# Digital Logic
# and
# Microprocessors

# Digital Logic and Microprocessors

Fredrick J. Hill
and
Gerald R. Peterson

John Wiley & Sons
New York   Chichester   Brisbane   Toronto   Singapore

*Library of Congress Cataloging in Publication Data:*

Hill, Fredrick J.
  Digital logic and microprocessors.

  Includes index.
  1. Logic design.   2. Microprocessors—Design and construction.   I. Peterson, Gerald R.   II. Title.

TK7888.4.H55   1984        621.3819′58        84-7377
ISBN 0-471-08539-1


Printed in the United States of America

10 9 8 7 6 5 4 3 2 1

# Preface

This volume has been designed to serve as a textbook for the first course in computer hardware. The topics have been chosen to introduce digital hardware to the student who will go on to study nothing more about computers than using a high-level language as well as to the student who plans to specialize in computer science or computer engineering. It begins with the time-honored topics in logic design, moves smoothly into microprocessors, and finishes with an in-depth treatment of interface design and assembly-language programming. The material should be suitable for a class of students including a broad range of majors or for a class of electrical engineering and computer science students. The authors use the book for a two-semester course given in the sophomore year to students in electrical and computer engineering.

Very little in the way of prior background has been assumed. Only the interest and desire to learn about computer hardware is required. Many students will have had prior experience with some high-level language, but even this is not essential. For a few pages in Chapter 5 the reader is presumed to have been previously introduced to electricity at the level of ohms law. In classes where this background is not uniformly present, the instructor need only take a more descriptive approach to the material on loading and delay.

Our intention has been to satisfy the need for a so-called "modern approach" to the topics of digital hardware. This need is easily justified by the observation that the two current action areas of digital design were unknown 15 or even 10 years before this writing. These two related, but separate, subjects are VLSI design and microprocessor-based system design. Chapters 1 through 10 of this book provide the fundamentals of digital logic common to both these areas. The potential VLSI designer will then go on to study MOS electronics, interactive graphics, exploitation of the pass transistor, and silicon compilers in other courses. A much larger portion of the readers will be involved at some level with microprocessor-based design. Chapters 11 through 16 continue with this topic. Mastery of these chapters should make it possible for students to apply this subject in the work place whether or not they subsequently take more advanced courses.

Chapters 2, 3, and 4 present the traditional subjects of Boolean algebra, number systems, and Karnaugh maps to develop students' background and catch their interest. Chapters 5 and 7 introduce the common off-the-shelf SSI and MSI parts and notation to facilitate their use. Chapter 6 treats arithmetic and coding, including such microprocessor-related topics as BCD addition and subtraction and overflow.

The treatment of circuits with memory, or sequential circuits, in Chapters 8, 9, and 10 has been somewhat abbreviated from that of the authors' previous book, *Introduction to Switching Theory and Logical Design.* Understanding the fundamental timing assumptions of sequential circuits is as important as ever, so this subject has not been neglected. State table minimization and state assignment have been deemphasized. The algorithmic state machine or ASM chart is introduced as an alternative notation to the state table. Only clock-mode sequential circuits are considered in these chapters. For the reader who wishes to explore the problems that appear when the clock-mode assumption cannot be satisfied, a comprehensive treatment is included at the end of the book as Chapter 17. A simplified RTL notation is introduced in Chapter 10 and is found helpful (in moderation) in Chapters 11, 13, 14, and 15.

Chapters 11 through 16 treat microprocessors at the hardware and assembly-language level. To come to grips with "real world" details and to include programming, it was necessary to select a particular microprocessor. We chose the 6502 as the simplest architecture still in (and likely to remain in) the marketplace. The instruction set of the TB6502, used in the book, is a proper subset of and includes nearly all the features of the 6502 instruction set. By not requiring the TB6502 to be cycle-by-cycle compatible with the 6502, we were able to provide a partial RTL description of the former at a level appropriate to this book. Simplified versions of the common parallel and serial interface chips were devised to familiarize the students most efficiently with the typical control features of these units.

Whether the students are freshmen, sophomores, juniors, or seniors, this book can serve as a first introduction to digital hardware. It may be used for either one or two semesters. The first 10 chapters can fit nicely into a one-semester course supplemented by a logic design laboratory. Chapters 11 through 16 can constitute a second semester, if supported by a comprehensive laboratory experience in assembly-language programming. With less emphasis on laboratory, Chapters 11 and 12 might be added to the first semester, thereby providing a brief introduction to microprocessors. Chapter 17 could be included at the end of either the first or the second semester.

For the second semester a laboratory using a 6502-based microcomputer, such as the AIM or SYM, is ideal but not essential. At the assembly-language level, there is considerable similarity between the various 8-bit microprocessors. It should be quite satisfactory to use this book for the lecture portion of a course in conjunction with a laboratory supported by a non-6502-based microcomputer and programming manuals for that computer.

Fredrick J. Hill
Gerald R. Peterson

# Contents

# Digital Logic and Microprocessors

# Between the Transistor and the High-Level Language

## 1.1 Prior Perceptions

As the title suggests, this book is organized to serve the reader without previous background in digital hardware. It would be surprising, however, to discover a reader who had no previous experience with the digital world. One encounters the *digital computer* in the office, in the home, in the elementary and high school classroom, in the game arcade, and on film. Bits and pieces of information about the computer are necessarily accumulated in the process of these encounters. Some readers may have sampled some of the material in the early chapters of this book in their work experience or in precollege course work.

The typical reader's perception of the inside of the computer is quite likely as characterized by Fig. 1.1. The reader may have learned that the active elements inside the box, labeled "computer," are *transistors,* as symbolized in the figure by the schematic diagram symbol for the transistor. Sometimes, a very large number of these transistors are found together in a tiny package called an *integrated circuit* (IC). Mysterious networks of these transistors "think and compute." Wires connect the computer box to other equipment such as the CRT (cathode ray tube) terminal and printer shown in Fig. 1.1. Through these peripheral devices, with outer covers that also conceal transistors and integrated circuits, the computer communicates with the outside world.

Many readers will have had some experience programming a computer in some *high-level language* such as BASIC, FORTRAN, or PASCAL. These programming tools are called *"languages"* simply because they are media for communicating orga-

1

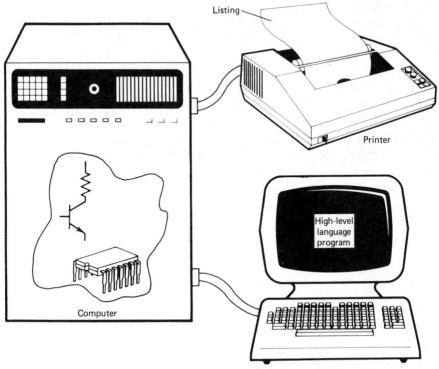

*Figure 1.1    The computer: inside and out.*

nized information between intelligent systems, in this case between the user and the computer. These languages are termed "high level" because they permit the user to express concepts in familiar formats. In general, the higher the level of a language the more convenient it is for the user. The notion of typing programs in a high-level language on a CRT terminal and receiving results from a printer as depicted in Fig. 1.1 is probably shared even by those readers without personal programming experience.

The scope of this book includes neither the design of transistor circuits for computers nor programming in a high-level language. Most readers will have the opportunity to learn more about transistor circuits in electronics courses. If any readers have not yet studied programming in a high-level language, they are almost certain to do so before they complete their formal education. This book will approach the spectrum of computer topics between these two extremes.

Transistors and other electronic components are connected to form the basic building blocks from which computers and other digital systems are constructed. The large majority of these building blocks fall into two classes, those that perform logic and those that store information. A *schematic* diagram showing how one such logic building block is constructed from transistors, resistors, and diodes is shown in Fig. 1.2. The rules for constructing such schematic diagrams can be thought of as a language. Text written in this language is two-dimensional graphics rather than a one-

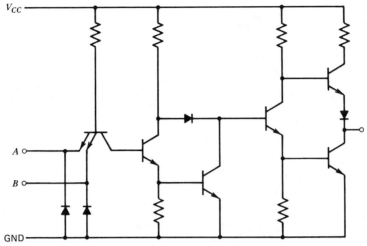

*Figure 1.2    TTL (transistor–transistor-logic) AND gate.*

dimensional sequence of symbols, but in the broadest sense it is a language nonetheless. A digital computer might consist of tens of thousands of logic elements, only one of which is shown in Fig. 1.2. It is easily seen that a representation of a computer as a composite diagram consisting of thousands of copies of Fig. 1.2 would be bewildering indeed. Electronic schematic diagrams are an inefficient language for designing and communicating the structures of digital computers.

## 1.2   A Spectrum of Digital Languages

A spectrum of languages, in addition to high-level languages and the lowest-level electronic diagrams, is both available and necessary for describing and using digital systems. Providing you with an understanding of these intermediate languages is the object of this book. In this chapter our goal is to provide a map showing the relationship between the various representations of digital systems and an indication of where these topics might be treated in later chapters. As was the case with Fig. 1.2, details will be displayed with which you may not be completely comfortable until you have considered subsequent chapters.

    The spectrum of languages available for describing digital systems and exploiting computers is summarized in Fig. 1.3. The lowest level, electronic schematic diagrams, is shown at the bottom of the chart; high-level languages are at the top. A variety of features distinguish the languages entered at various vertical levels in the chart. The single measure that applies to all the entries in the chart and increases from language to language moving upward through the chart may be characterized as efficiency of description. A higher-level language will be provided with more powerful and more complex building blocks or primitives and, therefore, will usually require fewer symbols, fewer lines, or less space on a page to describe a given system

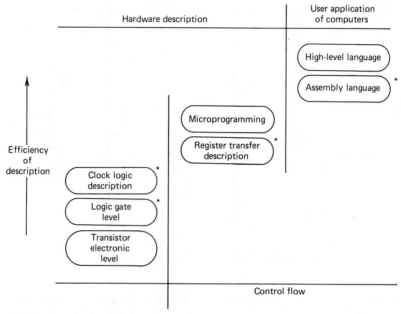

*Figure 1.3   Levels of computer languages.*

than would a lower-level language. The trick is to assure that the building blocks at one level are unambiguously defined in terms of lower-level languages so that accuracy of description is not sacrificed in favor of brevity.

The next level up from the electronic level in Fig. 1.3 is the *logic gate* level. The circuit of Fig. 1.2 may be represented by the gate level logic symbol shown in Fig. 1.4a. Figure 1.4b is borrowed from Chapter 7 to show a 4-bit binary adder realized in terms of logic gates. Chapter 2 will very simply define logic gates in terms of transistor electronics, and the next few chapters will develop a mechanism for realizing more complex logical operations such as addition in terms of networks of these gates. The concept of a binary bit will be clarified in Chapter 3.

It was mentioned in the first section that, in addition to logic gates, there is another class of building blocks that stores information. The treatment of *memory elements* is the distinction between the logic gate level and the next step up, the clock logic level. At the gate level, memory elements are modeled as networks of gates, an example of which is borrowed from Chapter 8 and shown in Fig. 1.5a. At the gate level designers have considerable freedom to configure memory elements as they wish, each element behaving a little differently from the others. For the sake of uniformity and easily synchronized timing, most digital systems with memory are designed at the clock logic level. At this level memory elements are primitive building blocks that change state synchronously with each occurrence of a pulse from a common clock source. These notions will be developed to your satisfaction in Chapters 8 and 9. Figure 1.5b shows the symbol for the primitive building block that will represent the clocked memory element of Fig. 1.5a for clock mode design.

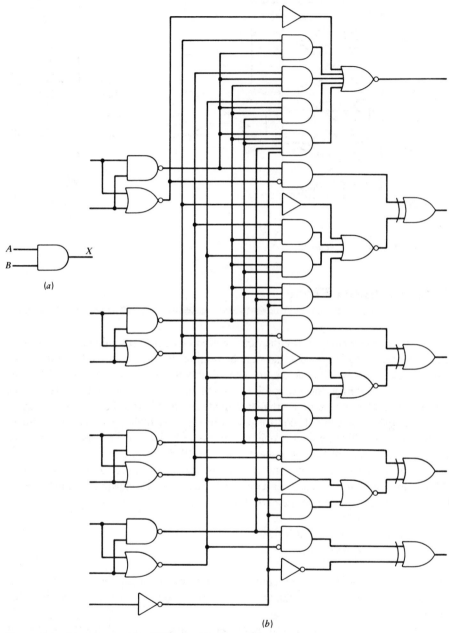

*(a)*

*(b)*

*Figure 1.4    Logic gates: (a) AND gate, (b) 4-bit adder.*

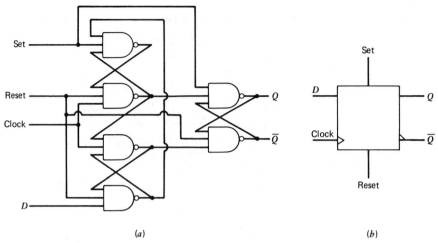

*Figure 1.5    Memory element.*

# 1.3    Control Flow

The languages described in the previous section bear little resemblance to the computer programming languages that may be familiar to some readers. Programs written in these high-level languages are lists of action statements together with some mechanism to control the order of execution of these action statements. In the BASIC language, for example, the statements are numbered. Following the execution of a BASIC statement, control passes to the next higher numbered statement unless the executed statement is one of a number of special control flow statements. Execution of a control flow statement will in itself determine the next active statement. The simplest control flow mechanism is illustrated in the following scrap of a BASIC program. In this case, statement 100 is executed following statement 90, if P1 < P2. Otherwise, the next active statement is 120.

        90. IF P1 >= P2 THEN 120
        100. X = X + 5

There is a large variety of such control flow statements in the many computer languages which have been developed over the years, but the general purpose is the same.

All the languages in the right-most two columns of Fig. 1.3 incorporate lists of action statements and control flow. This includes the center two languages, which are used to describe hardware. Chapter 10 will consider a hardware description language at the *register transfer level* (RTL). Hardware descriptions at this level very much resemble programs written in a high-level language. The RTL level is a formalization of clock logic description featuring vector operations and the separation of control and data unit realization. RTL descriptions have proved to be an extremely convenient medium for both the formulation and the communication of digital system designs.

The following are two lines taken from the middle of an RTL description of a digital system capable of arithmetic operations.

10. $C,A \leftarrow \text{ADD}(X,A)$.
11. $\rightarrow (C,\bar{C})/(12,15)$.

Step 10 specifies the addition of an input vector, $X$, to a register of memory elements, $A$. This is followed by step 11, which is executed one clock period later. This step effects a conditional branch to either step 12 or step 15, depending on whether a carry value, $C$, from the addition is 1 or 0. Although beyond the scope of this book, computer programs have been developed that can translate RTL descriptions to hardware realizations. Steps 10 and 11 in the RTL printout describe the hardware shown in Fig. 1.6. The control unit that determines the timing of execution of steps in the description is abbreviated as a single box in the figure. The control line shown coming from this box causes the results of the addition to be transferred into the register, $A$, and carry flip-flop (memory element), $C$. The output line from the carry flip-flop goes into the control unit to effect the branch. The two RTL steps given here are not independent of the rest of the RTL description of the system. Where, for example, is it specified that the register, $A$, consists of six clocked flip-flops? For answers to questions like this one, we await Chapter 10.

Those language levels marked by an asterisk in Fig. 1.3 are the subject matter of this course. Immediately above the RTL in the chart is *microprogramming*. A microprogramming language is a register transfer language constrained to conform to a predefined hardware structure. This topic is beyond the scope of this book; the reader is referred to Ref. [2] of Chapter 10.

The subject matter of Chapters 11 through 16 of this book will be *assembly*

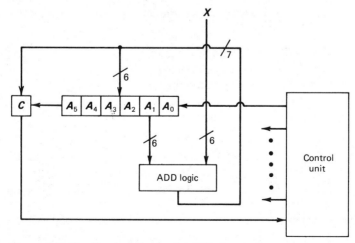

*Figure 1.6   Realization of a partial RTL description.*

*language* programming for an 8-bit *microprocessor*. Assembly language is a level above the hardware description languages in the chart, since it is a language for specifying the external function of a computer with an internal operation described in an RTL. The reader with prior experience with a high-level language will find that an assembly language for a particular computer is devised to reflect exactly the operations that the computer can accomplish rather than the operations most convenient for the user to express. For large computers, a program called a *compiler* is available to translate a user program written in a high-level language to assembly language for execution.

Compilers are sometimes available for microprocessors as well, but it is often desirable to program these computers directly in assembly language. One reason for this is that memory space is limited, particularly in 8-bit micrprocessors. A compiler will not always generate assembly language code as compactly as can be done manually. Most applications of 8-bit processors are as components of larger systems. As such, the timing of input and output operations must be very closely controlled. This is much more difficult if a high-level language is superimposed on the assembly language. Sometimes assembly language is used simply to provide increased speed of execution of programs; consider the arcade game.

Chapters 15 and 16 are specifically dedicated to the treatment of the microprocessor as a component. Where speed is not critical, a single microprocessor may often be used in place of a complicated special purpose logic network. Thus this device is the tool of the hardware designer as well as the programmer.

## 1.4  Terms and Roles

In the previous three sections we discussed the various language levels in terms of which computers can be designed and used. Corresponding somewhat to these language levels are areas of human activity. Most people working in the computer field will characterize their primary function as falling into one category in the following list.

> Circuit designer
> Logic designer
> System designer
> System programmer
> Programmer
> User

The circuit designer will use the transistor electronic level language. The logic designer will function at the logic gate level, the clock logic level, and, sometimes, the RTL level. The tools of the system designer's trade are the RTL and microprogramming level of description.

The systems programmer writes the programs such as compilers, *operating systems,* and input–output handlers, which make computer users more effective. These persons are involved with microprogramming and assembly language as well as high-level languages. Programmers write programs for their own use and the use of others, usually in a high-level language. The user may or may not be a programmer. Users who can do their job by calling on "off-the-shelf" programs are becoming increasingly common.

A person destined to function in any of these categories will be more effective having mastered the material contained herein. While in the book, however, he or she will usually wear the shoes of the logic designer. This will be the case even in the chapters on microprocessors, where the emphasis is on the use of these devices as powerful logic components. On occasion we shall move into the area of the system designer.

Although an individual's primary function may be one of the six areas in the list just presented, it would be surprising if he or she were not occasionally found wearing one of the other five hats. As just mentioned, the logic designer may be required to function as an assembly language programmer. To work effectively, the systems programmer must be familiar with the work of the systems designer. The circuit designer becomes a computer user as he simulates a design prior to its physical realization.

# 1.5 Some Digital History

This book will focus on the handling of information in discrete or digital form, in contrast to continuous or analog. With the exception of telegraphic transmission of discretely coded information beginning about 1840, the digital world has centered on computation. The history of mechanical aids to computation goes back many centuries. The development of the abacus apparently predates recorded history. In the seventeenth century, Pascal and Leibniz developed mechanical calculators, the ancestors through hundreds of years of development of the still-used mechanical desk calculator. The first device that was a computer in anything like the modern sense was proposed about 1830 by Charles Babbage.

In 1822 Babbage published a description of a *"difference engine"* for the computation of mathematical tables. The difference engine carried out a fixed sequence of calculations, specified by mechanical settings of levers, cams, gears, and such. Even data had to be entered mechanically, and the results were printed out immediately on computation. The difference engine was funded by the British government (the very first government-sponsored research project) and was partly completed when Babbage conceived the idea of using punched cards (invented by Jacquard in 1801) to provide the input and storage of results. He proposed that instructions be read from one set of cards and data from another set, with the results stored on still another set. Babbage started work on his second machine, the *"analytical engine,"* about 1830. Much of the remainder of his life (he died in 1871) was spent in a fruitless effort to get the machine built. His ideas were 100 years ahead of the technology of the day. The realization of Charles Babbage's dreams had to await the development of electronics.

Construction of an electromechanical computer based partly on Babbage's ideas was undertaken at Harvard University in 1939. This machine, the Mark I, was dedicated August 7, 1944, a date considered by many as the beginning of the computer era.

Borrowing concepts from a little-known computer built by Atanasoff and Berry at Iowa State University in the late 1930s, J. P. Eckert and J. W. Mauchly directed the development of ENIAC at the University of Pennsylvania, beginning in 1943. The ENIAC was very similar in logical organization to the Mark I, but it used vacuum tube technology. Experience gained on ENIAC and suggestions made by John von Neumann, the mathematical consultant to the ENIAC project, led to the design of EDVAC, the first stored-program computer. Because of the breaking up of the ENIAC team at the end of World War II, EDVAC was not completed until 1950. It was followed closely in time by several commercial computers, including Eckert and Mauchly's UNIVAC I.

Since 1950, computers have increased in capability and decreased in cost and size at a phenomenal pace. By 1960, a transistorized version of ENIAC, which occupied all of a very large room, could have been constructed to fit into a large suitcase. Today, much more powerful *microcomputers* are housed in integrated circuit packages smaller than a book of matches. The recent improvements in computers have been paralleled by progress in supporting digital equipment such as mass memories, input–output devices, and communications systems. Today the computer is truly pervasive in almost every aspect of life in the developed countries of the world.

# Boolean Algebra and Digital Logic

## 2.1  Computer Logic Circuits

In Chapter 1 it was pointed out that the electrical (usually) circuits within digital computers and related devices and systems fall primarily into two categories: those that perform logical operations on information and those that store information. It is the former category that we wish to treat in this chapter. Of all electrical circuits, logic circuits are the simplest to understand and the most invariant in their operation. For this reason we shall present simplified versions of these circuits first and then proceed to investigate how they are used. We shall use some elementary electrical concepts that we are confident are well within the background of our readers regardless of their fields.

The principal reason that digital circuits are simple is that they always deal with information represented in *binary* form. That is, the output wire of any properly functioning circuit within an all-digital system will always assume one of only two voltage values. If more than one binary bit of information is required, then more than one circuit is used. The possible binary output values for logic circuits are universally referred to as *logical 0* and *logical 1*. This does not mean that the actual voltages that can occur at the physical circuit outputs are 0 V and 1 V. The actual voltages that will represent logical 0 and logical 1 will vary with the technology in which the circuits are implemented and with the application of the circuits. For example, in emitter-coupled logic circuits 1.5 V might represent logical 0 and 0 V might represent logical 1. At the time of this writing 0 V most commonly represents logical 0 and 5 V most commonly represents logical 1 at measurable output points within digital circuits.

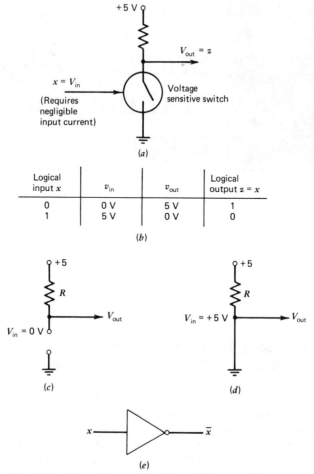

| Logical input $x$ | $v_{in}$ | $v_{out}$ | Logical output $z = x$ |
|:---:|:---:|:---:|:---:|
| 0 | 0 V | 5 V | 1 |
| 1 | 5 V | 0 V | 0 |

*(b)*

*Figure 2.1   Logical inverter circuit.*

Let us first consider the function of the logical value inverter shown in Fig. 2.1a. This circuit has been simplified for purposes of our analysis by substituting the voltage sensitive switch shown for a transistor and biasing network. This idealized circuit is actually a very good approximation of the real device. The switch shown will close whenever the input voltage, $V_{in}$, rises significantly above 0 V. As shown in Fig. 2.1b, the logical input, $x$, is represented by $V_{in}$, which may be either 0 or 5 V. When $V_{in}$ is 0, the inverter circuit assumes the form shown in Fig. 2.1c. The output, $V_{out}$, is connected through a resistor to the +5-V power source. Only a very tiny electric current (as determined by connecting circuits not shown) will actually flow through the resistor in the direction of the output arrow. Therefore, the output voltage will be almost the same as the 5-V source. When $V_{in}$ is 5 V, the switch is closed and the $V_{out}$ is connected to ground (0 V). These two values of $V_{out}$, together with the corresponding logical output values, are tabulated in the last two columns of Fig.

2.1b. It is easily seen that the logical output is the inversion (opposite value) of the input value.

Diagrams of logic networks are valuable communications aids. These diagrams seldom show electrical inverter circuits. Instead, the circuit is represented by the logical inverter symbol shown in Fig. 2.1e. The logical *inverter* may also be referred to as a NOT gate. In the next section we shall have occasion to use the NOT operation in algebraic expressions. We could write NOT($x$) for the inversion function of $x$, but the much more compact $\bar{x}$ is used instead. The use of the name NOT for this circuit arises from the meaning of the word NOT in natural language. The operation of the circuit could be described by the statement, "If the input, $x$, is at the 1-level, then the output, $\bar{x}$, will NOT be at the 1-level, and vice versa."

Let us next consider the circuit of Fig. 2.2a, which implements the logic function AND. A *diode* is connected in each input wire. (In this case only two inputs are shown, but *AND gates* may in general have two or more inputs.) The diode is a device through which electric current can flow in only one direction, the direction indicated by the diode arrow. In Fig. 2.2b we see a diode with current and voltage references indicated. Figure 2.1c depicts approximately the relation between voltage and current in a typical diode. Notice that if $V < 0$, then the current, $i$, is always 0. If a positive voltage is applied, the voltage drop across the diode will never exceed 0.2 V no matter how large the current through the diode may become.

Now we are ready to examine the AND gate itself. Suppose that one of the input voltages is +5 V while the other is 0 V as shown in Fig. 2.3a. A current of approximately $i = +5/R$ will flow through the resistor to the 0-V input. The output voltage, $V_z$, will be +0.2 V, the same as the voltage drop across the current-carrying diode. The situation will be essentially the same with a 0.2-V output, if the 0 and + 5 inputs are reversed or if both are 0. The output voltage, $V_z = 0.2$, is tabulated in the first three rows of Fig. 2.3c for these three cases. As the final case, suppose that both $V_x$ and $V_y$ are +5 V. In this case no current will flow through either diode or the resistor. Therefore, the output voltage, $V_z$, will be the same as the supply, or + 5 V. These values are tabulated in the last row of Fig. 2.3c. Now, as is usually the case, we let 0 and +5 V represent logical 0 and 1, respectively, to obtain the first two and last columns of Fig. 2.3c. One of the beauties of logic circuits is that we can

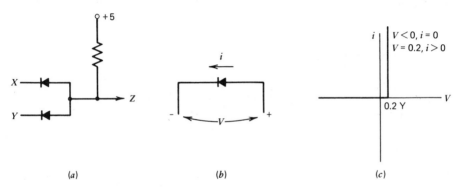

(a)         (b)         (c)

*Figure 2.2   Diode logical AND circuit.*

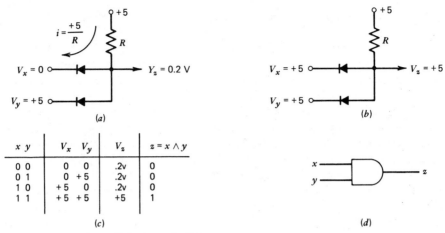

| x y | $V_x$   $V_y$ | $V_z$ | $z = x \wedge y$ |
|-----|-----------|------|-----------|
| 0 0 | 0      0 | .2v | 0 |
| 0 1 | 0    +5 | .2v | 0 |
| 1 0 | +5    0 | .2v | 0 |
| 1 1 | +5   +5 | +5 | 1 |

(c)                                      (d)

Figure 2.3   Analysis of the logical AND gate.

treat voltages that differ only slightly from the logical values the same as the logical values. Thus 0.2 V may be treated as 0. We see now that we have actually realized the logical AND function. The output, $z$, is 1 if and only if both inputs, $x$ AND $y$, are 1. Otherwise $z = 0$. As with the inverter, the circuit diagram for the AND gate is seldom used. Logical AND is symbolized as shown in Fig. 2.3d, and the AND function is written algebraically as $x \wedge y$.

Equally as important as the AND and NOT circuits is the OR circuit, one form of which is shown in Fig. 2.4a. Notice that the diodes have been turned around with respect to those in the AND circuit. The diodes are now there to isolate the circuits connected to the input lines, $x$ and $y$, from the OR circuit and do not seem to affect the function of the OR circuit itself. We leave it to the reader to analyze the circuit to obtain the output values tabulated in Fig. 2.4b. We see that this is, indeed, an OR gate, since the output is 1 whenever either input, $x$, OR input, $y$, is 1. The OR function is written $x \vee y$.

Fortunately we have been able to ignore small differences in voltage from the standard logic levels. We would not be able to do this if several diode AND and OR gates were connected in series output to input. For this reason diode AND and OR

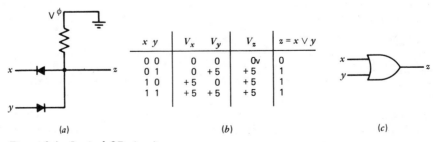

| x y | $V_x$   $V_y$ | $V_z$ | $z = x \vee y$ |
|-----|-----------|------|-----------|
| 0 0 | 0      0 | 0v | 0 |
| 0 1 | 0    +5 | +5 | 1 |
| 1 0 | +5    0 | +5 | 1 |
| 1 1 | +5   +5 | +5 | 1 |

(a)                                      (b)                                      (c)

Figure 2.4   Logical OR circuit.

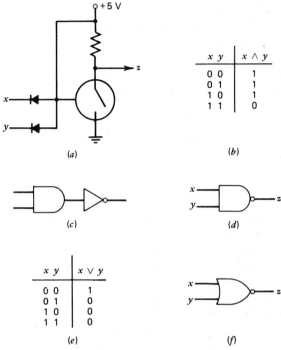

*Figure 2.5   NAND and NOR gates: (a) NOT–AND, (c) NOT–AND, (d) NAND, (f) NOR.*

gates are seldom used without at least one associated inverter to restore logic levels. The common practice is to rely on the NOT–AND or NAND circuit shown in Fig. 2.5a. The NOT–AND function as tabulated in Fig. 2.5b is easily seen to be an inversion of the output values given for the AND gate in Fig. 2.3c. We could symbolize the NOT–AND gate as shown in Fig. 2.5c, but for convenience we use only a circle to represent the inversion. Thus the standard NAND gate symbol is as given in Fig. 2.5d.

An OR gate output may also be connected to an inverter. The resulting NOR function is tabulated in Fig. 2.5e and is symbolized as in Fig. 2.5f.

We have now developed the set of basic building blocks for designing digital combinational logic networks. In the following sections we shall explore the use of these building blocks.

## 2.2   Application of Logic Circuits

In Chapter 1 we learned something about the characteristics of digital systems, but only in terms of very general qualitative descriptions. We also discussed some of the levels of description that can be used in dealing with such systems. At this point we wish to look at systems at the logic level. In order to work efficiently with logic circuits, either as designers or as users, you need to develop a variety of tools and techniques. High on the list of requirements is a form of mathematics suitable for dealing with digital systems. The conventional forms of mathematics with which we are all familiar are well suited for describing continuous (analog) systems, and the mathematical techniques available for the analysis and design of continuous systems are extensive and very powerful. But we need a different form of mathematics to deal efficiently with discrete systems.

While computers, a fairly recent development, are the most conspicuous example of digital systems, digital devices and systems were around long before computers. One early digital system was the dial telephone system, which goes back about 60 years. That part of the dial system that connects you to the other party is an enormous system of switches that open and close in an extremely complex manner to connect your phone to the one out of millions that you dialed. These days the system is mostly electronic, but it was originally constructed of actual switches, and the circuits were mostly designed by trial-and-error intuitive methods. As the system grew in size and complexity, the designers recognized the need for formal mathematical tools.

In 1938 a mathematician working for the Bell Telephone Laboratories, Claude Shannon, proposed a switching algebra for use in analyzing and designing discrete binary systems. Shannon's switching algebra was an outgrowth of an algebra of logic devised nearly 100 years earlier by George Boole, an English mathematician and logician. The AND, OR, and NOT building blocks derived in the previous section are the most convenient physical implementation of this algebra. Although switching algebra is only a special case of Boole's algebra, most people simply use the term *Boolean algebra* to refer to the mathematics used in dealing with digital systems and circuits.

Boole developed his algebra as a means of describing the complex logical relationships of natural language. This connection to natural language is very important, because it is the basis for a technique for moving from a verbal description of the function of the desired circuit to an unambiguous mathematical description. This procedure is very important in that designers are forced to continue consideration of their descriptions until they are certain they actually describe what they intend to build. It is surprising how much engineering time is wasted by attempts to bypass this step and to proceed from some vague mental picture of the circuit function directly to wiring circuits. Such a process rarely works and never produces good designs.

With the following example we shall illustrate both the algebraic formulation of a combinational logic problem and the application of the basic logic network building blocks developed in the previous section.

### Example 2.1

An elevator door control is to function as follows. When the elevator stops at a floor, the door will open and a timer will turn on a signal that will remain on for 10 sec to allow time for passengers to get on or off. In addition, there will be an electric eye to ensure that the doors do not close on someone in the doorway. After the 10 sec is up, if there is no one in the doorway, the doors will close if a call button has been pressed on another floor or if a passenger in the elevator has pressed a button for another floor. Obtain a logic network diagram of a network with an output that will be 1 whenever the elevator door is to close. As an intermediate step, determine a Boolean algebraic expression that will be logical 1 when the door is to close.

## Solution

As a beginning, let us delete extraneous information to reduce the above paragraph to a more concise logical statement.

The elevator door will close if and only if the timer signal is not on and the electric-eye circuit is closed and a call button at another floor has been pressed or a floor button in the elevator has been pressed.

Next, we observe that this compound statement is made up of simple declarative statements and assign a letter to represent each statement.

| | |
|---|---|
| The door will close | $D$ |
| The timer is on | $T$ |
| The electric-eye circuit is closed | $E$ |
| A call button has been pushed | $C$ |
| A floor button has been pushed | $F$ |

Each statement is true or false. The letters representing the statements are *logic variables*, which will be used as the names of binary signals in a logic circuit. A logical variable will be 0 if the corresponding statement is false and logical 1 if the statement is true. For example, the $E$ signal is a signal derived from the electric-eye circuit, taking on the logic 1 value if the beam is closed, the logic 0 value if the beam is interrupted by a passenger.

One drawback of the English language is that when a number of ANDs, ORs, and NOTs appear in a single sentence, the order in which they are to be applied is not always clear. Before writing the final logical expression, it is convenient to add some parentheses to the preceding statement to form a quasi-English expression emphasizing order.

((the timer is NOT on) AND (the electric eye signal is closed)) AND
((a call button has been pressed) OR (a floor button has been pushed))

We recognize that "the timer is not on" can be directly symbolized by $\overline{T}$. Replacing this and the remaining simple statements with the appropriate symbols results in a Boolean algebraic expression for $D$.

$$D = (\overline{T} \wedge E) \wedge (C \vee F) \tag{2.1}$$

We have represented the phrase "if and only if" by the symbol " $=$ ", which has the same meaning where the variables are restricted to the values 1 and 0. A circuit element is available to represent each of the operations appearing in Eq. 2.1, so we can translate it directly into the logic network realization shown in Fig. 2.6. ∎

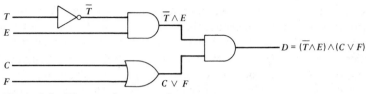

*Figure 2.6   Elevator door network.*

AND, OR and NOT are the operations of Boolean algebra in just the same way that addition, subtraction, multiplication, and so forth are the operations of conventional algebra. Operations in ordinary algebra are defined by rules. You can write an addition table for the decimal digits, but that does not completely define addition for all possible numbers. You need rules that tell how the addition table is extended to numbers other than the 10 digits. In Boolean switching algebra, however, each variable can take on only two values. For a finite number of variables we can then completely define an operation simply by listing the results of that operation for all possible combinations of the variables. Such listings are known as *truth tables*. The truth tables for the basic Boolean connectives AND, OR, and NOT together with NAND and NOR are summarized in Fig. 2.7. We shall see in the next example that truth tables for other functions may be expressed in terms of these connectives.

### Example 2.2

Determine a truth table for the Boolean function realized by the logic network given in Fig. 2.8.

### Solution

First we translate the network into a logic expression by reversing the process used in the elevator example.

$$f(x,y) = (\bar{x} \wedge y) \vee (x \wedge \bar{y}) \tag{2.2}$$

| $A$ | $\bar{A}$ | | $A$ | $B$ | $A \wedge B$ | $A \vee B$ | $\overline{A \wedge B}$ | $\overline{A \vee B}$ |
|---|---|---|---|---|---|---|---|---|
| 0 | 1 | | 0 | 0 | 0 | 0 | 1 | 1 |
| 1 | 0 | | 0 | 1 | 0 | 1 | 1 | 0 |
| | | | 1 | 0 | 0 | 1 | 1 | 0 |
| | | | 1 | 1 | 1 | 1 | 0 | 0 |

*Figure 2.7   Truth tables for Boolean connectives.*

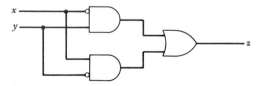

*Figure 2.8   Exclusive OR network.*

| $x$ | $y$ | $\bar{x}$ | $\bar{y}$ | $x \wedge \bar{y}$ | $\bar{x} \wedge y$ | $(x \wedge \bar{y}) \vee (\bar{x} \wedge y)$ | $x \oplus y$ |
|---|---|---|---|---|---|---|---|
| 0 | 0 | 1 | 1 | 0 | 0 | 0 | 0 |
| 0 | 1 | 1 | 0 | 0 | 1 | 1 | 1 |
| 1 | 0 | 0 | 1 | 1 | 0 | 1 | 1 |
| 1 | 1 | 0 | 0 | 0 | 0 | 0 | 0 |

*Figure 2.9   Evaluation of exclusive OR.*

Next we write the truth tables for $\bar{x}$ and $\bar{y}$ as shown in Fig. 2.9. We then obtain the truth table for $x \wedge \bar{y}$ by ANDing row by row the entries for $x$ and $\bar{y}$. The truth column for $\bar{x} \wedge y$ is obtained in a similar manner. Finally, we OR the entries for columns $x \wedge \bar{y}$ and $\bar{x} \wedge y$ row by row to obtain the truth column for $(x \wedge \bar{y}) \vee (\bar{x} \wedge y)$. ∎

In Example 2.2 we have conveniently defined the truth table for another very important binary connective, *exclusive OR*. The interpretation of OR given in Fig. 2.7 is clearly consistent with the desired operation of the elevator system. Assuming the other conditions are satisfied, we want to close the door if a call button has been pushed, if a floor button has been pushed, or if both events have occurred. However, the word "or" in English sometimes means one or the other but not both. For example, in the statement to a child,

You may have a cookie or a piece of cake,

the meaning is clearly one or the other but not both. This kind of "or" is known as *exclusive OR*, for which the symbol is "⊕", and the truth table is shown in the rightmost column of Fig. 2.9. The first type of OR is sometimes referred to as *inclusive OR*, but most people just use OR to denote the connective defined by the truth table of Fig. 2.7.

The symbol for the exclusive OR network is given in Fig. 2.10. Perhaps the most important distinction between the ⊕ and the other connectives is that there is no simple circuit for realization of ⊕. Even though we shall eventually see that four of the networks symbolized by Fig. 2.10 can be obtained in a single integrated circuit package, each is realized within the package by a network much the same as in Fig. 2.8.

*Figure 2.10   Symbol for exclusive OR.*

# 2.3 Evaluation of Boolean Functions

We have now seen how to formulate a Boolean equation corresponding to some desired system operation and how to translate that equation into a circuit. In Fig.

2.9 we saw our first example of an evaluation of a Boolean function. Let us examine more carefully what is meant by the evaluation of functions that can take on only the values 1 and 0. A function, in mathematical terms, is a relationship between variables. A function is evaluated by determining the values of a result variable corresponding to values of the other input variables. If you are given the function

$$y = x^2$$

you can evaluate $y$ as a function of $x$ by evaluating $x^2$ for various values of $x$. Since $x$ can take on an infinite number of values, $y$ cannot be fully evaluated in this manner, and the functional relationship between $y$ and $x$ must be stated as a rule. In switching algebra, however, there are only two possible values for each variable, so we can fully evaluate any function by tabulating the outputs corresponding to all possible combinations of the inputs in a truth table, as was done for Eq. 2.2 in Fig. 2.9. Equation 2.2 included parentheses to specify the order of evaluation. In that particular case the parentheses were actually unnecessary because of the existence of a universally accepted precedence rule specifying the normal order of evaluation of Boolean connectives. This rule may be stated as follows.

---

In the absence of parentheses, NOT is always evaluated first followed by AND. The OR operator is always evaluated last.

---

Equation 2.2 is repeated as follows without the parentheses.

$$x \oplus y = x \wedge \bar{y} \vee \bar{x} \wedge y \tag{2.3}$$

It can be seen that by following the rule just stated that the evaluation of Eq. 2.3 will yield the same results as given in Fig. 2.9.

We note that parentheses are still sometimes needed where exceptions to this rule are required. Consider again the elevator door equation.

$$D = (\bar{T} \wedge E) \wedge (C \vee F) \tag{2.4}$$

In this case the OR operation must be done before the AND. In addition, the first AND operation is enclosed in parentheses so that it may be evaluated in a truth table as a binary connective, relating only two variables. We group the variables in pairs and combine them to form single variables that can in turn be combined with other single variables.

Figure 2.11 is the truth table for the function of Eq. 2.4. Since there are four input variables, there are 16 possible input combinations. In the first column we evaluate $\bar{T}$. In the second column we evaluate $(\bar{T} \wedge E)$, which we equate to an intermediate variable, $X$. In the third column we evaluate a second intermediate variable, $Y$, equal to $(C \vee F)$. In the last column we evaluate $D = X \wedge Y$. Thus, by grouping the pairs and successively applying the definitions of the binary operations from Fig. 2.7, we have evaluated a function of four variables. The same method can be applied to any function of any number of variables.

| $T$ | $E$ | $C$ | $F$ | $\overline{T}$ | $X = \overline{T} \wedge E$ | $Y = C \vee F$ | $D = X \wedge Y$ |
|---|---|---|---|---|---|---|---|
| 0 | 0 | 0 | 0 | 1 | 0 | 0 | 0 |
| 0 | 0 | 0 | 1 | 1 | 0 | 1 | 0 |
| 0 | 0 | 1 | 0 | 1 | 0 | 1 | 0 |
| 0 | 0 | 1 | 1 | 1 | 0 | 1 | 0 |
| 0 | 1 | 0 | 0 | 1 | 1 | 0 | 0 |
| 0 | 1 | 0 | 1 | 1 | 1 | 1 | 1 |
| 0 | 1 | 1 | 0 | 1 | 1 | 1 | 1 |
| 0 | 1 | 1 | 1 | 1 | 1 | 1 | 1 |
| 1 | 0 | 0 | 0 | 0 | 0 | 0 | 0 |
| 1 | 0 | 0 | 1 | 0 | 0 | 1 | 0 |
| 1 | 0 | 1 | 0 | 0 | 0 | 1 | 0 |
| 1 | 0 | 1 | 1 | 0 | 0 | 1 | 0 |
| 1 | 1 | 0 | 0 | 0 | 0 | 0 | 0 |
| 1 | 1 | 0 | 1 | 0 | 0 | 1 | 0 |
| 1 | 1 | 1 | 0 | 0 | 0 | 1 | 0 |
| 1 | 1 | 1 | 1 | 0 | 0 | 1 | 0 |

*Figure 2.11   Truth table for elevator door function.*

As the number of variables increases, the size of the truth table will become unwieldy and the labor of obtaining it excessive. Computer programs are available that will generate the truth table from the Boolean equation. However it is obtained, the truth table is very useful for verifying that the equation is correct. When writing a verbal description it is quite natural to focus on expected situations and very easy to overlook unlikely combinations of events. With the truth table, the designer can look at every possible combination of inputs, evaluate what sort of situation each combination represents, and determine if the response is appropriate. The truth table is also very useful in troubleshooting the final circuit. If the circuit follows the form of Eq. 2.4 exactly, then each column of the truth table corresponds to the output of a gate. If the circuit does not work right, the truth table will make it easy to determine which gate is malfunctioning.

## 2.4  Boolean Algebra

At this point it might seem to the reader that we have pretty much solved the problem of combinational design. Just find an equation that represents what the circuit is supposed to do, verify it by truth-table analysis, and translate it directly into a circuit of gates. Providing the equation is correct and no mistakes have been made in verifying it (no small assumption), this process will produce a working design. But how good a design? Might there not be some other design, accomplishing the same function, but better, cheaper, simpler, more reliable? In a very simple problem, such as the elevator door control, the chances are that the first design found will be about as good as any other. As designs get more and more complex, this is less and less likely to be the case. The structure, or form, of the equations the designer finds initially reflect the way the designer thought through the problem. But that does not necessarily correspond to the best structure in a logic circuit. We need a way of manipulating Boolean equations to find other, possibly preferable, forms. Boolean equations, like equations of conventional algebra, can usually be expressed in many forms. For example,

$$y = (2ax^2)^3$$

can also be expressed

$$y = 8a^3x^6$$

This alternative form may or may not be preferable, but the important point is that we have rules that enable us to find alternative forms. If we are to exploit the potential of Boolean algebra, we need to develop some rules.

The development of Sections 2.1 and 2.2 may be taken as providing an informal definition of Boolean switching algebra. This is an algebra of two-valued variables, the two values being denoted 0 and 1, with three operations, NOT, AND, and OR, defined by the truth table of Fig. 2.7. We also defined three other operations, exclusive OR, NAND, and NOR, but these are usually considered derived operations because they can be expressed in terms of the other three. We could define the algebra rigorously, in terms of a set of postulates [4], but the informal definition just given will suffice for our purposes.

When you studied geometry, you started with a set of postulates that defined the system under study. You then developed theorems, rules that followed from the postulates and that were useful in solving problems. In the same way we can develop Boolean identities, rules that follow from the definitions of the operations and that will prove useful in working with logical equations. We start in Fig. 2.12 with a set of identities that follow by inspection from the defining truth tables.

The first identity, $A = A$, is the definition of inversion first developed in Section 2.1. The identity $A \wedge A = A$ can be demonstrated with reference to Fig. 2.7. If $A = 0$, we have $0 \wedge 0 = 0$; if $A = 1$, we have $1 \wedge 1 = 1$. The other identities can be similarly demonstrated with reference to the truth tables. Note that, except for the first identity, these identities occur in pairs. This arrangement is a

$$\overline{\overline{A}} = A$$

| | |
|---|---|
| $A \vee 0 = A$ | $A \wedge 1 = A$ |
| $A \vee 1 = 1$ | $A \wedge 0 = 0$ |
| $A \vee A = A$ | $A \wedge A = A$ |
| $A \vee \overline{A} = 1$ | $A \wedge \overline{A} = 0$ |

*Figure 2.12   Single-variable Boolean identities.*

Commutative laws
$$A \lor B = B \lor A$$
$$A \land B = B \land A$$

Associative laws
$$(A \lor B) \lor C = A \lor (B \lor C)$$
$$(A \land B) \land C = A \land (B \land C)$$

Distributive laws
$$A \lor (B \land C) = (A \lor B) \land (A \lor C)$$
$$A \land (B \lor C) = (A \land B) \lor (A \land C)$$

*Figure 2.13   Computational rules of Boolean algebra.*

consequence of an interesting characteristic of Boolean algebra, the principle of duality:

Given any Boolean equation that is true, a second equation that is also true can be obtained by replacing every $\land$ with $\lor$, and vice versa, and replacing every 0 with 1, and vice versa.

We shall see several interesting consequences of duality as we continue our investigation of Boolean algebra. An immediate consequence is that any time we prove one theorem, we have automatically proved its dual.

The next three identities (Fig. 2.13) are the commutative, associative, and distributive laws, which may be regarded as the "computational rules" of Boolean algebra. Each of these laws is also accompanied by its dual. The commutative laws follow directly from the defining truth tables for AND and OR. The associative laws tell us that when ANDing or ORing three or more variables, the order in which we combine them is immaterial. This in turn allows us to write expressions such as

$$A \lor B \lor C \lor D$$

without parentheses to group the variables into pairs. We will leave the demonstration of these laws by truth table as an exercise for you.

The second distributive law will look more familiar in the analogous form from ordinary algebra

$$a(b + c) = ab + ac$$

| $A$ $B$ $C$ | $B \lor C$ | $A \land (B \lor C)$ | $(A \land B) \lor (A \land C)$ | $A \land B$ | $A \land C$ |
|---|---|---|---|---|---|
| 0  0  0 | 0 | 0 | 0 | 0 | 0 |
| 0  0  1 | 1 | 0 | 0 | 0 | 0 |
| 0  1  0 | 1 | 0 | 0 | 0 | 0 |
| 0  1  1 | 1 | 0 | 0 | 0 | 0 |
| 1  0  0 | 0 | 0 | 0 | 0 | 0 |
| 1  0  1 | 1 | 1 | 1 | 0 | 1 |
| 1  1  0 | 1 | 1 | 1 | 1 | 0 |
| 1  1  1 | 1 | 1 | 1 | 1 | 1 |

*Figure 2.14    Truth table for second distributive law.*

Because of the analogy, application of this law is sometimes referred to as factoring, although it is not factoring in the arithmetic sense. The truth-table demonstration for this law is shown in Fig. 2.14.

Next, the two theorems, given in Fig. 2.15, can be derived algebraically using the identities provided in Fig. 2.14. We will derive the first of these as follows, leaving the rest as exercises for you. At each step the applicable identity that justifies that step is cited at the right. As you attempt to justify the remaining theorems, you are urged to adhere to the discipline of citing the justification for each step. As you become more adept at Boolean algebric manipulation, this practice will gradually be discarded.

$$
\begin{aligned}
x \lor (x \land y) &= (x \land 1) \lor (x \land y) & &A \land 1 = A \\
&= x \land (1 \lor y) & &\text{Distributive law} \\
&= x \land 1 & &A \lor 1 = 1 \\
&= x & &A \land 1 = A
\end{aligned}
$$

The final pair of theorems are generally referred to as *DeMorgan's laws,* after the mathematician Augustus DeMorgan, a contemporary of Boole, who first stated them. The first of DeMorgan's theorems is demonstrated by the truth table in Fig. 2.16*b*. The second then follows by duality.

These laws, which state the relationships between the complements of functions and the complements of variables, are probably the most useful of all the theorems, and we will see a lot of them. The attentive reader may have noted that we referred to truth-table *demonstrations* of the distributive laws and DeMorgan's laws, rather than *proofs.* The reason is that these theorems apply to any number of variables and cannot be proven by tabulation. A truth table such as that shown in Fig. 2.16*b* "proves" the law only for two variables. A proof for an arbitrary number of variables

$$A \lor (A \land B) = A \qquad A \land (A \lor B) = A$$
$$A \lor (\overline{A} \land B) = A \lor B \qquad A \land (\overline{A} \lor B) = A \land B$$

*Figure 2.15 Theorems for reducing expressions.*

$$\overline{A \lor B \lor C \cdots} = \overline{A} \land \overline{B} \land \overline{C} \cdots \qquad \overline{A \land B \land C \cdots} = \overline{A} \lor \overline{B} \lor \overline{C} \cdots$$

(a)

| $A$ | $B$ | $A \lor B$ | $\overline{A \lor B}$ | $\overline{A}$ | $\overline{B}$ | $\overline{A} \land \overline{B}$ |
|---|---|---|---|---|---|---|
| 0 | 0 | 0 | 1 | 1 | 1 | 1 |
| 0 | 1 | 1 | 0 | 1 | 0 | 0 |
| 1 | 0 | 1 | 0 | 0 | 1 | 0 |
| 1 | 0 | 1 | 0 | 0 | 0 | 0 |

(b)

*Figure 2.16 DeMorgan's laws.*

requires a proof by induction. As a further illustration of Boolean algebraic manipulation, we extend DeMorgan's theorem to three variables as follows.

$$\overline{A \lor B \lor C} = \overline{(A \lor B) \lor C} \qquad \text{Associative law}$$
$$= \overline{(A \lor B)} \land \overline{C} \qquad \text{DeMorgan's theorem}$$
$$= \overline{A} \land \overline{B} \land \overline{C} \qquad \text{DeMorgan's theorem}$$
$$\overline{A \land B \land C} = \overline{A} \lor \overline{B} \lor \overline{C} \qquad \text{Duality}$$

In writing Boolean expressions throughout the rest of the book, we shall rely on the precedence rules stated in the previous section and omit parentheses insofar as possible. We shall further simplify the text by omitting AND symbols and implying AND whenever two variables are written next to each other. This is the same practice followed for multiplication in ordinary algebra. Thus the expression

$$(A \land B) \lor (A \land C)$$

may be written

$$AB \lor AC$$

Note very carefully, however, that this convention requires that all variables be represented by single letters, or single letters with subscripts, such as

$$x_1, x_2, x_3, \ldots$$

In later chapters we shall be using variable names such as *ready* and *halt*. In such cases it is essential that all operation symbols be written out. It should also be noted that it is common in the literature to use the symbols $(+)$ and $(\cdot)$ for OR and AND. The problem with the use of these symbols is the possibility of confusion with addition and multiplication, especially when designing digital systems that perform arithmetic operations.

## 2.5  Simplification of Boolean Functions

Now that we have some theorems, let us see if we can use them to simplify some functions. First, let us consider an example that is similar to the proofs you studied in geometry. Recall that the usual problem in geometry is to try to prove the validity of some geometric construction or identity. The basic procedure is to apply the various theorems to modify one or both sides of the statement until they agree. We can use the same general procedure to prove additional theorems in Boolean algebra.

**Example 2.3**

Prove

$$ab \vee \bar{a}c \vee bc = ab \vee \bar{a}c \tag{2.5}$$

Solution

$$
\begin{aligned}
ab \vee \bar{a}c \vee bc &= ab \vee \bar{a}c \vee bc(a \vee \bar{a}) && (x \vee \bar{x} = 1; x \wedge 1 = x) \\
&= ab \vee \bar{a}c \vee abc \vee \bar{a}bc && \text{(Distributive law)} \\
&= ab(1 \vee c) \vee \bar{a}c(1 \vee b) && \text{(Distributive law)} \\
&= ab \vee \bar{a}c && (1 \vee x = 1)
\end{aligned}
$$

By duality we also have

$$(a \vee b)(\bar{a} \vee c)(b \vee c) = (a \vee b)(\bar{a} \vee c) \tag{2.6}$$

∎

In verifying identities we have the advantage that we know the goal of the manipulation. The usual problem in logic design is that of trying to find a simpler equivalent form for some logic function. In such cases we can only try various approaches, hoping to find one that will lead in the direction of simplification.

**Example 2.4**

The circuit of Fig. 2.17 has been designed by trial and error. Find a simpler form.

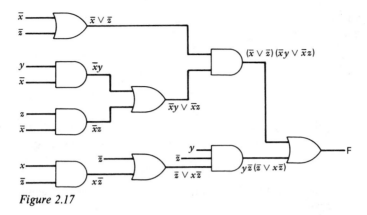

*Figure 2.17*

## Solution

First, we might wonder if there is a simpler form. We never know for sure, but in this case it seems fairly certain. Surely a function of only three variables does not need to be as complicated as this. In such cases it is sometimes useful to work on one part of the circuit at a time. Let us start with the output of the lowest OR gate,

$$\bar{z} \vee x\bar{z}$$

From the theorem $a \vee ab = a$ (Fig. 2.15) this reduces to $\bar{z}$, so the next AND gate forms

$$y \wedge \bar{z} \wedge \bar{z} = y\bar{z}$$

So one of the terms into the output OR gate is just $y\bar{z}$. Now consider the other input to the output OR gate

$$(\bar{x} \vee \bar{z})(\bar{x}y \vee \bar{x}z)$$

Applying the distributive law to factor out the common term we have

$$\bar{x}(\bar{x} \vee \bar{z})(y \vee z)$$

Applying $a(a \vee b) = a$ (Fig. 2.15) reduces this to

$$\bar{x}(y \vee z) = \bar{x}y \vee \bar{x}z$$

Combining with the other term at the output OR gate we have

$$f(x,y,z) = y\bar{z} \vee \bar{x}y \vee \bar{x}z$$

Rearranging terms,

$$f(x,y,z) = z\bar{x} \vee \bar{z}y \vee \bar{x}y$$

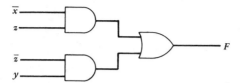

*Figure 2.18*

This may be recognized as the form of Eq. 2.5 with $a = z$, $b = \bar{x}$, and $c = y$, so that

$$f(x,y,z) = z\bar{x} \vee \bar{z}y \qquad (2.7)$$

leading to the circuit of Fig. 2.18, which is certainly a good deal simpler than the one we started with. ∎

The obvious problem with this procedure is that it is based on trial and error and largely dependent on intuition and just plain luck. How do we choose which theorem to use at each step? If we do find something simpler, how do we know that there may not be some other form that is simpler still? In general, we do not know. If Boolean simplification is to be a useful tool in digital design, we need systematic methods that can be applied to any logic function with reasonable assurance that the result will be in some sense optimal. As the next step toward developing such a method, let us explore an alternative interpretation of Boolean algebra.

## 2.6   Boolean Algebra as an Algebra of Subsets

So far we have discussed only one Boolean algebra, the two-valued switching algebra. This is our major interest, but there are other Boolean algebras. One is the Boolean algebra of sets. This algebra has more than the two values 1 and 0. Since some of the demonstrations in the previous sections assumed the existence of only two objects 1 and 0, we have not, strictly speaking, proved theorems for the Boolean algebra of *sets*. Nonetheless, all the identities given in Section 2.4 are also true for this algebra. We shall take the space to verify only a few of them in this section. We will, however, feel free to take advantage of the graphic properties of this example algebra wherever it will help to illustrate a point.

A set is simply a collection or group of objects that have some characteristics in common that enable them to be classified as a member of that set. An example would be the set of all cars. A subset is an identifiable group within the larger set, for example, the subset of all Ford cars, or all blue cars, or all Japanese cars. Subsets may be overlapping, having elements in common, for example, blue Fords, or may be *disjoint,* having no elements in common, for example, Ford cars and Japanese cars, since there are no Japanese Fords.

The set of all elements (all cars) is called the *universal set* and may be denoted by 1. Subsets are denoted by variables, such as $F$ for Fords, $B$ for blue cars, and $J$

for Japanese cars. There is also a special set, the null set, denoted by 0, that contains no elements at all.

We define three operations, union, intersection, and complementation. The *union* of two sets, denoted by $\vee$, is the set of all elements in either sets, for example,

$$B \vee F$$

is the set of all cars that are either blue or Fords, or both.

The *intersection* of two sets, denoted by $\wedge$, is the set of all elements common to both sets, for example,

$$B \wedge J$$

denotes the set of all blue Japanese cars.

The *complement* of a set is the set of all elements that are not in the set. For example, $\bar{J}$ signifies the set, all cars that are not Japanese.

A common pictorial device for representing set theory is the *Venn diagram*. In a Venn diagram the universal set is represented by a rectangle. In Fig. 2.19 the rectangle represents the set of all cars; the circle, the set of all Japanese cars; and the shaded area, the set of cars that are not Japanese. In Fig. 2.20 the circles represent cars that are blue and cars that are Japanese. The shaded area represents the intersection, blue Japanese cars. In Fig. 2.21, the shaded area represents the union, all cars that are either blue or Japanese or both.

To show that this is a Boolean algebra, we need to show that the various rules and identities are consistent with the pictorial interpretation shown in Figs. 2.19, 2.20, and 2.21. For example, consider the theorems

$$J \vee \bar{J} = 1 \quad \text{and} \quad J \wedge \bar{J} = 0$$

It is obvious that the union of all cars that are Japanese and all cars that are not is the set of all cars, the universal set. Similarly, the intersection of sets of cars that are Japanese and cars that are not is the set of no cars at all, the null set. From Fig. 2.20 we readily see that

$$B \vee (B \wedge J) = B$$

that is, the union of blue cars and blue Japanese cars is the set of blue cars. The other basic theorems can be similarly demonstrated.

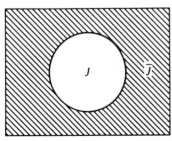

*Figure 2.19*

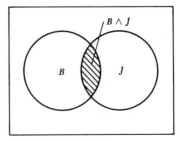

*Figure 2.20*

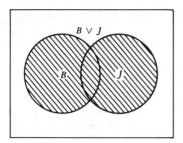

Figure 2.21

Consider the distributive law

$$A \lor (B \land C) = (A \lor B) \land (A \lor C)$$

In Fig. 2.22*a*, we see $A \lor (B \land C)$. In Fig. 2.22*b*, the horizontally lined area represents $A \lor B$, and the vertically lined area, $A \lor C$. The crosshatched area is thus the intersection

$$(A \lor B) \land (A \lor C)$$

and is seen to be the same area as in Fig. 2.22*a*, $A \lor (B \land C)$.

The value of this interpretation comes from the fact that relationships that may seem strange in algebraic form become almost self-evident when viewed in a pictorial form. For example,

$$B \lor (B \land J) = B$$

is quite obvious when viewed in terms of Fig. 2.20. Similarly, the identity

$$(B \land J) \lor (B \land \bar{J}) = B$$

is obvious in terms of Fig. 2.23. In words, the union of blue Japanese cars and blue non-Japanese cars is all blue cars. In both these examples, the result of the use of the theorem is simplification. In the complicated circuit problem of Example 2.4, many complex terms combined into a few simple terms. The pictorial interpretation will make it easier for us to recognize how complex terms can combine to form simpler terms.

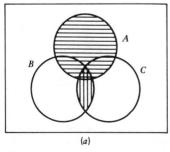

(a)

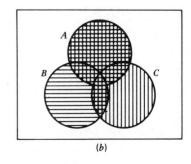

(b)

Figure 2.22

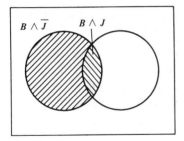

*Figure 2.23*

## 2.7   On the History of the Algebra of Logic

George Boole (1815–1864) was the English mathematician whose name is forever stamped on the algebraic underpinning of digital logic. Boole was the first to reduce the concepts of Aristotle's logic to algebraic notation. His work was an important step in mathematics and led to the symbolic logic more fully developed by Peano, Whitehead, Bertrand Russell, and others.

Interestingly, Boole's own formulation did not correspond exactly to the operations and identities presented in this chapter. For example, Boole used $x - y$ to represent what we would denote as $x \wedge \bar{y}$. His primary OR operation was the exclusive OR rather than the inclusive OR. This created problems that were eliminated by C. S. Pierce and others. An independent set of *postulates* for Boolean algebra, as presented in this chapter, was published by E. V. Huntington in 1904.

The same Claude E. Shannon who originated information theory was the first to observe that the two-valued Boolean algebra was useful in the design of switching networks. His 1938 article in the *Transactions of the American Institute of Electrical Engineers* is focused on the design of relay switching networks.

## Problems

**2.1**   A logic circuit is to be designed to produce a signal to open or close a valve controlling the flow of brine into a tank. The following statement describes the desired operation.

> The flow of brine is to be turned off if and only if (1) the tank is full or (2) the output is shut off, the salt concentration exceeds 2.5%, and the water level is not below a specified minimum level.

(a)   Break this statement up into the component simple statements and assign a letter variable to represent each one.

(b)   Write a Boolean equation for the valve control signal similar to Eq. 2.1

(c)   Draw a logic block diagram for this equation similar to Fig. 2.2.

**2.2**  A logic circuit is to be designed that will produce a signal to activate a burglar alarm system in a bank, subject to the following conditions.

> The alarm is to be operative only if a master switch at the police station has been turned on. Subject to this condition, the alarm will ring if the vault door is disturbed in any way, or if the door to the bank is opened unless a special switch is first operated by the security guard's key.

Repeat Problem 2.1 for this system.

**2.3**  In many cars the seat-belt alarm buzzer is also used to warn against leaving the key in the ignition or leaving the lights on. The following statement describes how such a system might operate.

> The alarm is to sound if the key is in the ignition when the door is open and the motor is not running, or if the lights are on when the key is not in the ignition, or if the driver belt is not fastened when the motor is running, or if the passenger seat is occupied and the passenger belt is not fastened when the motor is running.

Repeat Problem 2.1 for this system.

**2.4**  Set up a truth table for the system of Problem 2.1.

**2.5**  Using truth tables, verify
(a) $A \lor (B \land C) = (A \lor B)(A \lor C)$
(b) $\overline{A \land B \land C} = \overline{A} \lor \overline{B} \lor \overline{C}$
(c) $A\overline{B} \lor \overline{A}B = AB \lor AB$

**2.6**  Write the dual of the expression shown here and then prove its validity by truth-table analysis.

$$(a \lor b)(\overline{a} \lor c)(b \lor c) = (a \lor b)(\overline{a} \lor c)$$

**2.7**  Use the laws of Boolean algebra to verify algebraically that

$$\overline{A}\overline{B}\overline{C} \lor AC \lor BC = \overline{ABC \lor AC \lor BC}$$

**2.8**  Using the laws of Boolean algebra, simplify the following as much as possible.
(a) $(a \lor \overline{a}b)(ac \lor a\overline{c}(b \lor \overline{b}))$
(b) $AB\overline{C} \lor (AB\overline{C} \lor \overline{A}C)(B(A \lor C) \lor A\overline{B}\overline{C} \lor \overline{B}C)$

**2.9**  Use Venn diagrams to verify DeMorgan's law,

$$\overline{A \lor B} = \overline{A} \land \overline{B}$$

**2.10**  Professors Smith, Jones, and Green collect old books. Smith collects eighteenth-century English books and foreign-language fiction. Jones collects eighteenth-century books, except English fiction, and English nonfiction. Green collects English nonfiction and eighteenth-century foreign language books. Use Venn diagrams to determine the books for which there is competition, that is, those desired by two or more collectors.

**2.11** Three one-digit binary numbers, $a$, $b$, and $c$, are to be added together to form a 2-bit binary sum, $xy$, where $x$ is the more significant bit. Write Boolean expressions for $x$ and $y$ as functions of $a$, $b$, and $c$, and draw the logic block diagram for circuits realizing $x$ and $y$.

## References

1. Boole, G., *An Investigation into the Laws of Thought.* Dover Publications, New York, 1954.
2. Hill, F. J., and G. R. Peterson, *Introduction to Switching Theory and Logical Design,* 3rd ed. Wiley, New York, 1981.
3. Johnson, D. E., J. L. Hilburn, and P. M. Julich, *Digital Circuits and Microcomputers.* Prentice-Hall, Englewood Cliffs, N.J., 1979.
4. Quine, W. V., *Mathematical Logic.* Harvard University Press, Cambridge, Mass., 1955.
5. Roth, C. H., Jr., *Fundamentals of Logic Design,* 2nd ed. West Publishing, St. Paul, Minn., 1979.

# 0,1: Binary Numbers or Logical Values?

## 3.1 The Binary Number System

In the previous chapter we used 0 and 1 as symbols for the two logical values in our two-valued switching algebra. In working with this algebra, in manipulating and simplifying expressions, we made no assumptions about the significance or physical meaning of the variables. We simply followed abstract rules, in much the same way that we can add numbers without knowing whether the numbers represent dollars, people, or whatever.

Sometimes, as in the elevator door problem, individual logical variables may represent specific physical events, such as the closing of a door. In other cases, two or more logical variables may be treated as a unit, representing more complex information. For example, combinations of logical values (0's and 1's) on eight lines may be used to represent the letters of the alphabet.

Since numeric computation is a frequent requirement, it is common for sets of logical variables to be used to represent numbers. Because digital systems are inherently *binary* in the sense of two values, it is convenient to use the binary number system to carry out numeric or arithmetic operations in digital systems. In the *binary number system* we have only two numerals, 0 and 1, which we also use as symbols for logical values. Is there a conflict here? No, but there is an obvious source of confusion in the dual use of these symbols. Binary numbers and logical values are two distinct data types, each with its respective set of allowed operators, forming the binary number system on the one hand and switching algebra on the other. The fact that the same symbols are used to represent values in the two systems does not imply

**35**

any connection between them beyond the fact that they are both two-valued systems. Binary numbers are just as important in the world of digital computers as Boolean algebra, so you must take care to separate the two concepts in your mind.

In this chapter we shall learn just enough of the binary number system to support the numeric notation needed in Chapters 4 and 5 and to provide one or two design examples. In Chapter 6 we shall return to the topic of binary arithmetic and its implementation, as well as the more general topic of coding information into binary logical form. We shall use switching algebra as a means of implementing binary arithmetic. In the last section of this chapter we shall present an example of a logical design implementing a binary numeric process. This example will, we hope, clarify the distinction between the binary logical design process and binary arithmetic operations.

First, let us review what is involved in the definition of a number system. We normally write a number as a string of $n$ digits, for example,

$$d_{n-1}d_{n-2}d_{n-3} \cdot \cdot \cdot d_1 d_0$$

In a weighted number system the numeric value is given by

$$N = d_{n-1}r^{n-1} + d_{n-2}r^{n-2} + \cdot \cdot \cdot d_1 r + d_0$$

where $r$ is the radix, or base, of the number system. In the decimal system, for example, we write

$$7419$$

as a shortened notation for

$$7 \times 10^3 + 4 \times 10^2 + 1 \times 10^1 + 9 \times 10^0$$

Because the decimal number system is almost universal, the powers of 10 are implied without further notation. The fact that 10 was chosen as the base of the number system is commonly attributed to the 10 fingers on two hands. Any other relatively small number (a large number of symbols would be awkward) would have done as well.

Consider the number

$$N = 110101$$

and assume that the base of $N$ is 2. The number system with base 2 is called the *binary number system*. We may rewrite $N$ as follows:

$$N = 1 \times 2^5 + 1 \times 2^4 + 0 \times 2^3 + 1 \times 2^2 + 0 \times 2^1 + 1 \times 2^0$$

$$= 32 + 16 + 0 + 4 + 0 + 1$$

$$= 53 \text{ (decimal)}$$

thus obtaining the decimal equivalent. Note that the binary number system requires only two symbols, 0 and 1, whereas the familiar 10 symbols are employed in the decimal system. In general, the number of symbols in a number system must be equal to the radix of the system.

In contrast to people, digital computers do not have 10 fingers and are not biased toward the decimal system. They are constructed of physical devices that operate much more reliably when required to switch between only two stable operating points. These two points can, as just noted, be used to represent the binary digits (or bits) 0 and 1, with the result that the binary number system is widely used in digital systems.

## 3.2  Conversion between Bases

Not every binary string is a binary number, but binary numbers are so widely used in digital systems that it is important that the designer or user of such systems be thoroughly proficient in the use of these numbers. One important problem is conversion between the binary system and other systems. The conversion from binary to decimal was illustrated by the example in the last section; you simply add up the powers of two represented by the 1's in the binary number. In order to do this it is necessary to know the powers of two, which are listed for convenience in Appendix A.

**Example 3.1**

Convert $11010011_2$ to decimal

Solution

$$1 \times 2^7 + 1 \times 2^6 + 0 \times 2^5 + 1 \times 2^4 + 0 \times 2^3 + 0 \times 2^2$$
$$+ 1 \times 2^1 + 1 \times 2^0 = 128 + 64 + 16 + 2 + 1 = 211$$

   For conversion from decimal to binary, there are two methods. The first is just the opposite of the binary-to-decimal, subtracting the powers of two that "fit." ∎

**Example 3.2**

Convert $43_{10}$ to binary.

Solution

$$
\begin{array}{rl}
43 & \\
\underline{-32} & \qquad\qquad\qquad 1 \\
11 & \\
& 16 \qquad\qquad\qquad 0 \\
\underline{-8} & \qquad\qquad\qquad 1 \\
3 & \\
& 4 \qquad\qquad\qquad 0 \\
\underline{-2} & \qquad\qquad\qquad 1 \\
1 & \\
\underline{-1} & \qquad\qquad\qquad 1 \\
0 &
\end{array}
$$

$43_{10} = 101011_2$

   The largest power of two that will "fit" is 32, so we subtract it. The remainder,

11, is smaller than the next power of two, 16, so we insert a 0 in the 16's position.
The next power of two, 8, fits, 4 does not, and so forth.                          ■

An alternative method of conversion uses successive division by two.

## Example 3.3

Convert $43_{10}$ to binary.

Solution

$$
\begin{array}{rl}
156 & 43 \div 2 = 21 + \text{Rem } 1 \\
157 & 21 \div 2 = 10 + \text{Rem } 1 \\
158 & 10 \div 2 = 5 + \text{Rem } 0 \\
159 & 5 \div 2 = 2 + \text{Rem } 1 \\
160 & 2 \div 2 = 1 + \text{Rem } 0 \\
161 & 1 \div 2 = 0 + \text{Rem } 1
\end{array}
$$

$43_{10} = 101011_2$                                                            ■

The following example illustrates a more compact notation for this method.

## Example 3.4

Convert $653_{10}$ to base 2.

```
2|         653        Check
 2|        326   1    1010001101 = 512 + 128 + 8 + 4 + 1
  2|       163   0               = 653₁₀
   2|       81   1
    2|      40   1
     2|     20   0
      2|    10   0
       2|    5   0
        2|   2   1
         2|  1   0
            0   1
```

A disadvantage of the binary number system is the fact that so many bits are
needed for numbers of even modest magnitude. For example,

$$35465$$

a number easy enough to interpret in decimal, is equivalent to

$$111010101011001$$

Not only is it virtually impossible to get any sense of the magnitude of this string of
0's and 1's, you are very likely to make a mistake if you try to copy it. Because of

such problems, it is common to convert binary numbers to hexadecimal (base 16) form. The conversion is simple because 16 is a power of two. The process used in converting the binary number to base 16 may be informally derived as follows. Let $N$ be a number expressed in 12-bit binary form. Therefore,

$$N = d_{11}2^{11} + d_{10}2^{10} + d_9 2^9 + d_8 2^8 + d_7 2^7 + d_6 2^6 + d_5 2^5 + d_4 2^4 + d_3 2^3$$
$$+ d_2 2^2 + d_1 2^1 + d_0 2^0$$
$$= (2^4)^2 (d_{11}2^3 + d_{10}2^2 + d_9 2^1 + d_8) + 2^4 (d_7 2^3 + d_6 2^2 + d_5 2^1 + d_4)$$
$$+ 2^0 (d_3 2^3 + d_2 2^2 + d_1 2^1 + d_0)$$
$$= (d_{11}2^3 + d_{10}2^2 + d_9 2^1 + d_8) \cdot 16^2 + (d_7 2^3 + d_6 2^2 + d_5 2^1 + d_4) \cdot 16^1$$
$$+ (d_3 2^3 + d_2 2^2 + d_1 2^1 + d_0) \tag{3.2}$$

Each of the expressions in parentheses represents a 4-bit binary number in the range 0–15 (decimal). We noted in the last section that a number system requires a number of symbols equal to the radix. Thus, we will need 16 symbols to represent the terms in parentheses. Standard practice for hexadecimal is to use the 10 decimal symbols, plus the first six letters of the alphabet, as shown in Fig. 3.1.

Based on Fig. 3.1, we see that all we need to do to convert a binary number to hexadecimal is to divide the number into groups of four bits, starting at the the least significant bit, and convert each group of four bits to the equivalent hexadeximal digit given in Fig. 3.1.

### Example 3.5

Convert 10101010110011101 to hex.

Solution

```
0001  0101  0101  1001  1101

  1     5     5     9     D

         1559D
```

| Binary | Hex | Decimal | Binary | Hex | Decimal |
|--------|-----|---------|--------|-----|---------|
| 0000 | 0 | 0 | 1000 | 8 | 8 |
| 0001 | 1 | 1 | 1001 | 9 | 9 |
| 0010 | 2 | 2 | 1010 | A | 10 |
| 0011 | 3 | 3 | 1011 | B | 11 |
| 0100 | 4 | 4 | 1100 | C | 12 |
| 0101 | 5 | 5 | 1101 | D | 13 |
| 0110 | 6 | 6 | 1110 | E | 14 |
| 0111 | 7 | 7 | 1111 | F | 15 |

Figure 3.1   Hexadecimal equivalents of binary and decimal numbers.

If the number of bits is not a multiple of four, as is the case here, add leading zeros as required.                                                                                       ■

It should be emphasized that hex representation is used simply as a "shorthand" representation for long binary strings. Computers do not work in the hex system; this system has no particular computational merit. Indeed, the binary strings that we represent in hex need have no numeric significance at all. For example, the binary string

01001101

is used as the code for the letter "M" in a popular alphabetic code that we shall discuss later. The fact that the code has no numeric significance does not prevent us from writing it as

4D

in hex, simply to save writing out all those 0's and 1's. The hex system seems awkward at first, probably because we are not used to seeing letters mixed with numerals, but you will find that you soon get used to it.

The easiest way to convert hexadecimal numbers to decimal, or vice versa, is to first convert to binary and then complete the conversion by one of the methods given earlier. You can, if you prefer, add or subtract powers of 16, or repeatedly divide by 16. However, this will require some skill in hexadecimal arithmetic, which is probably more trouble than it's worth. Skill in binary arithmetic is important, but that topic will be delayed until Chapter 6.

## 3.3   Binary Coding of Decimal Digits

The most common procedure for handling numbers in computers is to convert them to binary on entry and to process them by means of binary arithmetic. The subject of binary arithmetic will be treated in Chapter 6. In some cases, however, it is desirable to preserve the decimal aspects of numbers throughout processing. The numbers have to be converted to binary in the sense of being represented internally by combinations of binary (two-valued) signals, but we can preserve the decimal character to the extent of converting each decimal digit individually to a binary form.

There are many situations in which it is desirable to maintain the decimal form. Probably the most common are those in which the decimal numbers are not numeric in the sense of representing countable or measurable quantities but are more or less arbitrary codes, for example, area codes or ZIP codes. Telephone numbers are a good example. An American telephone number is a 10-digit number, consisting of a three-digit area code, a three-digit exchange number, and a four-digit subscriber number. Suppose you had a list of telephone numbers and wanted to sort them on area code or exchange number. If the numbers were completely converted into 33-bit binary equivalents, the identity of the area codes and exchange numbers would be lost; that is, no particular bits would directly correspond to these parts of the telephone numbers, and sorting would be difficult. If each digit were individually converted, however, the identity of the parts of the telephone numbers would be retained and sorting would be relatively simple.

To represent decimal digits physically in a binary system, we use combinations of binary values on at least four wires; that is, we use a vector of four or more binary bits. If we use a 4-bit vector, these same four bits could represent a 4-bit binary number (with a value from 0000 to 1111) or a *hexadecimal* digit (with a value from 0 to F). If they are to represent decimal digits, we must assign a unique combination of 0's and 1's to each decimal digit. The simplest way to make such an assignment is to let the combinations of 0's and 1's forming the binary numbers with values from 0 to 9 (0000 to 1001) represent the corresponding decimal digits. The bit combinations corresponding to the binary numbers greater than 1001 (decimal 9) will not be used. Thus, the decimal digits are represented by their binary equivalents, that is, in *binary-coded decimal* (BCD) form, as shown in Fig. 3.2. To represent a multidigit number in BCD, we simply convert each digit to the corresponding code as listed in Fig. 3.2.

| **Binary number** | | | | **Decimal digit** |
|---|---|---|---|---|
| $b_3$ | $b_2$ | $b_1$ | $b_0$ | |
| 0 | 0 | 0 | 0 | 0 |
| 0 | 0 | 0 | 1 | 1 |
| 0 | 0 | 1 | 0 | 2 |
| 0 | 0 | 1 | 1 | 3 |
| 0 | 1 | 0 | 0 | 4 |
| 0 | 1 | 0 | 1 | 5 |
| 0 | 1 | 1 | 0 | 6 |
| 0 | 1 | 1 | 1 | 7 |
| 1 | 0 | 0 | 0 | 8 |
| 1 | 0 | 0 | 1 | 9 |
| 1 | 0 | 1 | 0 | None |
| 1 | 0 | 1 | 1 | None |
| 1 | 1 | 0 | 0 | None (Not used) |
| 1 | 1 | 0 | 1 | None |
| 1 | 1 | 1 | 0 | None |
| 1 | 1 | 1 | 1 | None |

*Figure 3.2  Binary-coded-decimal digits (BCD).*

**Example 3.6**

Convert $653_{10}$ to BCD

Solution

$$6 \quad 5 \quad 3$$

$$\overline{0110}\ \overline{0101}\ \overline{0111}$$

As a comparison, note that the direct binary equivalent, determined in Example 3.4 is

$$1010001101$$ ∎

Note that the BCD forms in Fig. 3.2 are also codes. They are not arbitrary codes, but they are alternative representations of the decimal digits, in much the same sense that area code (212) is an alternative representation of New York City. Codes are another of the subjects of Chapter 6. We have barely scratched the surface here.

At the beginning of this chapter we noted that 0 and 1 are logical values, representing the two possible signal levels in a binary system. In addition, they will also have numeric significance when the binary signals represent numbers in the binary number system. The BCD code represents an interesting mix of the two meanings. The codes for the decimal digits are chosen to be the equivalent binary numbers, but that does not mean that they should be treated as binary numbers in the design of circuits that process them. The following example will, we hope, clarify the relationship.

### Example 3.7

Binary-coded decimal digits are to be represented by the values on four wires, $x_3$, $x_2$, $x_1$, $x_0$. Design a combinational logic circuit that will have an output of 1 if and only if the code on wires $x_3$, $x_2$, $x_1$, $x_0$ actually represents a decimal digit. If, because of an error of some sort, one of the six "illegal" codes appears, the circuit output should be 0.

### Solution

A truth table defining the function $f(x_3,x_2,x_1,x_0)$ is given as Fig. 3.3. This table looks very much like Fig. 3.2, but now we are talking about logical values, not binary numbers! For the purposes of this problem, each 4-bit code in Fig. 3.3 is an aribitrary combination of values assigned to a particular decimal digit. The fact that these codes were chosen because they are also the equivalent binary numbers is irrelevant to the logical design problem. The nature of the logical design problem would be no different if we had chosen any arbitrary 10 of the 16 possible combinations. Thus, the logical operators AND, OR, and NOT are now valid on these logical variables whose values are tabulated in Fig. 3.3, and we may use these operators to realize the function, $f(x_3,x_2,x_1,x_0)$, which must be 1 if and only if the values of $x_3,x_2,x_1,x_0$ correspond to a binary-coded decimal digit.

Fortunately, we can easily obtain an expression for $f(x_3,x_2,x_1,x_0)$ by inspection of the truth table. Notice that $f$ is 1 whenever $x_3 = 0$ OR $x_3 = 1$ AND $x_2 = 0$ AND $x_1 = 0$. Otherwise, $f = 0$. We may, therefore, write

$$f(x_3,x_2,x_1,x_0) = \overline{x}_3 \vee x_3\overline{x}_2\overline{x}_1 = \overline{x}_3 \vee \overline{x}_2\overline{x}_1$$

The resulting logic circuit is shown in Fig. 3.4. ∎

In Chapter 6 and thereafter, we shall use logic circuits to implement arithmetic operations on binary numbers. We may not continue to remind the reader of the

| $x_3$ | $x_2$ | $x_1$ | $x_0$ | $f(x_3, x_2, x_1, x_0)$ |
|---|---|---|---|---|
| 0 | 0 | 0 | 0 | 1 |
| 0 | 0 | 0 | 1 | 1 |
| 0 | 0 | 1 | 0 | 1 |
| 0 | 0 | 1 | 1 | 1 |
| 0 | 1 | 0 | 0 | 1 |
| 0 | 1 | 0 | 1 | 1 |
| 0 | 1 | 1 | 0 | 1 |
| 0 | 1 | 1 | 1 | 1 |
| 1 | 0 | 0 | 0 | 1 |
| 1 | 0 | 0 | 1 | 1 |
| 1 | 0 | 1 | 0 | 0 |
| 1 | 0 | 1 | 1 | 0 |
| 1 | 1 | 0 | 0 | 0 |
| 1 | 1 | 0 | 1 | 0 |
| 1 | 1 | 1 | 0 | 0 |
| 1 | 1 | 1 | 1 | 0 |

Figure 3.3 BCD digit detection.

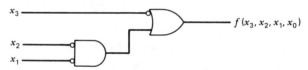

Figure 3.4 BCD digit detector.

transition between binary bits and logical values as we did in the previous example. In fact, the distinction may not be constantly on the mind of the practicing digital system designer. The underlying relationship is nonetheless important and should be understood. There is no automatic translation of an operation in binary arithmetic to a logic circuit realization. The design process relies on the careful enumeration of each logic functional value so that the Boolean expressions on logical values will yield the same results as the arithmetic operations on the binary bits.

## 3.4 Historical Note

Reference to a *dyadic* or binary number system has been uncovered in a Chinese book believed to have been written about 3000 B.C. The Western world of the sixteenth and seventeenth centuries did not have access to this work, so the concept had to be rediscovered. Notes left by John Napier (1550–1617), the inventor of the logarithm, indicated that he was aware of the value of the binary scale. A half-century later, Gottried Wilhelm von Leibniz, the inventor of calculus, more completely doc-

umented and extolled the virtues of the radix-two number system. Apparently he did
not make use of binary numbers in his calculator mentioned in Section 1.5.

Some algebra textbooks of the 1890s included a short chapter on scales of notation. The usefulness of base 2 was not apparent at that time, and it soon disappeared
from the textbooks. Fortunately, these concepts were well known to the computer
pioneers of the 1930s and 1940s.

## Problems

**3.1**  Convert each of the following decimal integers to their binary equivalents.
(a) 19
(b) 67
(c) 81
(d) 153

**3.2**  Convert the binary number 11100101011101 to hexadecimal.

**3.3**  Find the binary equivalent of the hexadecimal numbers
(a) A05F
(b) FFDE

**3.4**  Find the hexadecimal equivalent of each of the decimal numbers given in Problem 3.1.

**3.5**  Find the decimal equivalent of the hexadecimal numbers given in Problem 3.3.

**3.6**  (a) Determine the Boolean expression for a logic function, $f(x_3,x_2,x_1,x_0)$,
which will be be 1 if and only if a binary number $a_3a_2a_1a_0$ is greater than or
equal to 10 (decimal).
(b) Obtain a logic diagram of a realization of this function.

## References

1. Hill, F. J., and G. R. Peterson, *Introduction to Switching Theory and Logical Design,* 3rd
ed., Chap. 2. Wiley, New York. 1981.
2. Smith, D. E., and J. Ginsburg, "From Numbers to Numerals and from Numerals to
Computation," in *The World of Mathemtics,* J. R. Newman, ed., Vol. 1. Simon & Schuster, New York, 1956.

# Simplification of Boolean Functions

## 4.1 Standard Forms of Boolean Functions

We have seen that a problem with the direct application of Boolean theorems to the simplification of logic equations is that of deciding which theorems to apply in what order. This problem is made more difficult by the fact that the original functions to be simplified may be in almost any form. An important step in the process of developing a systematic procedure for simplification is the development of a standard Boolean form to serve as a starting point. A logical basis for such a standard form is the truth table, since it is unique to each function. Indeed, since the truth table lists the values of the function for all possible combinations of variables, the truth table defines the function. Thus, a Boolean form corresponding in some direct way to the truth table would seem a logical choice for a standard form.

Consider the function defined by the truth table of Fig. 4.1. We shall now show that the Boolean expression of Eq. 4.1 is a representation of this function.

$$f(A,B,C) = \overline{A}BC \vee A\overline{B}\overline{C} \vee ABC \tag{4.1}$$

First consider the AND term, $\overline{A}BC$. When $A = 0$, $B = 1$, and $C = 1$, this term takes on the value of $1 \wedge 1 \wedge 1 = 1$. Since 1 OR "anything" equals 1, the expression of Eq. 4.1 takes on the value 1 for this combination of variables, as specified by the truth table. Also note that this AND term will take on the value 0 for any other combination of input variables. In a similar fashion, the AND term $ABC$ will be equal to 1 for the combination of values $A = 1$, $B = 0$, and $C = 0$, and for no other, and the AND term $A\overline{B}\overline{C}$ will take on the value 1 only for the input combination $A$

| Row No. | $A$ | $B$ | $C$ | $f(A,B,C)$ |
|:---:|:---:|:---:|:---:|:---:|
| 0 | 0 | 0 | 0 | 0 |
| 1 | 0 | 0 | 1 | 0 |
| 2 | 0 | 1 | 0 | 0 |
| 3 | 0 | 1 | 1 | 1 |
| 4 | 1 | 0 | 0 | 1 |
| 5 | 1 | 0 | 1 | 0 |
| 6 | 1 | 1 | 0 | 0 |
| 7 | 1 | 1 | 1 | 1 |

*Figure 4.1*

$= 1, B = 1$, and $C = 1$. For each row in the truth table for which the function is equal to 1, one of the AND terms in Eq. 4.1 will equal 1 and the expression will therefore equal 1. For all other rows in the truth table, all three AND terms will equal 0 and the expression will equal 0. We see, then, that the logical equation of Eq. 4.1 represents the function of Fig. 4.1.

Because there is a certain similarity between arithmetic multiplication and Boolean AND, terms of the form $\overline{A}BC$ are often referred to as *product* terms. Furthermore, because there is some similarity between arithmetic additon and Boolean OR, a Boolean expression consisting of an ORing of AND terms is often referred to as a *sum-of-products (SOP)* form. An SOP form such as Eq. 4.1, in which each product contains all the variables of the function and corresponds to a single row in the truth table, is known as a *standard sum-of-products (SSOP)* form. It is considered a standard form because of its direct, one-to-one, correspondence to the defining truth table.

The procedure for finding the SSOP form is simple. For each row of the truth table for which the function is equal to 1, set up a product term containing all the variables of the function. Each variable should be true (uncomplemented) if the value of that variable on that row is 1 and false (complemented) if the value of the variable is 0 on that row. These products, containing all the variables of the function and corresponding to the 1-rows of the function, are known as the *minterms* of the function, for reasons that will be explained later.

### Example 4.1

Find the SSOP form for the elevator door function.

### Solution

We found the truth table for this function in Fig. 3.5. In Fig. 4.2 we show the rows of the truth table for which the function equals 1. On each row we have also shown the corresponding minterm. The SSOP form is

$$f(T,E,C,F) = \overline{T}E\overline{C}F \lor \overline{T}EC\overline{F} \lor \overline{T}ECF \qquad (4.2)$$

■

| Row no. | T | E | C | F | D | Minterm |
|---|---|---|---|---|---|---|
| 5 | 0 | 1 | 0 | 1 | 1 | $\overline{T}E\overline{C}F$ |
| 6 | 0 | 1 | 1 | 0 | 1 | $TECF$ |
| 7 | 0 | 1 | 1 | 1 | 1 | $\overline{T}ECF$ |

*Figure 4.2*

| Row No. | A | B | C | f(A,B,C) |
|---|---|---|---|---|
| 0 | 0 | 0 | 0 | 1 |
| 1 | 0 | 0 | 1 | 1 |
| 2 | 0 | 1 | 0 | 0 |
| 3 | 0 | 1 | 1 | 1 |
| 4 | 1 | 0 | 0 | 0 |
| 5 | 1 | 0 | 1 | 1 |
| 6 | 1 | 1 | 0 | 0 |
| 7 | 1 | 1 | 1 | 1 |

*Figure 4.3*

Next consider the truth table of Fig. 4.3. This function equals 1 for five rows of the table and 0 for three rows. We can determine the SSOP form,

$$f(A,B,C) = \overline{A}\overline{B}\overline{C} \vee \overline{A}\overline{B}C \vee \overline{A}BC \vee A\overline{B}C \vee ABC \tag{4.3}$$

but we see that it is somewhat complicated since it requires five minterms. Since there is a form that corresponds directly to the 1's of the function, it seems reasonable that there might be a form corresponding directly to the 0's. This form is shown in Eq. 4.4,

$$f(A,B,C) = (A \vee \overline{B} \vee C)(\overline{A} \vee B \vee C)(\overline{A} \vee \overline{B} \vee C) \tag{4.4}$$

which is an example of the *standard product-of-sums (SPOS)* form.

Consider first the sum term $(A \vee \overline{B} \vee C)$. For $A = 0, B = 1, C = 0$, this term equals $0 \vee 0 \vee 0 = 0$. Since 0 AND "anything" equals 0, the expression of Eq. 4.4 will equal 0 for this combination of inputs, as specified by the truth table. For any other combination of variables, at least one term in the sum will equal 1, and the entire sum term will equal 1. In a similar manner, the sum term $(\overline{A} \vee B \vee C)$ will equal 0 for the combination $A = 1, B = 0$, and $C = 0$, and for no other, and the sum $(\overline{A} \vee \overline{B} \vee C)$ will equal 0 only for the combination $A = 1, B = 1$, and $C = 0$. Thus, for each row of the truth table for which the function equals 0, one of the sums, and therefore the entire expression, will equal 0. For all other rows, all three sums, and therefore the entire expression, will equal 1.

The sum terms in an SPOS form are known as *maxterms* of the function. The procedure for finding the SPOS form is simple. For each row of the truth table for which the function equals 0, set up a sum term containing all the variables of the function. A variable should be uncomplemented if that variable has the value 0 on that row of the truth table, complemented if the variable is 1 on that row of the truth table. Note that the correspondence between complemented and uncomplemented variables and the 1's and 0's in the variable values is reversed from that for minterms.

### Example 4.2

Find the SPOS form for the function defined by the truth table of Fig. 4.4.

### Solution

$$f(A,B,C) = (A \lor B \lor C)(\overline{A} \lor B \lor \overline{C})(\overline{A} \lor \overline{B} \lor \overline{C}) \qquad (4.5)$$

∎

Because of the direct relationships to the 1's and 0's, it is common to say that the minterms "create" the 1's of the function and the maxterms "create" the 0's of the function. The reader should not interpret this terminology as implying that you need both, minterms for 1's and maxterms for 0's. The SSOP form directly specifies the values for which the function will be 1; it will equal 0 for all other values. The SPOS form directly specifies the values for which the function will be 0; it will be 1 for all other values. Since there are two standard forms, it is natural to wonder which is preferable, but we really cannot answer the question at this point.

You will note that the standard forms are not necessarily simple forms; an SSOP form with 14 minterms of five variables would be very tedious to write out. A shorter equivalent form would be useful. In Figs. 4.1 to 4.4 you will note a column labeled "Row No." Each row is numbered with the decimal equivalent of the variable combination interpreted as a binary number. For example, in Fig. 4.1 the variable combination $A = 0$, $B = 1$, $C = 1$ is interpreted as a binary number giving us

$$011_2 = 3_{10}$$

so that row is called Row 3. Now we can specify a function simply by listing the rows for which it equals 1.

| Row No. | $A$ | $B$ | $C$ | $f(A,B,C)$ | Maxterm |
|:---:|:---:|:---:|:---:|:---:|:---:|
| 0 | 0 | 0 | 0 | 0 | $A \lor B \lor C$ |
| 1 | 0 | 0 | 1 | 1 | |
| 2 | 0 | 1 | 0 | 1 | |
| 3 | 0 | 1 | 1 | 1 | |
| 4 | 1 | 0 | 0 | 1 | |
| 5 | 1 | 0 | 1 | 0 | $\overline{A} \lor B \lor \overline{C}$ |
| 6 | 1 | 1 | 0 | 1 | |
| 7 | 1 | 1 | 1 | 0 | $\overline{A} \lor \overline{B} \lor \overline{C}$ |

*Figure 4.4*

$$f(A,B,C) = \Sigma\, m(3,4,7) \tag{4.6}$$

This is known as the *minterm list form.* Even though we did not find the SPOS form for Fig. 4.1, we can specify the function by listing the rows for which the function equals 0, the maxterm list form.

$$f(A,B,C) = \Pi M(0,1,2,5,6) \tag{4.7}$$

In these forms, m indicates minterms, M indicates maxterms, the $\Sigma$ symbol suggests *sum of products,* the $\Pi$ symbol suggests *product of sums.* Since both these forms and the SSOP and SPOS forms correspond to the rows of the truth table on a one-to-one basis, it is obvious that there is a direct relationship between the forms, and conversion from one to the other is straightforward, as shown in Eqs. 4.8 and 4.9.

$$f(A,B,C) = \Sigma\, m(2,4,7) \tag{4.8}$$
$$010 \quad 100 \quad 111$$

$$f(A,B,C) = \overline{A}B\overline{C} \vee A\overline{B}\,\overline{C} \vee ABC$$

$$f(A,B,C) = \Pi M(1,5,6) $$
$$001 \quad 101 \quad 110 \tag{4.9}$$

$$f(A,B,C) = (A \vee B \vee \overline{C})(\overline{A} \vee B \vee \overline{C})(\overline{A} \vee \overline{B} \vee C)$$

Equation 4.8 illustrates the conversion between minterm list and SSOP form, and Eq. 4.9, the conversion between maxterm list and SPOS form. (Note that there is no connection between Eqs. 4.8 and 4.9; they are entirely different functions.) This relationship also leads to the identification of minterms and maxterms by number, namely,

$$\overline{A}B\overline{C} = m_2, \quad A\overline{B}\,\overline{C} = m_4, \quad (\overline{A} \vee B \vee \overline{C}) = M_5, \quad (A \vee \overline{B} \vee \overline{C}) = M_3, \ldots$$

When using minterm list or maxterm list forms, it is important to indicate the order in which the variables appear in the truth table. Referring to Fig. 4.2, the minterm list notation for the elevator door function is

$$f(T,E,F,C) = \Sigma\, m(5,6,7) \tag{4.10}$$

But suppose the original truth table had been set up with the variables in the order $C,E,F,T$. (There is no reason to list the variables in any particular order.) The 1's of the function would, of course, occur for the same combinations of input values, but the corresonding three rows of the truth table would now appear as shown in Fig. 4.5, from which we determine minterm list and SSOP forms are as shown in Eqs. 4.11 and 4.12.

$$f(C,E,F,T) = \Sigma\, m(6,12,14) \tag{4.11}$$

$$f(C,E,F,T) = \overline{C}EF\overline{T} \vee CEF\overline{T} \vee CEF\overline{T} \tag{4.12}$$

| Row no. | C | E | F | T | D | Minterm |
|---|---|---|---|---|---|---|
| 6 | 0 | 1 | 1 | 0 | 1 | $\overline{C}EF\overline{T}$ |
| 12 | 1 | 1 | 0 | 0 | 1 | $CE\overline{F}\overline{T}$ |
| 14 | 1 | 1 | 1 | 0 | 1 | $CEF\overline{T}$ |

*Figure 4.5*

A comparison of Eq. 4.12 and Eq. 4.2 will show that these are the same minterms, the variables are just written in a different order, which, from the commutative law, makes no difference. But it is clear that a notation such as

$$\Sigma m(4,12,14)$$

means nothing unless the order in which the variables are listed is known.

Determining any of these standard forms from the truth table obviously requires that we first find the truth table. Although finding the truth table can be a useful step in verifying the correctness of the logical expression, in many cases it just is not practical. As we saw in Fig. 3.5, the truth table for the elevator control function, we have to find tables for each part of the function and then combine them to find the final table. This can be a tedious job for functions of four variables; for functions of more variables it may be totally impractical. However, now that we know that there is a standard algebraic form corresponding to the truth table, we can apply algebraic manipulation to convert the original algebraic form to one of the standard forms. This is not as difficult as simplification since we know what we are looking for. Let us first consider the conversion of arbitrary Boolean forms to sum-of-products form.

**Example 4.3**

Express $f(A,B,C,D) = (AC \vee B)(CD \vee \overline{D})$ in SOP form.

Solution
First, let $AC = x$ and $CD = y$. Then

$$f(A,B,C,D) = (x \vee B)(y \vee \overline{D})$$
$$= (x \vee B)y \vee (x \vee B)\overline{D} \quad \text{(Distributive law)} \tag{4.13}$$
$$= xy \vee By \vee x\overline{D} \vee B\overline{D} \quad \text{(Distributive law)}$$
$$= ACD \vee BCD \vee AC\overline{D} \vee B\overline{D} \quad \blacksquare$$

**Example 4.4**

Repeat Example 4.3 for the function

$$f(A,B,C,D,E) = (\overline{AC} \vee \overline{D})(\overline{B} \vee CE)$$

Solution

$$f(A,B,C,D,E) = (\overline{AC} \vee \overline{D})(\overline{B \vee CE})$$
$$= [(\overline{A} \vee \overline{C}) \vee \overline{D}](\overline{B} \wedge \overline{CE})$$
$$= [(\overline{A} \vee \overline{C})\overline{B} \vee \overline{B}\,\overline{D}](\overline{C} \vee \overline{E}) \qquad\qquad (4.14)$$

$$= (\overline{A}\,\overline{B} \vee \overline{B}\,\overline{C} \vee \overline{B}\,\overline{D})(\overline{C} \vee \overline{E})$$
$$= \overline{A}\,\overline{B}\,\overline{C} \vee \overline{B}\,\overline{C} \vee \overline{B}\,\overline{C}\,\overline{D} \vee \overline{A}\,\overline{B}\,\overline{E} \vee \overline{B}\,\overline{C}\,\overline{E} \vee \overline{B}\,\overline{D}\,\overline{E} \qquad ∎$$

The basic procedure used in these examples, successive application of the distributive law, together with DeMorgan's law when complemented expressions are involved, will work for functions of any form or degree of complexity. Sum-of-products forms such as Eqs. 4.13 and 4.14 are often adequate for the purposes to be explored in the next sections, but they can be simply expanded to the SSOP form if desired.

**Example 4.5**

Convert the SOP form of Eq. 4.13 to SSOP form.

Solution

$$f(A,B,C,D) = ACD \vee BCD \vee AC\overline{D} \vee B\overline{D} = ACD(B \vee \overline{B})$$
$$+ BCD(A \vee \overline{A}) \vee AC\overline{D}(B \vee \overline{B}) \vee B\overline{D}(A \vee \overline{A})$$
$$(a \vee \overline{a} = 1, a \wedge 1 = a)$$
$$= ABCD \vee A\overline{B}CD \vee ABCD \vee \overline{A}BCD \vee ABC\overline{D}$$
$$\vee A\overline{B}C\overline{D} \vee B\overline{D}A \vee B\overline{D}\overline{A}$$

$$= ABCD \vee A\overline{B}CD \vee \overline{A}BCD \vee ABC\overline{D} \vee A\overline{B}C\overline{D}$$
$$\vee B\overline{D}A(C \vee \overline{C}) \vee B\overline{D}\overline{A}(C \vee \overline{C})$$
$$(a \vee a = a, a \vee \overline{a} = 1, a \wedge 1 = a)$$
$$= ABCD \vee A\overline{B}CD \vee \overline{A}BCD \vee ABC\overline{D} \vee A\overline{B}C\overline{D}$$
$$\vee ABC\overline{D} \vee AB\overline{C}\overline{D} \vee \overline{A}BC\overline{D} \vee \overline{A}B\overline{C}\overline{D} \qquad \text{(Distributive law)}$$
$$f(A,B,C,D) = ABCD \vee A\overline{B}CD \vee \overline{A}BCD \vee ABC\overline{D} \vee A\overline{B}C\overline{D}$$
$$\vee AB\overline{C}\overline{D} \vee \overline{A}BC\overline{D} \vee \overline{A}B\overline{C}\overline{D}$$
$$(a \vee a = a)$$

$$(4.15)$$

This form can in turn be converted to minterm list form.                          ∎

## Example 4.6

Convert the SSOP form of Eq. 4.15 to a minterm list.

Solution

$$f(A,B,C,D) = ABCD \lor A\overline{B}CD \lor \overline{A}BCD \lor ABC\overline{D} \lor A\overline{B}C\overline{D} \lor AB\overline{C}\overline{D} \lor \overline{A}BC\overline{D} \lor \overline{A}B\overline{C}\overline{D}$$

| 1111 | 1011 | 0111 | 1110 | 1010 | 1100 | 0110 | 0100 | (4.16) |
|------|------|------|------|------|------|------|------|--------|
| 15   | 11   | 7    | 14   | 10   | 12   | 6    | 4    |        |

$$f(A,B,C,D) = \Sigma m(4,6,7,10,11,12,14,15) \quad\blacksquare$$

Similar techniques can be applied to convert any Boolean function to the POS or SPOS form. However, most people find these techniques more difficult and prefer to work with SOP or SSOP forms. If the SPOS form is needed, it can be easily found from the SSOP form.

## Example 4.7

Find the maxterm list and SPOS forms for the function of Eq. 4.16.

Solution
The rows of the truth table represented by maxterms are the rows not represented by minterms, so we have

$$f(A,B,C,D) = \Pi M(0,1,2,3,5,8,9,13) \quad (4.17)$$

0000      0001      0010      0011

$$= (A \lor B \lor C \lor D)(A \lor B \lor C \lor \overline{D})(A \lor B \lor \overline{C} \lor D)(A \lor B \lor \overline{C} \lor \overline{D})$$

0101      1000      1001      1101

$$(A \lor \overline{B} \lor C \lor \overline{D})(\overline{A} \lor B \lor C \lor D)(\overline{A} \lor B \lor C \lor \overline{D})(\overline{A} \lor \overline{B} \lor C \lor \overline{D})$$

(4.18)

$\blacksquare$

# 4.2  Karnaugh Map Representation of Boolean Functions

The *Karnaugh map* [2] is one of the most powerful tools available to the logic designer. The power of the Karnaugh map does not lie in its application of any marvelous new theorems but rather in its use of the remarkable ability of the human mind to perceive patterns in pictorial representations of data. This is not a new idea. Any time we use a graph instead of a table of numeric data, we are using the human ability to recognize complex patterns and relationships in a graphical representation far more rapidly and surely than in a tabular representation.

A Karnaugh map may be regarded either as a pictorial form of a truth table or as an extension of the Venn diagram. First consider a truth table for two variables. We list all four possible input combinations and the corresponding function values; for example, the truth tables for AND and OR (Fig. 4.6).

| $A$ | $B$ | $A \wedge B$ |  | $A$ | $B$ | $A \vee B$ |
|---|---|---|---|---|---|---|
| 0 | 0 | 0 |  | 0 | 0 | 0 |
| 0 | 1 | 0 |  | 0 | 1 | 1 |
| 1 | 1 | 1 |  | 1 | 1 | 1 |
| 1 | 0 | 0 |  | 1 | 0 | 1 |

*Figure 4.6   Truth tables for AND and OR.*

As an alternative approach, let us set up a diagram consisting of four small boxes, one for each combination of variables. Place a "1" in any box representing a combination of variables for which the function has the value 1. There is no logical objection fo putting "0's" in the other boxes, but they are usually omitted for clarity. Figure 4.7 shows two forms of Karnaugh maps for $AB$ and $A \vee B$. The diagrams of Fig. 4.7a are perfectly valid Karnaugh maps, but it is more common to arrange the four boxes in a square, as shown in Fig. 4.7b.

As an alternative approach, recall the interpretation of the Venn diagram discussed in Section 3.6. We interpret the universal set as the set of all $2^n$ combinations of values of $n$ variables, divide this set into $2^n$ equal areas, and then darken the areas corresponding to input combinations in which the function has the value 1. We start with the universal set represented by a square (Fig. 4.8a) and divide it in half, corresponding to input combinations in which $A = 1$ and combinations in which $\overline{A} = 1$ (Fig. 4.8b). We then divide it in half again, corresponding to $B = 1$ and $\overline{B} = 1$ (Fig. 4.8c).

With this notation, the interpretations of AND as the intersection of sets and OR as the union of sets make it particularly simple to determine which squares should be darkened (Fig. 4.9).

We note that the shaded areas in Fig. 4.9 correspond to the squares containing

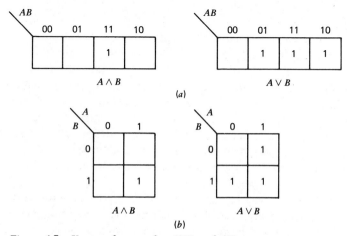

*Figure 4.7   Karnaugh maps for AND and OR.*

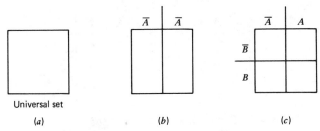

*Figure 4.8    Development of Karnaugh maps by the Venn diagram approach.*

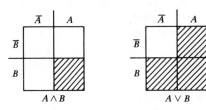

*Figure 4.9    Karnaugh maps of AND and OR (Venn form).*

1's in Fig. 4.7. Thus, both interpretations lead to the same result. We might say that the Karnaugh map is essentially a diagrammatic form of truth table and that the Venn diagram concepts of union and intersection of areas aid us in setting up or interpreting a Karnaugh map.

Since there must be one square for each input combination, there must be $2^n$ squares in a Karnaugh map for $n$ variables. Whatever the number of variables, we may interpret the map in terms of a graphical form of the truth table (Fig. 4.10*a*) or in terms of union and intersection of areas (Fig. 4.10*b*).

In Fig. 4.10*b*, we have changed the labeling slightly and placed 1's in the squares, rather than darkening them. The two types of map (Figs. 4.10*a* and *b*) are thus the same, except for labeling, and both types will be called Karnaugh maps or, for short, K-maps (see Ref. [2]). The K-maps for some other three-variable functions are shown in Fig. 4.11 to clarify the concepts involved.

Note particularly the functions mapped in Figs. 4.10*a* and 4.11*b*. These are both minterms, $m_7$ and $m_4$, respectively. Each is represented by one square; obviously, each one of the eight squares corresponds to one of the eight minterms of three variables. This is the origin of the name *minterm*. A minterm is the form of Boolean function corresponding to the minimum possible area, other than 0, on a Karnaugh map. A *maxterm,* on the other hand, is the form of Boolean function corresponding to the maximum possible area, other than 1, on a Karnaugh map. Figures 4.10*b* and 4.11*c* are maps of maxterms $M_0$ and $M_3$; we see that each map covers the maximum possible area—all the squares but one.

Since each square on a K-map corresponds to a row in a truth table, it is appropriate to number the squares just as we numbered the rows. These standard K-maps (as we shall call them) are shown in Fig. 4.12 for two and three variables. Now if a

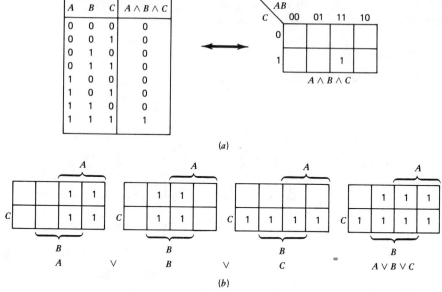

(a)

(b)

*Figure 4.10 Karnaugh maps for three-variable AND and OR.*

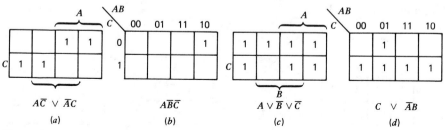

*Figure 4.11 Sample three-variable Karnaugh maps.*

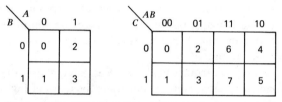

*Figure 4.12 Standard K-maps for two and three variables.*

function is stated in the form of the minterm list, all we need to do is enter 1's in the corresponding squares to produce the K-map.

### Example 4.8

Develop the K-map of $f(A,B,C) = \Sigma\, m(0,2,3,7)$.

Solution

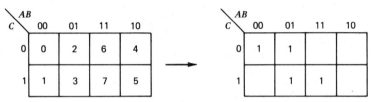

*Figure 4.13*

If a function is stated as a maxterm list, we can enter o's in the squares listed or 1's in those not listed.

### Example 4.9

Develop the K-map of $f(A,B,C) = \Pi M(0,1,5,6)$.

Solution

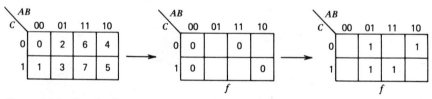

*Figure 4.14   Standard map.*

A map showing the 0's of a function is a perfectly valid K-map, although it is more common to show the 1's.

In developing these basic concepts, we have restricted ourselves to the simple two- and three-variable K-maps, but in practical cases we will more often be using maps for functions of more variables. The standard map for four variables is shown in Fig. 4.15 in both notations.

We have seen that one requirement for a Karnaugh map is that there must be a square corresponding to each input combination; the maps of Fig. 4.15 satisfy this requirement. Another requirement is that the squares must be so arranged that any pair of squares immediately adjacent to one another (horizontally or vertically) must

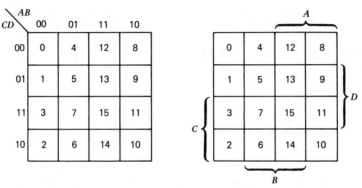

Figure 4.15 Standard K-maps for four variables.

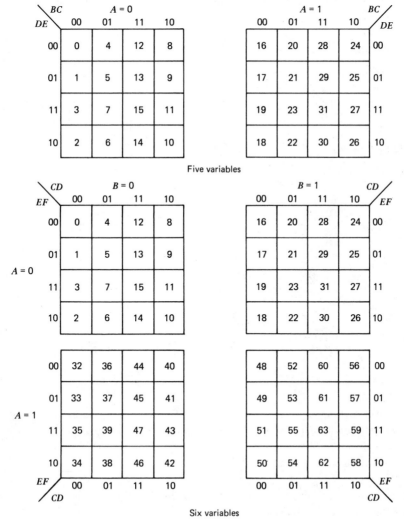

Figure 4.16 Standard K-maps.

**57**

correspond to a pair of input conditions that are *logically adjacent,* that is, differ in only one variable. For example, squares 5 and 13 on the maps in Fig. 4.15 correspond to input combinations $\overline{A}B\overline{C}D$ and $AB\overline{C}D$, *identical except in A. Note that squares at the ends of columns or rows are also logically adjacent.*

The standard K-maps for five variables and six variables are shown in Fig. 4.16. In the five-variable map we have two four-variable maps placed side by side. They are identical in *BCDE*, but one corresponds to $A = 1$, the other to $A = 0$. The standard four-variable adjacencies apply in each half. In addition, squares in the same relative position on the two halves, for example, 4 and 20, are also adjacent. Any maps that satisfy the requirements of two squares in proper adjacency may be considered Karnaugh maps. The six-variable map is formed by doubling the five-variable map and appropriately relabeling. Maps could be similarly formed for even more variables, but six variables is about the limit; maps for more variables are just too large to use efficiently.

### Example 4.10

Find K-maps for the following functions:
(a) $f(V,W,Y,Z) = \Sigma m(9,20,21,29,30,31)$.
(b) $f(A,B,C,D,E) = AB \vee \overline{C}D \vee DE$.

Solution
(a)  For the first function, we enter 1's in the listed squares.

| YZ \ WX | V = 0 |  |  |  | V = 1 |  |  |  | WX / YZ |
|---|---|---|---|---|---|---|---|---|---|
|  | 00 | 01 | 11 | 10 | 00 | 01 | 11 | 10 |  |
| 00 | 0 | 4 | 12 | 8 | 16 | 20 **1** | 28 | 24 | 00 |
| 01 | 1 | 5 | 13 | 9 **1** | 17 | 21 **1** | 29 **1** | 25 | 01 |
| 11 | 3 | 7 | 15 | 11 | 19 | 23 | 31 **1** | 27 | 11 |
| 10 | 2 | 6 | 14 | 10 | 18 | 22 | 30 **1** | 26 | 10 |

*Figure 4.17*

(b)  Referring to the standard map for five variables, we easily identify the maps of the individual product terms, as shown in Fig. 4.18. We then take the union of these to form the final K-map (Fig. 4.19). The minterm list of this function may now be read directly from the map.

$$f(A,B,C,D,E) = AB \vee \overline{C}D \vee DE$$

$$= \Sigma m(2,3,7,10,11,15,18,19,23,24,25,26,27,28,29,30,31) \qquad (4.19)$$

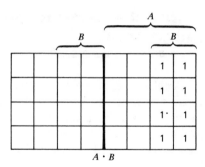

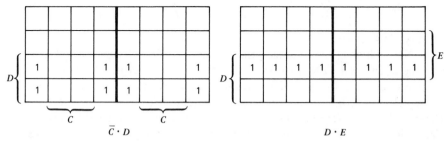

Figure 4.18

Figure 4.19

In part (*a*) of Example 4.10, note that the variables do not have to be *A,B,C,* . . . . Obviously, we can call the variables anything we want. The only precaution to be observed is that if we work with minterm numbers, the variables must appear on the K-map in the proper manner corresponding to the order of their listing in the function statement. Part (*b*) of Example 4.10 shows that it is not necessary to have the function of the SSOP or minterm list form to find the K-map. The Venn diagram interpretation makes it possible to move directly to the map from an SOP form.

# 4.3  Simplification of Functions on the Karnaugh Map

It has taken us a while to establish the necessary tools, but we are now ready to use them for simplifying functions. If we are to develop truly general-purpose methods, we need, in addition to a standard starting point, at least some idea of what sort of a circuit form is desired. The specification, "Find the simplest possible circuit," is too "open-ended" to have much meaning. Let us consider two different realizations of a four-variable function,

$$f(A,B,C,D) = (A \lor B)(C\overline{D} \lor \overline{C}D) \tag{4.20}$$

$$= AC\overline{D} \lor A\overline{C}D \lor BC\overline{D} \lor B\overline{C}D \tag{4.21}$$

as shown in Fig. 4.20.

Now, which of these forms is the simplest? At first glance, you will likely feel that the circuit of Fig. 4.20*a* is the simplest. It is certainly simpler in terms of the number of gates, but it has a drawback relative to the other circuit form. Note that the signals A and B, entering the left-most gate, pass through three gates, or three *levels of gating,* before reaching the output. By comparison, all the signals in the circuit of Fig. 4.20*b* pass through only two levels of gating. So far we have not concerned ourselves in any way with the speed at which gates function. This matter will be discussed in a later chapter; for now we shall simply point out that there is always some *delay* between the time that a gate input changes and the time when a resultant change is seen at the output of the gate. The delays are incredibly short in terms of a human frame of reference, but they can be significant. The sum of gate delays in a signal path within a computer will influence the speed at which that computer can operate. Since circuits with two levels of gating (such as Fig. 4.20*b*) will have less delay than circuits with more levels, a significant portion of the combinational logic circuits within any system will be of the two-level form.

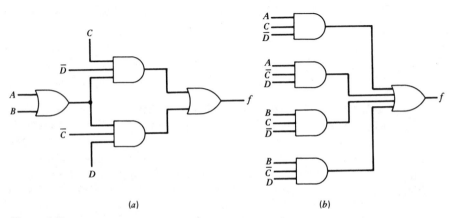

*(a)*                    *(b)*

*Figure 4.20*

Fortunately, simple procedures exist for finding the least costly two-level circuit directly realizing an SOP or POS form of any Boolean function. No such convenient process exists for finding multilevel realizations such as Fig. 4.20a. The least costly, or *minimal, two-level realization* may be defined as follows.

Definition: A two-level sum-of-products expression will be regarded as a *minimal* expression if there exists (1) no other equivalent espression with fewer products and (2) no other equivalent expression with the same number of products but a smaller number of variables in' the products. A minimal product-of-sums is defined in the same way with the word *products* replaced by the word *sums,* and vice versa.

Note that *a* minimal, rather than *the* minimal expression is characterized by this definition. As we shall see, there may be several distinct but equally complex expressions satisfying this definition.

In an SOP form, each product corresponds to a gate and each variable in each product to an input to that gate. The same holds for each sum in a POS form. The exact ratio between the cost of a gate and the cost of a gate input will depend on the type of gate, but the cost of an additional gate will usually be several times the cost of an additional input on an already existing gate. For that reason, the elimination of gates is usually the primary objective of minimization procedures. For the remainder of this section we shall concentrate on the SOP forms, because the K-map represents each minterm by a single square. We shall consider minimization of POS forms in the next section.

Simplification of functions on the K-map is based on the fact that sets of minterms that can be combined into simpler product terms will be logically adjacent on the K-map. Consider the function

$$f(A,B,C) = \Sigma m(0,6,7)$$

$$= \overline{A}\overline{B}\overline{C} \vee AB\overline{C} \vee ABC \tag{4.22}$$

By algebraic manipulation, we can simplify the function as follows:

$$f(A,B,C) = \overline{A}\overline{B}\overline{C} \vee AB(\overline{C} \vee C) \tag{4.23}$$

$$= \overline{A}\overline{B}\overline{C} \vee AB$$

On the map we have the pattern shown in Fig. 4.21. Note that the minterms that combined into a simpler term are adjacent on the K-map. Of course, that is no accident. That is the way we set up the K-map, so that adjacent squares represent min-

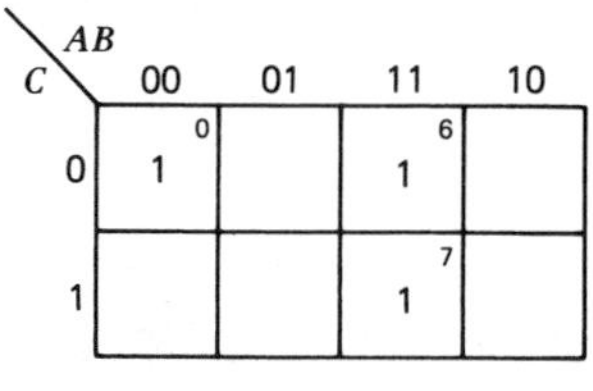

*Figure 4.21*

terms that differ in only one variable and may be combined by eliminating that variable. In this case, we see that the function takes on the value 1 if $A = 1$ and $B = 1$, regardless of the value of $C$, so $C$ can be eliminated from the two terms corresponding to $A = 1$ and $B = 1$. You may note that this is the same sort of simplification we discussed in connection with the Venn diagram of Fig. 3.19. In this case suppose that $A$ represents Ford cars, $B$ represents blue cars, and $C$ represents sedans. Then the simplification

$$A B \overline{C} \lor A B C = A B$$

corresponds to the statement that the union of the set of blue Ford sedans and the set of blue Ford nonsedans is the set of blue Fords.

Now consider the function $f(W,X,Y,Z) = \Sigma m(4,6,9,11,13,15)$, which is mapped in Fig. 4.22. This function may be manipulated algebraically as follows.

$$f(W,X,Y,Z) = \overline{W}X\overline{Y}\overline{Z} \lor \overline{W}XY\overline{Z} \lor W\overline{X}\overline{Y}Z \lor W\overline{X}YZ \lor WX\overline{Y}Z \lor WXYZ$$

$$= \overline{W}X\overline{Z}(\overline{Y} \lor Y) \lor W\overline{X}Z(\overline{Y} \lor Y) \lor WXZ(\overline{Y} \lor Y)$$

$$= \overline{W}X\overline{Z} \lor W\overline{X}Z \lor WXZ$$

$$(4.24)$$

Notice that the distributive law may be applied once again to the last two terms.

$$f(W,X,Y,Z) = \overline{W}X\overline{Z} \lor WZ(\overline{X} \lor X)$$

$$= \overline{W}X\overline{Z} \lor WZ$$

We see that the first product corresponds to a group of two minterms that are at opposite ends of a column rather than physically adjacent on the map. These two squares are, however, logically adjacent, since they represent minterms that are identical in three variables and differ only in $Y$. The second term represents the block of four adjacent squares in the $W = 1$, $Z = 1$ area of the map. Other sets of four adjacent squares that may be combined into single products are illustrated in Fig. 4.23.

Since sets of two minterms combine to eliminate one variable and sets of four combine to eliminate two, we would expect that sets of eight may combine to elim-

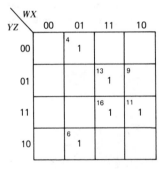

*Figure 4.22*

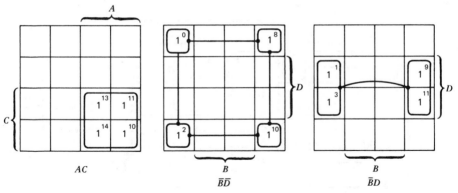

*Figure 4.23   Sets of four on the K-map.*

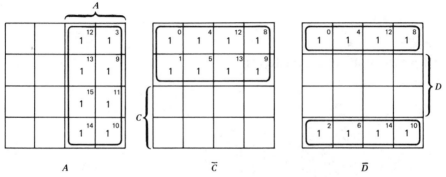

*Figure 4.24   Sets of eight on the K-map.*

inate three variables. Figure 4.24 illustrates some of these sets. The same general principles apply as we go to five-variable and six-variable maps, but we also have logical adjacency between squares, or sets of squares, in the same position on different sections of the map. Some groupings on a five-variable map are shown in Fig. 4.25.

It must be emphasized that not just any group of two, four, or eight, or more minterms will eliminate variables. The groups must be logically adjacent. We have shown some examples of logically adjacent groups, but we cannot show them all. When in doubt, the test of logical adjacency is whether or not enough variables remain constant over the entire set. On an $n$-variable map, a pair of minterms is adjacent if $n - 1$ variables remain constant over the pair; a set of four minterms is adjacent if $n - 2$ variables remain constant over a set; a set of eight minterms is adjacent if $n - 3$ variables remain constant over the set, and so forth.

The process of simplifying a function on a K-map consists of nothing more than determining the smallest set of adjacent groups that covers (contains) all the minterms of the function. Let us illustrate the process by a few examples.

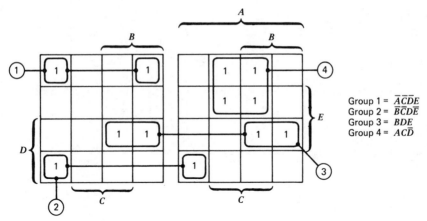

Group 1 = $\overline{A}\overline{C}\overline{D}E$
Group 2 = $\overline{B}\overline{C}D\overline{E}$
Group 3 = $BDE$
Group 4 = $AC\overline{D}$

*Figure 4.25   Sets on a five-variable map.*

### Example 4.11

Simplify

$$f(A,B,C,D) = \Sigma m(0,2,10,11,12,14)$$

Solution

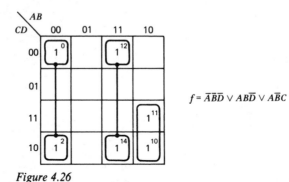

$f = \overline{A}\overline{B}\overline{D} \vee AB\overline{D} \vee A\overline{B}C$

*Figure 4.26*

In Example 4.11 there are choices. We could also combine $m_2$ and $m_{10}$, or $m_{10}$ and $m_{14}$. However, there is no reason to use these combinations since $m_2$, $m_{10}$, and $m_{14}$ are already covered by (contained in) the necessary pairings with $m_0$, $m_{11}$, and $m_{12}$, respectively, for which there is no choice. *Thus, a rule in using Karnaugh maps is to start by including those products (groups) that cover at least one minterm that can be covered in no other way (except by a group totally included in some larger group).*

### Example 4.12

Simplify

$$f(A,B,C,D) = \Sigma m(0,2,8,12,13)$$

Solution

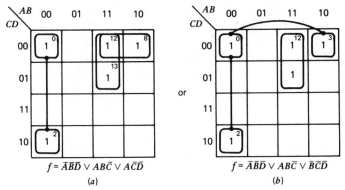

$f = \bar{A}\bar{B}\bar{D} \vee AB\bar{C} \vee A\bar{C}\bar{D}$

(a)

or

$f = \bar{A}\bar{B}\bar{D} \vee AB\bar{C} \vee \bar{B}\bar{C}\bar{D}$

(b)

Figure 4.27   Alternative maps for Example 4.12.

In this case there are two equally valid choices for $m_8$.   ■

Note that in both realizations for Example 4.12, there is some redundancy. In the first, $m_{12}$ is covered by terms $AB\bar{C}$ and $A\bar{C}\bar{D}$; in the second, $m_0$ is covered by $\bar{A}\bar{B}\bar{D}$ and $\bar{B}\bar{C}\bar{D}$. This procedure of covering a minterm more than once causes no trouble. In terms of an AND–OR realization, it simply means that, when the variable values are such that the particular minterm is 1, the output of more than one AND gate will be at 1. Since the OR is inclusive, it does not matter how many AND gates are at logic 1.

## Example 4.13

Simplify

$f(A,B,C,D) = \Sigma m(1,5,6,7,11,12,13,15)$

Solution

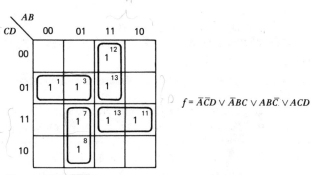

$f = \bar{A}\bar{C}D \vee \bar{A}BC \vee AB\bar{C} \vee ACD$

Figure 4.28

■

Example 4.13 illustrates a possible hazard in using K-maps. The temptation is great to use the group of four in the center. However, when we make the necessary pairings with the other four minterms, we find the four in the center have been covered. This illustrates the importance of the rule just stated in italics that we start with groups including minterms that can be covered in no other way.

We now present one more example without further comment. We suggest that you try to work it before looking at the answer. Once mastered, Karnaugh maps will seem almost second nature, but mastery requires practice.

**Example 4.14**

Simplify

$$f(A,B,C,D,E) = \Sigma m(0,1,4,5,6,11,12,14,16,20,22,28,30,31)$$

Solution

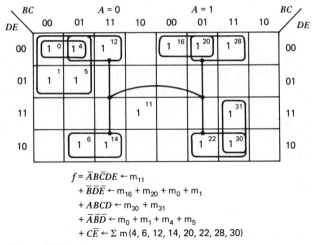

$$f = \overline{A}B\overline{C}DE \leftarrow m_{11}$$
$$+ \overline{B}D\overline{E} \leftarrow m_{16} + m_{20} + m_0 + m_1$$
$$+ ABCD \leftarrow m_{30} + m_{31}$$
$$+ \overline{A}\,\overline{B}D \leftarrow m_0 + m_1 + m_4 + m_5$$
$$+ C\overline{E} \leftarrow \Sigma\, m\,(4, 6, 12, 14, 20, 22, 28, 30)$$

*Figure 4.29*

# 4.4   Map Minimization of Product-of-Sums Expressions

With only minor changes, the procedure described in Section 4.3 can be adapted to product-of-sums design. We have seen that each 1 of a function is produced by a single minterm and that the process of simplification consists of combining minterms into products that have fewer literals and produce more than a single 1. We have also seen that each 0 of a function is produced by a single maxterm. It would seem reasonable, then, to expect that maxterms might combine in a similar fashion.

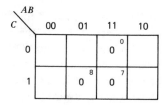

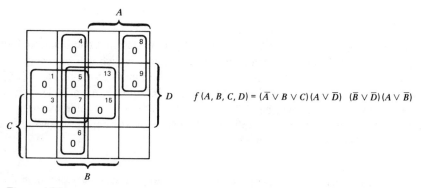

*Figure 4.30   K-Map of Eq. 4.25.*

Consider the function

$$f(A,B,C) = \prod M(3,6,7)$$

$$= (A \lor \bar{B} \lor \bar{C})(\bar{A} \lor \bar{B} \lor C)(\bar{A} \lor \bar{B} \lor \bar{C}) \qquad (4.25)$$

which is shown in Fig. 4.30. Maxterm 6 produces a 0 when $A = 1$, $B = 1$, $C = 0$, and maxterm 7 produces a 0 when $A = 1$, $B = 1$, $C = 1$. Taking them together, they produce 0's whenever $A = 1$ and $B = 1$, regardless of $C$. Furthermore, the single sum $(\bar{A} \lor \bar{B})$ will produce both 0's, since it will be 0 whenever $A = 1$ and $B = 1$, regardless of $C$. We also note that $M_3$ and $M_7$ together produce 0's for $B = 1$ and $C = 1$, regardless of A, as does the single sum $(\bar{B} \lor \bar{C})$. Algebraically, we have

$$f(A,B,C) = (A \lor \bar{B} \lor \bar{C})(\bar{A} \lor \bar{B} \lor C)(\bar{A} \lor \bar{B} \lor \bar{C})$$

$$= (A \lor \bar{B} \lor \bar{C})(\bar{A} \lor \bar{B} \lor \bar{C})(\bar{A} \lor \bar{B} \lor C)(\bar{A} \lor \bar{B} \lor \bar{C})$$

$$= (A\bar{A} \lor B \lor C)(A \lor B \lor C\bar{C})$$

$$= (\bar{B} \lor \bar{C})(\bar{A} \lor \bar{B})$$

To summarize, to design product-of-sums (POS) forms, select sets of the 0's of the function. Realize each set as a sum term with variables being the *complements* of those that would be used if this same set were being realized as a product, to produce 1's. Let us consider some example.

**Example 4.15**

Obtain a minimal product-of-sums realization of

$$f(A,B,C,D) = \Sigma m(0,2,10,11,12,14) = \prod M(1,3,4,5,6,7,8,9,13,15)$$

**Solution**

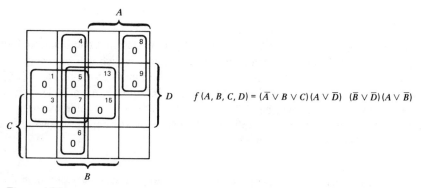

$$f(A, B, C, D) = (\bar{A} \lor B \lor C)(A \lor \bar{D})\ (\bar{B} \lor \bar{D})(A \lor \bar{B})$$

*Figure 4.31*

If we compare Examples 4.11 and 4.15 we see that the sum-of-products form is preferable to the product-of-sums forms for this function, since there are three products in the SOP form and four sums in the POS form. In other cases the POS form will be the more economical. It would be useful to have some means of determining in advance which form (SOP or POS) will be the best for a particular function. Unfortunately, no such method exists. Assuming there is no hardware preference, the designer should try both forms.

# 4.5   Incompletely Specified Functions

Recall that the basic specification of a switching function is the truth table, that is, a listing of the values of the function for the $2^n$ possible combinations of $n$ variables. Our design process basically consists of translating a (generally) verbal description of a logical job to be done into a truth table and then finding a specific function that realizes this truth table and satisfies some criterion of minimum cost. So far, we have assumed that the truth values were strictly specified for all the $2^n$ possible input combinations. This is not always the case.

Sometimes the circuit we are designing is a part of a larger system in which certain inputs occur only under circumstances such that the output of the cicuit will not influence the overall system. Whenever the output has no effect, we obviously *don't care* whether the output is a 0 or a 1. Another possibility is that certain input combinations never occur because of various external constraints. Note that this does not mean that the circuit would not develop some output if this forbidden input occurred. Any switching circuit will respond in some way to any input. However, since the input will never occur, we do not care whether the final circuit responds with a 0 or a 1 output to this forbidden input combination.

Where such situations occur, we say that the output is *unspecified*. This is indicated on the truth table by entering an "X" as the functional value, rather than 0 or 1.* Such conditions are commonly referred to as *"don't-cares"* and functions including "don't-cares" are said to be *incompletely specified*. A realization of an incompletely specified function is any circuit that produces the same outputs for *all input combinations for which output is specified*.

### Example 4.16

For some reason, of no interest to us, it is necessary to design a combinational logic circuit that we shall call a "divisible-by-three detector," which will be incorporated within a digital computer designed to operate in the binary-coded-decimal (BCD) mode. The circuit will have four input lines, $A,B,C,D$, representing a BCD digit, and a single output, $Z$, which will be 1 if and only if this digit is divisible by three. The most significant bit of the BCD number is $A$, $B$ is the next significant, and so forth. The circuit is illustrated in Fig. 4.32.

---

*The symbol for the unspecified output condition is unfortunately not standard. Other symbols used include $d$ and -.

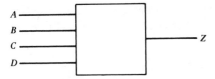

*Figure 4.32 Divisible-by-3 detector.*

This problem is interesting because there are only 10 BCD digits, 0000 to 1001, that can occur on lines $A$, $B$, $C$, $D$. The bit combinations 1010 through 1111 will never appear on these lines. Thus we do not care what output the circuit would have for these input combinations.

### Solution
The functional values for $Z$ may be entered directly on the Karnaugh map, since the BCD digits are the same as the minterm numbers. Thus 1's are entered in Fig. 4.33 in the squares corresponding to $m_3$, $m_6$, and $m_9$. Since $m_{10}$ through $m_{15}$ represent input combinations that will never occur, X's are entered in these squares.

Since the output is optional for the don't-cares, we assign them to be 0's or 1's in whatever manner will result in the simplest realization. On the K-map, this means that we group the X's with the 1's whenever this results in a larger group and ignore them if there is no advantage to be gained from their use. In this case we use the X's in squares 11, 13, and 15 together with the 1 in square 9 to form the product $AD$, use the X in square 14 with the 1 in square 6 to form the product $BC\overline{D}$, and reuse square 11 with the 1 in square 3 to form $\overline{B}CD$. The X's in squares 10 and 12 are not used because they cannot be grouped with any of the 1's to produce a larger group.

$$Z = AD \vee BC\overline{D} \vee \overline{B}CD$$

This expression will evaluate to 1 for each combination of values for which $Z = 1$ in Fig. 4.33 and to 0 wherever $Z = 0$ in this map. Should any of the don't-care conditions occur, the expression would evaluate to 1 for those that are included in groupings, to 0 for those that are not. ∎

It would be convenient to have some compact algebraic form for incompletely specified functions. Because each row in the truth table corresponds to an input com-

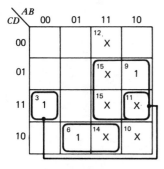

*Figure 4.33 K-Map for divisible-by-3 detector.*

bination, we simply add a list of the rows for which the output is unspecified, enclosed by parentheses and preceded by a "*d*" to signify "don't care." Thus a minterm list corresponding to Fig. 4.33 would be

$$f(A,B,C,D) = \Sigma m(3,6,9) + d(10,11,12,13,14,15)$$

## Example 4.17

Obtain a minimal sum-of-products representation for

$$f(w,x,y,z) = \Sigma m(0,7,8,10,12) + d(2,6,11)$$

### Solution

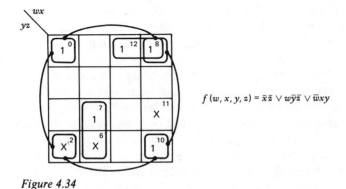

$$f(w, x, y, z) = \bar{x}\bar{z} \vee w\bar{y}\bar{z} \vee \bar{w}xy$$

*Figure 4.34*

We use the don't-cares in squares 2 and 6 to obtain larger groups than would otherwise be possible, but we ignore the don't-care in 11, since the only possible combination is with 10, which is already covered.  ∎

## Example 4.18

Obtain a minimal POS realization for the function of Example 4.17

### Solution

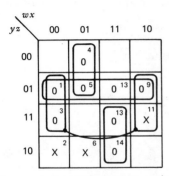

$$f(w, x, y, z) = (y \vee \bar{z})(x \vee \bar{z})(w \vee \bar{x} \vee y).(\bar{w} \vee \bar{x} \vee \bar{y})$$

*Figure 4.35*

Because we want the POS form in Example 4.18, we design on the 0's, using the don't-cares if they improve the groupings, ignoring them otherwise. Again we start with the necessary sets. For example, 4 combines only with 5, and 14 only with 15. We could combine 4 with the don't-care in 6, and 14 also with 6, but this would violate our rule of using don't-cares only if they make larger groups possible. The don't-care at 11 is used because it does place 3 in a larger group ($x \lor \bar{z}$) than could otherwise be obtained. ∎

### Example 4.19

Obtain a minimal sum-of-products representation for

$$f(A,B,C,D,E) = \Sigma m(1,4,6,10,20,22,24,26) + d(0,11,16,27)$$

Solution

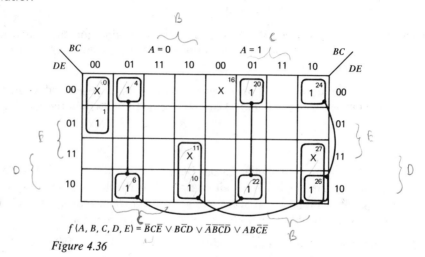

$$f(A, B, C, D, E) = \overline{BC}\overline{E} \lor B\overline{C}D \lor \overline{A}\,\overline{B}C\overline{D} \lor AB\overline{C}\overline{E}$$

Figure 4.36

Here the don't-care in 16 could be used in several ways, but it would not increase the size of any group. ∎

## Problems

**4.1** Convert the following Boolean forms to minterm list.
  (a) $f(w,x,y,z) = wy \lor x(w \lor y\bar{z})$
  (b) $f(U,V,W,X,Y) = \overline{V}(\overline{W} \lor \overline{U})(X \lor \overline{Y}) \lor \overline{U}\,\overline{W}\,\overline{Y}$
  (c) $f(V,W,X,Y,Z) = (X \lor Z)(\overline{Z} \lor \overline{WY}) \lor (VZ \lor W\overline{X})(\overline{Y} \lor Z)$

**4.2** By using the Karnaugh map, determine minimal sum-of-products realization of the following functions.
  (a) $f(A,B,C,D) = \Sigma m(0,4,6,10,11,13)$
  (b) $f(w,x,y,z) = \Sigma m(3,4,5,7,11,12,14,15)$

(c) $f(a,b,c,d) = \Pi M(3,5,7,11,13,15)$

(d) $f(V,W,X,Y,Z) = \Sigma m(0,2,3,4,5,11,18,19,20,23,24,28,29,31)$

**4.3**  A logic circuit is to be designed having four inputs $y_1$, $y_0$, $x_1$, $x_0$. The pairs of bits $y_1y_0$ and $x_1x_0$ represent two-bit binary numbers with $y_1$ and $x_1$ as the most significant bits. The only circuit output, $z$, is to be 1 if number $y_1y_0$ is greater than or equal to the binary number $x_1x_0$. Determine a minimal sum of products expression for $z$.

**4.4**  Determine a minimal sum-of-products expression equivalent to each of the following Boolean expressions.

(a) $f(A,B,C,D,E) = (\overline{C}\overline{E} \vee CE)(\overline{A} \vee B) D \vee \overline{\overline{A} \vee B)DCE}$

(b) $f(w,x,y,z) = (\overline{w} \vee x) \vee (\overline{x} \vee z) \wedge (\overline{y} \vee z)$

**4.5**  By using Karnaugh maps, determine minimal product-of-sums realizations for the functions of Problems 4.2 *b* and 4.2 *c*.

**4.6**  Determine the minimal product-by-sums realization for the function of Example 4.14 and compare the cost to that of the minimal sum-of-products realization.

**4.7**  A prime number is a number that is only divisible by itself and 1. Suppose the numbers between 0 and 31 are represented in binary in the form of the five bits

$$x_4x_3x_2x_1x_0$$

where $x_4$ is the most significant bit. Design a prime detector. That is, design a combinational logic circuit with output $Z$ that will be 1 if and only if the five input bits represent a prime number. Do not count zero as a prime. Base your design on obtaining a minimal two-level expression for $Z$.

**4.8**  Determine minimal sum-of-products realizations for the following incompletely specified functions:

(a) $f(A,B,C,D) = \Sigma m(1,3,5,8,9,11,15) + d(2,13)$

(b) $f(W,X,Y,Z) = \Sigma m(4,5,7,12,14,15) + d(3,8,10)$

(c) $f(A,B,C,D,E) = \Sigma m(1,2,3,4,5,11,18,19,20,21,23,28,31) + d(0,12,15, 27,30)$

(d) $f(a,b,c,d,e) = \Sigma m(7,8,9,12,13,14,19,23,24,27,29,30) + d(1,10,17, 26,28,31)$

**4.9**  The four lines into the combinational logic circuit depicted in Fig. P4.9 carry one binary-coded decimal digit. That is, the binary equivalents of the decimal

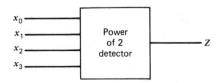

*Figure P4.9*

digits 0–9 may appear on the lines $x_3x_2x_1x_0$. The most significant bit is $x_3$. The combination of values corresponding to binary equivalents of the decimal numbers 10–15 will never appear on the lines. The single output $Z$ of the ciruit is to be 1 if and only if the inputs represent a number that is either 0 or a power of 2. Construct the logic block diagram of a minimal two-level realization of the circuit.

**4.10** A circuit receives two 3-bit binary numbers. $A = A_2A_1A_0$, and $B = B_2B_1B_0$. Design a minmal SOP circuit to produce an output whenever $A$ is greater than $B$.

**4.11** In a certain computer, three separate sections of the computer proceed independently through four phases of operation. For purposes of control, it is necessary to know when two of three sections are in the same phase at the same time. Each section puts out a 2-bit signal (00,01,10,11) in parallel on two lines. Design a circuit to put out a signal whenever it receives the same phase signal from any two or all three sections.

**4.12** A shaft-position encoder provides a 4-bit signal indicating the position as a shaft in steps of 30°, using a reflected (Gray) code as listed in Fig. P4.12. It may be assumed that the four possible combinations of four bits not used will never occur. Design a minimal SOP realization of a circuit to produce an output whenever the shaft is in the first quadrant (0–90°).

| Shaft position | Encoder output | | | |
|---|---|---|---|---|
| | $E_3$ | $E_2$ | $E_1$ | $E_0$ |
| 0–30° | 0 | 0 | 1 | 1 |
| 30–60° | 0 | 0 | 1 | 0 |
| 60–90° | 0 | 1 | 1 | 0 |
| 90–120° | 0 | 1 | 1 | 1 |
| 120–150° | 0 | 1 | 0 | 1 |
| 150–180° | 0 | 1 | 0 | 0 |
| 180–210° | 1 | 1 | 0 | 0 |
| 210–240° | 1 | 1 | 0 | 1 |
| 240–270° | 1 | 1 | 1 | 1 |
| 270–300° | 1 | 1 | 1 | 0 |
| 300–330° | 1 | 0 | 1 | 0 |
| 330–360° | 1 | 0 | 1 | 1 |

*Figure P4.12*

# References

1. Dietmeyer, D. L., *Logic Design of Digital Systems,* 2nd ed. Allyn & Bacon, Boston, 1978.
2. Karnaugh, M., "The Map Method for Synthesis of Combinational Logic Circuits." *Trans. AIEE,* 72, pt. I, 593–598 (1953).

3. Nagle, H. T., Jr., B. D. Carroll, and J. D. Irwin., *An Introduction to Computer Logic.* Prentice-Hall, Englewood Cliffs, N.J., 1975.
4. Roth, C. H., Jr., *Fundamentals of Logic Design,* 2nd ed. West Publishing, St. Paul, 1979.
5. Sloan, M. E., *Computer Hardware and Organization.* SRA Publishers, Chicago, 1976.
6. Veitch, E. W., *A Chart Method for Simplifying Truth Functions.* Proc. ACM, Pittsburgh, Pa., pp. 127–133 (May 2,3, 1952).

# Standard Digital Integrated Circuits

## 5.1  Introduction

In preceding chapters we have discussed the various logic functions primarily in terms of their basic logical characteristics, with little consideration of physical realizations. We have shown symbols for gates, devices that realize the basic logic functions, but have considered them to be idealized devices, with their behavior described in terms of the logic symbols 0 and 1. For example, an AND gate is any device in which the output voltage is at the level corresponding to logic 1 only if all inputs are at that same level. For purposes of abstract "paper" designs, that is really all we need to know about a gate. When we reach the point of translating a design into actual hardware, we need to know what sort of gates and other digital circuits are actually available and we need to know something about their physical limitations.

In Chapter 2 the critical circuit element in the NAND gate was the idealized switch as shown in Fig. 5.1*a*. If this circuit is to be physically realized, some available "real-world" component must be substituted for this ideal switch. In electric-circuit realizations of the NAND gate, the four most common physical approximations of the ideal switch have been the *relay*, the *vacuum tube*, the *bipolar transistor*, and the *MOS* (metal oxide semiconductor) *transistor*. Of these, the relay and the MOS transistor most closely approximate the properties of the switch. Because the bipolar transistor can change state (open and close) more rapidly, it has until recently been the most important switching device for digital circuit realization. In Fig. 5.1*b* we

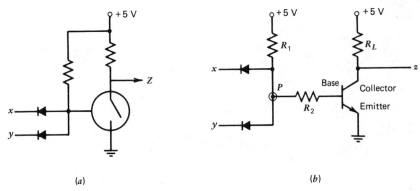

*Figure 5.1  Bipolar realization of a NAND gate.*

substitute a bipolar transistor* together with a resistor connected to its base (input) terminal to prevent damage due to excess current.

For now we restrict our discussion of the circuit of Fig. 5.1*b* to the static or dc situation in which the input values do not change. Suppose first that either of the inputs, *x* or *y*, shown in Fig. 5.1*b* is 0 V. In this case the transistor, like the switch of Fig. 5.1*a*, will be turned OFF. For this "turned-off" case the transistor may be modeled as shown in Fig. 5.2. This model is reasonably accurate for voltages at point *P* up to about 0.5 V depending on $R_2$. Notice that the only difference between this and the ideal switch model is that there is a very small current, $I_3$, into the transistor from point *P*.

Now suppose that both inputs, *x* and *y*, are at 5 V, so that the transistor NAND gate like the ideal switch is turned ON. For this case and for voltages at point *P* down to about 2.5 V the model of Fig. 5.3 approximates the transistor NAND gate. We see several differences between the model and the turned-on ideal switch NAND gate for which the output is always 0 V. Not only is there an input current from point *P*, but the output voltage is now some small value (about 0.2 V) rather than exactly zero. Most important is the box labeled current limiter. Regardless of the network connected to the output, *z*, the current through this box cannot exceed $\beta I_B$. This limit, $\beta I_B$, will vary with the physical size or power-handling capacity of the transistor. A typical value might be 10 mA. In any case, although the current, $I_C$, might be less than $\beta I_B$, it must never exceed this value. Whenever the current through this current limit box is less than $\beta I_B$, the voltage drop across the box is zero.

The bipolar "diode transistor logic" (DTL) NAND gate of Fig. 5.1*b* served only to illustrate its function in terms of the ideal switch version. Commercially available gates are complicated somewhat to improve their performance. A more realistic DTL NAND gate is shown in Fig. 5.4*a*. The second transistor has been added to increase its amplification or the constant $\beta$ referred to eariler. The models of Figs.

---

*To the reader with some experience in electronics, the transistor is obviously of the NPN variety. To hold this discussion to a reasonable length, we shall consider only a single type of bipolar transistor as well as a single example of a MOS device.

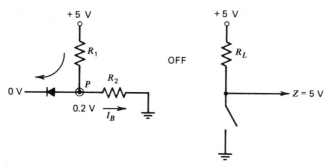

*Figure 5.2 "Turned-off" transistor NAND gate.*

5.2 and 5.3 remain valid for the overall circuit. Only the current limit, $\beta I_B$, is increased. A large family of DTL gates of various types with varying numbers of inputs are available in integrated circuit form, but this family is not the most commonly used bipolar logic family. More popular is the transistor–transistor logic or TTL or T²L family for which a three-input NAND is shown in 5.4b. This circuit may look quite different from the DTL NAND gate, but it functions essentially the same. The left-most transistor with three inputs to the emitter acts much the same as three separate diodes.

The other approximation of the ideal switch that we shall discuss in this section is the MOS field-effect transistor. This device is a closer approximation of the voltage controlled switch because the controlling input draws no current. The symbol for the MOS transistor is given in Fig. 5.5a. When a relatively low voltage representing logical 0 is applied to the input, x, in this circuit, we say that the device is open or OFF, as shown in Fig. 5.5c. Actually, there is a very high but not infinite resistance between points S and D. If a relatively high voltage representing logical 1 is applied to the input, x, the switch is ON and there is a much lower resistance between S and D as shown in Fig. 5.5c.

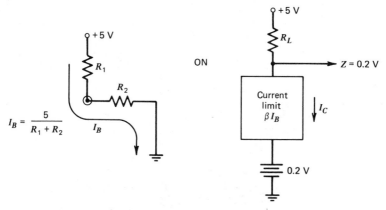

*Figure 5.3 "Turned-on" transistor NAND gate.*

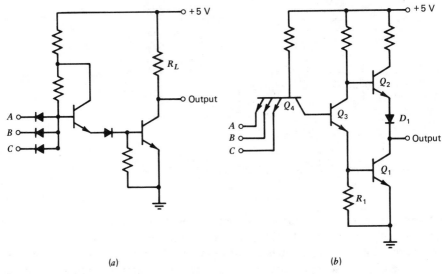

(a)　　　　　　　　　　　　　　　　　　(b)

*Figure 5.4　DTL (a) and TTL (b) NAND gates.*

Since diodes are not readily realized in metal oxide semiconductor integrated circuits and the turned on MOS transistor can be used as a resistor, it is customary to construct MOS logic circuits using only the MOS transistors themselves. A simple MOS NAND gate is shown in Fig. 5.6. Because of the high resitances involved, MOS circuits can be designed that consume less power than corresponding bipolar circuits. The fact that MOS devices draw almost no input current makes it possible to store logical values for short periods of time on naturally occurring capacitors at gate inputs. This makes it possible to replace the constant power supply with periodic supply signal. Under these conditions, individual gates will consume power less than 50% of the time. This low power consumption and the overall simplicity of MOS circuits make this technology an excellent choice for use in large-scale integrated circuits where very large numbers of devices are fabricated in a small area of semiconductor.

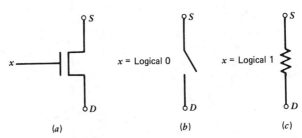

(a)　　　　　　　　　　(b)　　　　　　　　　　(c)

*Figure 5.5　MOS transistor switch.*

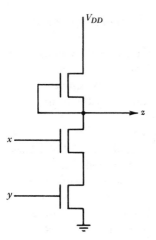

*Figure 5.6 MOS NAND gate.*

## 5.2 Small-Scale Integrated Circuits

Until about 1970, gate circuits such as those discussed in the last section were commonly realized from discrete components—resistors, transistors, diodes—mounted on plastic cards and wired together. Today logic circuits are almost invariably realized in integrated circuit form. An *integrated circuit* is a complete electronic circuit implanted by electrochemical methods on a single chip of silicon. The earliest integrated circuits typically realized a few gates on a single chip. As the technology advanced, the circuit density steadily increased, until today complete computers, comprising tens of thousands of components, can be placed on ¼-in. square chips.

Integrated circuits can be classified in a variety of ways. One classification is in terms of the type of electronic technology used. Currently popular technologies include TTL, ECL, CMOS, NMOS, and PMOS. NMOS and PMOS are two slightly different physical realizations of the MOS model discussed in the last section. These five technologies differ in such characteristics as speed, power consumption, and packing density. ECL is very fast but consumes a lot of power. CMOS is slower but consumes so little power that it is suitable for battery-powered applications, such as electronic watches. NMOS has the highest packing density and is used in very complex circuits, such as microprocessors. TTL, which falls about in the middle in all characteristics, is by far the most popular technology and is available in more different circuits than all the others put together.

Integrated circuits can also be classified in terms of circuit complexity. *Small-scale integration* (SSI) encompasses circuits with up to 10 gates per chip. *Medium-scale integration* (MSI) includes circuits with from 10 to 100 gates per chip. From 100 to about 5000 gates per chip we have *large-scale integration* (LSI), and above this we have *very-large-scale intergration* (VLSI). In SSI chips we have individual gates and flip-flops. MSI chips realize more complex logic functions, such as code conversion and arithmetic operations. LSI and VLSI chips realize complete digital systems, such as memory systems and microprocessors.

In terms of logic design the most important family of integrated circuits is 7400 TTL, which includes both SSI and MSI circuits. This family of logic circuits was introduced by Texas Instruments and is now manufactured by many companies. The various types of circuits in this family are denoted by numbers of the form 74xxx, where the xxx represents a two-digit or three-digit number, for example, 7410 or 74163. There are actually several families of TTL circuits, indicated by one or more letters between the 74 and the xxx, for example, 74508, 74L508, and 74H08. These families usually differ only in speed and power characteristics and can be mixed together, although some precautions must be observed. As of this writing the LS family is the industry standard, but new, improved families continue to appear. In this text we will designate all TTL circuits in the basic form, for example, 7486, and not worry about which family might be used for a specific realization, except for a few cases where the LS version is slightly different. In those cases we will include the LS in the chip number.

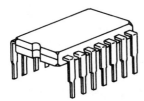

*Figure 5.7   Dual-in-line package.*

Each integrated circuit consists of a single chip of silicon, in the case of SSI about the size of the head of a pin. To make it possible to connect to these tiny circuits, they are mounted in plastic packages with the circuit leads brought out to pins that can be plugged into sockets or soldered to printed-circuit boards. The standard package for SSI TTL is the 14-pin *dual-in-line* package (DIP) shown in Fig. 5.7. Two of the pins are used for power supply connections ($+5$ V and ground for TTL), leaving 12 pins for gate inputs and outputs. The standard gate packages available are hex inverters, quad two-input, triple three-input, dual four-input, and single eight-input. Most of these packages are available in several types of gates. Figure 5.8 summarizes the available TTL gate packages. The pin diagrams have been shown only for NAND gates, but the pin arrangements will be exactly the same for any other types of gates, except for the 7402 Quad Nor chip. Note that the exclusive OR is available only in the quad two-input package.

## 5.3   Fan-out, Fan-in, and Noise Immunity

In the simplified logic circuits of the last section, the logic levels were assumed to be the same as the supply levels. In TTL the circuits are more complex and the nominal

## Part nos. for various gate types

| Chip type | Pin diagram | NAND | AND | OR | NOR | Ex-OR |
|---|---|---|---|---|---|---|
| Quad 2-input: | | 7400 | 7408 | 7432 | 7402 | 7486 |
| Triple 3-input: | | 7410 | 7411 | | 7427 | |
| Dual 4-input: | | 7420 | 7421 | | | |
| Single 8-input: | | 7430 | | | | |
| Hex inverter: | | 7404 | | | | |

Figure 5.8  Summary of TTL logic packages. For all chips: pin 14 is $V_{cc}$ = 5-V dc; pin 7 is $G_{nd}$ = 0-V dc.

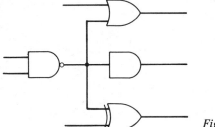

*Figure 5.9   Fan-out.*

logic levels are 0.2 V and 3.5 V, although the supply voltages are 0 V and 5 V. We say "nominally" because the voltages will vary from one circuit to the next and even in the same circuits, depending on operating conditions. We have defined a binary system as one in which the signal levels are restricted to two distinct values. Actually, no two devices ever operate at precisely the same values, and it is more precise to say that the signal levels are restricted to two distinct ranges of values. In TTL the output voltage ranges are 0 to 0.4 V for logic 0 and 2.4 to 5.0 V for logic 1. In addition to providing proper supply voltages, the designer must observe the *fan-out* limits of the gates. The fan-out is simply a measure of the number of other gates that a single gate can drive. Figure 5.9 shows a gate driving inputs of three other gates as a part of some larger circuit. If the gate had a rated fan-out of 10, it could drive seven more gates and still function properly. The fan-out is given as the number of inputs to gates in the same family that can be driven by a single gate. The fan-out for the 74LS family is 20. You will also encounter references to the *fan-in* of a gate; this is simply the number of inputs to the gate.

The fan-out limit of a gate is a function of the amount of current it can draw when it is turned on. If we restrict ourselves to the dc case where logical values are not changing, we can infer fan-out from models of the turned-on transistor such as the one given in Fig. 5.3. This is illustrated by the following example.

### Example 5.1 (Optional: Not essential for later material)

Assume that the elementary NAND gate of Fig. 5.1*b* is to be connected in a network composed of only NAND gates of this type (with various numbers of inputs). Assume that for all gates $R_1 = 5000 \ \Omega$, $R_L = 2500 \ \Omega$, $I_B = 12 \ 10^{-5}$ A when the gate is turned on, and $\beta = 100$. Determine a fan-out limit for the gate under these circumstances.

### Solution

Figure 5.10*a* depicts a single gate with output connected to inputs of *n* gates of the same type. Only the input diodes and the resistor, $R_1$, are shown for these *n* gates. The fan-out of the driving gate is *n*. Our goal is to determine the maximum allowable *n*, that is, the fan-out limit for the gate.

The fan-out limit depends on the current flowing into the collector (labeled as point *C* in Fig. 5.10*a*) when the gate is turned on. We therefore turn to the "switched-on" (saturated) model from Fig. 5.3. Since we already know the current

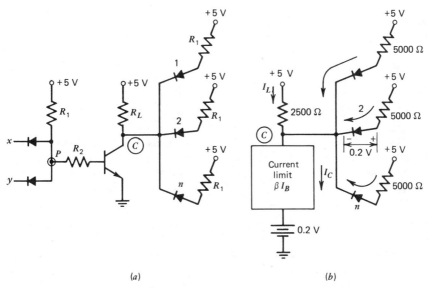

(a)                                                        (b)

*Figure 5.10   Fan-out computation.*

$I_B$ on the input side, only the output half of the model is shown in Fig. 5.10b. This is shown connected to the same $n$ gate inputs that are driven by this gate.

In determining limits the worst possible situation, or "worst case," must always be considered. Here the worst case is where all other inputs of the $n$ gates driven by our subject gate are at logical 1 except those connected to this gate. Therefore, all the current passing through $R_1$ of each of these gates must flow into point $C$ of the subject gate. The current through $R_L$ must also go into this point. As previously stated, there is no voltage drop across the current limit box unless the current $I$ through it exceeds $\beta I_B$. Allowing a 0.2 V drop across each conducting diode, we arrive at Eq. 5.1.

$$I_C = \frac{5 - 0.2}{2500} + n \cdot \left( \frac{5 - 0.4}{5000} \right) \tag{5.1}$$

$$I_C = 1.92 \times 10^{-3} + 0.92n \times 10^{-3} \tag{5.2}$$

The current determined from the simplified Eq. 5.2 must be less than the current limit, $\beta I_B$, if the model used in Fig. 5.10b is to be valid. If $I_C$ exceeds this value, the transistor is really not in the switched-on (saturated) mode. Instead it will act like an amplifier in the linear region, and the output voltage will range anywhere between 0.2 and 5 V. We do not actually care what this linear region is or what happens there. We need only realize that, if the gate is driven into that region, it will not be functioning as a digital logic gate. Therefore, this must not happen. Equation 5.3 must be satisfied.

$$I_C < \beta I_B = 12 \times 10^{-3} \tag{5.3}$$

Therefore,

$$1.92 \times 10^{-3} + 0.92n \times 10^{-3} < 12 \times 10^{-3}$$

$$0.92n < 10.08 \tag{5.4}$$

$$n < 10.95$$

Equation 5.4 suggests that the fan-out limit be established as 10.  ∎

If you proceed as in Example 5.1 for one of the gates listed in Fig. 5.8 using a circuit from a manufacturer's data manual, you may determine a fan-out limit larger than the one recommended by the manufacturer. One reason for this is simply a margin of safety. A second reason is that a driving gate will draw more current from a gate that it is causing to switch values than from one that is static with input 0 and output logical 1. Analysis of this non-steady-state situation is more complicated and will not be considered here. In your initial laboratory experience as well as later as a practicing logical designer you can confidently rely on simple tables in the manufacturers' data manuals for fan-out limits and other parameters. Circuit analysis, even in the simplified form of Example 5.1, will rarely be required.

In addition to output ranges, manufacturers of TTL also specify input ranges, 0 to 0.8 V for logical 0 and 2.0 to 5.0 V for logical 1. The differences in the rated input and output ranges provide *noise immunity. Noise* is unwanted signal that adds to the logic signals and may, if large enough, cause a logic signal to appear to have the incorrect value.

Consider Fig. 5.11, which shows one gate driving another over a transmission path in which some noise is added to the signal. Assume that the output of the driving gate is at logic 0, which the manufacturer guarantees to be no larger than 0.4 V. At the receiving end, the manufacturer guarantees that the input signal will be interpreted as logic 0 so long as it does not exceed 0.8 V. Therefore, up to 0.4 V of noise can be added to the logic signal and the circuit will still function properly. In a similar fashion, a logic 1 at the output of the drive gate will be no less than 2.4 V, but any voltage greater that 2.0 V at the receiving end will be interpreted as logic 1.

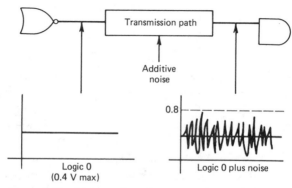

*Figure 5.11    Noise-on logic lines.*

Again, up to 0.4 V of noise can be added without malfunction, and we say that the noise margin of TTL is 0.4 V.

# 5.4   Switching Delay in Logic Circuits

We have mentioned differences in the *speed* characteristics of the different families of logic. Speed in logic circuits is usually characterized in terms of *delay time,* the time it takes for a change in logic level at the input of a circuit to cause a corresponding change in level at the output. The delay times in modern logic circuits are incomprehensibly short in terms of a human frame of reference. The typical delay for 74LS is 10 nsec, where a nsec (nanosecond) is $10^{-9}$ sec. One nanosecond is approximately the time it takes light to travel 1 ft. Short as these times may be, they are significant when building computer systems capable of millions of operations per second, and the higher costs of even faster circuits may well be justified in some cases.

The delay associated with a logic gate results from the fact that no real switching device can change from the "turned-on" condition to the "turned-off" condition instantaneously. This fact is due to a variety of physical phenomena including stored charge on the base of a bipolar transistor and effects that may be modeled as capacitors connected to gate inputs and outputs. A mathematical analysis of gate delay would require knowledge exceeding the current background of most of our readers. This is not a serious limitation, since gate delays for typical gates can be measured and the same delays are associated with all gates in the same logic family. Even in practice the engineer will usually depend on empirically determined gate-delay values published in manufacturers' data manuals and on simulations of simple models based on these measurements.

To consider the impact of gate delay, let us refer to the chain of bipolar inverters shown in Fig. 5.12a. The chain is driven by a signal source capable of generating pulses of varying duration. We ignore the waveform from the signal generator itself and begin our analysis at point $A$ (the output of gate 1). The signal at point $A$ is itself a delayed and inverted version of the signal generator output with less steep rising and falling edges.

In Fig. 5.12b we see graphs of the signals observed at points $A$, $B$, and $C$ of Fig. 5.12a expressed in 0–1 logic levels. We compare the signals at points $A$ and $B$ to determine the delay associated with gate 2. Similarly, the delay characteristics of gate 3 are found from comparing signals $B$ and $C$. Although two gates of the same logic family do not necessarily have exactly the same delays, the variation is usually less than $\pm 50\%$ from the average value. In this case the delay properties of gates 2 and 3 are the same, so we concentrate on gate 2. When $A$ goes from 1 to 0, it causes gate 2 to turn off so that its output goes from 0 to 1. The delay between the 1 to 0 transition at $A$ and the 0 to 1 transition at $B$ is called the *turn-off* delay of gate 2. The nominal value assigned to this delay is the time from the midpoint of the transition on $A$ ($t_1$) to the midpoint of the transition on $B$. This time interval is clearly labeled as the *turn-off delay* in Fig. 5.12b. Similarly, the interval from the midpoint

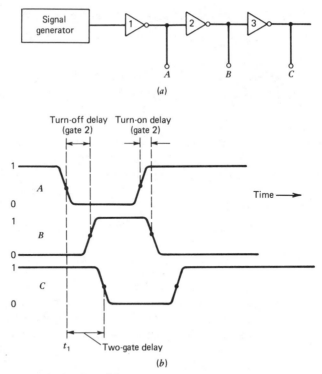

Figure 5.12    Gate delay.

of the 0 to 1 transition on $A$ to the midpoint of the 1 to 0 transition on $B$ is the *turn-on* delay for gate 2.

Also measured in Fig. 5.12 is the delay between the 1 to 0 transition on $A$ and the 1 to 0 transition on $C$. This two-gate delay is the sum of the turn-off delay of gate 1 and the turn-on delay of gate 2. This simple fact, that we can merely add the delays associated with two gates in series to obtain the overall delay, is invaluable in the determination of the delay through a path in a complex logic network.

As illustrated in Fig. 5.12, turn-off delay is usually longer than the turn-on delay of the same gate. Sometimes an average of the two is simply referred to as the gate delay and used in the calculation of delay through a path. This procedure gives a reasonably accurate result since turn-on and turn-off transitions alternate as a signal advances through a path.

### Example 5.2

Suppose the logic network of Fig. 5.13 is to be constructed using a logic family with a turn-off delay of 10 nsec and a turn-on delay of 6 nsec. Suppose input $A$ is initially 0 and inputs $B$ and $C$ are initially 1. If $B$ changes from 1 to 0, the output, $Z$, will change values. Determine the expected time delay between the change at $B$ and the change at $Z$.

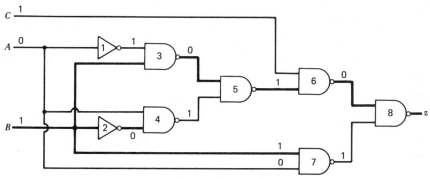

*Figure 5.13    Calculation of path delay.*

Solution

The initial values of every gate in the network are shown in Fig. 5.13. Notice in the figure that interconnecting lines from input $B$ toward the circuit output are darkened. The reader can verify that each gate along the darkened path will change values in reponse to a change on input $B$. Notice that the lower darkened path terminates at the inputs to gates 4 and 7, because these gates will have an output of 1 regardless of $B$. There are four gates on the darkened path to the output. The nominal gate delay, that is, the average of the turn-on and turn-off delays, is 8 nsec. Therefore,

$$\text{path delay} = 4 \times 8 = 32 \text{ nsec} \qquad \blacksquare$$

## 5.5    Circuit Implementation with NAND and NOR Gates

In Section 5.2 it was pointed out that the most commonly used SSI gates are NAND and NOR gates. For this reason we are interested in learning to convert Boolean expressions to forms that can be implemented using only these two types of gates. In the previous section we observed that the switching delay of a logic network is cummulative from gate to gate along each path through the network. In any digital system the delay in combinational logic paths will be the factor that determines how fast the system will function. We are, therefore, interested in minimizing logic path delay wherever possible. Consequently, we turn first to converting the sum-of-products and product-of-sums expressions developed in Chapter 4 to all-NAND or all-NOR forms.

We observed that any Boolean function, no matter how complex, can be expressed in AND–OR terms, that is, in expressions of the general form

$$f = AB \lor CD \lor A\overline{D}$$

or in the OR–AND form, such as

$$f = (A \lor \overline{B})(B \lor C)(\overline{C} \lor D)$$

We have also seen that the Karnaugh map provides a powerful design tool leading naturally to such forms. However, in terms of actual realization, the use of AND gates and OR gates may not be desirable. As shown in Fig. 5.1, the addition of a single stage of amplification to a diode gate inherently provides inversion, leading to NAND or NOR; AND or OR requires another level of inversion and, hence, higher cost.

Thus, it appears that using NAND or NOR gates might have some economic advantages, but how do we design for such gates? Fortunately, it turns out that the conversions from AND–OR or OR–AND forms, to which the Karnaugh maps naturally lead, are very simple. In Fig. 5.14a we show an AND–OR circuit that we assume is a minimal design arrived at by use of a Karnaugh map. In Fig. 5.14b we change all the AND gates to NAND gates and compensate for this change by inverting the inputs to the OR gate. In each line running from an AND gate to an OR gate the two successive inversions simply cancel, so the logical function of the circuit is not changed by this modification. We then note, from DeMorgan's law,

$$\overline{X} \vee \overline{Y} \vee \overline{Z} = \overline{XYZ}$$

that an OR gate with inverted inputs is equivalent to a NAND, leading to the final form of the circuit in Fig. 5.14c.

Thus, the conversion from AND–OR to all-NAND form is accomplished simply by replacing both the AND gates and the OR gates by NAND gates. Even more significant, this example shows that any function can be realized using only one type of gate, NAND. This fact makes possible significant economies through volume purchasing and simplification of inventory control. Ten thousand circuits of a single type will be considerably cheaper than five thousand each of two types. In a similar manner, an OR–AND circuit can be converted to all-NOR form. There is no theoretical preference between the all-NAND or all-NOR forms, but all-NAND forms are by far the most popular, probably because most designers find the AND–OR form more natural when setting up a problem. This popularity is reflected by the fact that NAND gates are available in more package configurations than any other gate type. Finally, it is important to remember that this direct one-for-one replacement works only for two-level circuits, that is, AND–OR for NAND, OR–AND for NOR. If there are more than two levels of gating, more complex methods are required, as will be discussed in the next section.

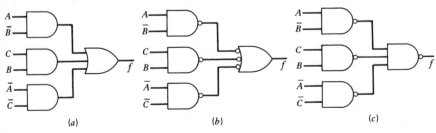

*Figure 5.14*

# 5.6  Multilevel ALL–NAND Realizations

Up to this point we have emphasized realizations in two-level form, that is, AND–OR or OR–AND. We can convert these to all-NAND or all-NOR form, but we still have two levels of gating. There are several reasons for this emphasis. First, the forms are completely general, any functions can be put in either form. Second, powerful design tools for handling these forms exist. Third, these are the fastest circuits, since any additional levels of gating will introduce additional delay. In spite of all these factors, there are times when designs with more levels of gating are preferable. In this section we wish to explore methods for finding multilevel forms.

The most common technique for multilevel design is to work initially in terms of AND, OR, and NOT and then convert to all-NAND. Consider the function

$$f = A\overline{C}D \lor AC\overline{D} \lor B\overline{C}D \lor BC\overline{D}$$

for which the direct realization is shown in Fig. 5.15a. By applying the distributive law, we can factor this function to the form

$$f = \overline{C}D(A \lor B) \lor C\overline{D}(A \lor B)$$

for which the realization is shown in Fig. 5.15b. This circuit, with three levels of gating, is clearly preferable if gate count is our only concern. The basic technique used to find such forms is, as suggested by the equations, algebraic factoring.

The circuit of Fig. 5.15b requires both two-input and three-input gates, which will not be available on the same chip in SSI form. When dealing with large production volumes, significant economies may be realized by reducing the number of

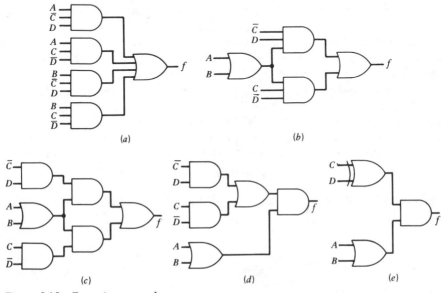

*Figure 5.15   Factoring networks.*

different types of SSI chips. For example, a manufacturer might specify that all functions are to be realized using only two-input gates. When AND and OR are involved, such a restriction is no problem, since AND or OR of three or more inputs can always be broken up into a cascade of two-input operations. For example, this function can be factored in alternative ways. Breaking the three-input ANDs into two-input ANDs leads to the form

$$f(A,B,C,D) = (A \vee B)(\overline{C}D) \vee (A \vee B)(C\overline{D})$$

leading to the realization of Fig. 5.15c. A different factoring leads to the form

$$f(A,B,C,D) = (A \vee B)(\overline{C}D \vee C\overline{D})$$

leading to the realization of Fig. 5.15d. Finally, in this equation we recognize the exclusive OR of $C$ and $D$, giving the form

$$f(A,B,C,D) = (A \vee B)(C \oplus D)$$

and the circuit of Fig. 5.15e.

The circuits of Fig. 5.15 c and d are both more complex than Fig. 5.15b in terms of numbers of gates but have the potential advantage of using only two-input gates. This sort of uniformity can be particularly important in all-NAND or all-NOR designs.

The circuit of 5.15e will almost certainly be the most economical form particularly if the other gates in a 7486 quad exclusive OR package are used elsewhere in the system. The savings in wiring costs will usually more than offset the higher cost of the 7486 package.

Up to this point we have always assumed that both the true and complement forms of every input variable have been available. Inputs in this form are often referred to as *double-rail* inputs. In many systems where inputs to logic networks originate from the outputs of memory elements within the same system, double-rail inputs are indeed available. By contrast, where variables pass from one system to another, they are frequently available only in one form, usually the true form. Such signals are known as *single-rail* inputs. The network of Fig. 5.15e has the additional advantage over 5.15d in that it requires only single-rail inputs.

Having found a multilevel design by factoring, it may then be desired to convert the design to all-NAND form. The basic technique is to break the circuit up into sections that have the AND-OR form, convert these to NAND–NAND form, and then reconnect the sections, adding inversions where necessary. Where the circuit does not divide conveniently into AND–OR sections, we use DeMorgan's law,

$$\overline{A \wedge B} = \overline{A} \vee \overline{B}$$

which, in terms of gate realization, tells us that a NAND gate is equivalent to an OR gate with inversion at the inputs (Fig. 5.16).

*Figure 5.16*

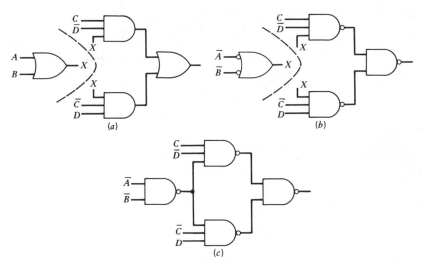

*Figure 5.17*

As a first example of AND–OR to NAND conversion, consider the circuit of Fig. 5.15*b*. In Fig. 5.17*a* we show this circuit divided into two sections, an AND–OR section and an isolated OR gate. Wherever we cut lines in dividing the circuit we assign symbols to those lines to keep track of them for future interconnec⸴ion. In Fig. 5.17*b* we convert the AND–OR section to NAND–NAND and replace the OR gate with one with inverted inputs, complementing *A* and *B* to compensate. Finally, in Fig. 5.17*c* we replace the OR gate with NAND and reconnect to complete the conversion.

As a second example, let us convert the alternate form of the same function shown in Fig. 5.18*c*. In Fig. 5.18*a* we first divide the circuit into three sections, an AND–OR section, an isolated OR gate, and an isolated AND gate. In Fig. 5.18*b* we convert the AND–OR to NAND–NAND, convert the OR as before, and convert the AND to NAND, giving $\bar{f}$, rather than *f* at the output. In Fig. 5.18*c* we convert the OR gate, add inversion at the output and reconnect. When we have an isolated AND, realization in NAND requires use of a NAND gate followed by inversion. We use a NAND with both inputs tied together rather than an inverter, since our objective is all-NAND design.

In comparing these two designs, the design of Fig. 5.17 would appear to be more economical because it uses only four gates compared to six in Fig. 5.18, but we should also consider chip count. Three-input gates come three to a chip, two-input gates come four to a chip, so both circuits will require two chips. Since mounting and wiring costs are usually dominant, there will be little or no difference in cost between these two realizations. Another factor is the gates that are not used. With the form of Fig. 5.17, there would be three gates left over for use in other circuits, whereas there would be only two left over with the form of Fig. 5.18. However, this advantage for the first circuit might well be outweighed by the economies due to using only one type of chip.

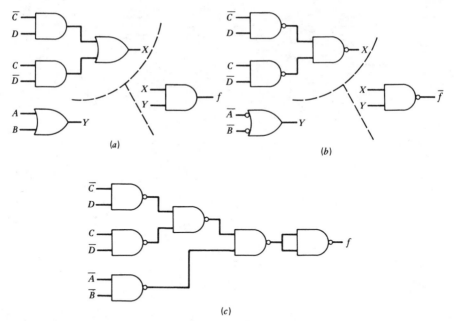

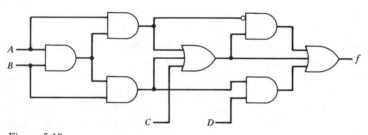

Figure 5.18

The basic procedures illustrated by these examples can be extended to circuits of any complexity. Break the circuit up into sections small enough that the conversion to NAND form is obvious, keep careful track of the identity of signals between sections, add inversions where required, and reconnect. The following example circuit is about as complicated as one is likely to see.

## Example 5.3

Convert the circuit shown in Fig. 5.19 to all NAND form.

Solution
In Fig. 5.20a we divide the circuit into three sections, two AND–OR sections and an isolated AND. Note that there are more connections between sections than in previous examples and that every line that has been cut is identified. In Fig. 5.20b

Figure 5.19

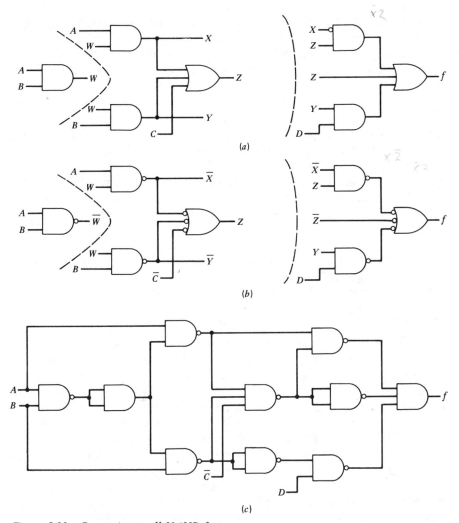

*Figure 5.20 Conversion to all-NAND form.*

we show the initial conversion to NAND form. In this case there are inputs to the OR gates in the AND–OR sections other than directly from the AND gates. We therefore initially convert these gates to the OR-with-inversion form, complementing the direct inputs to compensate. When we convert the upper-left AND to NAND we delete the input inverter and complement the signal. The outputs of the ANDs in the left-most AND–OR group go to other parts of the circuit, so we identify these outputs as complements of the signals developed by the corresponding AND gates. At this point we see the importance of keeping track of the identity of the intermediate signals. We know exactly what signals are required at the inputs of the various gates and we know what signals are available at the outputs of gates. For example, we need $\overline{x}$ at the input of the upper-right NAND gate and $\overline{x}$ is available from the upper-left NAND gate, so we can make a direct connection here. By contrast, we need $y$ at the input of the lower-right NAND gate but have $\overline{y}$ from the lower-left NAND, so an inverter will be needed here. In a similar manner we see that two other inverters are needed in connecting the final circuit, Fig. 5.20c.    ■

The conversion process discussed here involves, except for insertion or deletion of inverters, a direct gate-for-gate replacement, so that the basic form, or organization, of the circuit is preserved. In mathematical terms, it is an *isomorphic* translation. The procedure will always work, but there is no guarantee that a form that is optimal for realization in terms of AND, OR, and NOT will be optimal for all-NAND realization. This is particularly true in circuits with single-rail inputs, in which the inherent inversion of the NAND can sometimes be used to advantage to eliminate extra inverters. This notion will be explored in the next section.

## 5.7 Reducing Package Counts in Multilevel Realizations

Very often only a small part of an overall system design will be realized using SSI, that is, TTL packages. Although only a few such packages may be required, it remains important to use no more packages than necessary. To this end the designer will usually try to use all gates in every package even though this may mean that some gate inputs are wasted. For example, if one additional two-input gate is needed but only a three-input gate is available, the three-input gate will be used as a two-input gate by connecting one input to $+5$ V. In addition, where delay is not critical, package savings can sometimes be effected by resorting to multilevel realization.

A useful technique for direct design of multilevel NAND circuits is *map factoring*, an application of the K-map that takes advantage of the inverting characteristic of the NAND function. Map factoring makes use of the Venn diagram interpretation of intersection for the AND function. If two functions are ANDed on a map, 1's will be entered only in squares where both functions are 1. To apply this notion, we look for a large group of 1's that can be realized in terms of a single product intersected with complements of other products to eliminate squares containing 0's. This is convenient, since the inherent inversion of a NAND gate will make all squares in a product 0.

### Example 5.4

Find an all-NAND realization of the function in Fig. 5.21 assuming single-rail inputs. Use no more than two TTL packages.

Solution
The minimal SOP form of this function is

$$f(A,B,C,D) = \overline{C}\overline{D} \vee \overline{B}\overline{C} \vee \overline{A}BD \vee AB\overline{D} \qquad (5.5)$$

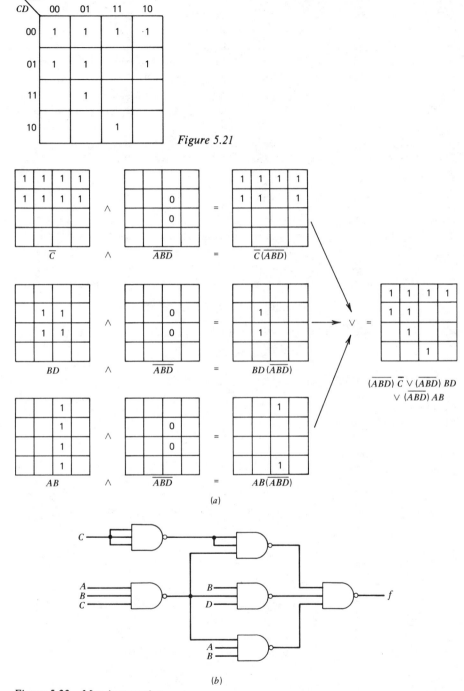

Figure 5.21

(a)

(b)

Figure 5.22   Map intersection.

One rather obvious factoring will lead to a three-chip realization. Before concluding that this is the best that can be done, let us attempt a map factoring. Assuming that the final NAND gate will realize an ORing of terms, we try to divide the 1's into three groups so that no four-input gates will be required. The maps of Fig. 5.22*a* show how this can be done. All but two of the 1's are realized by ANDing the term $\overline{C}$ with the complement of the three-variable product, *ABD*. The remaining two 1's are included by ANDing this same NAND gate output with *BD* and *AB*. The result is shown in Fig. 5.22*b*. By using a three-input gate to complement *C*, it is possible to realize the circuit of Fig. 5.22*b* using only two 7410 triple three-input NAND gate packages. Equation 5.5 cannot be realized using two packages, since all variables must be inverted; so at least one 7420 dual two-input NAND package is required. ■

The K-map can also be helpful in finding exclusive OR designs. As mentioned earlier, the exclusive OR leads to patterns of alternating rows or columns, or strongly diagonal patterns on the map. Figure 5.23 shows some typical patterns associated with exclusive OR. In Fig. 5.24 we see the map for the function discussed at the beginning of this section. Comparing this to Fig. 5.23*a*, we see that the pattern is *C* ⊕ *D* with the 1's in the first column masked off. Hence, a possible form is

$$f(A,B,C,D) = \overline{\overline{A}\,\overline{B}}(C \oplus D)$$

which, from DeMorgan's law, is equivalent to

$$f(A,B,C,D) = (A \lor B)(C \oplus D)$$

the form realized in Fig. 5.15*e*.

**Example 5.5**

Find a realization for

$$f(A,B,C,D) = \Sigma m(2,3,6,8,9,12)$$

Solution
The map of this function is shown in Fig. 5.25*a*. In direct two-level realization the best we can do is four pairs of minterms, requiring four three-input ANDs driving a

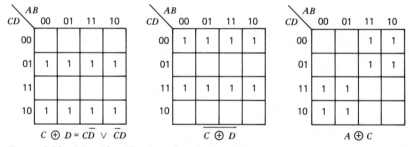

*Figure 5.23   Map identification of exclusive OR.*

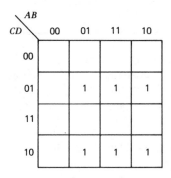

Figure 5.24

four-input OR. As shown in Fig. 5.25b, another possibility is to AND the grouping $(A \oplus C)$ with a block of four 0's on the center of the map, leading to the form

$$f(A,B,C,D) = \overline{BD}(A \oplus C)$$

and the circuit of Fig. 5.25c.

These map-factoring methods can be applied in many ingenious ways [3], but it is important to maintain perspective on the value of these or any other methods for multilevel design. People who have a knack for this sort of thing often find a challenge in trying to find the one best design. But these methods are often tedious and time-consuming and offer no guarantee of leading to a better design. It is important to balance the cost of the designer's time against possible economies in the final design. Unless a circuit is to be produced in very large volume, it is usually best to apply standard two-level techniques and use the first design you find that works.

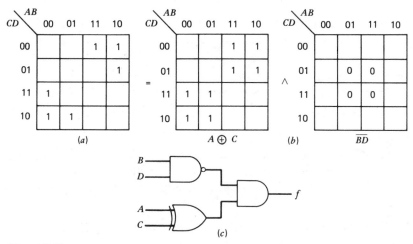

Figure 5.25

## Problems

**5.1** Assume that a higher power NAND gate with a circuit of the simple form given in Fig. 5.1*b* is to be used in a network composed only of NAND gates of the same type. Assume that for all gates $R_1 = 2000\ \Omega$, $R_L = 1000\ \Omega$, $I_B = 5 \times 10^{-4}$ A, when the gate is turned on, and $\beta = 200$. Determine a fan-limit for the gate subject to these conditions.

**5.2** Suppose two gates with the specifications given in Problem 5.1 are to be connected in the network shown in Fig. P5.2 where gate 1 drives only gate 2 and the low-current lamp. What is the minimum dc resistance allowed for the lamp if the circuit is to function properly?

*Hint:* **Current will flow through the lamp and $Z$ must be 1 when $A = B = C = 1$.**

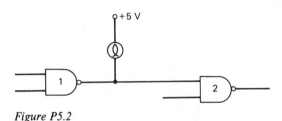

Figure P5.2

**5.3** Suppose the voltage waveforms, $A$ and $B$, shown in Fig. P5.3 were measured at the input and output of an inverter, respectively. Determine values for the turn-off delay and turn-on delay of the inverter.

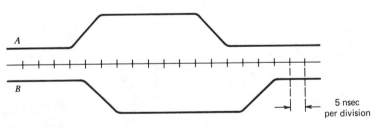

5 nsec per division

Figure P5.3

**5.4** The "turn-on" delay is 8 nsec and the "turn-off" delay is 12 nsec for all gates in the network of Fig. P5.4. Suppose that initially $A = B = C = 0$. Determine the time delay before the circuit output $Z$ changes following a change on input $C$ from 0 to 1.

**5.5** Determine the NAND equivalents of the circuits of Fig. P5.5.

**5.6** Convert the circuit of Fig. P5.6 to NOR form.

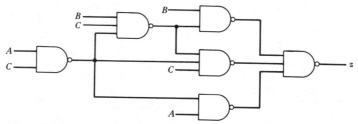

*Figure P5.4*

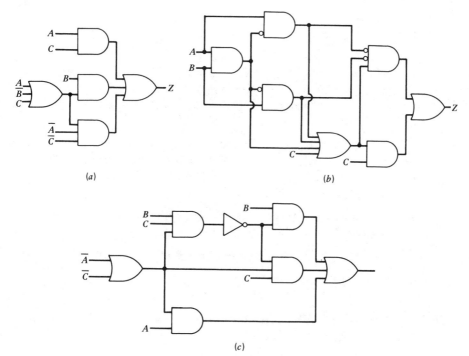

(a)                                                             (b)

(c)

*Figure P5.5*

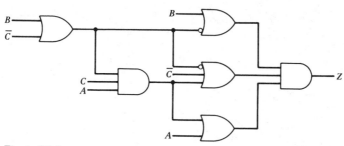

*Figure P5.6*

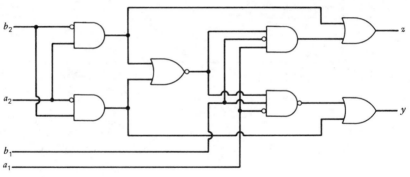

*Figure P5.7*

**5.7** The circuit of Fig. P5.7 has been designed to compare the magnitudes of two 2-bit binary numbers, $A = a_2a_1$, and $B = b_2b_1$. If $A > B$, $Z = 1$ and $Y = 0$. If $B > A$, $Z = 0$ and $Y = 1$. If $A = B$, $Z = Y = 0$. Convert this circuit to all-NAND form.

**5.8** Consider the all-NAND circuit developed as a result of Problem 5.2. Suppose the average of the turn-on and turn-off delays is 15 nsec. Determine the maximum time delay encountered in the change of either output in response to a change on any single input.

**5.9** Realize the following function in all-NAND form with a fan-in limit of two, double-rail inputs.

$$f(A,B,C) = \Sigma m(0,3,4,6,7)$$

*(Note:* **For the next three problems (5.10, 5.11, 5.12) assume that the only NAND packages available are those shown in Fig. 5.8.)**

**5.10** Determine an all-NAND circuit realizing the following function on a single chip, assuming single-rail inputs.

$$f(A,B,C) = \Sigma m(1,2,3,5)$$

**5.11** Find a two-chip realization of the function of Example 5.4, using NAND gates and exclusive OR gates.

**5.12** Find an all-NAND realization of the function that follows assuming a fan-in limit of three single-rail inputs. What is the chip count?

$$f(A,B,C,D) = \qquad f(A,B,C,D) = \Sigma m(0,1,2,3,7,8,9,10,13)$$

**5.13** Determine a minimal sum-of-products expression for the function realized by the circuit of Fig. P5.13.

**5.14** Realize the following functions, using exclusive OR wherever possible.
(a) $f(A,B,C,D) = \Sigma m(1,3,7,8,10,14)$
(b) $f(A,B,C,D) = \Sigma m(5,6,9,10)$

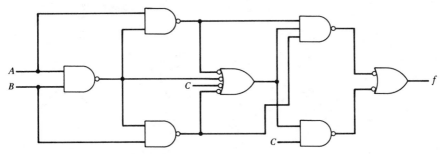

*Figure P5.13*

**5.15** Show how a single 7410 NAND gate package can be used to realize a five-input NAND function.

**5.16** Assume that AND gates and OR gates with four inputs cost $0.50 each and that AND gates and OR gates with two inputs cost $0.25 each. Determine a minimum cost two-level or three-level network to realize

$$f(A,B,C,D) = AC + BC + CD + AB$$

**5.17** A parity check function is one that emits a 1 when the number of 1's in a received word does not conform to the specified parity. Draw the map of a function that will receive a 5-bit word and put out a 1 if the number of 1's is odd. Determine a realization for this function, using exclusive OR gates wherever possible.

## References

1. Barna, A., and D. Porat, *Integrated Circuits in Digital Electronics.* Wiley-Interscience, New York, 1973.
2. Blakeslee, T. R., *Digital Design with Standard MSI and LSI,* 2nd ed. Wiley-Interscience, New York, 1979.
3. Maley, G. A., and J. Earle, *The Logic Design of Transistor Digital Computers.* Prentice-Hall, Englewood Cliffs, N.J., 1963.
4. Peatman, J. B., *Digital Hardware Design.* McGraw-Hill, New York, 1980.
5. Sandige, R. S., *Digital Concepts Using Standard Integrated Circuits.* McGraw-Hill, New York, 1978.

# Computer Arithmetic and Codes

## 6.1 Introduction

In Chapter 3 we reviewed the concept of a binary number and introduced the binary-coded decimal (BCD) code. We observed that the values 1 and 0 are also used as logical values in Boolean algebra, but with a different meaning than when used to form binary numbers. Our discussion of binary numbers was limited to what was necessary to support needed notation and some simple logic network examples.

In digital computers and related equipment, information is stored in the form of binary numbers or encoded in a variety of codes based on the digits 0 and 1. Since a goal of this book is to teach the design of logic networks for the processing of information within digital computers, we must look more deeply into binary numbers, binary arithmetic, and codes. In this chapter we shall consider the necessary operations on binary numbers and define the appropriate codes and code translation procedures. We shall then turn our attention to the implementation of combinational logic networks for operating on binary numbers and manipulating coded information. In the next chapter we shall introduce some of the available MSI parts that accomplish arithmetic and information-processing functions. The use of such special purpose parts can significantly reduce the design time associated with a digital network.

## 6.2 Conversion between Number Bases

In Chapter 3 we considered the conversion of integers expressed in binary form to decimal notation and vice versa. It is possible to convert numbers with digits to the right of the decimal or binary point in the same manner. For example, the binary number

10.101

is equivalent to

$$1 \times 2^1 + 0 \times 2^0 + 1 \times 2^{-1} + 0 \times 2^{-2} + 1 \times 2^{-3}$$
$$= 2 + 0.5 + 0.125 = 2.625$$

In the binary system, the "period" separating the integer and fractional parts of the number is known as the *binary point*.

For conversion from decimal to binary, there are two methods. The first is just the opposite of the binary-to-decimal, subtracting the powers of two that "fit."

**Example 6.1**

Convert $43.4_{10}$ to binary.

Solution

$$
\begin{array}{ll}
\begin{array}{r}
43.4 \\
-32 \\
\hline
11.4
\end{array} & \text{————— 1} \\[2ex]
& 16 \text{ ————— 0} \\
\begin{array}{r}
-8 \\
\hline
3.4
\end{array} & \text{————— 1} \\[2ex]
& 4 \text{ ————— 0} \\
\begin{array}{r}
-2 \\
\hline
1.4 \\
-1 \\
\hline
0.4
\end{array} & \text{————— 1} \\
& \text{————— 1} \\[2ex]
& 0.5 \text{ ————— 0} \\
\begin{array}{r}
-0.25 \\
\hline
0.15 \\
-0.125 \\
\hline
0.025
\end{array} & \text{————— 1} \\
& \text{————— 1}
\end{array}
$$

$43.4_{10} = 101011.011$

The largest power of two that will "fit" is 32, so we subtract it. The remainder, 11.4, is smaller than the next power of two, 16, so we insert a zero in the 16's position. The next power of two, 8, fits, 4 does not, and so forth.  ∎

Note that there is a remainder when we stop subtracting. The binary fraction will be nonterminating unless the decimal fraction can be expressed exactly as a sum of powers of two. If the process is nonterminating, how far should we carry out the conversion? Each decimal digit gives us a "precision" of 1 part in 10 in that position. When we write a number as 43.4, it is implied that the "actual" value is closer to

43.4 than to 43.3 or 43.5. In a similar manner, each binary digit (bit) gives us a precision of one part in two, each pair of bits, a precision of one part in four, and each group of three bits a precision of one part in eight. Thus, to obtain comparable precision, the binary fractions should have about 3 ⅓ bits per digit of the decimal fraction, rounded off to the nearest whole bit, of course. For one decimal digit in the fraction we should have 3 bits, for two decimal digits, 7 bits, for 3 decimal digits, 10 bits, and so on.

An alternative method of conversion uses successive division by two to convert the integer portion and successive multiplication by two to convert the fraction.

## Example 6.2

Convert $43.4_{10}$ to binary.

Solution

$$43 \div 2 = 21 + \text{Rem } 1$$
$$21 \div 2 = 10 + \text{Rem } 1$$
$$10 \div 2 = \phantom{0}5 + \text{Rem } 0$$
$$5 \div 2 = \phantom{0}2 + \text{Rem } 1$$
$$2 \div 2 = \phantom{0}1 + \text{Rem } 0$$
$$1 \div 2 = \phantom{0}0 + \text{Rem } 1$$

$43_{10} = 101011_2$

$$0.4 \times 2 = 0.8 \quad 0$$
$$0.8 \times 2 = 1.6 \quad 1$$
$$0.6 \times 2 = 1.2 \quad 1$$
$$0.2 \times 2 = 0.4 \quad 0$$
$$0.4 \times 2 = 0.8 \quad 0$$

$0.4_{10} = 0.01100_2$

(We have continued the process for more bits than needed to clarify the method.)■

For the integer conversion we repeatedly divide by two and list the remainders, starting with the last remainder as the last significant bit. For the fraction conversion we repeatedly multiply the fractional portion by two, with the overflows forming the bits of the binary fraction. There is no particular choice between the multiplication/division method and the subtraction method; use whichever you prefer.

We saw in Chapter 3 that integer binary numbers can be readily converted to hexadecimal form; the same is true for fractional binary numbers. We use the same process of grouping bits in groups of four, but we work right from the binary point for the fractional part.

**Example 6.3**

Convert $1010101011001.0101111_2$ to hex.

Solution

$$0001\ 0101\ 0101\ 1001\ \cdot\ 0101\ 1110$$

$$1\quad 5\quad 5\quad 9\quad \cdot\quad 5\quad E$$

$$1559.5E_{16}$$

If the number of bits on each side of the binary point is not a multiple of four, as is the case here, add leading or trailing zeros, as required.          ■

## 6.3  Coding of Information

A basic consideration in the design of digital systems is the manner in which information is represented within the systems, that is, the manner in which information is coded. A *code* is any assignment of specific combinations of bits to specific units of information. We may have codes for numbers, codes for letters, or codes for anything that may be of interest. We have already seen two examples of codes. Although the binary number system may seem to have some fundamental connection to the nature of digital systems, the use of binary numbers to represent numeric quantities is just one example of coding. The BCD code is another example of a code, in this case for decimal digits. Clearly, both codes are used to represent numbers, but why should we choose one rather than the other or in preference to other possible codes? The choice of codes is very important and far from arbitrary. At this point let us look at some factors that should be considered in choosing codes.

Suppose that an angular position indicator is to display the position of an airplane control surface with an accuracy of $\pm 1°$ over a range of $\pm 30°$ from the neutral position. Thus there are 61 (including 0) possible positions. We represent the positions by combinations of voltage and no-voltage on six lines. There are 64 possible combinations, such as NNVVNV, NVVNNV, and VNNNVV, where we have used N and V to represent the two possibilities, rather than the conventional 0 and 1, to emphasize that we are not dealing with numbers but simply combinations of values. Any assignment of 61 of the 64 available combinations to the 61 positions constitutes a coding of the control surface position. You could put the 64 combinations on slips of paper, draw them from a hat, and assign them to positions and the result would be a legitimate coding. It would probably not be a very good code, to be sure, but it would be a code.

Codes should not be assigned on a purely random basis. There are usually very good reasons for preferring some particular code to others. The most common situation is that there is some sense of numeric magnitude to the information to be coded, so that a numeric code is preferable. Note that numbers are basically derived from

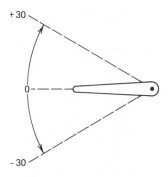

*Figure 6.1    Angular position indicator.*

counting. Far back in the mists of prehistory, when farmers traded apples for eggs or cows for pigs, they needed to know how many of each they had, so they counted, and counting in time led to number systems.* So numeric codes are appropriate any time we are using information that implies a count of anything. We also use numbers in cases where counting is not directly implied. When we assign an address of 436 E. Elm to a house, we do not usually mean that it is the 436th house on E. Elm. Rather, the number tells us something about the position of the house on E. Elm, for example, that it is farther east than number 415 and on the other side of the street. In other cases, numbers convey no numeric information at all. Telephone numbers and social security numbers do not imply counting in any sense, they are just identification codes, used because they are (presumably) unambiguous.

   In the case of the angular position indicator, the positional sense of numbering is clearly applicable. Suppose the positions are numbered as positive and negative numbers of degrees from the neutral, as shown in Fig. 6.1. If we had assigned totally arbitrary codes, positive or negative direction would not be apparent.

   Now that we have decided to number angular positions as shown in Fig. 6.1, is the coding position solved? Not at all, since we still have to decide what combinations of the six lines will represent the numbers. One obvious possibility is the binary number system. Of course, we will need a way to represent the sign of the number, but that is no problem, as we shall see later. Another possibility is to use the BCD code, but that would require some additional lines, since we have two decimal digits plus a sign. Straight binary would appear to be preferable, but there are problems with the use of coding directly in terms of binary numbers. As the indicator moves from +11 to +12, the binary code will change from 001011 to 001100. Note that three bits of the code are to change. No matter how carefully the encoder may be constructed, all three bits will not change at precisely the same time. The difference in time may be only a few milliseconds, but digital circuits are very fast and can easily respond in such intervals. Suppose that the 0 changes to 1 before either of the 1's change to 0. Then the code seen will momentarily be 001111, the code for

*Reference [6] provides a very interesting account of the evolution of numerals and number systems from counting.

+15. If this coded number were driving a physical controller such as a stepping motor, a serious problem could result.

This type of problem can be prevented by the use of a code in which only one bit changes as we move from one element of the code to the next. Such a code is the *Gray code,* or reflected code. The reason for the latter name is seen in the way the code is constructed. We start by writing down 0 and 1 in a column and then "reflecting" them about a horizontal line, that is, copy them in the opposite order below the line, and then put 0's in front of the entries above the line, 1's in front of the entries below the line. This gives us the four-element Gray code, shown in Fig. 6.2a. If we repeat the process, reflecting the four-element code and inserting 0's above the line, 1's below the line, we get the eight-element Gray code (Fig. 6.2b). Repeating the process again leads to the 16-element code (Fig. 6.2c), and so on.

If we use this code, as the indicator moves from +11 to +12, the code will change from 01110 to 01010. Only one bit changes, so the only uncertainty is the exact instant that it will change; there is no possibility of an erroneous signal being generated. We have solved one problem, but we have lost the advantage of a numeric code. Noting that we are using the code to represent positions such as +11 and +12, it might seem that it is a numeric code. But numeric codes, such as binary and BCD, are distinguished by positional weighting. Each digit or bit has a weight, a power of the radix, determined by its position. It is this weighting that makes it possible to perform arithmetic by the normal rules, to determine the relative magnitudes of numbers by subtraction. The Gray code is not a weighted code; individual bits have no numeric significance, so comparing magnitudes cannot be done other than by looking the codes up in a table to determine which represent larger quan-

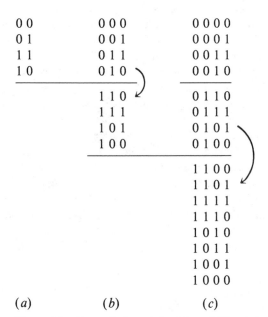

(a)                 (b)                 (c)

*Figure 6.2   Construction of the reflected (Gray) code.*

tities. In spite of these disadvantages, the Gray code is widely used, as in this example where codes represent continuously changing physical phenomena and momentarily erroneous codes might cause serious problems. If magnitude comparisons or other arithmetic operations are required, the Gray code can be sampled and then converted to regular binary. We shall see in Chapter 7 that the conversion from Gray to binary is fairly simple.  ·

Numeric data is the most common type of data encountered in digital systems, but there are many situations in which alphabetic data must be represented. There are several *alphanumeric* codes in use; by far most common in the United States is the *ASCII* code (American Standard Code for Information Interchange). The ASCII code is formally an 8-bit code, but only seven bits are specified in the official definition of the code. The use of the first bit is optional with the user. It may be arbitrarily set to 0 or 1 or used for parity checking, a concept to be discussed in the next section. With seven data bits, there are 128 possible codes, enough for upper- and lowercase letters, the decimal digits, various punctuation marks, and miscellaneous control codes, the function of which is usually optional with the user. Figure 6.3 shows the ASCII codes for the most common characters, with the leading bit set to 0.

The ASCII code perhaps deserves more attention than the other codes presented in this chapter, since it is the primary medium of interchange of textual information between digital systems. With the notable exception of IBM systems, which use the EBCDIC code, communications between computers and peripheral equipment such as keyboards, printers, magnetic tape drives, and magnetic disk drives are accomplished using this code. You have probably observed that the characters and symbols engraved on the keys of most computer terminals are those listed in Fig. 6.3. With the exception of letters and numbers that conform to the standard typewriter format, the location of symbols may vary from keyboard to keyboard, but those listed will almost always be available. Not all the 128 possible ASCII codes are shown in Fig. 6.3. Most of the other codes have been assigned to specific control functions that are meaningful for particular peripherals or particular information interchange formats. Three of the most common control codes, those for space, carriage return, and line feed, are included in the table. We shall not apply the ASCII code in this chapter, but you will find it to be used extensively in Chapters 11 through 16, which treat microprocessors.

In Fig. 6.4 we see an additional code that is widely used. All readers have seen the digital readout devices found, for example, on digital watches and pocket calculators. These devices display decimal digits as combinations of lighted and unlighted segments arranged as shown in Fig. 6.4b. The assignment of digits to display combinations may be considered a code. It is apparent that each decimal digit can be formed by lighting some subset of the seven segments. To control this display, we must generate a 7-bit code to indicate whether each segment should be on or off. If we let 0 correspond to OFF and 1 to ON, the *seven-segment code* will be as shown in Fig. 6.4a. Since the seven-segment code is highly redundant and nonweighted, it would be unsuitable for internal use in a computer. Thus, the use of this form of display will require logic to convert from some internal code, such as BCD, to the seven-segment code.

### ASCII codes for selected characters
### (most significant bit 0)

| Char | Binary | Hex | Char | Binary | Hex | Char | Binary | Hex |
|------|--------|-----|------|--------|-----|------|--------|-----|
| A | 0100 0001 | 41 | a | 0110 0001 | 61 | 0 | 0011 0000 | 30 |
| B | 0100 0010 | 42 | b | 0110 0010 | 62 | 1 | 0011 0001 | 31 |
| C | 0100 0011 | 43 | c | 0110 0011 | 63 | 2 | 0011 0010 | 32 |
| D | 0100 0100 | 44 | d | 0110 0100 | 64 | 3 | 0011 0011 | 33 |
| E | 0100 0101 | 45 | e | 0110 0101 | 65 | 4 | 0011 0100 | 34 |
| F | 0100 0110 | 46 | f | 0110 0110 | 66 | 5 | 0011 0101 | 35 |
| G | 0100 0111 | 47 | g | 0110 0111 | 67 | 6 | 0011 0110 | 36 |
| H | 0100 1000 | 48 | h | 0110 1000 | 68 | 7 | 0011 0111 | 37 |
| I | 0100 1001 | 49 | i | 0110 1001 | 69 | 8 | 0011 1000 | 38 |
| J | 0100 1010 | 4A | j | 0110 1010 | 6A | 9 | 0011 1001 | 39 |
| K | 0100 1011 | 4B | k | 0110 1011 | 6B |  |  |  |
| L | 0100 1100 | 4C | l | 0110 1100 | 6C | ! | 0010 0001 | 21 |
| M | 0100 1101 | 4D | m | 0110 1101 | 6D | " | 0010 0010 | 22 |
| N | 0100 1110 | 4E | n | 0110 1110 | 6E | # | 0010 0011 | 23 |
| O | 0100 1111 | 4F | o | 0110 1111 | 6F | $ | 0010 0100 | 24 |
| P | 0101 0000 | 50 | p | 0111 0000 | 70 | % | 0010 0101 | 25 |
| Q | 0101 0001 | 51 | q | 0111 0001 | 71 | & | 0010 0110 | 26 |
| R | 0101 0010 | 52 | r | 0111 0010 | 72 | ' | 0010 0111 | 27 |
| S | 0101 0011 | 53 | s | 0111 0011 | 73 | ( | 0010 1000 | 28 |
| T | 0101 0100 | 54 | t | 0111 0100 | 74 | ) | 0010 1001 | 29 |
| U | 0101 0101 | 55 | u | 0111 0101 | 75 | * | 0010 1010 | 2A |
| V | 0101 0110 | 56 | v | 0111 0110 | 76 | + | 0010 1011 | 2B |
| W | 0101 0111 | 57 | w | 0111 0111 | 77 | , | 0010 1100 | 2C |
| X | 0101 1000 | 58 | x | 0111 1000 | 78 | − | 0010 1101 | 2D |
| Y | 0101 1001 | 59 | y | 0111 1001 | 79 | . | 0010 1110 | 2E |
| Z | 0101 1010 | 5A | z | 0111 1010 | 7A | / | 0010 1111 | 2F |
|  |  |  |  |  |  | : | 0011 1010 | 3A |
|  |  |  |  |  |  | ; | 0011 1011 | 3B |
|  |  |  |  |  |  | < | 0011 1100 | 3C |
|  |  |  |  |  |  | = | 0011 1101 | 3D |
| Space | 0010 0000 | 20 |  |  |  | > | 0011 1110 | 3E |
| Carriage return | 0000 1101 | 0D |  |  |  | ? | 0011 1111 | 3F |
| Line feed | 0000 1010 | 0A |  |  |  | @ | 0100 0000 | 40 |

*Figure 6.3   ASCII code.*

| Decimal digit | \ | Seven-segment code | | | | | |
|---------------|---|---|---|---|---|---|---|
|  | a | b | c | d | e | f | g |
| 0 | 1 | 1 | 1 | 1 | 1 | 1 | 0 |
| 1 | 0 | 1 | 1 | 0 | 0 | 0 | 0 |
| 2 | 1 | 1 | 0 | 1 | 1 | 0 | 1 |
| 3 | 1 | 1 | 1 | 1 | 0 | 0 | 1 |
| 4 | 0 | 1 | 1 | 0 | 0 | 1 | 1 |
| 5 | 1 | 0 | 1 | 1 | 0 | 1 | 1 |
| 6 | 0 | 0 | 1 | 1 | 1 | 1 | 1 |
| 7 | 1 | 1 | 1 | 0 | 0 | 0 | 0 |
| 8 | 1 | 1 | 1 | 1 | 1 | 1 | 1 |
| 9 | 1 | 1 | 1 | 0 | 0 | 1 | 1 |

(a)                               (b)

*Figure 6.4   Seven-segment code and display.*

# 6.4  Parity

We commented earlier that one of the principal advantages of digital techniques is that the use of just two distinct signal levels reduces the likelihood of errors caused by noise. But complete immunity to noise is impossible. No matter how widely separated the two levels, it is possible to have so much noise that a transmitted 0 appears to be a 1, or vice versa. The basic method of combatting such errors is *redundancy,* sending more information than is necessary, with the hope that enough will get through correctly to provide reliable communications. Humans use redundancy every day in communicating with each other. Suppose you are talking on the phone and have a poor connection, so you are not sure the other person heard you correctly. So you repeat what you said, or ask him to repeat back what he heard.

Complete repetition is a rather extreme use of redundancy that is used in digital systems only in cases where absolute reliability is critical, as in military command signals. In most cases we add just enough redundancy to each message to reduce the probability of an undetected error to an acceptable level. Probably the most common technique is the use of a *parity bit*. For each data element to be transmitted we add one extra bit, selected to be 0 or 1 so that the total number of 1's in the code will be even (even parity) or odd (odd parity). As an example, assume we want to use the leading bit in the ASCII code to establish even parity. The codes for A and B would be as shown in Fig. 6.3,

$$0100\ 0001 \quad \text{and} \quad 0100\ 0010$$

since both contain an even number of 1's. The code for C would be

$$1100\ 0011$$

using a parity bit of 1 to give an even number of 1's. At the receiving end, if a code is received with an odd number of 1's, you know that something is wrong. You do not know what is wrong, but you can request retransmission or take other appropriate action. Two points should be noted. First, sender and receiver must agree ahead of time whether the parity is going to be odd or even. Second, parity is no guarantee against errors going undetected. If two errors occur in the same data element, parity will be unchanged. But the probability of two errors occurring in the same code is much less than the probability of one. If the addition of a single parity bit does not reduce the probability of error sufficiently, several parity bits can be added, each establishing parity of a different part of the data element. But there is no perfect system. No matter how elaborate the redundancy scheme, there is always a possibility, however remote, of some combination of errors occurring that will convert one "legal" code to another "legal" code, thus defeating all efforts to eliminate error.

So far we have inferred that parity bits are only used when information is transmitted from one system to another. Actually, parity bits can be employed within a single system. For example, even parity may be established on all characters as they are written on magnetic tape or magnetic disk. Parity can then be checked as they are read back from the storage medium.

The values to be assigned to parity bits at the sending end of a process can be generated using combinational logic. The logic necessary at the receiving end to determine if a parity error has occurred during transmission is very similar. Suppose we wish to transmit characters consisting of seven information bits, $X_0$, $X_1$, $X_2$, $X_3$, $X_4$, $X_5$, $X_6$, to which a parity bit, $P_7$, is to be added to form even parity characters. If parity for the eight-bit characters is to be even, then $P_7$ must be 1 if parity over the seven information bits is odd, and $P_7$ must be 0, if parity over the seven information bits is even. Therefore,

$$P_7 = \text{ODDPARITY}(X_0, X_1, X_2, X_3, X_4, X_5, X_6) \qquad (6.1)$$

where the function, ODDPARITY, is 1 if an odd number of the seven bits are 1. At the receiving end an error will have occurred, if an odd number of the eight received bits are 1. Therefore,

$$\text{Parity error} = \text{ODDPARITY}(X_0, X_1, X_2, X_3, X_4, X_5, X_6, P_7) \qquad (6.2)$$

If the eight-bit characters were to be odd parity, then the function, EVENPARITY, would be used to determine the parity bit at the sending end and to realize the error function at the receiving end.

The Boolean function for checking odd parity is the classic example of a function that is not efficiently realized in two levels. The function $f(X_1, X_2, X_3, X_4)$, which is 1 when an odd number of the variables $X_1$, $X_2$, $X_3$ or $X_4$ are 1, is depicted in Fig. 6.5. An odd-parity check over any number of variables will appear as a "checkerboard" pattern on a K-map. In such a map, there are no groupings of minterms into larger groupings, so the two-level realization will require an ORing of minterms. Furthermore, this is the maximum possible number of minterms, since the addition of any additional minterms to the map would create groupings. Thus, this is the worst possible function for two-level realization. A two-level parity check function over eight variables would require 129 NAND gates.

Let us consider an alternative arrangement. The circuit for detecting odd parity over two bits is simply exclusive OR. In Fig. 6.6a we show two exclusive OR gates with two distinct pairs of inputs. If exactly one of the four $X$ variables is 1, then exactly one of the $Z$ outputs must be 1. This will be true if three of the four $X$'s are 1. If all the $X$'s are 0 or all are 1, clearly both $Z$'s will be 0. If one gate has two 1's at the input and the other has two 0's, both $Z$'s will be 0. Finally, if both gates have one 1 at the input, both $Z$'s will be 1. This discussion is summarized in Fig. 6.6b. From this table, the number of $Z$'s equal to 1 is odd only if the number

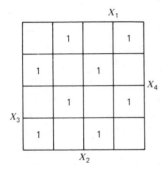

*Figure 6.5   Four-variable odd-parity check function.*

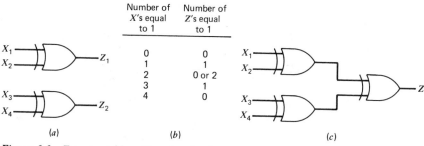

Figure 6.6  Four-variable odd-parity checker.

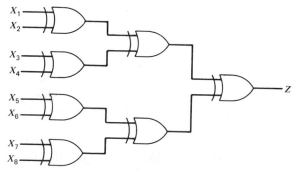

Figure 6.7  Eight-variable odd-parity checker.

of $X$'s equal to 1 is odd. Thus, a check of odd parity over all four $X$'s is provided by adding a third exclusive OR, as shown in Fig. 6.6c. Extending the argument to eight variables leads to the circuit of Fig. 6.7.

## 6.5  Binary Arithmetic

Digital systems do many things besides processing numeric data, but numeric processing is so common that every digital designer should have a basic understanding of binary arithmetic. Addition is derived originally from concepts of counting, but the operation is formally defined by the addition tables. When you were in grade school you learned your tables, you memorized $3 + 4 = 7$, and $7 + 5 = 12$, and so on. Binary addition is likewise formally defined by the addition table shown in Fig. 6.8. Note that there are two bits in the sum for the same reason that the addition of two decimal digits can produce a two-digit result; when the sum is equal to or larger than the radix, there is a carry into the next position.

The low-order bit of the result is known as the *sum bit,* the high-order bit, as the *carry bit*. The extension to the addition of multibit numbers proceeds in the same general manner as for decimal arithmetic, keeping in mind that the only numerals are 0 and 1 and there is a carry when the sum is 2 or larger.

| + | 0 | 1 |
|---|---|---|
| 0 | 00 | 01 |
| 1 | 01 | 10 |

*Figure 6.8   Binary addition table.*

**Example 6.4**

Add $7 + 6$ in binary.

Solution

$$
\begin{array}{rl}
 & 1\ 1 \quad \longleftarrow \text{Carries} \\
7 & 0\ 1\ 1\ 1 \\
+6 & +0\ 1\ 1\ 0 \\
\hline
13 & 1\ 1\ 0\ 1
\end{array}
$$

Note the third column from the right, where we have the addition of a carry to $1 +$ 1. Since $1 + 1 + 1 = 3_{10} = 11_2$, we have a sum bit and a carry.  ■

You may have wondered why we represented 6 and 7 as 4-bit numbers with leading 0's, since three bits would suffice. When converting decimal numbers to binary in the last section we used as many bits as needed, which is fine when writing the numbers on a sheet of paper. In digital systems, however, data are represented by voltages at a fixed, finite number of points. As a result, data in most digital systems are represented in the form of fixed-length words. A *word* in a digital system is simply a set of $n$ bits that can be taken together because of some common significance. For example, an 8-bit word might represent a single alphabetic character. A 16-bit word might be a four-digit decimal number in BCD form or simply a 16-bit number in the binary number system. The number of bits in a word is most commonly a multiple of 4, although any number is possible. In Example 6.5 we are assuming 4-bit words representing decimal digits. One problem with fixed-length words is that an operation may produce a result that is too long to fit in a single word. For example, if we add $8 + 9$ in a 4-bit system, we have

$$
\begin{array}{r}
1\ 0\ 0\ 0 \\
+\ 1\ 0\ 0\ 1 \\
\hline
1\ 0\ 0\ 0\ 1
\end{array}
$$

producing a result that will not fit in a 4-bit word. When this happens, we say that we have an overflow. Just what should happen as a result of an *overflow* depends on the situation. We shall consider the situation in more detail later.

Just as we can define binary addition by an addition table, we can define binary subtraction by a subtraction table. Before proceeding in that direction, however, we should note that allowing subtraction as a possible operation requires that we consider the representation of negative numbers. In "pencil and paper" arithmetic we indicate the signs of numbers by writing $(+)$ or $(-)$ in front of the number. Such a representation is known as *signed-magnitude representation*. We can do the same thing in a digital system by a sign bit in front of the magnitude, with the usual convention that 0 stands for (+), 1 for (−). We would then write +5 as 0,101 and −5 as 1,101. The comma is used in written notation to indicate that the first bit is a sign bit. There will be no "comma" in a physical system; the circuits will simply be designed to treat the first bit as the sign bit. The problem with signed-magnitude representation is that it requires the use of different rules for addition and subtraction, which in turn means separate circuitry in a physical system. The amount of hardware required can be reduced by the use of complement notation.

To illustrate the concepts of complement notation, consider the following decimal subtraction.

$$
\begin{array}{r}
56 \\
-\ 37 \\
\hline
19
\end{array}
$$

Simple enough, but let us try a different approach, replacing the subtraction (56 − 37) with the addition (56 + 63) in a two-digit calculator in which any overflow digits are lost.

$$
\begin{array}{r}
56 \\
+\ 63 \\
\hline
\end{array}
$$

Overflow →①19

With the overflow discarded, we see that the answer is correct, but why 63? This is a two-digit system, 10 = 100 and 63 = 10 − 37. The number 63 is the 10's complement of 37 in a two-digit system. In general, if we have any number $N$ represented in an $n$-digit decimal system, the 10's complement is given by

$$
\text{TENSCOMP}(N) = 10^n - N
$$

Given a number $N$ represented in an $n$-bit binary system, the 2's complement of $N$ is given by

$$
\text{TWOSCOMP}(N) = 2^n - N
$$

The 2's complement has the same characteristic as the 10's complement, making it possible to accomplish subtraction by addition. But there seems to be a flaw in the argument. The complement is defined in terms of subtraction. If we have to supply subtraction circuitry to form the complements, we are right back where we started, with separate circuits for addition and subtraction. Fortunately, it turns out that there are two methods for taking the 2's complements without subtraction.

**Method 1**   Reverse every bit (change 0's to 1's and 1's to 0's) and then add the least significant position.

**Method 2**   Starting with the least-significant bit, leave the bits unchanged up to and including the first 1, and then change all remaining bits.

### Example 6.5

Find the 2's complement representation of −37, using eight bits (seven numeric bits plus sign).

Solution

$$+37_{10} = 0,0100101_2 \quad \text{Reverse bits} \quad 1,1011010$$
$$\text{Add 1} \qquad \underline{\quad + 1 \quad}$$
$$(-37) \qquad 1,1011011$$ ∎

## Example 6.6

Find the 2's complement representation of $-56$, using eight bits.

Solution

$$+56_{10} = 0,0111000$$

Reverse   No change

$$(-56) \qquad 1,1001000$$ ∎

Note that we treat the sign bit like any numeric bit in forming the 2's complement. Method 1 is preferable for computer realization; use whichever you prefer for manual conversion.

Now let us investigate the use of the 2's complement in performing arithmetic.

## Example 6.7

Perform the addition $(+56) + (-37)$ in 8-bit binary.

Solution

$$(+56) \qquad 0,0111000$$
$$\underline{+(-37) \qquad +1,1011011}$$
$$(+19) \qquad 0,0010011$$

Note that the sign bits are treated as numeric bits, added by the normal rules of binary addition. In this case there is a carry into the sign position, which is added to 1 to produce a sum sign bit of 0 with a carry out that is discarded since the word length is eight bits. ∎

In Example 6.7 we have stated the subtraction problem in such a way as to emphasize that subtraction is simply the addition of numbers of unlike sign. Complement addition works no matter what the signs of the numbers or the sign of the sum. If the sum is negative, it will be in complement form.

## Example 6.8

Perform the addition $(-56) + (+37)$ in 8-bit binary.

Solution

$$(-56) \qquad 1,1001000$$
$$\underline{+(+37) \qquad +0,0100101}$$
$$\rightarrow (-19) \qquad 1,1101101$$

The answer indicated by the arrow is the correct answer ($-19$) in 2's complement form. To check this we can recomplement the number to obtain the positive value. Leaving the first 1 unchanged and reversing the other bits we obtain

$$0,0010011$$

which we see, by comparison to Example 6.7, is ($+19$).                                    ■

It is important to note that we have recomplemented the answer in Example 6.8 only to check our work. In a computer the answer would be left in the complement form, which is the correct representation for a negative number.

Complement addition will similarly work with addition of two negative numbers, producing the negative sum in complement form. The use of complements does not eliminate the problem of overflow. No matter what the system of representation, the addition of two numbers of like sign may produce a result that is too large to fit in a single word.

The problem of detecting overflow for signed addition is more complicated than for unsigned addition, as we shall see in a later section.

Signed binary arithmetic is the basic arithmetic capability required in digital systems. All other mathematical operations can be reduced to a series of additions. Multiplication can be implemented by repeated additions, division by repeated subtractions. In some cases specialized circuits may be developed for more complex operations in the interest of improved performance, but signed addition is the basic mathematical operation in most digital systems.

# 6.6  Implementation of Binary and BCD Addition

Since addition is the basic arithmetic operation in most computers, circuits for implementing addition are very important. Let us start by converting the binary addition table of Fig. 6.8 to truth table form, as shown in Fig. 6.9$a$. Here the outputs are labeled $S$ and $C$, for sum and carry. By inspection, we can see that

$$S = X \oplus Y \quad \text{and} \quad C = X \wedge Y$$

so that this simple binary addition can be accomplished with the circuit of Fig. 6.9$b$.

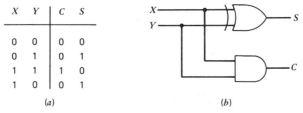

| X | Y | C | S |
|---|---|---|---|
| 0 | 0 | 0 | 0 |
| 0 | 1 | 0 | 1 |
| 1 | 1 | 1 | 0 |
| 1 | 0 | 0 | 1 |

(a)                                           (b)

*Figure 6.9   Truth table and circuit for binary addition.*

$$
\begin{array}{ccccc}
C_{n-1} \diagdown & & C_2 \diagdown & C_1 \diagdown & \\
X_{n-1} & \cdots & X_2 & X_1 & X_0 \\
Y_{n-1} & & Y_2 & Y_1 & Y_0 \\
S_n = C_n \leftarrow S_{n-1} & & S_2 & S_1 & S_0
\end{array}
$$

Figure 6.10   *Addition of* n-*bit numbers.*

This is certainly simple enough, but being able to add two 1-bit binary numbers is clearly not enough. Generally, we wish to add two $n$-bit binary numbers, where $n$ may be anywhere from 4 to 64. The basic process for binary addition of $n$-bit numbers is shown in Fig. 6.10. The first pair of bits is added, producing a sum bit and a carry that is then added to the next pair of bits, producing the second sum bit and another carry that is then added to the next pair of bits, and so on. We see that the circuit of Fig. 6.9b is not adequate except for the first bit position. In all the other positions we must add two bits plus an incoming carry. The circuit of Fig. 6.9b carries out only "half" the addition process and is known as a *half adder*. A circuit that adds two bits plus a carry is known as a *full adder*. The truth table and K-maps for the full adder are shown in Fig. 6.11. The sum bit is seen to be identical to the odd-parity function, that is, a three-input exclusive OR,

$$S_i = X_i \oplus Y_i \oplus C_i \tag{6.3}$$

Also, you should be able to verify that the carry term can be realized in the form

$$C_{i+1} = ((X_i \oplus Y_i) \wedge C_i) \vee (X_i \wedge Y_i) \tag{6.4}$$

leading to the circuit of Fig. 6.12, a realization of the full adder by interconnection of two half adders.

This circuit is fairly simple, but it is not a two-level circuit and therefore not as fast as it might be. If speed is critical, we can realize the K $=$ maps of Fig. 6-11 in the two-level AND-OR form,

$$S_i = X_i \overline{Y_i}\,\overline{C_i} \vee \overline{X_i}\,\overline{Y_i} C_i \vee X_i Y_i C_i \tag{6.5}$$
$$C_{i+1} = X_i Y_i \vee X_i C_i \vee Y_i C_i \tag{6.6}$$

leading to the circuit of Fig. 6.13.

With the availability of full adders, construction of $n$-bit adders simply requires cascading $n$ full adders, as shown in Fig. 6.14. This form of adder is known as a *ripple-carry adder,* because the carries "ripple through" from one stage to the next.

| $X_i$ | $Y_i$ | $C_i$ | $C_{i+1}$ | $S_i$ |
|-------|-------|-------|-----------|-------|
| 0 | 0 | 0 | 0 | 0 |
| 0 | 0 | 1 | 0 | 1 |
| 0 | 1 | 1 | 1 | 0 |
| 0 | 1 | 0 | 0 | 1 |
| 1 | 1 | 0 | 1 | 0 |
| 1 | 1 | 1 | 1 | 1 |
| 1 | 0 | 1 | 1 | 0 |
| 1 | 0 | 0 | 0 | 1 |

K-map for $S_i$:

| $C_i$ \ $X_i Y_i$ | 00 | 01 | 11 | 10 |
|-------------------|----|----|----|----|
| 0 | | 1 | | 1 |
| 1 | 1 | | 1 | |

K-map for $C_{i+1}$:

| $C_i$ \ $X_i Y_i$ | 00 | 01 | 11 | 10 |
|-------------------|----|----|----|----|
| 0 | | | 1 | |
| 1 | | 1 | 1 | 1 |

*Figure 6.11   Truth table and maps for full adder.*

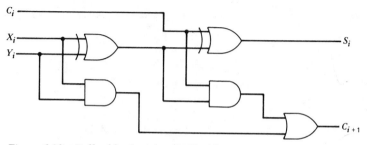

Figure 6.12 Full adder based on half adders.

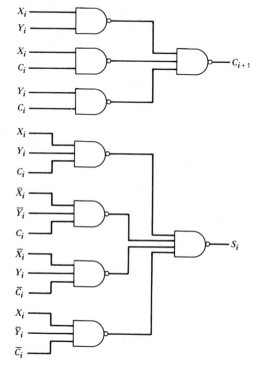

Figure 6.13 Two-level realization of full adder.

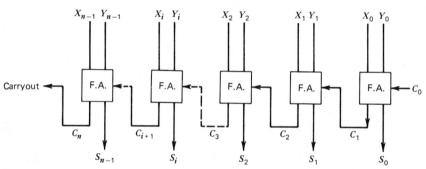

Figure 6.14 n-bit binary adder.

**119**

A half adder could be used in the first bit position, but it is usual to use full adders, both for uniformity and for the flexibility provided in signed arithmetic by an input carry. For example, the input carry can be used to provide the $+1$ required for Method 1 of taking the 2's complement.

Because addition is so important in computers, a great deal of effort has been devoted to trying to find better designs. The two designs shown in Figs. 6.12 and 6.13 are the "classic" designs, but many others have been used. Particularly in integrated circuits, other designs may be better suited to the special characteristics of a particular technology. A special concern in adder design is delay. Even if the two-level full adder is used, the $n$-bit adder is not a two-level circuit. As we can see from Fig. 6.14, a carry may propagate through the entire adder, and the highest-order output bits may be functions of the lowest-order inputs. Thus, the $n$-bit adder of Fig. 6.14 is not a two-level circuit, but a $2n$-level circuit. Theoretically, a high-order sum bit can be made a two-level function of all lower-level inputs, but the amount of logic required is totally impractical for more than just a few bits. Several ingenious designs have been developed that strike a compromise between the relatively slow ripple-carry adder and the impractically expensive two-level adder. One design, the *carry-look-ahead* adder, achieves approximately a 10-to-1 speed advantage over the ripple-carry adder for a 2-to-1 increase in cost, for a 64-bit adder.

As discussed in Chapter 3, numeric data are sometimes processed in BCD form, in which each decimal digit is individually converted to binary form. Although BCD is normally used for reasons other than computational convenience, if data are in BCD form, the capability to perform arithmetic on BCD data will be needed. It can be programmed, but many computers provide for BCD arithmetic in hardware. We can best illustrate the difference between binary addition and BCD addition by an example.

**Example 6.9**

Perform the addition $37_{10} + 48_{10}$ in binary and BCD.

Solution

$$
\begin{array}{lll}
\text{Binary} & 0100101 & \text{BCD} \quad 0011\ 0111 \\
& 0110000 & \quad\quad\ \ 0100\ 1000 \\
& 1010101 & \quad\quad\ \ 1000\ 0101 \\
\text{Binary 85} & & \text{BCD 85}
\end{array}
$$

Note that the addition of $7 + 8 = 15$ produces 5 (0101) plus a carry into the next digit, in accordance with the rules of decimal arithmetic. As we saw in Chapter 3, the fact that we are performing operations derived from the rules of decimal arithmetic poses no basic problems in logical design. The rules of decimal arithmetic specify the functional relationships between the binary codes, and logical design can then proceed in the usual way. In this case, the design could be derived directly from truth tables defining the rules of decimal addition for BCD codes, but it is usually found simpler to start with a binary adder and then modify it. The procedure followed is to add the two BCD codes in a 4-bit binary adder, just as if they were 4-bit binary numbers. If the sum is between 0 and 9, it will be correct. If the sum is greater than 9, we then add 6 to produce the correct code for the sum and an output carry. Thus, the addition of $7 + 8$ in Example 6.9 would be carried out in two steps, with the correction factor of 6 informally justified as follows.

$$
\begin{array}{r}
0111 \\
+ \quad 1000 \\
\hline
1111 \leftarrow \text{Binary 15} \\
+ \quad 0110 \leftarrow \text{Add 6 since sum} > 9 \\
\hline
\text{Decimal carryout} \qquad 1\,0101 \qquad \text{BCD 5}
\end{array}
$$

A carry occurs when the sum is equal to or greater than the radix, 10 in the case of the decimal system. A 4-bit binary adder is effectively a hex adder, that is, its radix is 16. In this case, 15 exceeds radix 10 by 5; if we add 6 to it, it will exceed radix 16 by 5, thus producing the correct sum plus carry.

Signed arithmetic poses special problems in BCD. We can use 10's complement arithmetic, but we must recall that complement arithmetic requires that we know the sign of each number, because each digit is handled separately in BCD, it would be necessary to have a sign bit with each digit or else provide some special means of keeping track of the sign of a string of BCD digits. As a result, computers that implement BCD arithmetic commonly include a subtract command and treat all BCD numbers as positive.

## 6.7  Carry and Overflow

In Section 6.5 we commented on the problem of overflow, the situation that occurs when the addition of two $n$-bit numbers produces an $n + 1$-bit number in a system in which numbers are limited to $n$ bits. Referring to the $n$-bit ripple-carry adder of Fig. 6.14, we see that an $n + 1$-bit sum would be indicated by $C_n = 1$. Does this mean that overflow is always indicated by output carry (carry-out) from the adder? Not necessarily, it depends on the character of the two numbers applied to the adder. Let us consider some examples, in all of which we will assume 4-bit data and a 4-bit adder. First, we will assume that the 4-bit words represent unsigned integers, so that the largest possible number is 1111 (15 decimal). Suppose we perform the addition $(8 + 9)$

$$
\begin{array}{r}
1000 \\
+ \quad 1001 \\
\hline
\text{Carry-out} \rightarrow 1\,0001
\end{array}
$$

In this case the carryout clearly corresponds to overflow. Next, assume that the numbers are represented in signed 2's complement notation, in which case there are only three numeric bits, so that the largest number that can be represented is 111 (7 decimal). Suppose we perform the addition $((+4) + (+5))$.

$$0100$$
$$+\ 0101$$

No carry-out $\rightarrow$ 0 1001

Note that we have omitted the commas used to separate the sign bits in the previous discussion of complement arithmetic. There are no commas inside the computer; the commas are a graphical convention used in print to remind the reader that we are dealing with complement forms. The adder simply sees two 4-bit numbers and treats the "sign bits" like any numeric bits. You may at first think the answer is correct, since 1001 is binary 9. But it is not correct; that first bit is a sign bit and it appears that we got a negative sum. We did not; what we got was overflow, since nine cannot be represented in a 4-bit signed number system. From this result, it would be natural to conclude that overflow in a signed system can be provided by detecting carry-out from the most significant numeric bits. Before we accept that conclusion, let us try the addition $((-4) + (-5))$

$$1100$$
$$+\ 1011$$

Carry-out $\rightarrow$ 1 0111

There is overflow, and there is a carry-out, but there is no carry from the most significant numeric bits. This is, in fact, a general principle—when adding two negative numbers in 2's complement form, it is the *absence* of carry from the most significant numeric bits that indicates overflow. There is, however, a more general rule for detecting signed overflow. In both the preceding examples we have added two numbers of the same sign and have, apparently, obtained a sum with the opposite sign, which is obviously wrong and is the result of overflow. On the other hand, if the signs are opposite, we do not have to worry about overflow, since the addition of oppositely signed numbers will produce a sum with magnitude smaller than either input.

From the preceding discussion, we see that overflow and carry-out are not necessarily the same thing, and the detection of overflow requires knowing what kind of numbers are being added. All binary adders work the same way, their logic being based on the assumption that the inputs are $n$-bit binary numbers, producing an $n$ + 1-bit sum, including the carry-out. In most computers, a carry-out is stored in a $C$ flag, a storage location that can be tested. But the meaning of this $C$-flag depends on the meaning of the data fed to the adder, and that is up to the programmer. Because signed 2's complement arithmetic is so common, many computers also have logic to detect overflow from 2's complement addition. This logic does not change the functioning of the adder; it still performs $n$-bit binary addition. The logic will compare the leading bits of the two inputs and the leading bit of the sum. If the input "sign bits" are the same but the sum "sign bit" is the opposite, an overflow flag (often designated as the $V$-flag) will be set. For example, if we have an 8-bit adder of the form of Fig. 6.14, the equation for $V$ will be

$$V = X_7 Y_7 \overline{S_7} \lor \overline{X_7}\, \overline{Y_7} S_7 \tag{6.7}$$

But remember, the setting of this flag does not mean overflow unless the two numbers being added were in signed 2's complement form.

For computers having BCD arithmetic, the carry-out will indicate decimal carry-out. For example, in an 8-bit computer, each word can accommodate two BCD digits, so the largest number that can be represented in a single word is $99_{10}$. If we perform the BCD addition (35 + 28 = 63), there will be no carry-out, but there will be a carry-out if we perform the addition (45 + 57 = 102).

## Problems

**6.1** Convert the following decimal numbers to binary.
   (a) 11.6
   (b) 23.47
   (c) 193.175
   (d) 727.413

**6.2** Convert the following binary numbers to decimal.
   (a) 1101.011
   (b) 110010.10101
   (c) 1010110.011101

**6.3** Convert the following binary numbers to hexadecimal.
   (a) 10110.011010
   (b) 1101100.01
   (c) 101011.010111001

**6.4** Convert the binary numbers of Problem 6.2 to hex.

**6.5** Convert the following hexadecimal numbers to binary.
   (a) 4A.3
   (b) C2.D6
   (c) 4F7.9A2
   (d) 7D2C.A49

**6.6** Convert the following hexadecimal numbers to decimal.
   (a) 34.A
   (b) 6D.2C
   (c) 27F.49A
   (d) DC29.4A7

**6.7** Convert the hex numbers of Problem 6.5 to decimal.

**6.8** Convert the following decimal numbers to BCD.
   (a) 492
   (b) 34.62
   (c) 391.76
   (d) 214.63
   (e) 4.9157

**6.9** Convert the following BCD numbers to decimal.
   (a) 10010110.0011
   (b) 100001000101.10000011

(c) 00010100.00101000

(d) 001010000001.01001001

**6.10** Consider the numbers in Problem 6.14 as straight binary and convert them to decimal.

**6.11** Convert the following negative decimal numbers to 2's complement binary form. In each case use as few bits as possible, subject to the restriction that the number of bits (including sign) must be a multiple of 4.
(a) $-23$
(b) $-47$
(c) $-171$
(d) $-256$
(e) $-511$

**6.12** In the following, all numbers are decimal. Carry out the indicated additions in binary, using signed 2's complement arithmetic. Choose the number of bits by the same criterion as in Problem 6.11.
(a)    $(+23)$
    $+(+14)$
(b)    $(+67)$
    $+(-31)$
(c)    $(-23)$
    $+(+19)$
(d)    $(-29)$
    $+(-61)$
(e)    $(+125)$
    $+(+89)$
(f)    $(+58)$
    $+(-91)$
(g)    $(-213)$
    $+(+214)$

**6.13** A message has been stored in a computer memory in ASCII code. Shown here is a listing of the message in hex. What is the message in English?
43 4F 4D 50 55 54 45 52 53 2C 20 52 41 48 21

**6.14** A system transmits a message in ASCII code with odd parity. Shown following is a list of codes as received. For each character, indicate if it is a legal or illegal code. If it is a legal code, indicate what the character is as received. In spite of any errors, can you figure out what the message probably is?

11000100
01001001
11010111
01001001
01010100
11000001
01001101

```
11010011
11001001
11010011
01010100
01000101
10001101
11010011
```

**6.15** Construct the logic diagram of a combinational logic circuit that checks for odd parity over seven inputs. The circuit output is to be 1 for odd parity. Now modify the circuit so that the output will be 1 for even parity.

**6.16** Carry out the following multiplications in binary arithmetic. The numbers are given in 2's complement form.
(a)   01001101   01010111
(b)   10101100   01010111

**6.17** Design a combinational logic circuit that will accomplish the multiplication of a 2-bit binary number, $X_1 X_0$, by a 2-bit binary number, $Y_1 Y_0$. The circuit will have four outputs representing the product, $P_3 P_2 P_1 P_0$.

**6.18** Design a combinational logic circuit that will convert BCD-coded numbers to seven-segment code. The circuit will have four input lines and seven outputs.
(a) Assume that only valid BCD-coded characters will appear.
(b) Provide for displaying the $E$ for illegal input characters.

**6.19** Consider a computer with 8-bit words, with a *C*-flag and a *V*-flag. The *C*-flag is a storage location that is set to 1 if there is a carry-out on addition, and set to 0 if there is not. The *V*-flag is set to 1 if addition produces an overflow in 2's complement addition and is set to 0 if there is no overflow. For each of the following additions, determine the 8-bit sum and the contents of *C* and *V* after addition.

(a)   01010110
     +00011100

(b)   11001100
     +10110101

(c)   01101100
     +10100011

(d)   10011100
     +10111001

**6.20** Assume the numbers in Problem 6.19 each represent two decimal digits in BCD and assume the computer performs BCD addition. Determine the sums and the contents of the *C*-flag after addition.

## References

1. Givone, D. D., and R. P. Roesser, *Microprocessors/Microcomputers: An Introduction.* McGraw-Hill, New York, 1980.
2. Hill, F. J., and G. R. Peterson, *Introduction to Switching Theory and Logical Design,* 3rd ed., Chap. 2. Wiley, New York, 1981.
3. Johnson, D. E., J. L. Hilburn, and P. M. Julich, *Digital Circuits and Microcomputers.* Prentice-Hall, Englewood Cliffs, N.J., 1979.
4. Roth, C. H., *Fundamentals of Logic Design,* 2nd ed. West Publishing, St. Paul, Minn., 1979.
5. Sloan, M. E., *Computer Hardware and Organization,* 2nd ed. SRA, Chicago, 1983.
6. Smith, D. E., and J. Ginsburg, From Numbers to Numerals and from Numerals to Computation, in *The World of Mathematics,* J. R. Newman, ed., Vol. 1. Simon & Schuster, New York, 1956.

# Combinational MSI Parts, ROMs, and PLAs

## 7.1 Perspective

Up to now we have approached the realization of combinational logic functions with a bag of SSI parts composed of one to four independent gates. In the not too distant past this was the only option available. Now, however, much more complex logic circuits can be found in single IC packages. To take advantage of the parts available, designers must give up some of their freedom to decide every design detail precisely according to their individual tastes. Will it be worth it? Usually, but not always, the answer will be yes. Occasionally a purely SSI realization will be most cost effective in every respect. In each individual case the decision to adopt a particular design approach will be based on consideration of production costs, parts costs, and design costs. The relative importance of each of these three cost categories will depend on the number of copies of the given design expected to be manufactured. Very often this is only one. In this case minimizing design cost or engineering time is probably most important. Where quantity production is anticipated, parts cost and production cost will determine the design approach.

Using the most complex packages available, and thereby minimizing the overall package count, will almost always result in the lowest production cost. Fewer packages means fewer interconnecting wires and smaller boards on which the IC's are mounted. Unless a complex package is new to the market, it will almost always cost less than the several less complex parts required to do the same job. The impact of a design approach on engineering time is less obvious. As you will no doubt agree as you begin the next section, the time required to understand a complex part and to integrate it into an overall design may well be greater than the time required if simpler and more familiar parts were used. Just as with mastery of a new technique, the first application of a new MSI or LSI part may be painful; but the part quickly becomes a convenient tool for subsequent designs.

In the next section we shall consider the use of MSI parts for the realization of arbitrary logic functions. In Section 7.3 we shall introduce some special parts for realizing the important arithmetic operations discussed in the last chapter. The last two sections of the chapter will treat logic realization using *read-only memories* (ROMs) and *programmed logic arrays* (PLAs).

## 7.2  Combinational MSI Parts

We have seen that any logical function, of any degree of complexity, can be realized using logic gates and thus in terms of SSI packages. However, there are certain more complex functions that occur so frequently in design problems that manufacturers have found it worthwhile to manufacture IC's specifically to implement these functions. Such circuits generally fall into the category of MSI, integrated circuits with 10 to 100 gates on a single chip. Significant economies result from the replacement of several chips with a single chip. Even if the single chip should cost as much as the set of chips it replaces, the economies will still be significant because the cost of mounting and connecting IC chips usually exceeds the cost of the chips themselves. In this section we want to look at some of the combinational functions that are available in MSI form.

Suppose we have serial signals coming in on four input lines to a computer and we want to route the signals from one of the lines onto a common line, with another pair of lines indicating, by the four possible combinations of 0 and 1 possible on two lines, which of the four lines is to be selected. The circuit shown in Fig. 7.1 will do the job. This is an example of a data selector, or *multiplexer.*

Next, suppose we have a pair of logic lines in a computer system that indicate which one of four output devices is to be used. Our problem is to put a logic 1 on one of four lines running to the output devices to indicate the selected device. The circuit shown in Fig. 7.2a will accomplish this task. This circuit is an example of a *decoder,* a circuit that accepts input on $n$ input lines and puts a 1 on one out of $2^n$ output lines. Finally, assume that we have a single signal line and we want to route the signal to one of four lines going to the four output devices. The circuit of Fig. 7.2b will do the job. This is an example of a *demultiplexer,* which is seen to be only a minor variant of the decoder. Indeed, such circuits are generally listed as decoder/demultiplexers, since they can be used for either function.

The basic function of a decoder is to accept an $n$-bit code and decode, or convert, it to a 1-out-of-$2^n$ code. It consists of AND gates that realize all the minterms of $n$ input variables, and each output is designated by the number of the minterm it represents. For example, if $B,A = 10$, line 2 will be high, since this corresponds to minterm 2.

The structures of the multiplexer and demultiplexer are very similar. In each

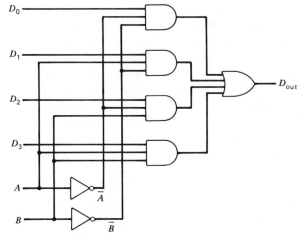

*Figure 7.1   Multiplexer.*

we have a gate for each possible minterm of the control variables, but each gate has an additional data input, so that the occurrence of a particular minterm combination on the control variables routes the data through the corresponding gate. In Fig. 7.1, if $B,A = 10$, data $D2$ is routed to DOUT. In Fig. 7.2$b$, if $B,A = 10$, DATA is routed to output 2. Because they are widely used in digital systems, decoder, multiplexer, and demultiplexer circuits are available as MSI parts in a wide variety of forms, and standard symbols have been developed so that designers will not have to draw out the complete circuits.

Before introducing these symbols we need to introduce some new terminology. Many signals in digital systems indicate that some specific situation or condition

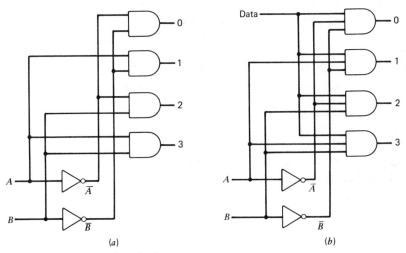

*Figure 7.2   Decoder/demultiplexer.*

exists. For example, the $E$ signal in the elevator example (Fig. 2.11) indicates that the electric eye circuit is closed. In that example we chose to let logic 1 indicate that the circuit is closed and logic 0 indicate that the circuit is open (beam interrupted). But this assignment says nothing about the actual voltage levels that will represent the circuit being closed or open. *Positive logic,* in which logic 1 is indicated by the more positive (high) voltage is assumed in specifying the functions of all standard TTL gates; for example, a TTL AND gate is one in which the output is at the high level only if all inputs are at the high level. But the use of this convention in defining gate functions does not mean that we must associate the high level with the fact that some condition has occurred and the low level with the fact that it has not occurred. There may be very good reasons, usually having to do with circuit requirements, for making the opposite assignment. To deal with this ambiguity, we say that a line is *asserted,* or *active,* when it is at the voltage level indicating that the situation exists, or the event has occurred, that is associated with that line. Thus, the $E$ line is asserted when it is at the voltage level indicating that the electric eye circuit is closed. If a line goes to the high (more positive) level to indicate assertion, it is an *active-high* line. If a line goes to the low (more negative) level to indicate assertion, it is an *active-low* line.

We have refrained from introducing the concepts of assertion and active lines to this point because they can lead to a lot of confusion with the concepts of positive logic and negative logic. "Active-high" and "active-low" specify only the polarity of a signal that is asserted. They do not change the function of a gate to which lines may be connected. For example, if lines $A$ and $B$ are the inputs to a TTL AND gate, the output will be at the high level only if both inputs are at the high level, regardless of whether $A$ and $B$ are active-high or active-low.

Although the use of active-high and active-low designations does not alter the physical functioning of a gate, it does permit the designer to view the logical function in a different way. For example, some designers prefer to think of a NAND gate as an AND gate with an active-low output, that is, the output is asserted (at the 0 level) only if both inputs are asserted (at the 1 level). Similarly, the NAND equivalent of an OR gate with inverted inputs (Fig. 5.16) may be viewed as an OR gate with active-low inputs and active-high outputs. The output will be asserted (logic 1) if either input is asserted (logic 0). If you think about it a minute, you will see that this operational description is consistent with both the meaning of the OR connective (output true if any inputs true) and the NAND truth table of Fig. 2.5a. Confusing as all this may be, the concept of active lines is essential in dealing with MSI circuits, and the confusion will (we hope) lessen in due time. With this background we are now ready to consider the block diagram symbol for the decoder.

The standard symbol for a two-to-four-line decoder is shown in Fig. 7.3. The identifying symbol X/Y indicates that the basic function is to convert from code X to code Y. The inputs are assigned binary weights, 1,2 in this case. The outputs are also assigned weights; the output with weight equal to the sum of the weights of the active inputs will be active. For example, if inputs 1 and 2 are both high, output three $(1 + 2 = 3)$ will be high.

Decoders often include an *enable line,* which allows the decoder to be turned off. Figure 7.4a shows a two-to-four-line decoder with an enable input, $E$. Also illus-

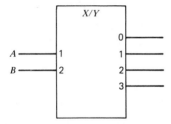

Figure 7.3    Decoder symbol.

trated in this figure is the alternative standard notation for listing the input weights. In this case the numbers are not the weights themselves, but are the logarithms of the weights in base 2. Thus the weight of $B$ is $2^1$, whereas the weight for $A$ is $2^0$. This notation is valid only when accompanied by the brackets and range of input sums in the form

$$\Big\rvert \frac{0}{3}$$

Because of the inverter in the enable line, this is an active-low enable, that is, the input must go low (to 0) to enable the decoder. If $E = 1$, a logic 0 will be applied to all the AND gates, driving all the outputs to 0. If $E = 0$, the decoder will function in the normal manner, with one input high, all the others low. The symbol for this decoder is shown in Fig. 7.4$b$. The notation EN indicates an enable input; the small triangle indicates that it is active-low. Outputs can also be active-low. Suppose we realize the decoder with NAND gates instead of AND gates, as shown in Fig. 7.5$a$. The output of the selected gate will go low if enabled, all others will be high. This is indicated in the symbol of Fig. 7.5$b$ by the triangles on the outputs.

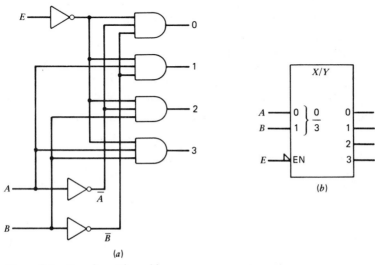

(a)

(b)

Figure 7.4    Decoder with enable.

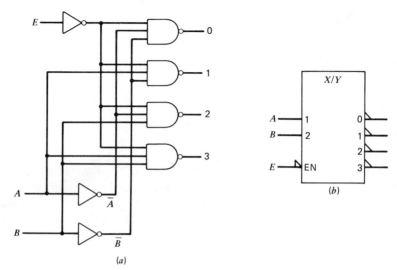

*(a)*

*Figure 7.5   Inverting outputs.*

Note that, except for the inverter, the circuit of the decoder with enable in Fig. 7.4*a* is identical to the circuit of the demultiplexer in Fig. 7.2*b*. Thus, if we use the enable input as a data input, the decoder with enable is a demultiplexer. If the control inputs *B,A* are set to select a particular gate, the data (enable) input will control the output of that gate. In the circuit of Fig. 7.5, because of the double inversion (in the enable line and in the NAND gate), the output of the selected NAND gate will be equal to the value at the enable (data) input. Because of this dual character, most IC decoders are provided with an enable or data input and are classified as decoder/ demultiplexers.

Before showing the symbols for multiplexers, we must first introduce a basic concept used in block diagram notation, AND dependency. A label G followed by a number (e.g., G2) indicates a gating input, one that is ANDed with any input or output labeled with the same number. Consider the symbol in Fig. 7.6. Whatever the function of this device may be, input G1 is ANDed with input 1 and input G2 is ANDed with output 2. Thus, input 1 will have no effect on the operation of the circuit unless G1 is active, and output 2 will be inactive unless G2 is active. By contrast, input 0 cannot be disabled because there is no G0 input. Gating inputs function

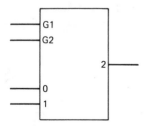

*Figure 7.6   Gate dependency symbol.*

in a manner similar to enable inputs, except that they affect only specified lines and can affect both inputs and outputs.

The block diagram symbol for the multiplexer of Fig. 7.1 is shown in Fig. 7.7. The label MUX identifies this device as a multiplexer. The symbol

$$G \frac{0}{3}$$

is shorthand for

$$G0,G1,G2,G3$$

The active gating signal is determined by the inputs enclosed in brackets, with weights equal to the powers of two indicated, in this case $2^0$ and $2^1$. For example, if $A$ and $B$ are both active, the total weight is $2^1 + 2^0 = 2 + 1 = 3$, so G3 is active, in turn enabling data input 3. If only $A$ is active, $2^0 = 1$, so G1 is active, enabling data input 1.

The use of gating inputs does not preclude the use of enable inputs. Figure 7.8 shows the symbol for a multiplexer that functions in the same way as that of Fig. 7.7 except that a 1 on the active-low enable line will hold the output inactive regardless of the signals at any other inputs.

In some cases more than one multiplexer may be included on a single chip. Figure 7.9 shows the actual circuit and the block diagram symbol for the 74153 dual four-to-one multiplexer. Note from the circuit that the $A,B$ inputs control both sections of the multiplexer. This is indicated in the block diagram symbol by the "indented box" at the top that represents a *control block*. Inputs to a control block affect (or control) all lower sections in the same way. In this case the inputs to the control block are gating inputs. If the sum of the weights of the active gating inputs is 2, for example, the number 2 data inputs in both sections will be enabled. By contrast, there is a separate enable input for each section, so that either may be completely turned on or off independently of the other. In the circuit diagram, note the two inverters in series in the $A$ and $B$ lines. These do not change the logical function of these inputs but provide the necessary amplification to drive the eight AND gates. The $A$ and $B$ inputs thus appear as normal one-gate loads to any driving circuit.

Decoder, multiplexer, and demultiplexer circuits were originally developed for

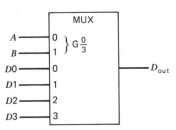

*Figure 7.7   Four-input multiplexer.*

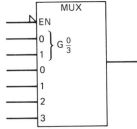

*Figure 7.8   Multiplexer with enable.*

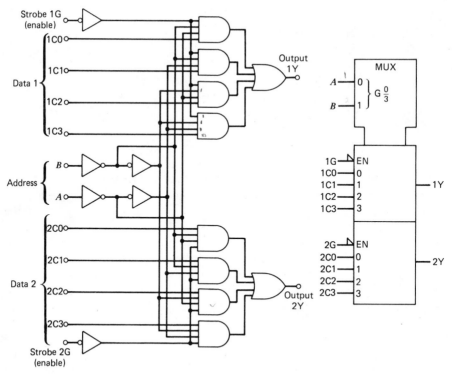

*Figure 7.9   Dual multiplexer.*

and continue to be used primarily for the basic purposes suggested at the beginning of this section. In a computer, a decoder might receive a binary word as an address and put a 1 on a single line to select the addressed location in memory. A multiplexer might be used to route data from one of several possible inputs to a main register. A demultiplexer might be used to route data from a main register to one of several possible outputs. Designers have also found ingenious ways to use these circuits as general-purpose logic blocks. Noting that a decoder realizes all the minterms of the input variables, we see that a decoder with an external OR gate could be used to realize any function. We will explore this idea in some detail in Section 7.3.

Multiplexers can also be used very effectively for realization of arbitrary functions. For this type of application, we regard the multiplexer as a general-purpose AND–OR circuit with considerable flexibility as to what is connected to the AND gates.

### Example 7.1

Implement the following function, using one half of a 74153 multiplexer.

$$f(X,Y,Z) = \Sigma\, m(3,4,6,7)$$

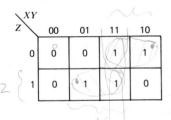

$$X\bar{Z} + YZ$$

*Figure 7.10*

## Solution

We start by drawing the K-map of the function, as shown in Fig. 7.10. We then connect $X$ to the $B$ address line and $Y$ to the $A$ address line. With that connection, each column of the K-map corresponds to one AND gate in the multiplexer. For example, if $B,A = X,Y = 0,0$, the top AND gate (Fig. 7.9) will be selected and the output will be determined by 1C0. For this set of values for $X$ and $Y$, the function is to be 0, so we connect logic 0 to 1C0. For $XY = 01$, the second gate down is selected and the output is determined by 1C1. For this set of values of $X$ and $Y$, we see that the output is equal to $Z$, so we connect $Z$ to 1C1. In a similar manner, for $XY = 10$, the output is equal to $\bar{Z}$, so we connect $\bar{Z}$ to 1C2, and we connect logic 1 to 1C3 to provide the outputs required for $XY = 11$. The complete circuit is shown in Fig. 7.11, with the strobe tied low so that the output will follow the address and data inputs.   ∎

It can be seen from Example 7.1 that any function of three variables can be realized with a 4-input mulitplexer. In the example we have all four possible com-

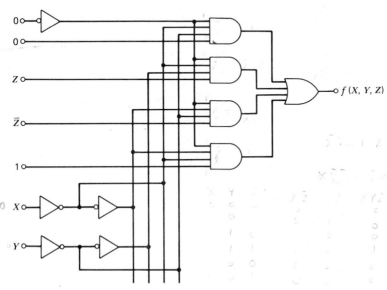

*Figure 7.11*

binations of 1 and 0 in the K-map columns, so this basic circuit form can be used no matter where the 1's and 0's appear on the map. In the same manner, any four-variable function can be realized with an 8-input multiplexer, any five-variable function with a 16-input multiplexer, and so on. For an $n$-variable function realized with a multiplexer, $n - 1$ variables are connected to the address line, and the data lines are driven by the remaining variable or its complement, or 0 or 1.

These rules provide for realizing any function of $n$ variables with a multiplexer with $2^{n-1}$ inputs. Sometimes functions can be realized with a smaller multiplexer and a small amount of external logic.

### Example 7.2

Realize the following function using a 4-input multiplexer.

$$f(W,X,Y,Z) = \Sigma m(1,2,5,7,9,11,13,15)$$

Solution

The procedure described here would require an 8-input miltiplexer, but let us see if we can do it with half of a 74153. We connect $W$ to $B$ and $X$ to $A$, and draw the map of the function with these two variables across the top, as shown in Fig. 7.12.

Assume we are going to use the top half of the 74153 (Fig. 7.9). The first column of the map corresponds to $B,A = 00$, for which the top AND gate is selected and the output is controlled by 1C0. From the map we see that the output should be 1 in this column if $YZ = 10$ or if $YZ = 01$, so that these 1's can be realized by connecting $(Y \oplus Z)$ to 1C0. The second column represents $B,A = 01$, for which 1C1 is the controlling input. In this column the output is to be 1 if $Z = 1$, so we connect $Z$ to 1C1. In a similar manner we see that $Z$ should also be connected to 1C2 and 1C3. The complete circuit, using one 4-input multiplexer and one exclusive OR gate is shown in Fig. 7.13. ∎

An obvious question in connection with this design is why we chose to use $W$ and $X$ to drive the address lines. Why not $Y$ and $Z$, or some other combination? If we chose some other combination, would the resultant realization be any simpler? The only way to find out which variables should be connected to the address lines is to try all possibilities. We can try one other possibility on the same map (Fig. 7.12).

| BA | 1C0 | 1C1 | 1C3 | 1C2 |
|---|---|---|---|---|
| WX | 00 | 01 | 11 | 10 |
| YZ | | | | |
| 00 | | | | |
| 01 | 1 | 1 | 1 | 1 |
| 11 | | 1 | 1 | 1 |
| 10 | 1 | | | |

*Figure 7.12*

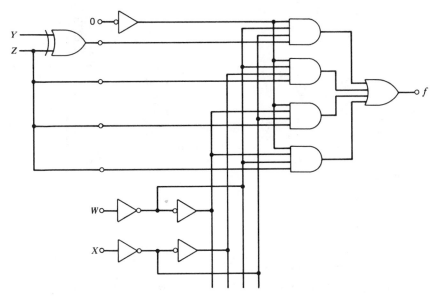

*Figure 7.13*

If we connect $Y,Z$ to $B,A$, then each row of the map corresponds to one of the gates. The top row corresponds to 1C0, so we connect 0 to that input. The second row is all 1's, so we connect 1 to input 1C1. The third row corresponds to 1C3, so we connect $W \lor X$ to that input. Finally, we connect $\overline{W \lor X}$ to 1C2. We see that this realization is not as simple as the first one found. We can try other possibilities by redrawing the map, first with $WY$ along the side and $XZ$ along the top, then with $WZ$ along the side and $XY$ along the top. We shall not explore this matter further, as it is of somewhat marginal utility. Realizing a function on a single chip will usually result in significant economies rélative to using discrete (SSI) gates, but the savings become minimal when we have to add extra gates.

## 7.3 Read-Only Memory

In the previous section we discussed decoders as they appear by themselves as MSI parts. An even more important use of decoders occurs within LSI memory and read-only memory (ROM) parts. These memories are organized with as many as $64 \times 1024 = 65,536$ (usually referred to as 64K) or even 256K words, sometimes with only one bit per word in each package. Decoders are used to select a word electronically to be read from or written in memory based on an input vector called an *address*. Treatment of read–write memories must await our introduction of information storage elements in the next chapter. In this section we shall restrict our discussion to read-only memories. The application of decoders is much the same in either type of memory.

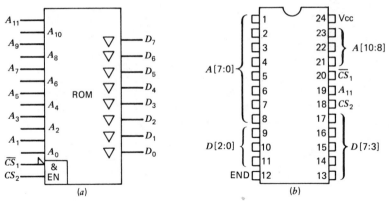

*Figure 7.14   Package and logic symbols for 4K $\times$ 8 ROM.*

In Fig. 7.14 we see the logic symbol and a package representation of a currently available ROM. This ROM has 12 address input lines, labeled $A_{11}$ to $A_0$, and eight data output lines, labeled $D_7$ to $D_0$. Since the part is found in a 24-pin package, four pins in addition to these inputs and outputs are available. Two of these are used for the supply voltage and ground, and the remaining two as chip select lines, a concept to be discussed shortly.

A read-only memory is really only a combinational logic device. Each of the lines, $D_7$ to $D_0$, is a combinational logic function of the 12 inputs. For each of the 4096 combinations of input values a particular set of values will appear on the eight output lines. From a slightly different viewpoint, each combination of input values may be thought of as an address that points within the ROM at one of 4096 8-bit data words, which is then routed to the output. By either interpretation, the result is the same.

The usefulness of the ROM lies in the fact that its implementation can be separated into two parts as shown in Fig. 7.15. The more complicated left-most box, the decoder, will be common to all ROMs of this type. The output matrix is specified independently for each application. The manufacturer will not usually make available the details of the logic circuits within either of these two boxes. This is not critical, because we can easily develop our own logic network representation of both

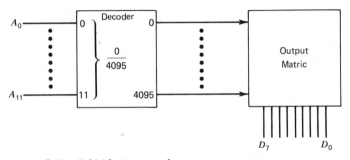

*Figure 7.15   ROM logic network.*

boxes that will relate package outputs to inputs in exactly the same way as the actual MSI part.

Let us turn our attention first to the rather large decoder that must translate the address information on the 12 input lines to 4096 output lines, only one of which is logical 1 for each input combination. So far we have seen decoders with only two input lines. The result of extending the approach of Fig. 7.2a to three inputs is shown in Fig. 7.16.

To extend the approach of Fig. 7.16 directly to 12 inputs would require 4096 12-input AND gates. So large a number of gates with so many inputs would be difficult to implement in any technology. A better way is shown in Fig. 7.17. To keep the diagram readable, only a few connections are actually shown. The 3-bit decoders may be considered to be copies of the circuit of Fig. 7.16. There are 64 pairs of output lines, one from each of the upper two 3-bit decoders. These pairs form the inputs to the upper 64 second-level AND gates. The outputs of these gates are the 64 possible minterms of the variables $a_1$, $a_2$, $a_3$, $a_4$, $a_5$, and $a_6$. The lower 64 second-level gate outputs are the minterms of $a_7$, $a_8$, $a_9$, $a_{10}$, $a_{11}$, and $a_{12}$. The $2^{12}$ 12-bit minterms are formed by using all possible pairs of outputs of the second-level gates (one from the upper 64 and one from the lower 64) to form inputs to the final $2^{12}$ AND gates.

Assuming the fan-in level permits, it might seem that it would be better to combine the outputs of all four decoders in 4-input AND gates, thus eliminating the second level of gates. Let us assume the cost is proportional to the number of gate inputs. The number of gate inputs, exclusive of the 3-bit decoders, is given by Eq. 7.1. Clearly the dominant cost is the output gates. No matter what the form of the decoder, the cost will always be dominated by the $2^n$ output gates, so these should have the minimum number of inputs—two.

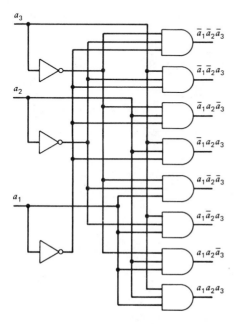

*Figure 7.16   Three-input decoder.*

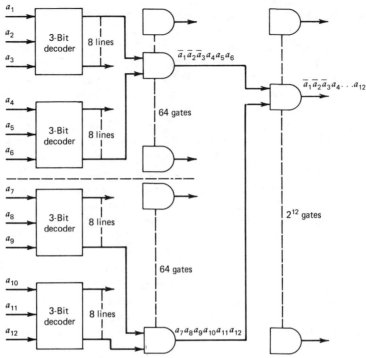

*Figure 7.17    Twelve-line to 4096-line decoder.*

$$\text{Number of inputs} = \overbrace{2 \cdot 2 \cdot 64}^{\text{2nd level}} + \overbrace{2 \cdot 2^{12}}^{\text{Output gates}} \tag{7.1}$$
$$= \quad 2^8 \quad + 2^{13} \approx 2^{13}$$

It may have been observed in Fig. 7.17 that each level of AND gates following the 3-bit decoders doubled the number of inputs handled by the overall decoder. The symmetry of this particular design resulted from the fact that the number of inputs (12) was divisible by a power of two, that is $2^2$, with the resulting small integer quotient, 3. Suppose the number of inputs to the decoder was not divisible by a power of 2, for example, 13. In spite of a lack of symmetry, a 13-input decoder quite similar to Fig. 7.17 could be constructed. The first level would consist of three 3-input and one 4-input decoder. The second level would be a bank of 64 AND gates and another bank of 128 AND gates. The final level would consist of 8192 AND gates.

Let us now turn our attention to implementing the output matrix portion of Fig. 7.15. This matrix is logically nothing more than a vector of OR gates, one for each output. A specification of the ROM is merely a tabulation of those decoder outputs that should be connected to each OR gate input. Each OR gate or ROM output will be represented by a column in the table. Ones are entered into the column corresponding to decoder outputs that are to be connected to the OR gate. Zeros represent no connection. When several outputs are tabluated together, the rows of the ROM table resemble memory words and are referred to as such. To illustrate this, let us turn to an example.

## Example 7.3

Consider the problem of converting from a BCD code to a seven-segment code, as tabulated in Fig. 6.4. Design a ROM circuit with outputs that are the signals that will drive the seven lamp segments. The BCD inputs are to be connected to the four four address lines.

## Solution

It can be seen that a ROM is applicable to our problem by noting that a full decoder produces all minterms of the input variables and that any function can be obtained by an ORing of minterms. Since the inputs are in BCD code, the decimal digits are the minterm numbers. A tabulation of all the bits in the ROM is shown in Fig. 7.18. For the six input combinations that will never occur, the output entries are all 0's. Actually, these values are "don't-cares" and could be entered as either 1 or 0. Because the outputs will effectively be an ORing of minterms, nothing is gained by postponing the assignment of these don't-care values as 0.

Only those decoder outputs that are actually connected to at least one of the output OR gates are shown in the partial logic diagram of the BCD to seven-segment ROM given in Fig. 7.19. The AND gates implementing the remaining six decoder outputs will be on the ROM chip but will have no effect. The connections to only two of the output OR gates are actually shown in the figure. From the table in Fig. 7.18 it can be seen that these outputs are given by the minterm lists

$$g = \Sigma m(2,3,4,5,6,8,9) \qquad (7.2)$$

| $A_3$ | $A_2$ | $A_1$ | $A_0$ | x | a | b | c | d | e | f | g |
|---|---|---|---|---|---|---|---|---|---|---|---|
| 0 | 0 | 0 | 0 | 0 | 1 | 1 | 1 | 1 | 1 | 1 | 0 |
| 0 | 0 | 0 | 1 | 0 | 0 | 1 | 1 | 0 | 0 | 0 | 0 |
| 0 | 0 | 1 | 0 | 0 | 1 | 1 | 0 | 1 | 1 | 0 | 1 |
| 0 | 0 | 1 | 1 | 0 | 1 | 1 | 1 | 1 | 0 | 0 | 1 |
| 0 | 1 | 0 | 0 | 0 | 0 | 1 | 1 | 0 | 0 | 1 | 1 |
| 0 | 1 | 0 | 1 | 0 | 1 | 0 | 1 | 1 | 0 | 1 | 1 |
| 0 | 1 | 1 | 0 | 0 | 0 | 0 | 1 | 1 | 1 | 1 | 1 |
| 0 | 1 | 1 | 1 | 0 | 1 | 1 | 1 | 0 | 0 | 0 | 0 |
| 1 | 0 | 0 | 0 | 0 | 1 | 1 | 1 | 1 | 1 | 1 | 1 |
| 1 | 0 | 0 | 1 | 0 | 1 | 1 | 1 | 0 | 0 | 1 | 1 |
| 1 | 0 | 1 | 0 | 0 | 0 | 0 | 0 | 0 | 0 | 0 | 0 |
| 1 | 0 | 1 | 1 | 0 | 0 | 0 | 0 | 0 | 0 | 0 | 0 |
| 1 | 1 | 0 | 0 | 0 | 0 | 0 | 0 | 0 | 0 | 0 | 0 |
| 1 | 1 | 0 | 1 | 0 | 0 | 0 | 0 | 0 | 0 | 0 | 0 |
| 1 | 1 | 1 | 0 | 0 | 0 | 0 | 0 | 0 | 0 | 0 | 0 |
| 1 | 1 | 1 | 1 | 0 | 0 | 0 | 0 | 0 | 0 | 0 | 0 |

*Figure 7.18   BCD to seven-segment ROM.*

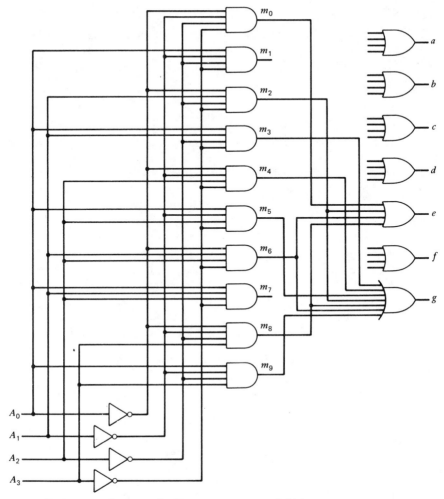

*Figure 7.19   Logic diagram of BCD to seven-segment ROM.*

and

$$e = \Sigma m(0,2,6,8) \tag{7.3}$$

In each case the listed minterms are connected to the corresponding output OR gate in Fig. 7.19.                                                                  ■

The code converter of Fig. 7.19 is a very simple example of a ROM. The decoder is evident, and the vector of OR gates on the right is the output matrix of Fig. 7.15. This circuit is actually available as an MSI part. Clearly the ROM package presented in Fig. 7.14 is capable of realizing much more complicated logic functions or storing much more information than the amount tabulated in Fig. 7.18.

Figure 7.19 is not the minimal two-level realization of the seven logic functions tabulated in Fig. 7.18; and this would not be the approach if it were necessary to realize these functions with discrete gates. The notion of minimal is not relevant to read-only memories. The advantage of these devices is convenience of manufacture. A variety of processes have been developed whereby the ouptut matrix of a ROM from a uniformly manufactured batch can be custom-programmed with user information.

The large fan-in requirement for the output matrix OR gates of a ROM necessitates a special approach to circuit design. One form of bipolar ROM realization is illustrated in Fig. 7.20. The wires in the output matrix are known as *word lines* and

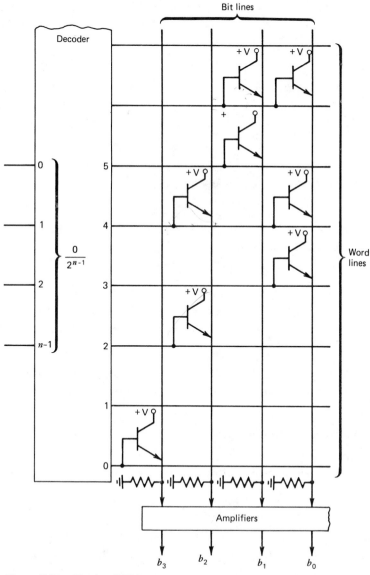

*Figure 7.20  Bipolar ROM.*

*bit lines.* The word lines are driven by an $n$-to-$2^n$ decoder, and the outputs are taken from the bit lines. Wherever a connection is to be made between a word line and bit line, a transistor is connected, emitter to the bit line and base to the word line. All collectors are tied to a common supply voltage.

In the absence of any input, all word lines are held at a sufficiently negative level to cut off the transistors so that the bit lines are at 0 V. When an input appears, the corresponding word line is raised to a sufficiently positive level to turn on the transistors, thus raising the connected bit lines to a positive level. For example, in Fig. 7.20, if the input were 100 ($m_4$), word line 4 would be turned on and the output would be $b_3b_2b_1b_0 = 0101$. Comparing this to Fig. 7.19, we see that the bit lines with their associated transistors make up the array of OR gates. (Please note that we have described here only the basic principles involved; the precise details of the technology vary widely.)

In large quantities, ROMs are available as special purpose parts with the stored information entered as a last step of the manufacturing process. A variety of sizes of user programmable ROMs called PROMs are also available. Some PROMs are eraseable, electrically or using ultraviolet light through a window in the package, others are not. A noneraseable ROM of the type shown in Fig. 7.20 would be manufactured with a transistor in each bit position. The PROM programming process would then consist of literally blowing out with high-current pulses the fusible links in the transistors corresponding to 0 bits. This type of PROM might be competitive with the more flexible eraseable variety for reasons of either speed or economy.

Smaller ROMs may be used in any application in place of SSI logic. The ROM becomes a more likely choice as the number of functions to be realized with the same set of variables increases. Also influencing the decision is the cost of programming the ROM versus that of interconnecting the SSI logic on a board.

Larger ROMs are used as permanent memory. They are widely applied for storing system software and other types of semipermanent programs. For example, many microcomputer systems come equipped with monitors and assemblers in ROM. This approach provides great advantages in reliability over the more traditional approach of loading software into regular read–write memory, since ROMs are immune to accidental loss of programs through power failure or user errors.

ROMs are also used as control storage in microprogrammable computers. Microprogramming is closely related to the digital system control sequences to be discussed in Chapters 9 and 10 but is itself beyond the scope of this book.

## 7.4  Chip-Select, Buses, and Three-State Switches

In this section let us view the ROM as a permanent memory rather than a set of logic functions. In memories the data vector called forth by each input address is referred to as a *word*. The number of bits per word is the *word length*. One particularly important word length is eight bits, which has come to be known as a *byte*. We noticed in Chapter 6, for example, that an ASCII-coded character occupies a byte. The ROM of Fig. 7.14 was quite consciously organized with byte-length words.

Suppose now that some application requires a large read-only memory consisting of 8192, or 8K, bytes of permanently stored information. At the time of this writing the designer would search the manufacturers' data manuals unsuccessfully for this size ROM in a single package. However, two of the packages shown in Fig. 7.14 contain just this amount of information. How can they be combined together as a single read-only memory?

As shown in Fig. 7.21, the system requiring a byte of information from the 8K ROM will indicate the address of the desired byte on a vector of 13 address wires, labeled **ADDR**[12:0]. The numerical addresses corresponding to the 13-bit vectors range from 0 to 8191. The bits of the address vector are numbered according to their weights in determining the numerical values of the addresses. For example, the most significant bit, **ADDR**[12], has a weight of $2^{12}$, or 4096. Half the data bytes, those with addresses from 0 to 4095 will be found in ROM A, and bytes 4096 to 8191 will be found in ROM B. Each of the ROMs has only 12 address input lines, so that the 13 **ADDR** lines cannot be connected directly to both ROMs. To see the solution, consider the address ranges expressed in hexadecimal.

| | |
|---|---|
| ROM A addresses | 0000 to 0FFF |
| ROM B addresses | 1000 to 1FFF |

We note that all addresses of bytes stored in ROM B have a most significant hex digit of 1, whereas this digit is 0 for addresses in ROM A. Only the single most significant of the 13 system address wires is needed to indicate whether this digit is 1 or 0. The remaining 12 wires represent the least significant three hex digits and may be connected directly to the 12 address inputs of both ROMs.

Also shown in Fig. 7.21 is a data bus. This is a vector of eight wires connected to the corresponding output wires of both ROMs and to the data input lines of the system interested in reading from the ROMs. Somehow it must be assured that data

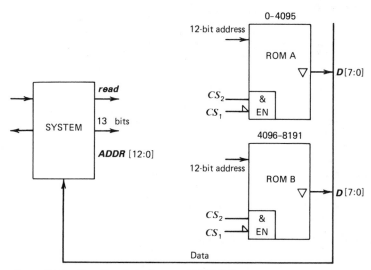

*Figure 7.21   Reading data from an 8K ROM.*

from ROM A will be routed to these wires whenever it is desired to read from an address in the range 0000 to 0FFF and that data from ROM B will be routed to the wires whenever data from the range 1000 to 1FFF is desired. To provide for this we return our attention to the *chip select* inputs cited only briefly in the previous section.

In Fig. 7.22*a* we see a more detailed internal configuration of ROM A, including an output switching network enabled by the two chip-select inputs. The downward pointing triangles included in Fig. 7.14 indicated that these outputs were three-state controlled. To interpret the function of the three-state switches we refer to the more detailed illustration of Fig. 7.22*b*. The physical implementation of the switches shown will vary with the technology, but the resultant operation is always very much like actual switches. In the case of Fig. 7.22*b* the switches are controlled by the output of the AND gate shown. If the AND output is 1, the switches will be closed. If it is 0, the switches will be open. When the switches are closed, the selected output of ROM A is connected to the system data bus. When the switches are open, ROM A is completely disconnected from the data bus and has no effect whatever on these bus wires. With respect to the ROM A switches, this latter situation is referred to as the *high-impedance state.*

A single three-state switch is sometimes depicted as shown in Fig. 7.22*c.* The three output states are 0, 1, and high impedance. The latter state will occur whenever the control input is 0. If the control input is 1, the output will be 0 or 1 according to the data input.

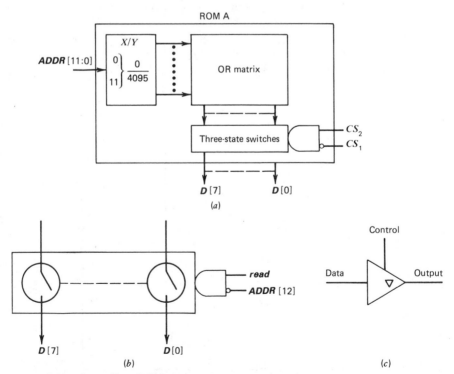

*Figure 7.22   Controlling ROM A output.*

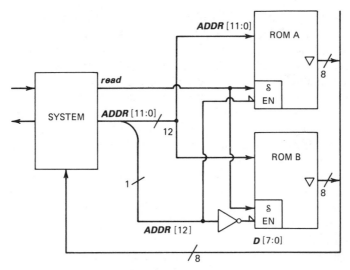

*Figure 7.23 Completed read-only memory system.*

The output **read** from the block labeled SYSTEM in Fig. 7.21 has not yet been discussed. This output will be a 1, if and only if the system actually desires to read a byte from read-only memory. Thus it is desired that a word from ROM A be connected to the data bus, when **read** = 1 and the most significant address bit, **ADDR**[12] = 0. Figure 7.22*b* shows these two signals connected to the enabling AND gate of the three-state switches so that the ROM A output is connected to the data bus at the proper time. Similarly, a byte from ROM B should be placed on the data bus, when **read** = 1 and **ADDR**[12] = 1. Figure 7.23 shows a complete logic block diagram of the system. The notation **ADDR**[11:0] denotes a vector consisting of bits 11 through 0 of **ADDR**. The corresponding wires are connected to the 12 address inputs of each of the 4096 byte ROMS. The symbol $\overline{\quad}_{12}$ represents a vector of 12 wires.

## 7.5 Programmed Logic Arrays

The decoder and OR array (ROM) is not an efficient approach to the realization of a single Boolean function. The reason for the inefficiency of the ROM is simple; the ROM realization implements the canonic form, the ORing of the minterms of a function. As we saw in Chapter 4, the canonic form is usually the worst possible realization in terms of gate count. A ROM realization of a single function may be more economical than an SSI realization in terms of package count, but a multiplexer form, as discussed in Section 7.2, will probably be still better.

The decoder (ROM) is most appropriate when realizing multiple functions. Since the decoder realizes all the minterms, it can be used to realize any number of functions, just by adding another OR gate for each function. By contrast, a multiplexer can realize only one function at a time. Consider the seven-segment encoder of Fig. 7.19. Here we make very efficient use of the decoder since each minterm is used several times. Each individual function could be realized more efficiently by minimization, but the total number of gates for all functions would be considerably greater since there would be few shared terms. The choice between a single-chip ROM and a multichip ROM (decoder plus OR gates) will usually depend strictly on production volume. To justify a ROM, the production volume must be enough so that the initial cost of programming will be offset by the savings in wiring costs.

Summarizing the discussion to this point, if we are interested in single-chip or "few-chip" realizations, the multiplexer form will usually be preferable for single functions of up to six variables. For multiple functions of up to six variables the decoder/ROM form is likely to be more economical. Above 10 variables, the situation becomes more complex. Suppose we have a system with 10 input variables in which 10 different functions are to be realized. None are functions of more than six variables, but every variable appears in several functions in such a way that there is no way to partition the functions into groups dependent on a smaller number of variables. We could use 10 multiplexers to realize the functions independently, but a single-chip realization might be more economical. For that purpose, one could use a 10-input/10-output ROM. Such a ROM would require a decoder realizing 1024 minterms, in turn requiring at least 1024 AND gates, but many of these minterms, probably the majority, would not be needed.

In order to obtain more efficient realization in such cases, the programmed logic array (PLA) has been developed. In a ROM, we have a fixed array of AND gates (the decoder) and an array of OR gates, with programmable connections between the AND gate outputs and the OR gate inputs. In a PLA, we have an array of inputs and an array of AND gates with programmable connections between the inputs and the AND gates, plus the array of OR gates with programmable connections to the AND gates. Figure 7.24 shows the basic organization of a PLA with three input variables and three output lines. The three input variables and their complements provide the inputs to the AND array. There are four AND gates, each with six inputs, for the three variables and the complements, so that any AND function of the variables can be realized. The outputs of the AND gates are connected to the three OR gates through another array. In each of the arrays, an X marks a point where a connection is made.

In this example, AND gate 1 realizes $A\overline{B}$, AND gate 2 realizes $\overline{A}B\overline{C}$, AND gate 3 realizes $\overline{B}C$, and $AND$ gate 4 realizes $A\overline{C}$. OR gate 1 is connected to AND gates 1 and 2 and so realizes $A\overline{B} \vee \overline{A}B\overline{C} = f_1$. Similarly, the other two OR gates realize $f_2 = \overline{A}B\overline{C} \vee \overline{B}C$ and $f_3 = \overline{A}B\overline{C} \vee A\overline{C}$.

Note that, in both arrays, lines that are not connected have no effect. The diagram of Fig. 7.24 is intended only to show the basic logical character of a PLA. The AND and OR functions are generally realized directly in the connection matrices, as the case in the ROM of Fig. 7.20, rather than in physically distinct gates. Figure 7.24 does represent accurately the basic logical character of the PLA, but it is unwieldy, and the simplified form of Fig. 7.25 is commonly used. Here we have just one input line to each gate, representing all the input lines, with X connections at as many intersections as required to indicate the inputs of that gate. The diagram shown in Fig. 7.25 is thus exactly equivalent to that in Fig. 7.24.

Practical PLAs will be much larger than the simple example we have shown

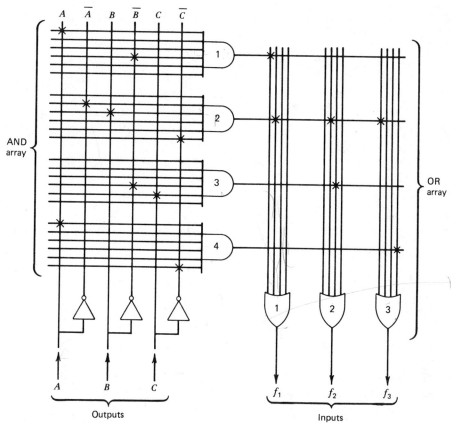

Figure 7.24   Basic logic model of PLA.

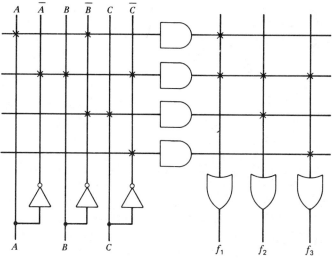

Figure 7.25   Simplified diagram of PLA.

here. Typical PLAs have 10 to 20 inputs, 30 to 60 AND gates, and 10 to 20 OR gates, so that a single PLA can realize a considerable number of complex functions. Like ROMs, PLAs can be mask-programmable, with the interconnection matrix established during manufacturing, or field-programmable (FPLAs). An FPLA is delivered with all the connections made (X at every intersection), and the undesired *links*, or *fuses*, are "blown out" to program the desired functions.

As mentioned, PLAs are usually employed in cases of large numbers of functions of many variables. We do not have space for such examples here, but let us consider a simpler example to illustrate that obtaining an optimum PLA realization is not a simple matter.

**Example 7.4**

The box labeled TRANSLATOR in Fig. 7.26 is to be realized as a PLA. The function of this circuit is to translate simple coded output characters from digital system

*Figure 7.26*

1 into a form understandable by system 2. The first four letters of the alphabet must be intermittently transmitted from system 1 to system 2. In system 1, these letters are coded on three lines, $x_1$, $x_2$, $x_3$, as shown in Fig. 7.27$a$. In system 2, they are coded on two lines, $y_z$ and $y_1$, as shown in Fig. 7.27$b$.

|        | $a$      | $b$ | $c$ | $d$ |
|--------|----------|-----|-----|-----|
| $x_1$  | 0        | 1   | 1   | 0   | 1 |
| $x_2$  | 1 or 0   | 1   | 0   | 0 |
| $x_3$  | x        | 0   | x   | x   | 1 |

$(a)$

|        | $a$ | $b$ | $c$ | $d$ |
|--------|-----|-----|-----|-----|
| $y_1$  | 0   | 0   | 1   | 1   |
| $y_2$  | 0   | 1   | 0   | 1   |

$(b)$

*Figure 7.27*

**Solution**

The translating functions $y_1(x_1, x_2, x_3)$ and $y_2(x_1, x_2, x_3)$ are easily compiled directly on Karnaugh maps, as in Fig. 7.28. To accomplish this, we determine from Fig. 7.27$b$ which letters of the alphabet require 1 for either $y_1$ or $y_2$. We then determine from Fig. 7.27$a$ which input combinations correspond to these letters. For example, $y_1$ is to be 1 for $c$ or $d$ (Fig. 7.27$b$). From Fig. 7.27$a$, we see that $c$ or $d$ is represented by $x_1x_2x_3 = 000$ or 001 or 101, so we enter 1 in the corresponding squares of Fig. 7.28$a$.

From these maps, the minimal realizations of $y_1$ and $y_2$, when considered indi-

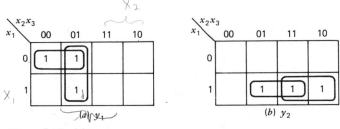

**Figure 7.28**

vidually, are easily seen to be those of Fig. 27.29. Alternatively, we note that $y_1$ and $y_2$ could be expressed as

$$y_1 = \bar{x}_1\bar{x}_2 \lor x_1\bar{x}_2x_3 \qquad \text{and} \qquad y_2 = x_1x_2 \lor x_1\bar{x}_2x_3 \qquad (7.4)$$

Taking advantage of the common term, $x_1\bar{x}_2x_3$, permits the implementation found in Fig. 7.30. Only three rows of the PLA are used. A PLA realization of Fig. 7.29 would have required four rows.                                                          ■

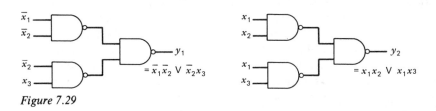

**Figure 7.29**

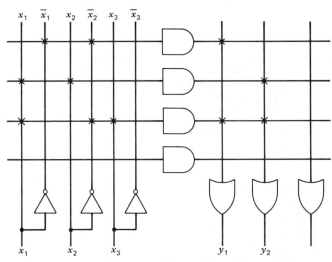

**Figure 7.30**   *PLA realization of the code translator.*

The savings of a single row in the simple PLA example just given may not have seemed overly impressive, but the underlying principle is most important. We have observed that the minimal PLA realization of a set of Boolean functions cannot be obtained if the functions are considered independently. In order to take maximum advantage of products that may be shared by two or more functions, the functions must be minimized together in a common procedure. Such a procedure is very difficult to employ by hand. For some additional insight, see Section 7.5 of Ref. [1]. Most organizations that make heavy use of PLAs will have available a program that will determine PLA realizations that are at least approximately minimal.

In comparing ROMs and PLAs we should keep in mind that these devices may be used within more complicated VLSI realizations of large systems as well as in the form of individual field programmable IC parts. In the former case, a PLA version of a realization of a set of Boolean functions would almost always require less chip area than the ROM version. The question would in each case be whether the savings would justify the increased engineering time required by the PLA approach. On-chip microprogram control store would still be implemented using ROMS.

What are the factors involved in comparing costs so that a choice can be made between a PROM and a field programmable PLA? Costs depend on technology, production volume, and type of programming. Given a specific application, with a certain volume and type of programming (i.e., mask, field, and erasable), the basic production costs of the two types of circuits are likely to be comparable, in which case programming cost becomes the determining factor. Again, assuming other factors are similar, programming cost will be proportional to the number of links (fuses), so let us consider the number of links in the two types. In all cases we assume $n$ input variables and $m$ output functions. In a ROM, we have $2^n$ decoder outputs to be connected to $m$ OR gates, so the number of links is given by

$$\text{ROM links} = 2^n \times m \tag{7.5}$$

In a PLA the number of AND gates will always be less than in a ROM, so let the number of AND gates be

$$2^n/k$$

where $k$ is a number greater than 1, usually in the range of 10 to 100. These gates must be linked to $2n$ inputs (true and complemented values of each variable) and $m$ OR, so the total number of links is given by

$$\text{PLA links} = \frac{2^n}{k}(2n + m) \tag{7.6}$$

To make it easier to compare them, we show the ratios of these numbers.

$$\frac{\text{ROM}}{\text{PLA}} = \frac{2^n \times m}{\frac{2^n}{k}(2n + m)} = \frac{k}{\frac{2n}{m} + 1} \tag{7.7}$$

In many cases, $n = m$, in which case the programming cost of a ROM is approximately $k/3$ times that of a corresponding PLA. Since $k$ is essentially a measure of how much we can simplify a set of functions relative to the canonic form, this confirms our earlier observation that ROMs are economical only if the minterms are needed. In the BCD-to-seven-segment code converter, discussed earlier, there are only 16 minterms and all are needed, so the ROM is clearly preferable. By contrast, consider the conversion of Hollerith card code to 6-bit ASCII code. The standard Hollerith code is a 12-bit code with 46 characters, so the conversion to ASCII requires the realization of six functions of 12 bits. Since there are only 46 characters, a maximum of 46 AND-gates will suffice, even if no simplification is possible. A 12-input ROM, if one were available, would have 4096 AND gates, so that a PLA would clearly be preferable. A standard PLA suitable for this application is the 82S100, with 16 inputs, 48 AND-gates, and 8 outputs, a total of 1920 links, compared with 24,576 links in a 12-input/6-output ROM.

# 7.6 Describing Multiplexers with a Graphic Vector Notation

The multiplexer packages that we have examined thus far have had at most two output lines. Very often multiplexers are used to realize system buses, in which cases the outputs may be vectors of 4, 8, 16, or more bits. Corresponding to each output there must be two or more alternative inputs as well. To show that some special notation is required for these many-input, many-output multiplexers let us consider the simple case of the quad two-line-to-one-line multiplexer. The term *quad* indicates four outputs, each of which can be connected to two alternative inputs.

In Fig. 7.31*a* we see the detailed logic diagram of a 74157 quad two-line-to-1-line multiplexer. This representation is tractable, but suppose there were 8 or 16 outputs. Somewhat simpler is the dependency representation shown in the example of Fig. 7.31*b*. Here one of two 4-bit vectors, **A** or **B**, is to be connected to the 4-bit output vector, **C**. If $x = 0$, the outputs are to be **A**; and if $x = 1$, the outputs are to be **B**. In the notation of Fig. 7.31*b*, each bit of an input vector or output vector must be represented by a separate line, even though all bits of a vector are treated the same.

In Fig. 7.31*c* we see a vector representation of Fig. 7.31*b* with only one fourth as many data lines. This notation is defined to indicate that bits **A**[3] and **B**[3] are inputs to the first bit of the multiplexer with **C**[3] as the output, and so on with **A**[0] and **B**[0] as inputs to the last bit. This type of notation will be useful wherever logic networks are merely replicated.

Occasionally, individual bits of vectors are treated repetitively by part of a logic network but in distinct ways elsewhere in the same overall network. In these cases it may be necessary to fork a single line representing a vector into two lines representing two subvectors as illustrated in Fig. 7.32*a*. Notation for convergence of two vectors or a vector and a scalar to form a single vector including all bits is illustrated in Fig. 7.32*b*. As shown, labels indicating the ordering and destination of individual bits must appear with each fork or convergence. Notice the distinctive curved lines of the

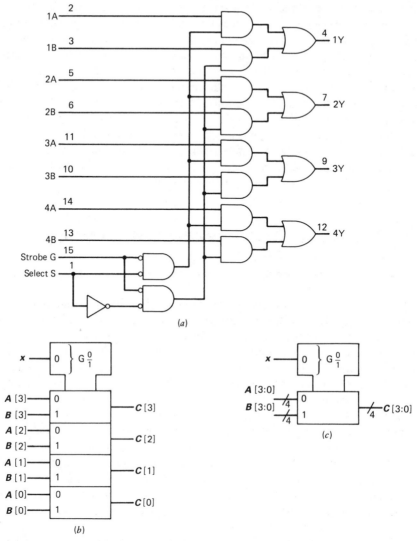

*Figure 7.31   Quad two-line-to-one-line multiplexer (74157).*

fork in Fig. 7.32a, indicating division of the vector into separate components, as distinguished from Fig. 7.32c, which represents fan-out of the same vector to different destinations.

## 7.7   Special Purpose MSI Parts

A digital designer's ideal world would include a huge catalog of available off-the-shelf MSI parts together with a powerful indexing system to permit the designer to

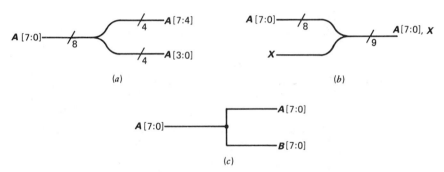

*Figure 7.32   Fork and convergence notation.*

find a part of interest easily. The designer would formulate Boolean expressions or perhaps logic block diagrams of a proposed design and then turn to that catalog with a good expectation of finding parts precisely tailored to do the job. In the "real world" this is too much to ask. A manufacturer will make available a standard part only if the projected market is sufficient to provide a profit from selling the part at a reasonable price. Consequently, only a few standard combinational logic chips have been developed to accomplish some of the most commonly used arithmetic and logic operations on a small number of bits, usually four. A designer would usually expect to use several of these parts together with some SSI logic to accomplish a design goal. Even so, the savings in package count and wiring cost of this approach over the use of only SSI parts can be dramatic.

It will be possible to make similar statements about sequential MSI parts as we study this topic in the next two chapters. While designers and managers were trying to decide which sequential MSI parts to make and market, the single-chip microprocessor arrived. Very early it was observed that microprocessors could be programmed to accomplish tasks that would otherwise require considerable special-purpose logic. Hence much of the creative output of semiconductor manufacturers has recently centered around the design and support of microcomputers. These devices are the subject matter of Chapters 11 through 16.

For a survey of available combinational MSI parts, refer to the catalogs of the various semiconductor manufacturers. In this section we shall only illustrate how such parts might be incorporated into a design. Nonstandard but comprehensible logic symbols for three such typical devices, the 7485 magnitude comparator, the 74280 parity generator/checker, and the 74283 4-bit binary adder, are shown in Fig. 7.33. Verbal descriptions of the functions of these devices are given in Table 7.1. The logic symbols, together with the verbal descriptions, provide sufficient information to allow us to accomplish the goals of the design examples in this section. The logic symbols represent functions operating on vectors of inputs that are symbolized by single lines to make the logic diagrams more readable.

The internal logic diagram of one of the packages, the 74283 4-bit adder is included in Fig. 7.33c to illustrate the logic operations on the individual bits of the input vectors. This network is an improvement over the 4-bit version of the ripple-carry adder discussed in Section 6.6. Additional gates are added to propagate carries ahead so that all four sum bits and the carry output are available after a maximum

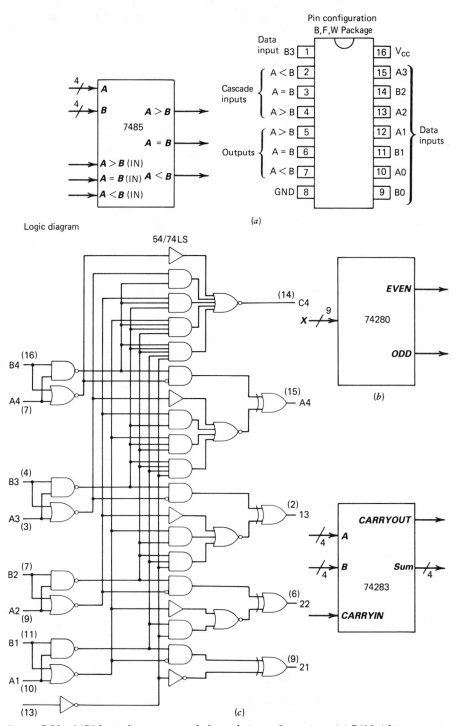

Figure 7.33   MSI logic diagrams, symbols, and sin configurations: (a) 7485 4-bit magnitude comparator, (b) 74280 9-bit odd/even parity generator/checker, and (c) 74283 4-bit binary adder.

**156**

**TABLE 7.1    Part Descriptions**

I.   **7485 4-bit magnitude comparator:**   This device compares the magnitude of two 4-bit binary numbers $A_3$, $A_2$, $A_1$, $A_0$ and $B_3$, $B_2$, $B_1$, $B_0$. If the first is greater, the output labeled $A > B$, will be 1. If the second is greater, the output labeled $A < B$, will be 1. If the two numbers are equal, the ouputs will have the same values as the similarly labeled inputs. These inputs may reflect the comparison of less significant bits in the overall comparison of two many-bit vectors.

II.  **74280 9-bit odd/even parity generator/checker:**   This device has two outputs, labeled **ODD** and **EVEN**. **ODD** is 1 if there is odd parity over the nine input bits. **EVEN** is 1 if parity over these bits is even.

III. **74283 4-bit binary adder:**   This part handles four bits of a binary addition. A carry input is provided that passes the carry from the less significant bits. The outputs are four sum bits and a carry to the next more significant bits. Some special logic circuitry not discussed in the previous chapter is included so that there are only three gate delays between the carry input and carry output. The most significant input bits are $A4$ and $B4$.

---

of four gate delays. The reader may accept the fact that this circuit carries out addition properly and is not required to analyze the network in detail.

In this section we shall consider our examples complete once a vector form of a logic block diagram is obtained. To accomplish the physical wiring of a design, the pin configuration from the manufacturers catalog is needed. An example of a package pin configuration diagram is given in Fig. 7.33$a$ for the magnitude comparator.

## Example 7.5

Construct the logic block diagram of a combinational logic network that will check parity over an 8-bit vector $X$ and then add the binary number $X$ to another 8-bit binary number, $Y$, if and only if parity was odd. If parity was even, then a third 8-bit binary number, $W$, is to be added to $Y$ instead. Use MSI parts in so far as possible.

Solution:
The desired network, which requires only one 74280 parity checker, two 4-bit adders, and two quadruple 2-input multiplexers is shown in Fig. 7.34. The inputs to the upper multiplexer are the most significant bits of the vectors $X$ and $W$. If the parity checker output **EVEN** is 0, the most significant bits of $X$ are routed to the multiplexer outputs. If **EVEN** $= 1$, the multiplexer outputs are the bits from $W$. The lower multiplexer handles the least significant bits of $X$ and $W$ in the same manner. The lower 74283 adds the least significant 4-bits of the two 8-bit arguments. The carryout from this adder package serves as the carry input to the upper 74283, which accomplishes the addition of the most significant 4-bits.

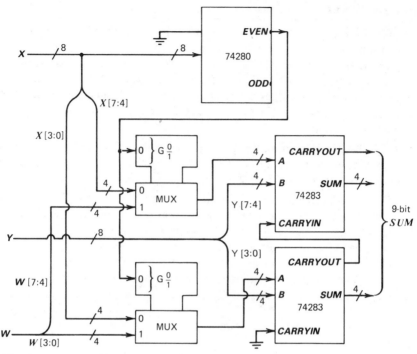

*Figure 7.34   Logic block diagram of adder with multiplexed input.*

Note that the available 9-bit parity generator/checker is used to check parity over 8-bits. This is done by simply connecting the ninth input bit to 0 so that it will not affect the outcome of the parity check.

## Problems

**7.1**  Given the function

$$f(A,B,C,D) = \Sigma\, m(1,2,4,5,6,7,8,9,10,11,14,15)$$

  (a)  Realize this function using an 8-input multiplexer.
  (b)  Realize this function using a 4-input miltiplexer plus external gating as required.

(*Hint:*  **Use *A* and *B* as the control variables to the multiplexer.**)

**7.2**  Realize the following two functions using both halves of a 74153 multiplexer and external gating as required.

$$f(A,B,C,D) = \Sigma m(2,4,6,10,12,15)$$

$$f(A,B,C,D) = \Sigma m(3,5,9,11,13,14,15)$$

(*Hint:*  **Use *A* and *D* as the control inputs to the multiplexer.**)

**7.3** Use half a 74153 dual four-line-to-one-line multiplexer in the design of a circuit that will check parity over three input bits. The output should be 1 if parity is odd. Construct a logic block diagram of your network.

**7.4** Use four single four-line-to-one-line multiplexer packages (each similar to one half a 74153) to design a network that will check parity over nine input bits. The output is to be 1 if parity is odd. Extend the results of Problem 7.3.
(a) Construct a logic block diagram of your network.
(b) Could both halves of two 74153 dual multiplexers be used to accomplish this function? Why or why not?

**7.5** A table of the valid two-out-of-five coded characters for the BCD digits is given in Fig. P7.5. All 5-bit characters not included in the list are invalid. Tabulate (in a table similar to Fig. 7.18) the bits in a ROM that can be used to translate the two-out-of-five code to the seven-segment code. A PROM with 32 8-bit words is available. In each of the following cases the solution will be a table consisting of 32 rows and eight columns of 1's and 0's. Enter 0's where the entries could actually be "don't cares."
(a) Assume that only valid two-out-of-five coded characters will occur.
(b) Assume that all 5-bit combinations are possible. Add an additional output that will be 1 whenever an invalid 5-bit character appears.

| Decimal digit | 2 out of 5 | | | | |
| --- | --- | --- | --- | --- | --- |
| | $X_4$ | $X_3$ | $X_2$ | $X_1$ | $X_0$ |
| 0 | 0 | 0 | 0 | 1 | 1 |
| 1 | 0 | 0 | 1 | 0 | 1 |
| 2 | 0 | 0 | 1 | 1 | 0 |
| 3 | 0 | 1 | 0 | 0 | 1 |
| 4 | 0 | 1 | 0 | 1 | 0 |
| 5 | 0 | 1 | 1 | 0 | 0 |
| 6 | 1 | 0 | 0 | 0 | 1 |
| 7 | 1 | 0 | 0 | 1 | 0 |
| 8 | 1 | 0 | 1 | 0 | 0 |
| 9 | 1 | 1 | 0 | 0 | 0 |

*Figure P7.5  Two-out-of-five code.*

**7.6** Tabulate the bits of a ROM that will translate seven-segment-coded characters into the two-out-of-five code. Assume that a PROM with 128 8-bit words is available and that only valid coded characters will appear.
(a) How many of the bits in your table are actually don't-cares?
(b) What is the ratio of don't-care bits to the total number of bits?
(c) How does this compare with a similar ratio for Problem 7.5a?

**7.7** Suppose that four 1024-word (eight bits per word) PROMs are to be used to provide a system with 4K of ROM. Each of the ROMs has two chip select inputs as shown in Fig. 7.21. Construct a logic block diagram similar to Fig.

7.23 for this system. Assume that there are only 12 address lines from the system and no read line. That is, some addressed word will appear on the data bus at all times.

**7.8** Repeat Problem 7.7 except that now the system has 16 output address lines, so that it can address a total of 64K bytes of memory. There is also a read line functioning in the same manner as in Fig. 7.23. The read-only memory packages will have the following address ranges in octal.

> 0000 to 03FF
> 0400 to 07FF
> 0800 to 0BFF

and      0C00 to 0FFF

(*Hint:* **Additional logic gates will be required to select the proper ROM as a function of the six most significant address lines.**)

**7.9** Suppose that a field programmable PLA of the form depicted in Fig. P7.9 is available. Mark X's at the proper points in the condensed input and output arrays to obtain a realization of the translation function of Problem 7.5a. Do not attempt to obtain a minimal realization. Merely find a realization that will fit into PLA given.

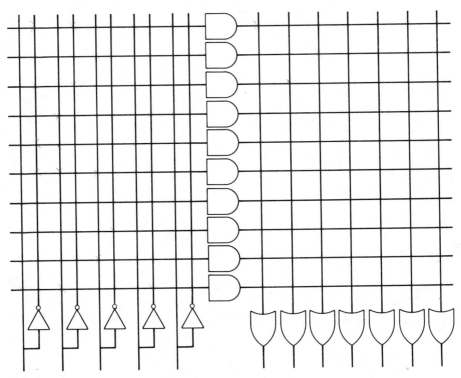

*Figure P7.9*

**7.10** Use Eqs. 7.5 and 7.6 to determine the total number of fusable links in the ROM of Problem 7.5a and the PLA of Problem 7.9. Which approach might be the most economical solution? Why?

**7.11** Use the PLA of Fig. 7.9 to obtain a realization of the three Boolean functions the minterm lists of which are given here. Mark X's at the appropriate points in the input and output arrays. Not all PLA inputs and not all output columns will be required. Use a minimal number of rows in the PLA. This may be accomplished by inspection of the Karnaugh maps of the three functions simultaneously.

$$f(A,B,C,D) = \Sigma\, m(0,2,7)$$

$$f(A,B,C,D) = \Sigma\, m(0,1,4,5,6,8,14)$$

$$f(A,B,C,D) = \Sigma\, m(6,8,12,13,14)$$

**7.12** Construct the logic block diagram of a 16-bit adder composed of four 74283 4-bit adder packages.

**7.13** Construct the logic block diagram of a network that will check parity over 10 bits using one 74280 and any additional NAND, NOT, or exclusive OR gates required.

**7.14** Use two 74283 4-bit adders and any additional NAND, NOT, or exclusive OR gates required to obtain a combinational logic network that will generate the two's complement of an 8-bit binary number.

**7.15** Modify the network of Problem 7.13 so that it will find the absolute value of an 8-bit two's complement number. The input number may be positive or negative.

**7.16** Use two 7485 comparators and any additional gates necessary to construct the logic block diagram of a network with a single output, **Z**, which will be 1 if and only if the 8-bit number **X** is greater than or equal to the 8-bit number **Y**.

**7.17** Suppose the maximum switching delay through the 74280 is 25 nsec. Suppose each gate in the quad 2-input multiplexer of Fig. 7.32 and the 74283 4-bit adder has a switch delay of 5 nsec. Use Fig. 7.13 and 7.31 to determine the maximum overall swtiching delay through the network in Fig. 7.32.

# References

1. Hill, F. J., and G. R. Peterson, *Introduction to Switching Theory and Logical Design,* 3rd ed. Wiley, New York, 1981.
2. *Intel Component Data Book.* Intel Corporation, Santa Clara, Calif., 1982.
3. Peatman, J. B., *Digital Hardware Design.* McGraw-Hill, New York, 1980.
4. *Signetics Data Manual.* Signetics Corporation, Sunnyvale, Calif., 1976.

# 8 Sequential Circuits

## 8.1  Storage of Information

Our attention up to this point has been directed to the design of combinational logic circuits. Consider the general model of a switching system in Fig. 8.1. Here we have $n$ input, or *excitation,* variables, $X_k(t)$, where $k = 1, 2, \ldots , n$ and $p$ output, or response, variables, $Z_i(t)$, $i = 1, 2, \ldots , p,$ all assumed to be functions of time. If at any particular time the present value of the outputs is determined solely by the present value of the inputs, we say that the system is *combinational.* Such a system can be totally described by a set of equations of the form of Eq. 8.1.

$$A = f(X_1, X_2, \ldots X_n) \tag{8.1}$$

The dependence on time need not be explicitly indicated, since it is understood that the values of all the variables will be those at some single time. If, on the other hand, the present values of the outputs are dependent not only on the present value of the inputs but also on the past history of the system, then we say that it is a *sequential* system.

   If the output of a sequential system is to be dependent on past inputs, there must exist within the system some mechanism for remembering these past inputs. In electronic sequential circuits or systems, the most common memory element is the *flip-flop.* (The origin of the term flip-flop is uncertain, but it has become standard.) Most sequential circuits will contain several flip-flops as depicted in Fig. 8.2. Each of these devices will remember a single binary bit of information. Together the set of binary values stored in these memory elements will constitute what we shall call

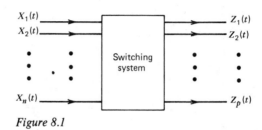

Figure 8.1

STATE

$X_1$
$X_2$

$X_n$                    LOGIC

FF  $Y_1$              $Z_1$

FF  $Y_2$              $Z_2$

FF  $Y_m$              $Z_p$

Figure 8.2   *Memory elements in a sequential circuit.*

the *state* of the sequential circuit. When the values stored in the flip-flops change, we say that the sequential circuit *changes state*.

Figures 8.3*a* and *c* show the circuit for a flip-flop constructed from two NOR gates and the timing diagram for a typical operation sequence. We have repeated the truth table for NOR, as Fig. 8.3*b,* for convenience in explaining the operation. At the start, both inputs are at 0, the $Y$ output is at 0, and the $X$ output at 1. Since the outputs are fed back to the inputs of the gates, we must check to see that the assumed conditions are consistent. Gate 1 has inputs of $R = 0$ and $X = 1$, giving an output $Y = 0$, which checks. Similarly, at gate 2, we have $S = 0$ and $Y = 0$, giving $X = 1$. At time $t_1$, input $S$ goes to 1. The inputs of gate 2 are thus changed from 00 to 01. After a delay (as discussed in Chapter 5), $X$ changed from 1 to 0 at time $t_2$. This changes the inputs of gate 1 from 01 to 00, so $Y$ changes the inputs of gate 2 from 01 to 11, but this has no effect on the outputs. Similarly, the change of $S$ to 0 at $t_4$ has no effect. When $R$ goes to 1, $Y$ goes to 0, driving $X$ to 1, thus "locking-in" $Y$ so that the return of $R$ to 0 has no further effect.

So far, nothing has been said about the case where $R = 1$ and $S = 1$ simultaneously. In Fig. 8.3*a* we see that, if this is the case, that outputs $X$ and $Y$ will both be 0 and will have no effect on each other. This behavior is undesirable for two reasons. First, a flip-flop should have only two stable states, Set and Reset, in both of

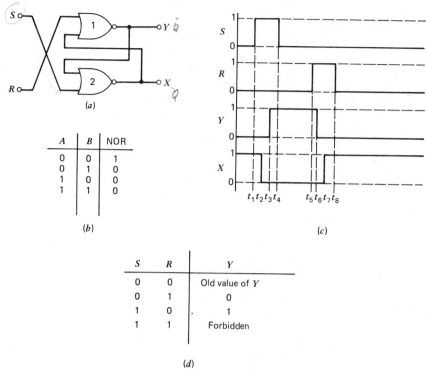

| A | B | NOR |
|---|---|-----|
| 0 | 0 | 1 |
| 0 | 1 | 0 |
| 1 | 0 | 0 |
| 1 | 1 | 0 |

(b)

| S | R | Y |
|---|---|---|
| 0 | 0 | Old value of Y |
| 0 | 1 | 0 |
| 1 | 0 | 1 |
| 1 | 1 | Forbidden |

(d)

Figure 8.3   Operation of a flip-flop.

which output $X$ will always be the complement of output $Y$. A device that does not behave in this manner is simply not a flip-flop. Second, the behavior of the flip-flop when the inputs return to 0 may not be predictable. If $R$ returns to 0 first, the eventual output values will be $Y = 1$, $X = 0$. If $S$ returns to 0 first, the resulting outputs will be the opposite. If both $R$ and $S$ appear to return to 0 at the same time, then both gates will try to go to 1 at the same time, which will cause both to try to go back to 0 at the same time, and so on. The final stable state will depend on random delays in the gates. Since even the slightest possibility of unpredictable operation must be avoided if the device of Fig. 8.3a is to be used as a memory element, the simultaneous existence of $R = 1$ and $S = 1$ is *forbidden*. That is, in any application in which this cross-coupled gate configuration is employed as a memory element, the designer must take some action to assure that $R = S = 1$ will never occur.

If the device in Fig. 8.3a is used subject to the constraints just discussed, it is known as a reset–set or $R$–$S$ flip-flop. It may be called a memory element, because it remembers which of its inputs was most recently 1. If $R$ was 1 most recently, $Y = 0$. If $S$ was 1 most recently, $Y = 1$. This memory continues indefinitely (unless the power is turned off) while the inputs remain at $R = S = 0$. The function of the $R$–$S$ flip-flop may be summarized by the table in Fig. 8.3d, which we shall call a *transition table*.

## 8.2   Clocking

The elementary *R-S* flip-flop of Fig. 8.3 can be used directly as a basic component in the design of digital systems, but it is not very well suited to this purpose. The problem lies in the fact that inputs and outputs can change at any time, and it is not easy to relate changes in the output of one such flip-flop to those of other flip-flops within the same system. The problem becomes much more manageable if we restrict the operation of the system in such a manner that changes in the internal state of the system may take place only at specified points in time separated by discrete time intervals. To enforce such a restriction requires the existence of a special signal that can be used to synchronize all changes in memory element values throughout a system. A *synchronizing signal* of this sort is almost universally called a *clock*. Three examples of clock signals are shown in Fig. 8.4.

    In Fig. 8.4*a* we see a periodic sequence of narrow (proportionately short time period) pulses. If this signal is used as a synchronizing clock for a system, changes in flip-flop values will occur in the system only at the time of the occurrence of a clock pulse. Between clock pulses, signals propagate through the combinational logic, so that the appropriate values are available at the flip-flop inputs in time for the next clock pulse. Synchronizing implies that all flip-flops that change value in response to a particular clock pulse change value at the same time. Thus, **the new value of any one flip-flop will be determined only by the old values of that and other flip-flops.** The effect of the new flip-flop values resulting from a particular clock pulse will only reach the input of other flip-flops in time to influence their response to the next clock pulse. A digital system that is synchronized by a clock so that it functions in the manner just described is said to be operating in the *"clock mode."*

    Clock signals will usually, but not always, be periodic as in Fig. 8.4*a*. Whether the clock is periodic or aperiodic as in Fig. 8.4*b,* there must always be a sufficient time interval between two successive clock pulses to allow signals to propagate through the combinational logic path in the system with the longest delay. This interval will, of course, depend on the type of logic circuit technology with which the system is implemented.

    The clock signal will be generated by a semianalog circuit, the design of which

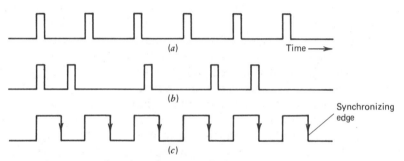

*Figure 8.4   Examples of synchronizing clocks.*

will be specified in a form different from that of the digital systems it drives. It is not always convenient to provide a clock with pulses that are very narrow with respect to the time interval between pulses (Fig. 8.4a). Often a clock will more nearly resemble the square wave of Fig. 8.4c. When the active portion of a clock period (the high value in Fig. 8.4c) extends over half the period, the precise time at which all flip-flops will change values is not very precisely defined and our assumption that all flip-flop value changes take place at the same time is called into question. This delemma is avoided by using one of the clock edges as the synchronizing event. Although the transitions in clock values are not actually instantaneous in practice, the conditions for *clock-mode* operation set forth in the previous paragraph will usually be satisfied if all changes in flip-flop values take place during the time of this *clock transition*. It is the option of the designer to choose whether flip-flops should change value on the rising edge or the falling edge of the clock. In most cases the authors' preference will be *falling-edge synchronization* as illustrated by the arrowheads in Fig. 8.4c.

Let us now add two AND gates, an inverter, and a clock line to the *R-S* flip-flop of Fig. 8.2a. The result is the *D-latch* configuration of Fig. 8.5a. Each time the clock line in the *D*-latch goes to 1, the output line, *Y*, will be updated to the value found on the input line, *D*. As illustrated in Fig. 8.5b, *S* will go to 1 and then the output *Y* will go to 1, if *D* = 1 when the clock goes high. If *D* = 0 when the clock goes high, *R* will go to 1 and then *Y* will go to 0.

At first glance it might seem that connecting a synchronizing clock line to the clock inputs of all *D*-latches in a digital system composed of these memory elements would result in a clock-mode system. This might be the case if clock pulses were

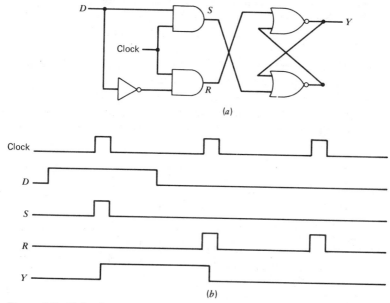

(a)

(b)

*Figure 8.5* D-*latch.*

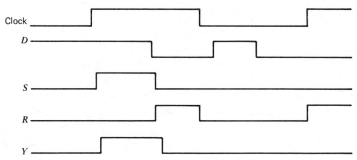

*Figure 8.6    Double output change for* D-*latch.*

made appropriately narrow as suggested by Fig. 8.5. If the clock remains 1 for rel-atively long periods of time as in Fig. 8.4c, clock-mode operation cannot be achieved using the *D*-latch. This is partially illustrated in Fig. 8.6, which shows a new set of input signals applied to the same *D*-latch given in Fig. 8.5a. Notice that *D* changes in value while the clock signal is 1. The output *Y* is first set to 1 since *D* = 1 when the clock first goes to 1. When *D* returns to 0 while the clock is still 1, *R* goes to 1 and the output, *Y*, returns to 0. The output of the *D*-latch has changed twice as a result of a single clock pulse. Why such a non-clock-mode operation is undesirable in a digital system will be made clear in the next section where we examine a simple 2-bit shift register.

Let us now consider a different type of *D* flip-flop, represented by the symbol in Fig. 8.7a. This is the standard block diagram symbol for the edge-triggered *D* flip-flop. The output, labeled *Q*, corresponds to the output *Y* in Figs. 8.2a and 8.4a; that is, it is the output that is 1 when the flip-flop is set. The lower output, as indicated by the inversion bubble, is the complementary output that is 0 when the flip-flop is set, 1 when it is reset. A timing diagram for a typical operation is given in Fig. 8.7b. Note that *Q* goes to 1 immediately following the first positive transition on the clock line, as marked by the arrow. In this respect the operation is similar to that of the *D*-latch in Fig. 8.4. Note, however, that a change in *D* occurs while the clock is still high, but no change occurs in *Q* until the next positive transition in the clock signal, also marked by an arrow. For the flip-flop to work reliably, the *D* input must be stable throughout a short interval around each positive clock transition. The reason

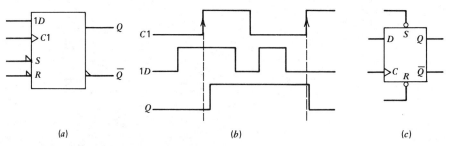

*Figure 8.7    Rising edge-triggered* D *flip-flop.*

for the name "edge-triggered" is now apparent; the flip-flop can change state only on the positive *edge*, or positive-going transition of the clock signal. No other changes on either the clock or D lines will cause state transitions.

A *D flip-flop*, which functions as suggested in Fig. 8.7, may be used in the design of clock-mode systems. The flip-flop will change state no more than once for each clock pulse, and all transitions are synchronized by the rising edge of the clock signal. In Fig. 8.8a we see the logic-block diagram of one circuit that will satisfy the specifications given for the *rising-edge D flip-flop* of Fig. 8.7a. A detailed analysis of this circuit is beyond the scope of this book. We must be content with the discussion of the input–output behavior of the circuit as just discussed. Figure 8.8b specifies the period of stability around the rising edge of the clock for this circuit. The transition period during which the input must remain stable is a factor of both the circuit parameters and the rise time of the clock pulse. The value of D must be "set up," that is, established at a stable value a specified time before the clock transition starts, and it must "hold" at that value until a specified time after the clock transition is complete. The manufacturer will provide specifications for these various time periods. The total time during which the D input must be held stable is the sum of the setup time, the rise time, and the hold time and is typically about 25 nsec for 74LS flip-flops.

The lines labeled S and R in Fig. 8.7a are the direct Set and Reset (sometimes called preset and clear) inputs, respectively. It can be seen in Fig. 8.8a that these lines are applied directly to the output NAND gates. These lines are normally held high, in which case they have no effect on the operation of the flip-flop. If the Set line goes low, it drives gate 1 high, setting the flip-flop; if the Reset line goes low, it drives gate 2 high, resetting the flip-flop. The Set and Reset lines are also applied to gates 3, 5, and 6 to prevent the flip-flop from switching to the opposite state if the clock is high when the Set or Reset signal goes back to 1. The Set and Reset lines are primarily used for intializing flip-flop values before a sequence of clocked operations begins.

Figure 8.8c shows the complete transition table for the rising-edge-triggered D flip-flop including the Set and Reset lines. As noted in the preceding discussion and indicated by the inversion bubbles on the block diagram symbol, the Set and Reset lines are active-low. The notations $C1$ and $1D$ on the symbol are an example of control dependency. An input line labeled $Cn$, where $n$ can be any digit, is a control line, normally a clock line, used to control data inputs to flip-flops. Those data lines that are controlled by the $Cn$ line will have labels starting with $n$, such as $nD$, $nS$, or $nR$. In this case, $C1$ controls $1D$, that is, the $1D$ input will have no effect unless $C1$ is active. Since the S and R inputs have no "1" in front of them, they are not controlled by the clock. This is shown in the first two lines of the transition table, indicating that S and R override the other inputs. The next three lines indicate that the flip-flop acts as a normal D flip-flop when the S and R lines are high, with the "up arrows" indicating that the transition occurs on the positive edge of the clock pulse. On the block diagram symbol, the "arrowhead" on the $C1$ input indicates positive-edge-triggering. Flip-flops can also be built that trigger on the negative (falling) edge. This is indicated by an inversion bubble on the clock input, preceding the "arrowhead."

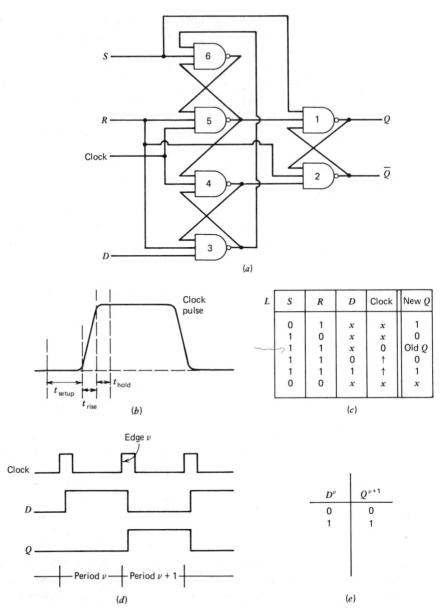

Figure 8.8   *Positive edge-triggered* D *flip-flop (7474). (a) Logic-block diagram. (b) Transition time specifications. (c) Complete transition table. (d) Timing. (e) Clocked transition table.*

The block diagram symbol shown in Fig. 8.7a is an example of the flip-flop symbols recommended by the revised IEEE Std 91/ANSI Y32.14. One requirement of this new standard (illustrated by this example) is that all inputs must be on the left side of the symbol, all outputs must be on the right side. The earlier version of this standard (1973) had only rudimentary symbols for flip-flops, with the result that various manufacturers developed their own symbols. Figure 8.7c shows a typical earlier symbol for the edge-triggered $D$ flip-flop. Note several differences. The control dependency of $D$ on the clock is not indicated. The outputs are labeled $Q$ and $\overline{Q}$, and the inversion bubble is omitted on the $\overline{Q}$ output. The $S$ and $R$ inputs appear at the top and bottom of the symbol. This form of symbol is not as precise as the new form, but it does allow more flexibility in arranging drawings so we shall use it occasionally where the meaning is clear and it makes simpler drawings possible.

If we assume that $S = R = 1$, the table in Fig. 8.8c may be reduced to its fifth and sixth rows. Now let us eliminate the clock column by relabeling the data input column as $D^v$, where $v$ is the number of the clock period that is terminated by synchronizing transition $v$, as illustrated in Fig. 8.8d. The output column may then be labeled $Q^{v+1}$, which represents the flip-flop output during the period immediately following the transition $v$. The resulting clock-mode transition table is given in Fig. 8.8e. This simple table is the usual definition of a $D$ flip-flop, where all output changes occur in response to a clock transition.

The table in Fig. 8.8e can easily be expressed in equation form as given by Eq. 8.2

$$Q^{v+1} = D^v \tag{8.2}$$

This is the standard next-value expression for the clocked $D$ flip-flop.

## 8.3  Registers

When a set of flip-flops is organized (at least conceptually) in vector form as shown in Fig. 8.9 the result is called a *register*. Usually the clock inputs to the individual flip-flops of a register are connected together as shown in the figure. A register will typically store a vector of binary bits representing a more complex item of information such as a binary number or an ASCII character. In the particular 4-bit reg-

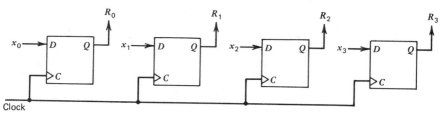

*Figure 8.9  Register.*

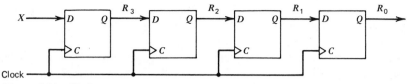

*Figure 8.10  Shift register.*

ister of Fig. 8.9 the current value of the data vector $X_0$, $X_1$, $X_2$, $X_3$ is transferred into (stored in) the register $R$ with each synchronizing transition on the clock line.

Often a digital system will have many more flip-flops than input lines. Such systems will effectively accumulate in internal storage registers the information appearing on the input lines over time. The configuration in Fig. 8.10, called a *shift register,* is designed to assemble information arriving from a single serial source. It is assumed that a new bit of information may be sampled from line $x$, with the arrival of each clock pulse. If $x$ is 1 at the time of the first clock pulse, the first flip-flop will be set to 1; if $x$ is 0, the first flip-flop will be reset. The second clock pulse will similarly shift the next value of $x$ into the first flip-flop while simultaneously shifting the first value of $x$ into the second flip-flop, and so on as each bit of $x$ is successively sampled.

A shift register constructed using edge-triggered $D$ flip-flops or otherwise operating in the clock mode will function as just described. To further point out the problems associated with the use of $D$-latches in clock-mode design, consider the 2-bit shift-register network given in Fig. 8.11. Note that the absence of the "arrowhead"

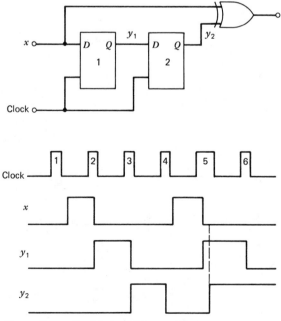

*Figure 8.11   Double-transition problem.*

on the clock inputs indicates that these are latches. Note that clock pulse 5 is significantly wider than the other pulses, which were tuned to the delays in the flip-flops. We see that $x = 1$ at the beginning of clock pulse 5. This value is shifted into flip-flop 1 and appears as output $y$ while clock pulse 5 is still present. Unfortunately, this causes $y$ to go to 1, also. This is not the desired behavior, the value of $x$ was shifted two places in the shift register by a single clock pulse. As noted earlier, clock-mode operation requires that the circuit state change only once for each clock pulse.

> To summarize one last time, design in the clock mode requires that the designer take steps to assure that there can be no more than one change in the system state in response to each clock pulse. This is usually accomplished by causing all flip-flops to trigger on the same transition of each pulse.

## 8.4   Memory Element Input Logic

The organization of a set of flip-flops as a register into which separately generated information can be triggered (Fig. 8.9) is easily the most common application of D flip-flops. However, a memory element is sometimes used in such a way that its next value following a clock pulse is an immediate function of its present value (before the pulse), as well as of external input information. In this case the analysis is slightly more complicated than for registers. Consider the network of Fig. 8.12.

To understand the operation of this network, let us consider separately the combinational logic network with output (labeled $f(a,b,y^\nu)$) leading to the D input of the flip-flop. This network may be represented by the Boolean sum of products,

$$f(a,b,y^\nu) = ab \lor a\overline{y}^\nu \qquad (8.3)$$

We now recall from Eq. 8.2 that the value at the D input of a D flip-flop will be the next value of that flip-flop following the next clock pulse. This leads to

$$y^{\nu+1} = f(a,b,y^\nu) = ab \lor a\overline{y}^\nu \qquad (8.4)$$

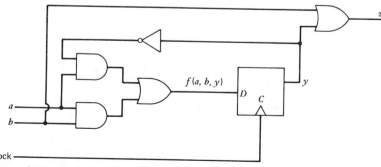

*Figure 8.12   Analysis of input logic.*

Now let us enter the information represented by Eq. 8.4 in the specially arranged Karnaugh map of Fig. 8.13a. Notice that the values represented on the Karnaugh map are the next values of the flip-flop. All external inputs are arranged along the top of the map while the present values of the flip-flop are arranged vertically. In this form the Karnaugh map is known as the *network transition table*.

We next consider the output of the network, given by z in Fig. 8.12. This output is characterized by

$$z = b \vee y'' \tag{8.5}$$

For completeness we tabulate Eq. 8.5 in the Karnaugh map of Fig. 8.13b. On this output map or output table the external inputs and the present flip-flop values are arranged in the same way as they were in the transition table.

In this section we have only analyzed an existing sequential circuit. A greater challenge would be to design such a circuit provided only with some sort of English-language statement of the problem. This will be the topic of the next section.

## 8.5  A First Design Example

In the previous section we analyzed a simple digital network designed using $D$ flip-flops and obtained an alternative description in the form of a transition table and output table. A more interesting problem is to synthesize a digital network beginning only with a word statement of the problem. One commonly used design approach is to first obtain a transition table for all memory elements in the network and an output table for all outputs. For $D$ flip-flops, the transition and output tables, which are in Karnaugh map form, may be translated directly to network form, taking advantage of Eq. 8.2 in the case of the transition table. In the next section we shall introduce another type of clocked memory element, the *J-K flip-flop*. For this type of flip-flop there will be another step between the transition table and the final network. For now we illustrate the synthesis approach for the simpler $D$ flip-flop case.

---

### Statement of Design Problem

A clock-mode digital system called a *start-signal generator* is to be designed having two input lines, **ready** and **go,** and one output line, **z.** The output is to remain 0 until the following sequence of events takes place. First, **ready** must go to 1 for an unspecified number of clock periods. Following this the output, **z,** will go to 1 for exactly one clock period as soon as **go** goes to 1. It may be assumed that **ready** and **go** will never be 1 simultaneously. Once a one-period output has occurred, **ready** must appear before another one-period output can be generated. A typical timing diagram for the network is given in Fig. 8.14.

---

You will note in Fig. 8.14 that the two input signals, *ready* and *go,* only change value just after the falling edge of a clock pulse, which in this case is the synchro-

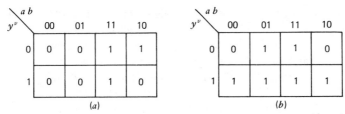

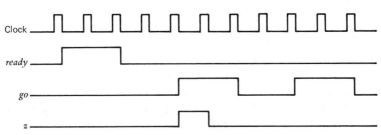

Figure 8.13   *Network transition table and output table: (a)* $y^{v+1}$ *and (b) Z.*

Figure 8.14   *Timing diagram for start signal generator.*

nizing edge for any state changes in the system. This relationship will hold in cases where the inputs are supplied by some system with an internal function that is synchronized by the same clock. In general, inputs to a clock-mode system may not change values during some short interval of time prior to the synchronizing clock edge. Controlling input changes with the system clock is the easiest approach to assuring that this constraint is satisfied. The effect of gate delays and memory-element response times in the input generating circuitry is reflected in Fig. 8.14. Often these delays will be neglected in timing diagrams for clock mode systems, and input transitions will seem to be exactly synchronized with clocking edges.

To the novice designer the gap between the English-language statement of a design problem and a final circuit may seem large. A variety of techniques have been developed for bridging this gap, all of which call, to a greater or lesser extent, on the designer's experience and engineering insight. A certain self-confidence and willingness to accept a challenge is also required. If the latter is present, you almost certainly have the insight, and the experience factor will grow surprisingly quickly.

One technique that is helpful here as well as in more complicated problems is simply to

> Ask oneself what must be remembered (how much information must be stored) and then identify a sufficient number of flip-flops to store this information.

In the problem at hand, it is only necessary to remember whether or not

> The *ready* signal has gone to 1 since the last time there was an output signal.

If so, a *go* signal will generate an ouptut. If not, no output will appear, when *go* goes to 1. Therefore, let us include only one flip-flop in our circuit, label this flip-flop *Y*, and use it to store whether or not the preceding statement is true. We then proceed to tabulate the next value of $Y$ ($Y^{\nu+1}$) following clock periods in which the inputs assume the various possible combinations of variables. We see such a table in Fig. 8.15a. Notice that $Y$ is initially 0 and remains at this value until *ready* becomes 1. For *ready, go* = 1,0 and $Y$ = 0, the next value of $Y$ is 1. $Y$ remains 1 until we have *go* = 1, after which $Y$ returns to 0 for the next clock period.

Nothing has yet been said about the output. Since $Y$ will always be 1 following *ready,* the output will be 1 when *go* = $Y$ = 1. $Y$ will return to 0 immediately following the first period for which *go* = 1. Therefore, the output **z** will be 1 for only one clock period as desired. A tabulation of output values as a function of the input and flip-flop values is given in Fig. 8.15b.

Notice that in both Figs. 8.15a and b that the entries in the columns for *ready,go* = 1,1 are all don't-cares. In the problem statement it was made clear that this combination of input values would never occur. Hence we use don't-care entries in much the same way as was done for combinational logic circuits.

We have only scratched the surface of the general problem of obtaining a transition table from a word description. We will return to the overall synthesis problem at various points in the book. If you are interested in a "state diagram"/"state table" approach, consult Ref. [4]. For now we return to our more modest objective, which is translation of the transition and output tables to a logic block diagram for the circuit. We remember that Fig. 8.15b is essentially a Karnaugh map of $Y^{\nu+1}$ as a function of $Y^{\nu}$, *ready,* and *go.* Obtaining a minimal sum of products from the map yields

$$Y^{\nu+1} = ready \lor (Y^{\nu} \land go) \tag{8.6}$$

To wire a network implementing Eq. 8.6 using clocked D flip-flops, we refer to the defining relation for the clocked D flip-flop given by Eq. 8.2. It is necessary to determine a logical expression for $D$ so that the proper network of gates can be connected to the $D$ input of the flip-flop realizing $Y$. Since $Y^{\nu+1} = D^{\nu}$, the right-hand side of Eq. 8.3 may be used to realize $D$ as a function of input and present flip-flop values all available at time $\nu$. Therefore we may write

$$D^{\nu} = ready \lor (Y^{\nu} \land \overline{go}) \tag{8.7}$$

This equation is used to define the gate network leading to the $D$ input of the flip-flop shown in Fig. 8.16.

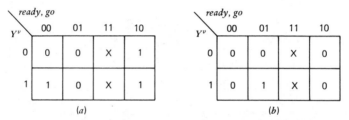

*Figure 8.15   Next value and output tables.* (a) $Y^{\nu+1}$ *and* (b) Z.

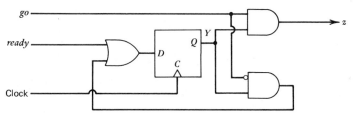

*Figure 8.16    Start signal generator.*

The expression for the output, **z,** may be obtained by treating Fig. 8.15*b* as a Karnaugh map. It is easily seen **z** is given by

$$z = Y^{\nu} \wedge go \tag{8.8}$$

We observe from the discussion of the start-signal generator that obtaining the gate network realizing the *D* input to a *D* flip-flop is very easy. The next value of the flip-flop output, which is the same as the current value of *D*, is specified directly in Karnaugh-map form in the next-value table for the flip-flop. A realization of this table is thus a realization of the network leading to the *D* input.

Most activity in digital systems consists of transferring computed data into registers as illustrated in Fig. 8.8. The clocked *D* flip-flop is ideally suited for this function. Therefore, the *D* flip-flop is by far the most commonly used memory element in digital system design.

There are a few special applications for which the *D* flip-flop is not best suited. Typically these are applications in which the next values of the flip-flops are primarily a function of their own present values. The start-signal generator is such an application. For such systems the *J-K* flip-flop, which will be discussed in the next section, is a better choice.

## 8.6    The *J-K* Flip-Flop

Let us reexamine the transition table for the *S-R* flip-flop as given in Fig. 8.3. Notice once again that the input combination $S = R = 1$ is forbidden. Let us relabel the *S* and *R* columns with *J* and *K*, respectively, and replace the entry "forbidden" by the complement of the old value of *Y*. In doing so we have developed a version of the transition table of the *J-K* flip-flop, as shown in Fig. 8.17*a*. For this transition table to have meaning it is necessary to insist that the *J-K* flip-flop be clocked. Otherwise we would have no idea how often the output would be replaced by its complement if both *J* and *K* remained 1 for a period of time. Thus the *J-K* flip-flop is always a clocked device and the old output is replaced by a new value once each clock period.

The symbol for a clocked *J-K* flip-flop is given in Fig. 8.17*b*. A complete transition table with the clock represented in a separate column is given in Fig. 8.17*c* for a falling edge-triggered *J-K* flip-flop.

The information in Fig. 8.17*a* can be rearranged to form the Karnaugh map

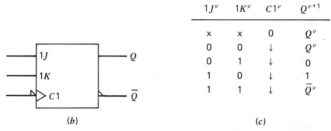

| J | K | $Y^{\nu+1}$ |
|---|---|---|
| 0 | 0 | $Y^{\nu}$ |
| 0 | 1 | 0 |
| 1 | 0 | 1 |
| 1 | 1 | $\overline{Y}^{\nu}$ |

(a)

| 1$J^{\nu}$ | 1$K^{\nu}$ | C1$^{\nu}$ | $Q^{\nu+1}$ |
|---|---|---|---|
| x | x | 0 | $Q^{\nu}$ |
| 0 | 0 | ↓ | $Q^{\nu}$ |
| 0 | 1 | ↓ | 0 |
| 1 | 0 | ↓ | 1 |
| 1 | 1 | ↓ | $\overline{Q}^{\nu}$ |

1J — Q

1K

C1 — $\overline{Q}$

(b)                                              (c)

**Figure 8.17   Transition table and symbol for J-K flip-flop.**

of Fig. 8.18*a*. Taking a minimal sum of products expression from this map yields Eq. 8.9, which may be called the transition equation for the clocked *J-K* flip-flop.

$$Y^{\nu+1} = (\overline{K} \wedge Y^{\nu}) \vee (J \wedge \overline{Y}^{\nu}) \tag{8.9}$$

To determine logical expressions for the *J* and *K* inputs of a flip-flop, it is convenient to arrange the information of Fig. 8.18*a* in a slightly different form. There are four combinations of new and old values for the output of a *J-K* flip-flop. In Fig. 8.18*b* we list the values of *J* and *K* that will effect each of the old-to-new transitions: $0 \rightarrow 0$, $0 \rightarrow 1$, $1 \rightarrow 0$, and $1 \rightarrow 1$. A don't-care entry in the *J* or *K* column indicates that either a 0 or a 1 will do.

The left column in Fig. 8.18*b* lists the four possible combinations of present value and desired next value of the output *Q*. The entries give the necessary values

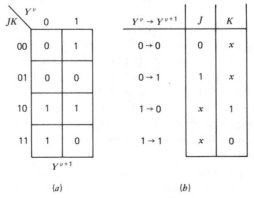

| JK \ $Y^{\nu}$ | 0 | 1 |
|---|---|---|
| 00 | 0 | 1 |
| 01 | 0 | 0 |
| 10 | 1 | 1 |
| 11 | 1 | 0 |

$Y^{\nu+1}$

(a)

| $Y^{\nu} \rightarrow Y^{\nu+1}$ | J | K |
|---|---|---|
| $0 \rightarrow 0$ | 0 | x |
| $0 \rightarrow 1$ | 1 | x |
| $1 \rightarrow 0$ | x | 1 |
| $1 \rightarrow 1$ | x | 0 |

(b)

**Figure 8.18   J-K flip-flop transitions: (a) K-map and (b) transition list.**

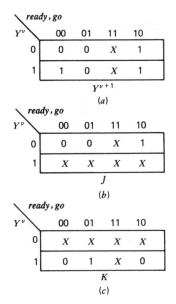

Figure 8.19   Karnaugh maps for J and K inputs.

is 0 and it is to remain 0, then *J* must equal 0 and *K* can be either 0 or 1 (don't-care). If *Q* is now 0 and is to go to 1, that is, the flip-flop is to be set, then *J* = 1 and *K* is a don't-care. The other entries follow in a similar manner.

Let us now reconsider the start-signal generator using *J-K* flip-flops in an attempt to determine an even simpler realization than the one given in Fig. 8.16. We repeat the next value of Fig. 8.15*a* as Fig. 8.19*a*.

Figure 8.19*a* specifies only the next flip-flop value. To construct a *J-K* flip-flop realization, we must obtain logical expressions for *J* and *K*. We begin by drawing Karnaugh maps for *J* and *K* as functions of the same variables $Y^{\nu}$, *ready,* and *go*. Next we directly enter the two don't-cares from Fig. 8.19*a* directly in these maps. If we don't care about the value of $Y^{\nu+1}$ for a particular combination of inputs, then we don't care about the corresponding values of *J* and *K*. Next we compare the individual transitions in Fig. 8.19*a* with the *transition lists* of Fig. 8.18*b*. For example, we see that, for the first two entries in the first row of the map for $Y^{\nu+1}$, both the present values and the next values are 0. The transition list indicates that the corresponding entries in the *J* and *K* maps should be *J* = 0 and *K* = *X*. These are the values shown in Figs. 8.19*b* and *c*. The final entry in the first row of the map for $Y^{\nu+1}$ shows present value 0 and next value 1. Therefore we must have *J* = 1 and *K* = *X*, also as given in Figs. 8.19*b* and *c*. The remaining entries in these Karnaugh maps are determined using the transition lists in a similar fashion.

Once we have the maps for *J* and *K*, we determine the minimal sum of products realizations given by Eqs. 8.10 and 8.11. In this case the sum of products forms for both *J* and *K* require no gates at all. That is,

$$J = ready \tag{8.10}$$

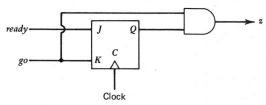

*Figure 8.20*  J-K *flip-flop version of start-signal generator.*

and

$$K = go \qquad\qquad (8.11)$$

Ordinarily we will not be so fortunate. The expression for the output is the same as given previously as Eq. 8.8, so the *J-K* flip-flop realization of the start-signal generator is as shown in Fig. 8.20.

Edge-triggered versions of *J-K* flip-flops are available. More commonly available in IC form are devices called *master–slave J-K flip-flops*. A master–slave flip-flop is nothing more than a cascade of two flip-flops, as shown in Fig. 8.21. The two *R-S* flip-flops are elementary flip-flops, such as shown in Fig. 8.3. The clock is applied directly to the first flip-flop, the *master,* and inverted before being applied to the second flip-flop, the *slave.* When the clock is low the master is disabled, so that changes in *J* and *K* will have no effect on the matter. The slave is enabled at this time but cannot change since it is driven by the master, which is disabled. When the clock goes high, the master is enabled and will behave as determined by the values of *J* and *K* at that time. At the same time, the inverted clock disables the slave, so it will not be affected by any changes at the output of the master. When the clock goes back down, the master is again disabled, isolating the flip-flop from changes in *J* and *K.* The inverted clock enables the slave, and the state of the master, as determined by *J-K* when the clock went high, is copied into the slave. Because the outputs of the complete flip-flop are taken from the slave, we are thus assured that the outputs cannot change until after the clock pulse has ended.

The standard block diagram symbol for a *J-K* master–slave flip-flop is shown in Fig. 8.21*b.* This is not a true edge-triggered device since the device does not make a single-state change on a clock transition. Instead, the master triggers on the rising edge of the clock, the slave on the falling edge, and the device is sometimes classified as a pulse-triggered flip-flop. Since the flip-flop is not edge-triggered, the "arrowhead" is not present at the clock input. The " ⌐ " symbol at the outputs indicates that the outputs change on the falling edge of the clock pulse. The standard TTL 7473 integrated circuit package contains two of the master–slave flip-flops shown in Fig. 8.21.

The master–slave flip-flop is synchronized by the falling edge of the clock. Unfortunately, there is one important disadvantage of the master–slave flip-flop with respect to a genuine falling edge-triggered *J-K* flip-flop. For reliable operation, the propagation of all signals through combinational input logic must be accomplished prior to the rising edge of the clock. Thus the clock pulse must be narrow with respect

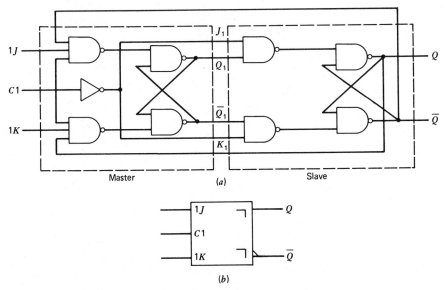

Figure 8.21   Master–slave J-K flip-flop.

to the clock period or a significant portion of the clock period is wasted. Some changes in $J$ and $K$ will not be recognized by the master–slave flip-flop while the clock is high. Suppose for example that $J = 1$ and $K = 0$ at the time of the onset of a clock pulse. If the inputs change to $J = K = 0$ while the clock is on, these new values will not be accepted. The $J = 1$ will have already set the master to 1, and this value will be passed on to the slave by the falling edge of the clock.

## 8.7   Design of Counters

One of the most useful and probably the most common type of sequential circuit is the counter. A counter would rarely be designed using $D$ flip-flops, since its next state is primarily a function of its present state. For these reasons we have chosen counters as our primary set of examples to illustrate design using $J$-$K$ flip-flops.

  An $n$-bit counter has $n$ flip-flops, which generally drive the $n$ output lines directly. In its elementary form the counter has only a clock input, and the function of the counter is to count the pulses appearing on the input line. Each time the input line is pulsed, the count, represented by $n$ binary bits, advances. Counters can be designed to count in any desired sequence; binary counters are most common. In a 2-bit binary counter, the output sequence would be 00, 01, 10, 11, 00, 01, and so on. Note that the count starts over after the maximum count is reached. The capacity of a counter is often stated in terms of its *modulus,* the number of counts that must be received before the count cycle starts to repeat. The 2-bit counter with the sequence given here would thus be known as a *modulo*-4 counter.

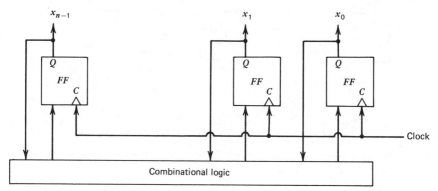

*Figure 8.22   General model for synchronous counter.*

In a synchronous counter, the line carrying the pulses to be counted is applied directly to the clock inputs of the flip-flops. With the flip-flops directly connected to the outputs, the general model of the synchronous counter is as shown in Fig. 8.22. Counters are quite simple to design. Since the circuit outputs are the same as the flip-flop outputs, the count sequence is the same as the sequence of internal states. The output sequence thus directly specifies the transition table.

Figure 8.23 shows the transition table for the 2-bit counter with the sequence just given. Note the difference between this transition table and the more general one given in Fig. 8.13. First, no output section is needed, since the outputs are equal to the flip-flop outputs. Second, since there is no input other than the clock input, which does not appear explicitly in the table, there is only one column of entries in the next state section. Third, since there are two flip-flops, the states have to be coded in terms of two state variables (flip-flop outputs are generally referred to as state variables). Thus, the present state column lists all possible counts immediately prior to a count pulse, the next-state column shows the corresponding counts to which the counter should go as the result of a count pulse. The count sequence is completely defined by the progression from each present state to the corresponding next state; the order in which the present states are listed is irrelevant. However, most people find it simplest to list the present states in the desired count order.

For the next step in the design procedure, we need the transition list for the *J-K* flip-flop (Fig 8.18*b*). Using the counter transition table and the *J-K* flip-flop tran-

| **Present state** | | **Next state** | |
|---|---|---|---|
| $y_1^\nu$ | $y_0^\nu$ | $y_1^{\nu+1}$ | $y_0^{\nu+1}$ |
| 0 | 0 | 0 | 1 |
| 0 | 1 | 1 | 0 |
| 1 | 0 | 1 | 1 |
| 1 | 1 | 0 | 0 |

*Figure 8.23   Transition table for modulo-4 binary counter.*

| $y_1$ $y_0$ | |
| --- | --- |
| 0 0 | 0 |
| 0 1 | 1 |
| 1 1 | $x$ |
| 1 0 | $x$ |

$J_1$

| $y_1$ $y_0$ | |
| --- | --- |
| 0 0 | $x$ |
| 0 1 | $x$ |
| 1 1 | 1 |
| 1 0 | 0 |

$K_1$

| $y_1$ $y_0$ | |
| --- | --- |
| 0 0 | 1 |
| 0 1 | $x$ |
| 1 1 | $x$ |
| 1 0 | 1 |

$J_0$

| $y_1$ $y_0$ | |
| --- | --- |
| 0 0 | $x$ |
| 0 1 | 1 |
| 1 1 | 1 |
| 1 0 | $x$ |

$K_0$

Figure 8.24  Karnaugh maps for modulo-4 counter.

sition list, we next obtain the Karnaugh maps for the $J$ and $K$ inputs of the two flip-flops as given in Fig. 8.24.

The first row of the four maps were generated by noting that, in the first row of Fig. 8.23, $y$ remains 0 so that $J_1 = 0$, whereas $K_1$ may be a don't-care. The second state variable, $y_0$, changes from 0 to 1 so that $J_0$ must be 1 but $K_0$ may be a don't-care. The remaining rows of the maps are generated in the same fashion. It is important to note, however, that Fig. 8.24 is in Karnaugh map form and there the third row of the maps ($y_1, y_0 = 1,1$) corresponds to the last row of the transition table in Fig. 8.23. From the Karnaugh maps we determine the following Boolean expressions.

$$J_1 = K_1 = y_0 \tag{8.12}$$

$$J_0 = K_0 = 1 \tag{8.13}$$

These functions describing the control input signals to the counter flip-flops are known as *excitation functions*. In this simple example they are trivial, but they may be quite complex in more complex counters. The corresponding circuit for the 2-bit binary counter is shown in Fig. 8.25. Note that we have included a RESET line. In any counter it is likely to be desirable to start the count at some particular value, most commonly zero. This is easily accomplished by tying the RESET inputs of the flip-flops together, so that pulling the RESET line low will reset the count to 0,0. In

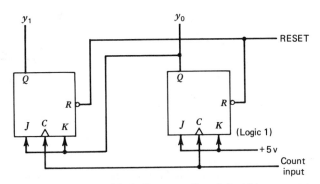

Figure 8.25  Logic block diagram of modulo-4 binary counter.

some applications this may not be required, since the counter can always be advanced to 0,0 with a maximum of three input pulses.

If a counter has $n$ flip-flops, the maximum possible modulus is $2^n$, but it is obvious that not all counters will have moduli equal to powers of two. To see what difference this makes in the design process, let us consider the design of a modulo-6 binary counter, that is, one that counts from 0 to 5 in binary and then starts over at zero. Clearly, three flip-flops are needed, and the transition table is shown in Fig. 8.26. The reader will note that it is of the same basic form as the table for the modulo-4 counter, Fig. 8.22; the only difference is that there are three variables and six counts.

The Karnaugh map for one of the excitation functions is shown in Fig. 8.26b, with two of the state variables treated as column variables for convenience. Here we see the effect of the two unused states out of eight possible with three variables. The transition table shows only the states used, but the Karnaugh map, being a map of three variables, has eight entries. Two of these are for the unused states, corresponding to the counts 110 and 111, which will never occur. Since the states will never occur, the corresponding next states are don't-cares. Since the next states are don't-cares, all six values of $J$ and $K$ will be don't-cares. Thus all six maps will contain don't-cares in the squares marked with an asterisk in Fig. 8.26b. This is general; if the modulus of the counter is not a power of two, there will be don't-care's on the maps of the excitation functions corresponding to the unused counts.

The counters discussed to this point have been fixed-sequence counters, in which the next count is completely determined by the present count, but this need not always be the case. Inputs can be added to the logic section to control the count sequence. The most common example of this sort of counter is the up–down counter. Such a counter will have an input, $X$, to control the direction of count. If $X = 0$, the counter will count up; if $X = 1$, the counter will count down. Figure 8.27 shows the

| Present state | | | Next state | | |
|---|---|---|---|---|---|
| $y_2^{\nu}$ | $y_1^{\nu}$ | $y_0^{\nu}$ | $y_2^{\nu+1}$ | $y_1^{\nu+1}$ | $y_0^{\nu+1}$ |
| 0 | 0 | 0 | 0 | 0 | 1 |
| 0 | 0 | 1 | 0 | 1 | 0 |
| 0 | 1 | 0 | 0 | 1 | 1 |
| 0 | 1 | 1 | 1 | 0 | 0 |
| 1 | 0 | 0 | 1 | 0 | 1 |
| 1 | 0 | 1 | 0 | 0 | 0 |
| 1 | 1 | 0 | $x$ | $x$ | $x$ |
| 1 | 1 | 1 | $x$ | $x$ | $x$ |

(a)

| $y_0$ \ $y_2 y$ | 00 | 01 | 11 | 10 |
|---|---|---|---|---|
| 0 | | | $x$ | $x$ |
| 1 | | 1 | $x$ | $x$ |

$J_2 = y_1 y_0$

(b)

Figure 8.26   Transition table and excitation map for modulo-6 binary counter: (a) transition table and (b) one of six excitation maps.

| Present state | | Input $X$ | |
|---|---|---|---|
| $y_1^v$ | $y_0^v$ | **0** | **1** |
| 0 | 0 | 0 1 | 1 1 |
| 0 | 1 | 1 0 | 0 0 |
| 1 | 0 | 1 1 | 0 1 |
| 1 | 1 | 0 0 | 1 0 |
| | | $y_1^{v+1}, y_0^{v+1}$ | |
| | | Next state | |

*Figure 8.27   Transition table for modulo-6 up–down counter.*

transition table for a modulo-4 up–down counter, and Fig. 8.28 shows the maps of the excitation functions. Note that this transition table looks more like that of Fig. 8.13. Since there is an input, there are two columns of next-state entries. The maps are three-variable maps, since the excitation functions are functions of the state variables and the input. The circuit of the up–down counter is shown in Fig. 8.29.

A special problem in the design of counters is that of making them self-starting. When we apply power to a counter it may come on in any state. We can always provide reset circuitry to take it to the desired starting state, but this may be undesirable in some situations. Instead, it may be preferable that the circuit be so designed that some number of "conditioning" pulses on the count line will always bring the counter to the desired condition. If the modulus of the counter is a power of two, there is no problem, since any possible state is a state in the count sequence. But this is not the case if the modulus is not a power of two. For example, suppose we have the modulo-6 counter of Fig. 8.26a, and it comes on in the 111 state. If we start pulsing the count input, will the counter go to the desired count sequence, or will it stick at 111 or go into a loop between 111 and 110? The only way to find out is to determine the values of the excitation functions for "illegal" counts.

In designing such counters, the best procedure is to first carry out the design as discussed here, assuming all the unused counts are truly don't-cares. Then, from the maps of the excitation functions, determine the values actually assigned to these

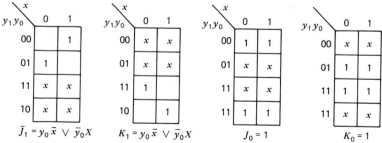

*Figure 8.28   K-maps for up–down counters.*

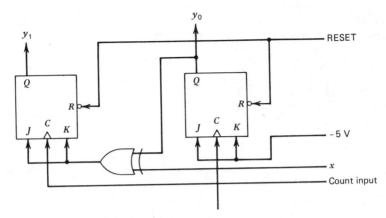

*Figure 8.29   Modulo-4 up–down counter.*

don't-cares. With this information and the excitation table, you can then determine what transitions will take place. If these transitions do not lead back to the main count sequence, you must then modify the excitation functions as necessary to obtain the desired transitions.

## Example 8.1

Obtain a realization of a modulo-3 counter that is self-starting.

Solution
The transition table of a modulo-3 counter is given in Fig. 8.30a. Suppose that, as the result of the don't-care assignments, the next state, entry for $y_1 y_0$ was 1,1, as shown in Fig. 8.30b. In this case, if the counter started in state $y_1 y_0 = 11$, it would remain forever in state 11. Thus it would not be self-starting. On the other hand, if the don't-cares were assigned to make the next-state entry for present state 11 either

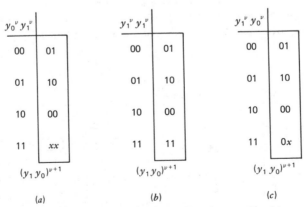

| $y_0^\nu y_1^\nu$ | |
|---|---|
| 00 | 01 |
| 01 | 10 |
| 10 | 00 |
| 11 | xx |
| $(y_1 y_0)^{\nu+1}$ | |

| $y_1^\nu y_1^\nu$ | |
|---|---|
| 00 | 01 |
| 01 | 10 |
| 10 | 00 |
| 11 | 11 |
| $(y_1 y_0)^{\nu+1}$ | |

| $y_1^\nu y_0^\nu$ | |
|---|---|
| 00 | 01 |
| 01 | 10 |
| 10 | 00 |
| 11 | 0x |
| $(y_1 y_0)^{\nu+1}$ | |

(a)

(b)

(c)

*Figure 8.30   Modulo-3 counter: (a) transition table, (b) not self-starting, and (c) always self-starting.*

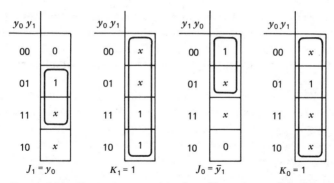

Figure 8.31    Excitation equations for self-starting modulo-3 counter.

00, 01, or 10, then the counter will be self-starting even if the initial state is 11 when power is applied. One simple way to assure that the next state in the last row will not be 11 is to make one of the next state variables 0, as is done in Fig. 8.30c. This might not lead to the absolutely optimal realization, but at least one of the state variables is a don't care, and the next state in the last row will be 00 or 01.

The excitation maps and expressions corresponding to Fig. 8.30c are given in Fig. 8.31. The don't-care in Fig. 8.30c translates to a don't care in the maps for both $J$ and $K$.                                                                          ■

## 8.8   MSI Registers and Counters

In Chapter 5 we saw that we could acheive economies in combinational design through the use of MSI parts containing many gates in place of separate gate chips. In a similar manner, we can put a number of flip-flops on a single MSI chip with resultant economies in sequential design. The simplest type of sequential MSI part

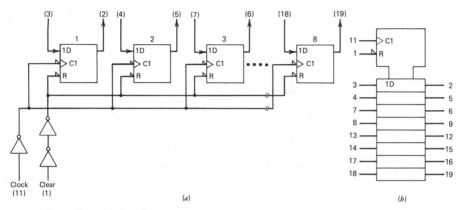

Figure 8.32    74273 8-bit register.

is the *register chip,* consisting of nothing more than a set of flip-flops intended to be used as a group to realize a register, rather than as independent flip-flops. Because the flip-flops are to be used as a group, there is no need for separate clock or reset lines for each. Also, it is usual to bring out only the true output of each flip-flop in order to reduce pin count.

Figure 8.32*a* shows the internal organization of the SN 74273 octal *D* flip-flop chip, and Fig. 8.32*b* shows the standard block diagram symbol for this chip. Note that this is simply a set of eight positive edge-triggered flip-flops with common clock and clear lines. The internal inversions of the clock and clear lines are provided so that these lines will appear as single TTL loads rather than eight loads. In the block diagram symbol, the clock input (C1) and the clear input (R) appear in the control block since they affect all flip-flops in the same way. Note that the 1D input is labeled only in the top block of eight. This is a convention of block diagram symbols; when there are identical blocks under control of a common control block, the inputs and outputs need be labeled only in the top block, although they may be labeled in all blocks if desired.

An important variant of the register chip is the chip with three-state output. Three-state outputs can be provided on any type of TTL circuit and provide a means of logically disconnecting the output from whatever it may be wired to. These three-state output switches function in the same way as in combinational logic parts, as discussed in Section 7.4. As before, in a block diagram symbol, a three-state output is indicated by an "upside-down triangle" next to the output line. An enable input will be labeled EN. In an ordinary TTL device without three-state output, an EN signal in the inactive state will hold all outputs at the inactive level, as was the case with the decoder of Fig. 7.2. In devices with three-state outputs, the outputs will be in the high-impedance state when the EN input is inactive. Figure 8.33*a* shows the block diagram symbol for the SN 74LS374 octal *D* flip-flop with output control. This chip is exactly equivalent to the SN 74LS273 except for the three-state outputs.

The EN signal does not affect the operation of the flip-flop in any way. Any

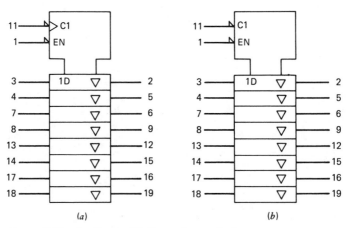

*(a)*                                 *(b)*

*Figure 8.33   Registers with three-stage output.*

time they are clocked, they will store what is at the $D$ inputs, regardless of the state of the EN input. Registers with three-state output find use in situations where it is desired to select the contents of one of several registers to be sent over a set of communication lines. With three-state outputs, this can be accomplished simply by connecting all the registers to the set of lines (often called a bus) and enabling the one with contents to be sent and disabling all the others.

Figure 8.33*b* shows another type of register chip, the 74LS373 octal $D$-type latch with output control. This device is the same as the 374 except that each flip-flop is a latch, exactly as shown in Fig. 8.5. This type of device is sometimes referred to as a *transparent latch,* because when the clock is high, the outputs simply follow the inputs, that is, the device is "transparent." When the clock goes low, the current values at the $D$ inputs are latched (stored). In the block diagram symbol, the only difference is the absence of the "arrowhead" that is used to indicate edge-triggering on a clock point.

Another important type of MSI sequential part is the *shift register.* There are about 15 different types of MSI shift registers, offering various combinations of serial and parallel at input and output, right shift or left shift, or both and various numbers of bits. One of the simplest of these devices is the 74LS164 8-bit parallel-out serial-in shift register. This device accepts input in serial at the left-most flip-flop and provides output in parallel from all eight flip-flops. Figure 8.34*a* shows the equivalent internal block diagram of this device, and Fig. 8.34*b* shows the block diagram symbol. In the block diagram symbol, SRG identifies a shift register, followed by 8 indicating the number of bits. It is seen from Fig. 8.34*a* that the serial input is formed by ANDing two inputs. This is indicated on the block diagram symbol by the box with the "&" symbol. The C1 and 1D inputs together indicate that a positive clock transition on C1 will load the AND of the two input signals into the first $D$ flip-flop. The "/→" symbol following C1 indicates that each positive edge of the clock will shift the contents of the register down (right). An arrow in the opposite direction

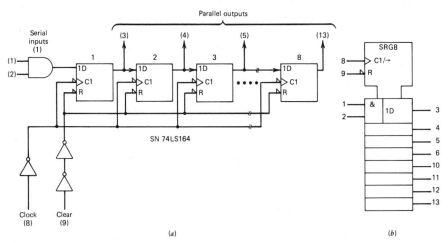

*Figure 8.34   Eight-bit shift register (74LS164).*

would indicate an upward (left) shift. Serial output would be indicated by having output only from the bottom block, indicating data bits are available only as they are shifted into the right-most flip-flop.

Figure 8.35a shows the simplified block diagram of the 74LS295B 4-bit shift register with both serial and parallel input. This register differs from the one previously discussed in that it has two modes of operation, shifting and loading. Note that the D inputs to the flip-flops are driven by AND–OR networks. When MODE = 0, the left AND gate of each pair is enabled, so that the left-most flip-flop receives the serial input and the register shifts right when clocked. When MODE = 1, the right AND gate of each pair is enabled and the register loads parallel data when clocked.

Figure 8.35b shows the block diagram symbol for this register, in which we introduce a new form of dependency, *mode dependency*. Recall that in the notation for multiplexers we used gate dependency, in which the symbol

$$G \frac{0}{1}$$

meant that the line so labeled was interpreted as a binary number taking on the values 0 or 1, enabling other input lines labeled 0 or 1. In mode dependency we have a similar interpretation. When the line labeled

$$M \frac{0}{1}$$

is at the 0 level, the chip is operating in mode 0; when the line is at the 1 level, the chip is operating in mode 1. Next, consider the label $C2/0\rightarrow$ on the clock line. We use C2 instead of C1 because 1 has already been used to identify a mode. The "$/0\rightarrow$" indicates that the chip will shift right when clocked if in mode 0. In the top flip-flop block, the notation 0,2D indicates that the serial input will be loaded on the C2 clock if in mode 0. In all four flip-flop blocks, the notation 1,2D indicates that the parallel input data will be loaded on the C2 clock if in mode 1.

Probably the most versatile of all shift registers is the SN 74194 universal shift register, the block diagram symbol for which is shown in Fig. 8.36. This chip has two mode inputs, allowing for four modes of operation. Mode 0 is a no-op mode, nothing happens if the chip is clocked while in mode 0. The notation $C4/1\rightarrow/2\leftarrow$ on the clock line indicates that the chip will shift right when clocked in mode 1, left when clocked in mode 2. The clock is labeled C4 because 0-to-3 are used to indicate modes. Each mode number is treated as if it were a separate input line. If $n$ of the actual physical input lines are assigned to specify the mode, the first $2^n$ input labels are assigned to the various modes. This enables mode numbers to be referred to in the same sense as control inputs.

The notation 1,4D on the top flip-flop indicates loading of the right serial input on C4 clock if in mode 1, and 2,4D on the bottom flip-flop indicates loading of the left serial input on C4 clock if in mode 2. The notation 3,4D indicates loading of the parallel input on C4 clock if in mode 3. Finally, the active-low $R$ input clears the register to 0000.

Counters make up the third important category of MSI sequential chips. MSI

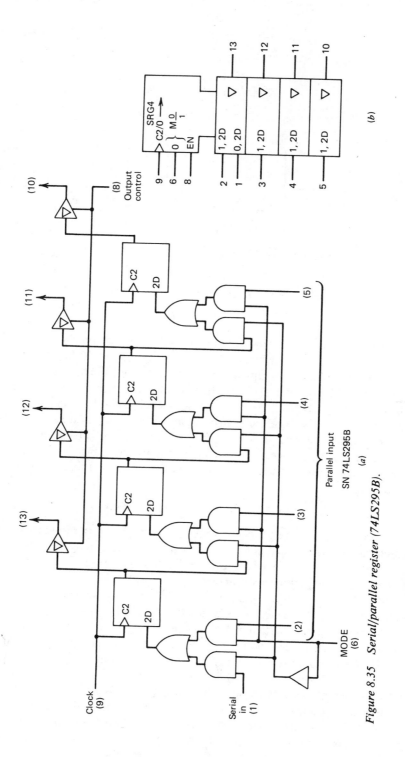

Figure 8.35  Serial/parallel register (74LS295B).

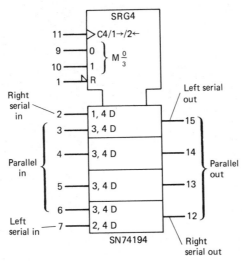

*Figure 8.36   Left/right shift register (74194).*

counters are available in two basic types, asynchronous and synchronous. In *asynchronous counters*, the count input (clock) is applied only to the first flip-flop, with changes in the output of this flip-flop triggering the second flip-flop, and so on. *Synchronous counters* are the type we studied in the last section, in which the clock is applied simultaneously to all flip-flops, and logic determines the order in which they change state. Given the same technology, asynchronous counters are generally faster than synchronous counters. However, they have a possible disadvantage in that the output bits do not all change at the same time but instead change one at a time. For example, when an asynchronous counter counts from 0111 to 1000, it will change in the sequence

0111      0110      0100      1000

so that incorrect counts are temporarily seen at the outputs. As a result, most designers prefer to use synchronous counters except in special cases where the speed of the asynchronous counter is needed and the intermediate transitions are no problem. We shall consider only synchronous counters.

Most MSI counters are 4-bit counters. All count in the binary sequence but are classified as binary (modulo-16) or decode (modulo-10).

Before considering particular MSI counters, let us introduce the basic elements of counter notation by considering the simple modulo-4 up–down counter of Fig. 8.29. The block diagram symbol for this counter is shown in Fig. 8.37. The indentifying label CTR2 specifies this block as a 2-bit binary counter (modulo-4). This counter has two modes of operation, count up (mode 0) and count down (mode 1). Since the clock (count) input does not control any inputs to the counter, it is not necessary to label it with a symbol such as C1. Instead, the notation $0+/1-$ indicates that a positive transition on that input will cause a count up if in mode 0, a count down if in mode 1. The active-low $R$ line clears the counter to 00. The individual flip-flop blocks are labeled with their numeric weight in the count sequence.

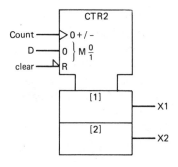

Figure 8.37 *Block notation for modulo-4 counter.*

Figure 8.38 shows the block diagram symbol for the SN 74LS163A binary counter with synchronous load and clear. The label CTR4 identifies it as a 4-bit binary counter (modulo-16). It has two modes of operation, load (mode 0) and count (mode 1). There are also two control inputs, G2 and G3. These could be considered additional mode control lines, but it is simpler to consider them separately. In complex parts with several control inputs, it is often a matter of choice as to whether to consider these inputs as M lines or G lines. The notation C4/1,2,3+ indicates that the counter will count up on positive transitions of C4 if the counter is in mode 1 and G2 and G3 are both active. All three conditions must be satisfied for counting to occur; just putting the chip in the count mode, or activating G2 or G3, is not sufficient. The 0,4D notation on the flip-flop indicates that a new count will be loaded on C4 in mode 0. Here we see the reason for considering the G2 and G3 inputs separately from the mode input; in mode 0, the G2 and G3 inputs have no effect. The upper output is the ripple-carry output, used to cascade these 4-bit counters to form larger counters. The notation 3CT = 15 indicates that this output will be active

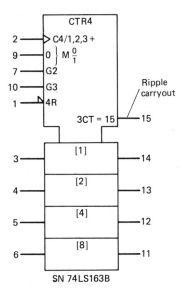

Figure 8.38 *Parallel load counter (74LS163B).*

when the count equals 15, provided G3 is active. Finally, the notation 4R indicates a clocked reset; the counter will clear to 0000 on C4 if this line is low (active).

Many other types of counters are available in MSI form, with a variety of features. Some are even more complex than the one in Fig. 8.38 in terms of their control functions, and a full consideration of them is not necessary for our purposes. One important note, counters with moduli that are not powers of two will be labeled CTRDIV*n*, where *n* is the modulus, for example, CTRDIV10 for a decade counter. This notation derives from the fact that counters are often used as frequency dividers. For example, the ripple-carry output in Fig. 8.38 will provide a pulse out for every 16 count pulses; that is, it divides the frequency by 16.

The parallel load feature on shift registers and counters can be used in a variety of ways to expand the versatility of these chips. For example, the counter of Fig. 8.38 can be converted to a shift register by using it in the load mode, with the outputs connected to the inputs, shifted one position left or right. The load inputs can also be used to change the modulus of a counter.

### Example 8.2

Connect a 74LS163B as a modulo-11 counter.

### Solution

The necessary connections are shown in Fig. 8.39. We connect the inverted ripple-carryout to the mode input. Except when the count is 15, the counter will thus be in mode 1, the count mode. When the count is 15, the ripple-carry goes to 1, driving the counter to mode 0, and the clock pulse then loads the counter to 5 (0101). The counter thus continually counts from 5 to 15 and back to 5, providing a modulo-11

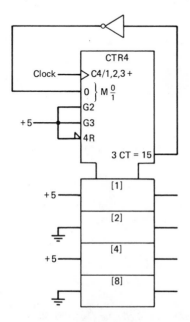

*Figure 8.39  Modulo-11 counter.*

counter. It is not the "usual" 11 counter, in the sense of counting from 0 to 10 and back, but it is nonetheless a modulo-11 counter. ■

## Problems

**8.1**  Refer to the complete transition table of the edge-triggered $D$ flip-flop of Fig. 8.8 to determine the value of flip-flop output, $Q$, at time $T_1$ in the timing diagram of Fig. P8.1.

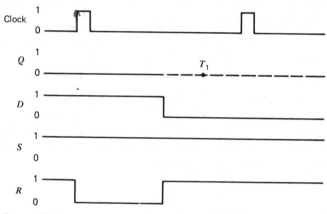

*Figure P8.1*

**8.2**  In the circuit of Fig. 8.10, $R_3 = R_2 = R_1 = R_0 = 1$ during clock period $n$. If $x$ is held constant at 0, what will be the values of $R_3$, $R_2$, $R_1$, and $R_0$ during clock period $n + 2$?

**8.3**  Analyze the network of Fig. P8.3 to determine a transition table and an output table for the network.

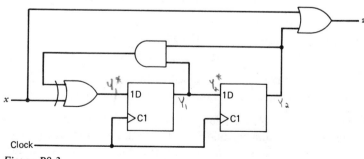

*Figure P8.3*

**8.4** A certain sequential circuit has two inputs, $a$ and $b$, and a single output, $z$, and will require only one flip-flop to implement. The flip-flop, $y$, is to store the value of $a \vee b$ that existed at the time of the most recent clock pulse. The output is to be $ay \vee b\overline{y}$. Obtain a transition table and an output table for this circuit.

**8.5** Obtain a logic-block diagram of the $D$ flip-flop realization of the transition table and output table obtained in Problem 8.4.

**8.6** Obtain the logic-block diagram of a $J$-$K$ flip-flop realization of the transition table and output table obtained in Problem 8.3.

**8.7** Obtain the logic-block diagram of a $J$-$K$ flip-flop realization of the transition table and output table given in Problem 8.4.

**8.8** Obtain the logic-block diagram of a $D$ flip-flop realization of the modulo-4 counter with the transition table given in Fig. 8.23. Compare the $D$ flip-flop with the $J$-$K$ flip-flop realization. Which is most economical?

**8.9** A $J$-$K$ flip-flop realization is to be generated for the sequence generator with the transition table given in Fig. P8.9. Note that one sequence will be generated if input $x = 0$, another if input $x = 1$. The circuit outputs will be the same as the flip-flop outputs. Determine sum-of-products expressions for the $J$ and $K$ inputs of each flip-flop.

| $y_1^v$ | $y_0^v$ | $x = 0$ | $x = 1$ |
|---------|---------|---------|---------|
| 0 | 0 | 1 1 | 1 0 |
| 0 | 1 | 0 0 | 1 1 |
| 1 | 0 | 0 1 | 0 1 |
| 1 | 1 | 1 0 | 0 0 |
| | | $y_1^{v+1}, y_0^{v+1}$ | |

*Figure P8.9*

**8.10** Design a modulo-6 counter to count in the following sequence. Use $J$-$K$ flip-flops.

$$000 \rightarrow 010 \rightarrow 001 \rightarrow 101 \rightarrow 110 \rightarrow 011$$

Show the transition table, the maps of the excitation functions, and the final circuit. You need not concern yourself with reset.

**8.11** Determine if the counter designed in Problem 8.10 is self-starting. If it is not, modify the design as required to make it self-starting.

**8.12** Design a variable-modulus binary counter. If control input $M = 0$, the counter is to count modulo-5 in binary. If control input $M = 1$, the counter is to count modulo-7 in binary. Use $J$-$K$ flip-flops. Show the transition table, the maps of the excitation functions, and the final circuit. Do not worry about reset.

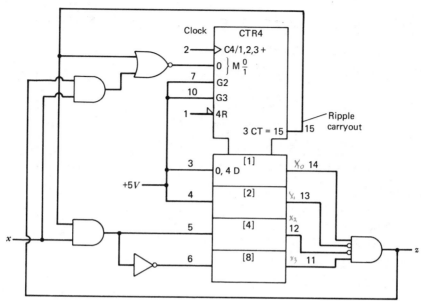

*Figure P8.13*

**8.13** Construct a complete 16-row transition table for the counter shown in block notation in Fig. P8.13. The circuit has only one input, $x$, in addition to the clock and one output, $z$. The counter chip is a 74LS163B. Let the state variables by $y_8$, $y_4$, $y_2$, $y_1$ with $y_8$ representing the most significant bit.

**8.14** Using the standard block diagram symbols, show the connections for two 74194 shift register chips to function as an 8-bit left-shifting register with serial input, parallel output.

**8.15** Show the connections of two 74LS163B counter chips to function as an 8-bit binary counter with reset.

**8.16** Show the connections for two 74LS163B counter chips to function as a divide-by-24 frequency divider. The output need be 1 for only one period out of every 24 clock periods.

**8.17** Show how appropriate connections to the load inputs can be used to cause the SN 74LS295B to function as a left-shifting register.

**8.18** Show how two 74LS163B counter chips can be connected to function either as an 8-bit counter with reset or as an 8-bit left-shifting register, controlled by a COUNT/SHIFT control line. Some external gating logic may be required.

# References

1. Barna, A., and D. Porat, *Integrated Circuits in Digital Electronics,* Wiley-Interscience, New York, 1973.

2. Blakeslee, T. R., *Digital Design with Standard MSI and LSI,* 2nd ed. Wiley-Interscience, New York, 1979.
3. Givone, D. D., *Introduction to Switching Circuit Theory.* McGraw-Hill, New York, 1970.
4. Hill, F. J., and G. R. Peterson, *Introduction to Switching Theory and Logical Design,* 3rd ed. Wiley, New York, 1981.
5. Johnson, D. E., J. L. Hilburn, and P. M. Julich, *Digital Circuits and Microcomputers.* Prentice-Hall, Englewood Cliffs, N.J., 1979.
6. Nagle, H. T., Jr., B. D. Carroll, and J. D. Irwin, *An Introduction to Computer Logic.* Prentice-Hall, Englewood Cliffs, N.J., 1975.
7. Peatman, J. B., *Digital Hardware Design,* McGraw-Hill, New York, 1980.
8. Roth, C. H., Jr., *Fundamentals of Logic Design,* 2nd ed. West Publishing, St. Paul, Minn., 1979.
9. Sandige, R. S., *Digital Concepts Using Standard Integrated Circuits.* McGraw-Hill, New York, 1978.
10. Sloan, M. E., *Computer Hardware and Organization,* 2nd ed. SRA Publishers, Chicago, 1983.

# Synthesis of State Machines

## 9.1  A Language Is Needed

In Chapter 8 we learned to design counters and more general sequential circuits with only one flip-flop. The reader is surely aware that digital designers will be confronted by more difficult problems than these. How should they proceed?

Recall from the design of the start-signal generator in Section 8.5 that merely communicating specifications for the design in English was a major problem. Despite our best efforts, you may not have been completely certain what was to be designed until after the next value table or transition table was finally derived. It would be surprising if this language problem were not compounded as the complexity of the design task increased. Quite naturally, then, we look to other language forms to bridge the gap between English and the transition table.

Historically, the earliest computers were developed with the availability of very few language tools, probably little more than English and Boolean algebra. For this reason, as well as the primitive state of available hardware, the development cycle for these machines was long and replete with error and redesign. When these machines were first used, the languages available for their programming were equally limited. The great potential of these machines was recognized, nonetheless, and much thought was devoted to ways to ease the burden of their programming. *Flow charts* came into use quickly and, through the years, a number of high-level programming languages appeared. It was eventually realized that variations of these same tools could also be used in the process of designing digital systems.

The flow chart has been an informal tool for digital design for many years.

More recently it has been formalized as the ASM or *algorithmic state machine* chart, see Refs. [4, 8]. The use of high-level languages as a means of describing hardware is also a relatively recent innovation, see Refs. [1–3]. We shall adopt both these techniques in this book. In this chapter we shall use the ASM chart to describe more complex sequential circuits than we were able to deal with in Chapter 8. In Chapters 10 and 11 we shall open our field of view to still more complex systems, including a computer central processor. For this description task we shall turn to the use of a *hardware description language*. The structure of hardware description languages has much in common with that of high-level programming languages. But the syntax is usually adjusted somewhat to reflect hardware structure. The flowchart is often used in the process of writing a program in a high-level language. Similarly, the ASM chart will prove helpful in deriving the representation of a digital system in a high-level language. It will also continue to serve as a formal specification for a realization of the control portion of the more complicated digital system.

## 9.2   Standard Symbols for the ASM Chart

An ASM chart is a flowchart that describes the behavior of a sequential circuit. A precise set of rules has been formulated to govern the construction of ASM charts. Each individual state of the sequential circuit is represented by an ASM block. An *ASM block* is a small configuration of symbols that represent the name of the state, the flip-flop values for that state, the circuit outputs as a function of the state, any inputs that can occur while the circuit is in this state, and the next states as they depend on circuit inputs. When the ASM blocks representing all states of a sequential circuit are connected together, as a single network, the resulting ASM chart is a behavioral description of that circuit.

Only three symbols may appear in an ASM block. Always included is the rectangular *state box* illustrated in Fig. 9.1. Within this box are listed the circuit outputs that can occur whenever the circuit is in the corresponding state regardless of input values. At the left of this box is the name of the state. Immediately above the box is a list of the flip-flop values that define that state.

Two additional symbols may appear within a block depending on the inputs

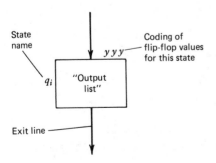

*Figure 9.1   ASM state box.*

that may occur while the circuit is in the corresponding state. Inputs that will always be inactive while the circuit is in a particular state are completely ignored in the corresponding state box. This feature is particularly convenient if the circuit has many inputs, only a few of which can be active at any given time. Each input wire that can possibly be active when the circuit is in a particular state must be represented in a *decision diamond* associated with that state, as illustrated in Fig. 9.2a. Sometimes Boolean expressions of input wires may be entered in a decision diamond. Each value that may be assumed by an input wire or expression in a diamond is associated with an exit line from that diamond. These lines lead to the blocks corresponding to the next states of the circuit following the next clock pulse. If output values are also functions of these inputs, these exit lines pass through conditional output boxes of the form illustrated in Fig. 9.2b. The outputs that occur when the path to a conditional output box is satisfied are listed within the box. Outputs that are not listed in either the state box or a conditional output box in a particular ASM block are always inactive when the circuit is in that state.

An example of an ASM block is given in Fig. 9.3. The sequential circuit that this state partly represents includes three flip-flops, $y_2, y_1, y_0$. In this state, which is labeled $q_1$, these flip-flops assume the values 0,1,0. While the circuit is in this state, the circuit output, $Z_1$, will always be 1. If $X_1 = 0$, the circuit output $Z_2$ will also be 1, and the next state will be $q_2$. If $X_1 = 0$, $Z_2$ will be 0 and the next state will depend on the input, $X_2$. For $X_1 X_2 = 1,0$, the next state will be $q_3$. For $X_1 X_2 = 1,1$, the next state will be $q_4$.

The ASM chart is primarily a tool in terms of which complex designs can be initially formulated prior to the development of a transition table that will lead to a hardware realization. For an example to illustrate a first complete ASM chart we turn, however, to a sequential circuit that has already been realized. This example, the start-signal generator of Section 8.5, conveniently has only two states.

**Example 9.1**

For convenience, the transition and output tables of the start-signal generator of Fig. 8.15 are given below as Figs. 9.4a and b. We have replaced the don't-care entries in the column under input combination 11 with the values that will allow the next state

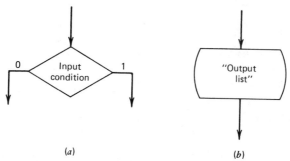

(a)                                                              (b)

*Figure 9.2   Decision and conditional output symbols.*

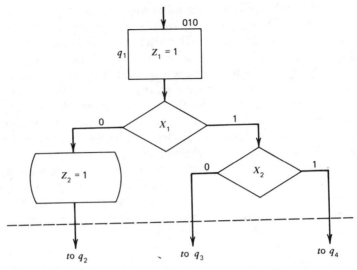

Figure 9.3   ASM block.

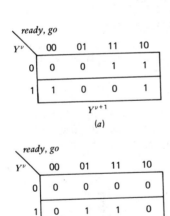

ready, go

| $Y^v$ | 00 | 01 | 11 | 10 |
|---|---|---|---|---|
| 0 | 0 | 0 | 1 | 1 |
| 1 | 1 | 0 | 0 | 1 |

$Y^{v+1}$

(a)

ready, go

| $Y^v$ | 00 | 01 | 11 | 10 |
|---|---|---|---|---|
| 0 | 0 | 0 | 0 | 0 |
| 1 | 0 | 1 | 1 | 0 |

$z$

(b)

ready, go

| Present state | 00 | 01 | 11 | 10 |
|---|---|---|---|---|
| $Q0$ | $Q1$ | $Q0$ | $Q1$ | $Q1$ |
| $Q1$ | $Q1$ | $Q0$ | $Q0$ | $Q1$ |

Next state

(c)

Figure 9.4   Specification of start-signal generator.

always to be a function of a single input wire. We may, of course, assign don't-cares to be any values we wish. Each state in the ASM chart must have a name, so we designate the state for which $Y = 0$ as $Q0$ and the state for which $Y = 1$ as $Q1$. We then replace the values of $Y^{\nu+1}$ in Fig. 9.4$a$ with $Q0$ and $Q1$ to form Fig. 9.4$c$. This convenient table is actually a state table, but more on that later.

In Fig. 9.5 we see a partial ASM chart for the start-signal generator that includes only state $Q0$. From the output table in Fig. 9.4 we see that the only output, $z$, is always 0 for this state. Therefore, the box in Fig. 9.5$a$ is empty. Figure 9.4$c$ tells us that from $Q0$ the next state is $Q1$, if $ready$ is 1. Otherwise the circuit will remain in $Q0$. This is reflected in the decision diamond in Fig. 9.5$a$.

Now we add the state $Q1$ to form the complete ASM chart of Fig. 9.5$b$. Again, the box for $Q1$ is empty, this time since the value of $z$ depends on the input go. Notice from Figs. 9.4$b$ and $c$ that for $Q1$, $z = 1$, and the next state is $Q0$. This is reflected

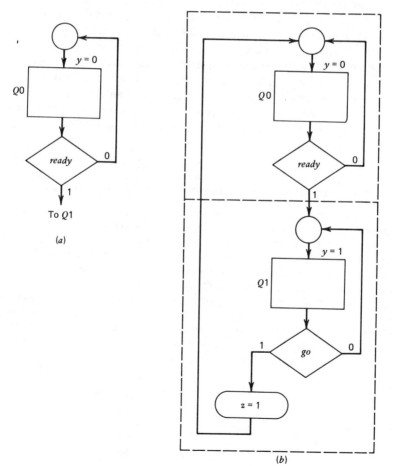

*Figure 9.5   ASM chart for start-signal generator.*

in the ASM chart by the arrow from the *go* decision diamond through the conditional output block, $z = 1$, to $Q0$. If $go = 0$, the circuit remains in $Q1$. ∎

In the next section we shall present an example that will illustrate the use of the ASM chart as a tool for the initial formulation of a design specification for a sequential circuit. In the process we shall extend our overall experience in sequential circuit synthesis.

# 9.3   Vending-Machine Control

We are now ready to consider an example in which we begin with only an English language statement of generally what the system to be designed must do. For this we choose a vintage vending-machine controller.

A vending machine is an example of a product for which the cost must necessarily increase at a rate faster than the overall rate of inflation. As with other manufactured products, component and labor costs increase in proportion to the overall rate of inflation. An additional complication is the fact that inflation shortens useful lifetime and increases the complexity of successive models. Newer models must keep track of larger amounts of money and, therefore, more combinations of coins than older models. For an example of the appropriate complexity, we select a vending machine model that will, regrettably, never be seen again. It should be clear to the reader that the approach will be the same for versions that must accept greater numbers of coins.

---

### Statement of Vending-Machine Controller Problem

A controller that keeps track of the value of the coins accepted by a vending machine is to be designed. The actual coin sensors are sampled once each clock period to generate two inputs, $I_1$ and $I_2$, to the controller. A third input line, **R**, will be on for one clock period each time the coin return button is pushed. This line, **R**, will also activate the coin return mechanism, so the circuit we are designing need produce no output for that situation. One output line, **C**, of the controller is to be 1 for one clock period following the deposit of a quarter to activate a change-return mechanism. Change will be made only if a quarter is deposited. The price of all items is $0.20. A second output, **A**, which enables the merchandise selection mechanism, is to be 1 for one clock period whenever the value of coins received has reached this amount. During any clock period, the values of $I_1$ and $I_2$ may be interpreted as follows.

| $I_1$ | $I_2$ | Received coin |
|-------|-------|---------------|
| 0     | 0     | none          |
| 0     | 1     | quarter       |
| 1     | 0     | nickel        |
| 1     | 1     | dime          |

Each coin will cause nonzero inputs for only one clock period.

We first determine that there must be a quiescent state in which the controller will rest when no coins have been received since completion of the last transaction. We refer to this as an *initial*, or *reset*, state and label it $Q0$. When in this state any of the input combinations in the table may appear. The possible next states to which the controller will go in response to each of these inputs are shown in the partly completed ASM chart of Fig. 9.6. We find it convenient to label each state with the cumulative value of coins received so far when the machine is in that state. Thus we are able to anticipate all the states of the controller in advance. In many design problems the designers are not so fortunate. They may gradually add states as they work their way through the problems, ending uncertainty regarding the number of states in the designs only when they are completed.

From the $Q25$ state two outputs are issued. $C$ causes a nickel in change to be returned to the customer while $A$ enables the dispensing of a product. The next state following $Q25$ is always $Q0$, since no inputs are expected during the single clock period while the machine is in $Q25$. The boxes are blank corresponding to states $Q5$ and $Q10$, since all outputs are 0 when the controller is in these states.

As we complete the flowchart as given in Fig. 9.7, you will notice that we are assuming that the customer will never deposit a quarter after having first deposited a dime or a nickel. You are free to question the wisdom of this assumption on the grounds that the cost of physical damage to machines in the field resulting therefrom may far exceed anything saved by simplifying the design. Our purpose here is to provide an understandable first example of the synthesis of an ASM chart. We therefore make the assumption to avoid an even more complicated ASM chart than the one given in Fig. 9.7.

When in states $Q5$ or $Q10$, the input may be a return or may indicate a nickel

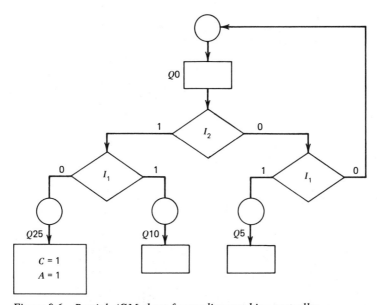

*Figure 9.6  Partial ASM chart for vending-machine controller.*

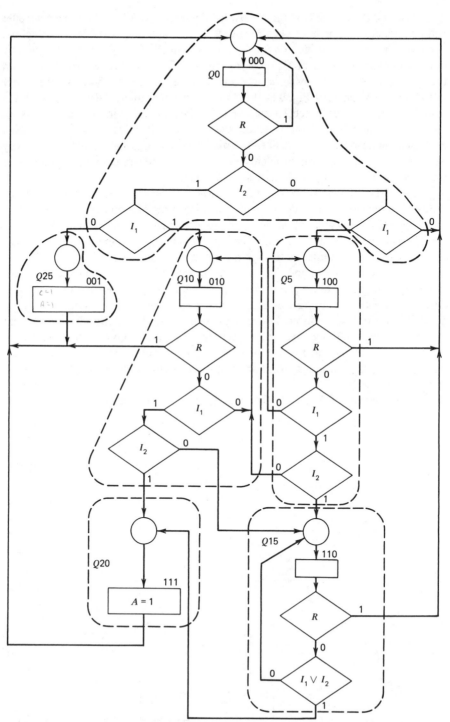

*Figure 9.7   ASM chart for vending-machine controller.*

or dime but not a quarter. Therefore, $I_2$ will not be tested if $I_1$ is 0, in which case the circuit remains in the current state. If an $R$ occurs at anytime, the machine will return to the initial state, $Q0$. We even allow for the possibility that the coin-return lever may be pulled when the machine is in state $Q0$, an option not provided for in Fig. 9.6. Indeed, return levers may occasionally be operated by noncustomers of the machine in the hope of recovering unclaimed coins. It may be assumed that $I_1 = I_2 = 0$, whenever the return button is pushed. From $Q5$ a nickel takes the circuit to $Q10$ and a dime to $Q15$. From $Q10$ a nickel takes the controller to $Q15$ and a dime takes it to $Q20$. From $Q15$ the next state is $Q20$, if any coin is received. (If an individual with two dimes puts in a nickel first, he is out of luck.) The output is $A = 1$ for $Q20$, and the next state is always $Q0$.

We observe from Fig. 9.7 that there are in all six possible states for the vending-machine controller. Each of these states must correspond to some unique combination of flip-flop values. Sometimes, but not always, we will want to use as few flip-flops as possible. If three flip-flops are used, there are eight combinations of flip-flop values, six of which could be assigned to the six states of the controller. Clearly two flip-flops with four combinations of values would be insufficient. In general the minimum number of flip-flops, $r$, necessary for a state assignment must satisfy Eq. 9.1, where $n$ is the number of states in the ASM chart.

$$2^{r-1} < n \le 2^r \tag{9.1}$$

The particular way chosen to assign circuit states to combinations of flip-flop values will influence the complexity of the eventual circuit realization. No satisfactory method exists that will always lead to a best state assignment, see Refs. [6, 7]. In this book we shall rely on information about the structure of the circuit that may present itself during the development of the ASM chart to suggest a reasonable state assignment. In the case of the vending-machine problem, there is relatively little information available on which to base a state assignment. Notice that the assignment given in Fig. 9.8 does have the advantage that relatively few of the flip-flop values will change for each state change. For example, only one state variable changes during each of the transitions $Q0 \rightarrow Q5$, $Q0 \rightarrow Q10$, $Q0 \rightarrow Q25$, $Q5 \rightarrow Q15$, $Q10 \rightarrow Q15$, and $Q15 \rightarrow Q20$. This can be seen in Fig. 9.7 where the assignment of flip-flop values is also recorded. A small number of flip-flop state changes may result in an economical realization, particularly if $J$-$K$ flip-flops are used.

| State | $y_1$ | $y_2$ | $y_3$ |
|-------|-------|-------|-------|
| $Q0$ | 0 | 0 | 0 |
| $Q25$ | 0 | 0 | 1 |
| $Q10$ | 0 | 1 | 0 |
| $Q5$ | 1 | 0 | 0 |
| $Q15$ | 1 | 1 | 0 |
| $Q20$ | 1 | 1 | 1 |

*Figure 9.8  Assignment of flip-flop states for controller.*

# 9.4    From ASM Charts to Transition Tables

Two different forms of realization can be obtained for sequential circuits described by ASM charts. One approach will realize the input logic to the $D$ flip-flops representing the state variables as well as the input logic using an ROM. The second form will feature a minimal realization of the logic functions implemented by networks of individual gates or by PLAs. With this approach, either $D$ or $J$-$K$ flip-flops may be used.

In ROM realizations don't-care input/present state combinations are of no advantage. It is, therefore, possible to translate the ASM chart directly to a tabular representation of the ROM that will implement the combinational logic. We shall return to this approach in the following section.

For minimal logic realizations, the process is more complicated and incorporates the realizing of sequential circuits from transition tables, as discussed in the last chapter. In all, the realization process will consist of three steps.

1. Translate the ASM chart to a state table.
2. Convert the state table to a transition table.
3. Develop Boolean expressions for circuit outputs and memory element inputs.

As we observed in Chapter 8, the final step will vary somewhat depending on the type of memory elements used.

As a first example of translation of an ASM chart to a state table and a transition table, let us return to the vending-machine controller of Fig. 9.7. Notice that not all input lines are necessarily referenced in the various decision diamonds associated with each particular state in the ASM chart of Fig. 9.7. This feature is one of the advantages of the ASM charts as a means of representation of sequential circuits. In a complex system, only a few of a large number of input lines may have meaning at a given time. The disadvantage of this form of notation is that it does not specifically say that combinations of the values of state variables not referred to at a specific point in the ASM chart can actually occur at that point. Consider, for example, the case where the vending machine is in state $Q10$ and inputs $R$ and $I_1$ are both 0. What values can occur for $I_2$? The ASM chart does not tell us. In this case, $I_2$ must be 0, since we have assumed that a quarter will not be inserted into the machine after a dime. Thus for state $Q10$ input combination $R\,I_1\,I_2 = 001$ is a don't-care. This fact will be recorded in the state table.

In laying out a state table we must allow for every combination of input values that might ever occur. In Fig. 9.9 the five possible combinations of input values that can occur in the vending-machine controller are listed at the top of the five columns in the table. The last two columns are headed by the two circuit outputs. Notice that we assume that the "return" signal can occur when the other two inputs are 0 or when no coin is being deposited.

Each of the states of the ASM chart is entered at the right of a row in the table of Fig. 9.9. Thus there is a square in the table for each combination of present state and inputs. We have already indicated that some of the input/present state combinations in the five columns of possible inputs will never occur. It is necessary

| $R, I_1 \cdot I_2$ Present state | 000 | 001 | 011 | 010 | 100 | Outputs $A$ | $C$ |
|---|---|---|---|---|---|---|---|
| $Q0$ | $Q0$ | $Q25$ | $Q10$ | $Q5$ | $Q0$ | 0 | 0 |
| $Q25$ | $Q0$ | X | X | X | $Q0$ | 1 | 1 |
| $Q10$ | $Q10$ | X | $Q20$ | $Q15$ | $Q0$ | 0 | 0 |
| $Q5$ | $Q5$ | X | $Q15$ | $Q10$ | $Q0$ | 0 | 0 |
| $Q15$ | $Q15$ | X | $Q20$ | $Q20$ | $Q0$ | 0 | 0 |
| $Q20$ | $Q0$ | X | X | X | $Q0$ | 1 | 0 |

Next states

*Figure 9.9   State table.*

to refer back to the original problem statement to find this information, since once again the ASM chart will not tell us. For example, a quarter will never be inserted in the machine if some other coin has previously been deposited. For a don't-care combination, we enter an $X$ in the table. Thus, the last five entries in the column headed by 001 are $X$'s. With the aid of the discussion in the previous section, you should be able to verify that the remaining $X$ entries in Fig. 9.9 also correspond to combinations of values that will not occur.

For each input/present state combination that can occur, the ASM chart identifies the next state that the circuit will assume following a clock pulse. These next states may be determined by following the arrows from the present state through the decision diamonds, as specified by the particular input values. Each of the next states, entered in a square in Fig. 9.9 corresponding to an input/present state combination, was determined from Fig. 9.7. A table of the form just developed is commonly refered to as a *state table*.

We recall that the values of the state variables for each state were already assigned in the ASM chart of Fig. 9.7. Replacing both the present state and next state entries in the state table of Fig. 9.9 with the corresponding combinations of state variable values, as given in the ASM chart, results in the transition table of Fig. 9.10. This table contains two additional rows representing the two combinations of state variable values that are not assigned to any state. The next state and output entries in these rows are left blank but may be interpreted as don't-cares.

We saw no cases of output values being conditioned on inputs in the vending-machine controller example. We were, therefore, able to display the outputs in two separate columns with the output entries corresponding to particular states listed on the respective rows in these two columns. A somewhat different form of state table will be required if the output is a function of the input as well as the present state.

| $y_1^\nu\, y_2^\nu\, y_3^\nu$ | $R, I_1, I_2$ 000 | 001 | 011 | 010 | 100 | 101 | 111 | 110 | Outputs $A$ | $C$ |
|---|---|---|---|---|---|---|---|---|---|---|
| $Q0$   000 | 000 | 001 | 010 | 100 | 000 | X | X | X | 0 | 0 |
| $Q25$   001 | 000 | X | X | X | 000 | X | X | X | 1 | 1 |
|     011 | | | | | | | | | | |
| $Q10$   010 | 010 | X | 111 | 110 | 000 | X | X | X | 0 | 0 |
| $Q5$   100 | 100 | X | 110 | 010 | 000 | X | X | X | 0 | 0 |
|     101 | | | | | | | | | | |
| $Q20$   111 | 000 | X | X | X | 000 | X | X | X | 1 | 0 |
| $Q15$   110 | 110 | X | 111 | 111 | 000 | X | X | X | 0 | 0 |

$$y_1^{\nu+1}\, y_2^{\nu+1}\, y_3^{\nu+1}$$

*Figure 9.10   Transition table for vending-machine controller.*

In example 9.2 we shall be asked to generate one of four output waveforms depending on the values present on two input lines. In order to express this in ASM chart form it will be necessary to use the conditional output notation defined in Fig. 9.2*b*.

**Example 9.2**

Develop an ASM chart and a transition table for a controllable waveform generator that will output any one of the four waveforms given in Fig. 9.11, as determined by the values of its two inputs, $x_1$ and $x_2$. The period of the first two waveforms is four

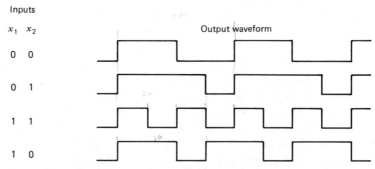

*Figure 9.11   Specification of controllable waveform generator.*

clock cycles, the period of the third is two, and the period of the final waveform is three clock periods. Changes in input values will be infrequent. When an input change does occur, the new waveform may begin at any point in its period.

Solution

Our approach to the problem will be to lay out an ASM chart with four states, one for each clock cycle of the waveforms with the longest period. For each state the output will be conditional on the values of the input lines in effect at that time. In addition, branching must be provided so that the last state may be skipped in the case of the fourth waveform. It is easiest to treat the third waveform as if the period were four with the third and fourth clock cycles identical to the first and second.

The ASM chart that describes a realization of the waveform generator is given in Fig. 9.12. We notice that all three of the waveforms are logical 1 for the first clock

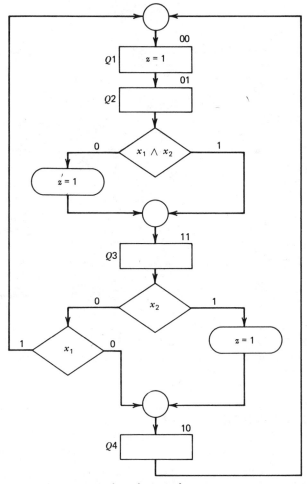

Figure 9.12   ASM chart for waveform generator.

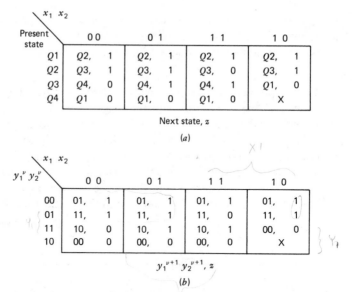

Figure 9.13   State and transition tables for waveform generator.

cycle. Therefore, the output, $z = 1$, is listed in the state box for the first state, $Q1$. During the second cycle the output is 1 except for the waveform corresponding to inputs $x_1 x_2 = 11$. Thus we place the expression $x_1 \wedge x_2$ in the decision diamond in the $Q2$ ASM block. The line corresponding to this expression being 0 leads to a conditional output box, containing $z = 1$, and then to state $Q3$. If the expression is 1, there is no output. During the third clock cycle the output will be 1, if $x_2 = 1$. This is specified with a conditional output box in the $Q3$ ASM block. For $x_1 x_2 = 10$, the fourth state is not used and the line leads back to $Q1$. Otherwise the circuit goes to $Q4$ for which the output is always 0. From $Q4$ the circuit returns to $Q1$ so that the waveform may be repeated.

We next express the ASM chart as a state table in Fig. 9.13$a$. In this case the outputs are a function of the inputs as well as the present states, so we list the outputs along with the next states in the columns corresponding to each combination of input values.

The ASM chart indicates that the values of the state variables $y_1 y_2$ assigned to $Q1$, $Q2$, $Q3$, and $Q4$ are 00, 01, 11, and 10, respectively. We simply substitute the state-variable values wherever we see the respective states in the state table to form the transition table of Fig. 9.13$b$. ∎

## 9.5   Circuit Realization

Once a transition table has been obtained, the process for developing the circuit realization may proceed in the same manner as for the simpler examples of Chapter 8.

The transition table for the waveform generator from Fig. 9.13 may be translated directly to the three separate Karnaugh maps for the next values of the state variables and the output as given in Fig. 9.14.

Using $D$ flip-flops in our realization allows us to use these maps directly to obtain the Boolean expressions for $z$ and for the $D$ inputs of the flip-flops as given by Eqs. 9.2, 9.3, and 9.4. We leave the more complicated $J$-$K$ flip-flop realization as a problem for the reader.

$$z = \bar{y}_1\bar{y}_2 \vee \bar{y}_1\bar{x}_1 \vee \bar{y}_1\bar{x}_2 \vee y_1y_2x_2 \tag{9.2}$$

$$y_1^{v+1} = y_2(\bar{y}_1 \vee \bar{x}_1 \vee x_2) \tag{9.3}$$

$$y_2^{v-1} = \bar{y}_1 \tag{9.4}$$

In Chapter 7 we saw that the ROM (read-only memory) provides an alternative means of realizing combinational logic. It permits the designer to bypass the task of obtaining minimal Boolean expressions, which we have just accomplished, since the ROM is merely an ORing of all the minterms as might be listed in a truth table. Once a transition table has been obtained, the process of translating this information to a tabulation of the words in a ROM realization is straightforward. It is only necessary to consolidate the columns of the transition table into truth-table form. The input lines and state variables become the address bits. A ROM realization of the waveform generator would appear as depicted in Fig. 9.15a with the decoder and ROM in the same IC package. A tabulation of the ROM words for each combination of address bits is given in Fig. 9.15b. This is, of course, nothing more than a truth table for the three functions to be realized and a rearrangement of the information given in the transition table of Fig. 9.13. The single don't-care entry in the transition table has been filled in with 0's in Fig. 9.15b.

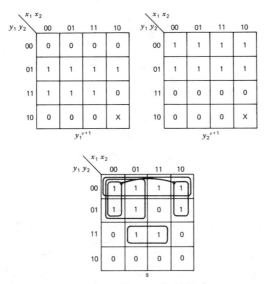

*Figure 9.14  Karnaugh maps for waveform generator.*

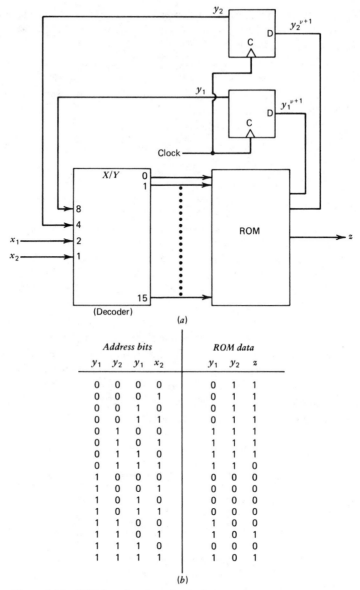

Figure 9.15  ROM realization of waveform generator.

| Address bits | | | | ROM data | | |
|---|---|---|---|---|---|---|
| $y_1$ | $y_2$ | $y_1$ | $x_2$ | $y_1$ | $y_2$ | $z$ |
| 0 | 0 | 0 | 0 | 0 | 1 | 1 |
| 0 | 0 | 0 | 1 | 0 | 1 | 1 |
| 0 | 0 | 1 | 0 | 0 | 1 | 1 |
| 0 | 0 | 1 | 1 | 0 | 1 | 1 |
| 0 | 1 | 0 | 0 | 1 | 1 | 1 |
| 0 | 1 | 0 | 1 | 1 | 1 | 1 |
| 0 | 1 | 1 | 0 | 1 | 1 | 1 |
| 0 | 1 | 1 | 1 | 1 | 1 | 0 |
| 1 | 0 | 0 | 0 | 0 | 0 | 0 |
| 1 | 0 | 0 | 1 | 0 | 0 | 0 |
| 1 | 0 | 1 | 0 | 0 | 0 | 0 |
| 1 | 0 | 1 | 1 | 0 | 0 | 0 |
| 1 | 1 | 0 | 0 | 1 | 0 | 0 |
| 1 | 1 | 0 | 1 | 1 | 0 | 1 |
| 1 | 1 | 1 | 0 | 0 | 0 | 0 |
| 1 | 1 | 1 | 1 | 1 | 0 | 1 |

(b)

The address bits and the functions to be realized are so few in the case of the waveform generator that a ROM realization would probably not be the method selected. The vending-machine controller is more complex and would most likely be actually realized using a ROM.

It is not necessary first to translate an ASM chart to state-table form prior to obtaining a table specifying a ROM realization. Recall that indication of don't-care

combinations of values of inputs and present states was the only information added
to the state table that was not found in the ASM chart. These don't-cares are important if a minimal realization is to be obtained. A word is included in a ROM for
every combination of values of the address bits whether this combination can actually
occur or not. Thus there is no penalty attached to assuming that all combinations of
input values are possible at each present state even though some input lines might
not be referenced at particular points in the chart. To illustrate this approach, a
ROM realization of the vending-machine controller is obtained in Example 9.3
directly from the ASM chart, ignoring the state table of Fig. 9.9.

**Example 9.3**

Obtain a ROM table for a realization of the vending-machine controller directly
from the ASM chart of Fig. 9.7

Solution
We note that there are three inputs to the vending-machine controller and three state
variables, combinations of values of which were already assigned to the six states in
Fig. 9.7. These six variables thus become the address bits of the ROM. Each combination of values of these address bits is tabulated in Fig. 9.16. For convenience, the
state variables are the most significant address bits. The ROM must have five outputs, the next values of the three state variables as well as circuit outputs $A$ and $C$.
The values of these five variables are tabulated for each combination of address bits
in Fig. 9.16.

The ROM output entries in Fig. 9.16 were obtained by a state-by-state consideration of the ASM chart. For example, the first eight rows (0–7) correspond to
present state $Q0$ or $y_1y_2y_3 = 000$. First notice that, if $R = 1$, no reference is made
to the other two inputs in the ASM block for $Q0$. A strict interpretation of the ASM
chart is that, if $R = 1$, the next state is to be $Q0$ regardless of the values of the other
two inputs. This is precisely what is entered for ROM output values in rows 4
through 7 of Fig. 9.16. The fact that $I_1I_2$ can only be 00, if $R = 1$, is not considered.
For each of the first eight rows of this truth table, $A$ and $C$ are both 0, since outputs
are only functions of the present state of the vending-machine controller and not the
inputs. In fact, we duplicate the same output values for each input combination for
all six states in the table.

No inputs at all are referenced in the ASM block for state $Q25$. Thus for each
combination of input values on rows 8 through 15 of the table the next values of the
state variables are 000 and the outputs are 11. The combinations of state variable
values 011 and 101 are not assigned to states in the ASM chart. Since these states
will never occur, next state and output values are all don't-cares, as observed previously. These don't-care entries are specified as 0's in the ROM. Again, this is no
disadvantage, since the ROM realizes all minterms without minimization.

A logic-block diagram of the vending machine controller is given in Fig. 9.17.
It appears quite simple, since all the logic is collected in the single block labeled
ROM. If MSI parts and a user programmable ROM (PROM) are used, this ROM-
based approach may be most economical in terms of both engineering time and production cost. If the design were to be realized as a special purpose VLSI part, it

| Address bits | | | | | | ROM data | | | | | Address bits | | | | | | ROM data | | | | |
|---|---|---|---|---|---|---|---|---|---|---|---|---|---|---|---|---|---|---|---|---|---|
| $y_1$ | $y_2$ | $y_3$ | $R$ | $I_1$ | $I_2$ | $y_1^{n+1}$ | $y_2^{n+1}$ | $y_3^{n+1}$ | $A$ | $C$ | $y_1$ | $y_2$ | $y_3$ | $R$ | $I_1$ | $I_2$ | $y_1^{n+1}$ | $y_2^{n+1}$ | $y_3^{n+1}$ | $A$ | $C$ |
| 0 | 0 | 0 | 0 | 0 | 0 | 0 | 0 | 0 | 0 | 0 | 1 | 0 | 0 | 0 | 0 | 0 | 1 | 0 | 0 | 0 | 0 |
| 0 | 0 | 0 | 0 | 0 | 1 | 0 | 0 | 1 | 0 | 0 | 1 | 0 | 0 | 0 | 0 | 1 | 0 | 0 | 0 | 0 | 0 |
| 0 | 0 | 0 | 0 | 1 | 0 | 1 | 0 | 0 | 0 | 0 | 1 | 0 | 0 | 0 | 1 | 0 | 0 | 1 | 0 | 0 | 0 |
| 0 | 0 | 0 | 0 | 1 | 1 | 0 | 1 | 0 | 0 | 0 | 1 | 0 | 0 | 0 | 1 | 1 | 1 | 1 | 0 | 0 | 0 |
| 0 | 0 | 0 | 1 | 0 | 0 | 0 | 0 | 0 | 0 | 0 | 1 | 0 | 0 | 1 | 0 | 0 | 0 | 0 | 0 | 0 | 0 |
| 0 | 0 | 0 | 1 | 0 | 1 | 0 | 0 | 0 | 0 | 0 | 1 | 0 | 0 | 1 | 0 | 1 | 0 | 0 | 0 | 0 | 0 |
| 0 | 0 | 0 | 1 | 1 | 0 | 0 | 0 | 0 | 0 | 0 | 1 | 0 | 0 | 1 | 1 | 0 | 0 | 0 | 0 | 0 | 0 |
| 0 | 0 | 0 | 1 | 1 | 1 | 0 | 0 | 0 | 0 | 0 | 1 | 0 | 0 | 1 | 1 | 1 | 0 | 0 | 0 | 0 | 0 |
| 0 | 0 | 1 | 0 | 0 | 0 | 0 | 0 | 0 | 1 | 1 | 1 | 0 | 1 | 0 | 0 | 0 | 0 | 0 | 0 | 0 | 0 |
| 0 | 0 | 1 | 0 | 0 | 1 | 0 | 0 | 0 | 1 | 1 | 1 | 0 | 1 | 0 | 0 | 1 | 0 | 0 | 0 | 0 | 0 |
| 0 | 0 | 1 | 0 | 1 | 0 | 0 | 0 | 0 | 1 | 1 | 1 | 0 | 1 | 0 | 1 | 0 | 0 | 0 | 0 | 0 | 0 |
| 0 | 0 | 1 | 0 | 1 | 1 | 0 | 0 | 0 | 1 | 1 | 1 | 0 | 1 | 0 | 1 | 1 | 0 | 0 | 0 | 0 | 0 |
| 0 | 0 | 1 | 1 | 0 | 0 | 0 | 0 | 0 | 1 | 1 | 1 | 0 | 1 | 1 | 0 | 0 | 0 | 0 | 0 | 0 | 0 |
| 0 | 0 | 1 | 1 | 0 | 1 | 0 | 0 | 0 | 1 | 1 | 1 | 0 | 1 | 1 | 0 | 1 | 0 | 0 | 0 | 0 | 0 |
| 0 | 0 | 1 | 1 | 1 | 0 | 0 | 0 | 0 | 1 | 1 | 1 | 0 | 1 | 1 | 1 | 0 | 0 | 0 | 0 | 0 | 0 |
| 0 | 0 | 1 | 1 | 1 | 1 | 0 | 0 | 0 | 1 | 1 | 1 | 0 | 1 | 1 | 1 | 1 | 0 | 0 | 0 | 0 | 0 |
| 0 | 1 | 0 | 0 | 0 | 0 | 0 | 1 | 0 | 0 | 0 | 1 | 1 | 0 | 0 | 0 | 0 | 1 | 1 | 0 | 0 | 0 |
| 0 | 1 | 0 | 0 | 0 | 1 | 0 | 0 | 0 | 0 | 0 | 1 | 1 | 0 | 0 | 0 | 1 | 0 | 0 | 0 | 0 | 0 |
| 0 | 1 | 0 | 0 | 1 | 0 | 1 | 1 | 0 | 0 | 0 | 1 | 1 | 0 | 0 | 1 | 0 | 1 | 1 | 1 | 0 | 0 |
| 0 | 1 | 0 | 0 | 1 | 1 | 1 | 1 | 1 | 0 | 0 | 1 | 1 | 0 | 0 | 1 | 1 | 1 | 1 | 1 | 0 | 0 |
| 0 | 1 | 0 | 1 | 0 | 0 | 0 | 0 | 0 | 0 | 0 | 1 | 1 | 0 | 1 | 0 | 0 | 0 | 0 | 0 | 0 | 0 |
| 0 | 1 | 0 | 1 | 0 | 1 | 0 | 0 | 0 | 0 | 0 | 1 | 1 | 0 | 1 | 0 | 1 | 0 | 0 | 0 | 0 | 0 |
| 0 | 1 | 0 | 1 | 1 | 0 | 0 | 0 | 0 | 0 | 0 | 1 | 1 | 0 | 1 | 1 | 0 | 0 | 0 | 0 | 0 | 0 |
| 0 | 1 | 0 | 1 | 1 | 1 | 0 | 0 | 0 | 0 | 0 | 1 | 1 | 0 | 1 | 1 | 1 | 0 | 0 | 0 | 0 | 0 |
| 0 | 1 | 1 | 0 | 0 | 0 | 0 | 0 | 0 | 0 | 0 | 1 | 1 | 1 | 0 | 0 | 0 | 0 | 0 | 0 | 1 | 0 |
| 0 | 1 | 1 | 0 | 0 | 1 | 0 | 0 | 0 | 0 | 0 | 1 | 1 | 1 | 0 | 0 | 1 | 0 | 0 | 0 | 1 | 0 |
| 0 | 1 | 1 | 0 | 1 | 0 | 0 | 0 | 0 | 0 | 0 | 1 | 1 | 1 | 0 | 1 | 0 | 0 | 0 | 0 | 1 | 0 |
| 0 | 1 | 1 | 0 | 1 | 1 | 0 | 0 | 0 | 0 | 0 | 1 | 1 | 1 | 0 | 1 | 1 | 0 | 0 | 0 | 1 | 0 |
| 0 | 1 | 1 | 1 | 0 | 0 | 0 | 0 | 0 | 0 | 0 | 1 | 1 | 1 | 1 | 0 | 0 | 0 | 0 | 0 | 1 | 0 |
| 0 | 1 | 1 | 1 | 0 | 1 | 0 | 0 | 0 | 0 | 0 | 1 | 1 | 1 | 1 | 0 | 1 | 0 | 0 | 0 | 1 | 0 |
| 0 | 1 | 1 | 1 | 1 | 0 | 0 | 0 | 0 | 0 | 0 | 1 | 1 | 1 | 1 | 1 | 0 | 0 | 0 | 0 | 1 | 0 |
| 0 | 1 | 1 | 1 | 1 | 1 | 0 | 0 | 0 | 0 | 0 | 1 | 1 | 1 | 1 | 1 | 1 | 0 | 0 | 0 | 1 | 0 |

*Figure 9.16   ROM for vending-machine controller realization.*

might be best to obtain minimal sum-of-products functions and to realize them using PLAs.   ∎

# 9.6  The State Diagram, an Alternative Notation*

The ASM chart is a convenient graphic notation in terms of which digital systems can be initially formulated. It is a particularly useful form in systems with large

*The material of this section and the following section is not essential to the developments in succeeding chapters. It may, therefore, be skipped at the option of the instructor.

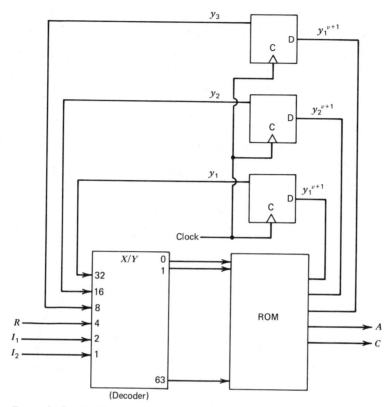

*Figure 9.17   Realization of vending-machine controller.*

numbers of input lines, only a few of which are of interest at any time, or where not all aspects of the problem are completely understood at the outset. The ASM chart is bulky, but it exerts a variety of pressures on designers that help them to avoid errors. A disadvantage is that, in order to conform to the simplest configuration of decision diamonds on the ASM chart, designers might occasionally specify next state or output values that could be left as don't-cares.

The *state diagram* is in most cases a more compact graphic notation that conveys the same information as the ASM chart. As with the state table, the state diagram represents combinations of input values rather than individual inputs or functions of inputs. To illustrate the state diagram, we turn once again to the vending-machine controller. For easy reference, we repeat the state table for the vending-machine controller as Fig. 9.18a. In the state diagram each state is represented by a circle. Each combination of input values and present state that can actually occur (i.e. excluding don't-cares) is represented by an arrow. This arrow originates at the present state and terminates with the arrow head at the next state resulting from this input combination. For convenience, the inputs are usually coded so that each combination of input values is represented by a single symbol. The appropriate input symbols are then entered adjacent to the corresponding arrows in the state diagram.

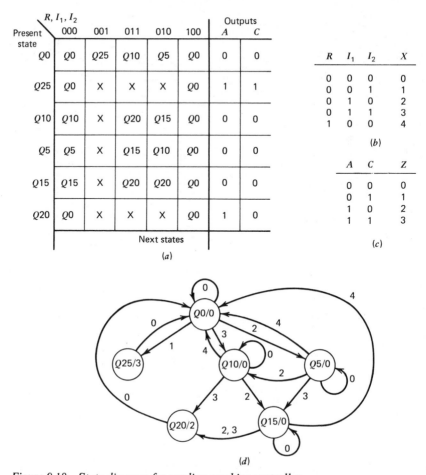

| $R, I_1, I_2$ | | | | | | Outputs | |
| Present state | 000 | 001 | 011 | 010 | 100 | A | C |
|---|---|---|---|---|---|---|---|
| Q0 | Q0 | Q25 | Q10 | Q5 | Q0 | 0 | 0 |
| Q25 | Q0 | X | X | X | Q0 | 1 | 1 |
| Q10 | Q10 | X | Q20 | Q15 | Q0 | 0 | 0 |
| Q5 | Q5 | X | Q15 | Q10 | Q0 | 0 | 0 |
| Q15 | Q15 | X | Q20 | Q20 | Q0 | 0 | 0 |
| Q20 | Q0 | X | X | X | Q0 | 1 | 0 |

Next states

(a)

| R | $I_1$ | $I_2$ | X |
|---|---|---|---|
| 0 | 0 | 0 | 0 |
| 0 | 0 | 1 | 1 |
| 0 | 1 | 0 | 2 |
| 0 | 1 | 1 | 3 |
| 1 | 0 | 0 | 4 |

(b)

| A | C | Z |
|---|---|---|
| 0 | 0 | 0 |
| 0 | 1 | 1 |
| 1 | 0 | 2 |
| 1 | 1 | 3 |

(c)

(d)

*Figure 9.18   State diagram for vending-machine controller.*

The input coding for the vending-machine controller is given in Fig. 9.18b. The state diagram for this system is given in Fig. 9.18d. You will find an arrow corresponding to each entry in the next-state portion of Fig. 9.18a that is not a don't-care.

The representation of outputs in the state diagram will depend on whether the outputs are functions of the inputs as well as the present states. In Fig. 9.18d the outputs are a function of the present states only and are, therefore, denoted within the present-state circle, separated by a slash from the present state symbol. The outputs are coded in a manner similar to the inputs, as shown in Fig. 9.18c. If the outputs were a function of the inputs as well as the present state, a separate set of output values would be associated with each arrow in the state diagram. The output symbols would be placed next to the appropriate arrows following the input symbols and a slash. This will be illustrated in the next example.

| Present state | $x_1 x_2$ 0 0 | | 0 1 | | 1 1 | | 1 0 | |
|---|---|---|---|---|---|---|---|---|
| $Q1$ | $Q2,$ | 1 | $Q2,$ | 1 | $Q2,$ | 1 | $Q2,$ | 1 |
| $Q2$ | $Q3,$ | 1 | $Q3,$ | 1 | $Q3,$ | 0 | $Q3,$ | 1 |
| $Q3$ | $Q4,$ | 0 | $Q4,$ | 1 | $Q4,$ | 1 | $Q1,$ | 0 |
| $Q4$ | $Q1$ | 0 | $Q1,$ | 0 | $Q1$ | 0 | X | |

Next state, $z$

Figure 9.19   State table of waveform generator.

## Example 9.4

Obtain a state-diagram representation of the waveform generator designed in Example 9.3.

Solution
We begin by reproducing the state table for this circuit as given in Fig. 9.13a as Fig. 9.19.

We see from Fig. 9.19 that the outputs are indeed a function of the inputs as well as the present state in this example. Each square in the state table corresponds to a unique combination of present state and input values. In general, the output values may be assigned independently in each square.

In the state diagram we have the opportunity to include a separate arrow from present state to next state for each combination of present state and input values. Where the outputs are also functions of the present state and inputs, it would seem natural to represent an output value by associating it with the appropriate arrow on the state diagram. The state table of Fig. 9.19 was translated to the state diagram of Fig. 9.20 in this fashion. Where more than one arrow would have connected the same present state to the same next state with the same output value, these multiple arrows have been combined into single arrows with all the individual input values separated by commas. The list of input values is followed by a slash and the common output values. The combinations of input values $X_1 X_2$, are represented in the state diagram by their binary-coded decimal integer equivalents. ∎

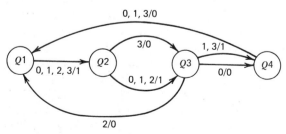

Figure 9.20   State diagram for waveform generator.

The state diagram will be used from time to time in the rest of the book for the sake of compactness. Sometimes the designer may elect to use this form in place of the ASM chart for the same reason. As we have noted earlier, the ASM chart does not represent all the information that can be included in a state diagram or a state table. The ASM chart cannot indicate whether a particular combination of input and present-state values can actually occur or is a don't-care combination. If a sequential circuit has only a few inputs with numerous don't-care combinations, it might be most easily formatted as a state diagram. If there are many inputs only a few of which affect the circuit at any given time, then an ASM chart formulation is probably preferred.

We conclude with a final example in which we first formulate a system as a state diagram and then convert it to an ASM chart.

**Example 9.5**

A sequential circuit is to have a single input, $x$, and a single output, $z$. Input, $x$, will be 1 for short intervals of one, two, or three clock periods. Any two intervals in which $x = 1$ will be separated by ten or more clock periods in which $x = 0$. Immediately following each $x = 1$ interval the output, $z$, will be 1 for a short interval. If $x$ was 1 for one period, $z$ is to be 1 for three periods. If $x$ was 1 for two periods, $z$ is to be 1 for two periods. If $x$ was 1 for only one period, $z$ should remain 1 for three periods. $z$ is to be 0 at all other times. Develop a state diagram representation of this circuit.

Three typical input and output waveforms are given in Fig. 9.21c. Note that all transitions of both input and output are synchronized by transitions on a common clock. The waveform of the clock is not shown in the figure.

Solution
Since there is only one input line, a state-diagram approach seems appropriate. We assume a starting state labeled, $a$, in Fig. 9.21a. We shall arrange for the circuit to return to state $a$ at the end of each output signal. As long as the input, $x$, is 0, the circuit may remain in state $a$. For the first period in which $x = 1$, the output will remain 0; but the circuit must be sent to a new state, labeled $b$ in Fig. 9.21a, by the clock pulse at the end of the period. From state $b$ the behavior will depend on the input. If the input remains 1, the output will remain 0 and the circuit will go to state $c$ to wait for the eventual input change. If the input is 0, the output must be 1 for this and the following two clock periods. Thus, the output is listed as 1 and the circuit goes to state $d$ to continue this behavior. All this is depicted in Fig. 9.21b.

The input may be either 1 or 0, when the circuit is in state $c$. If the input continues as 1, the corresponding output will be 0; and the circuit will go to state $e$. If the input is zero, the output should be 1, as enforced by the arrow leading to state $f$ in Fig. 9.21b. Once the input has returned to 0, it will remain there for several clock periods. Therefore, the only arrow from state $d$ in Fig. 9.21c is for $x = 0$ and $z = 1$, leading to state $g$.

Since the input cannot be 1 for more than three consecutive clock periods, only one arrow is needed from state $e$ for $x = 0$ and the single period output of 1. For state $f$ the input must be 0 to generate the second of two output 1's. For state $g$ the input must also be 0 to generate the third of three periods in which $z = 1$. This

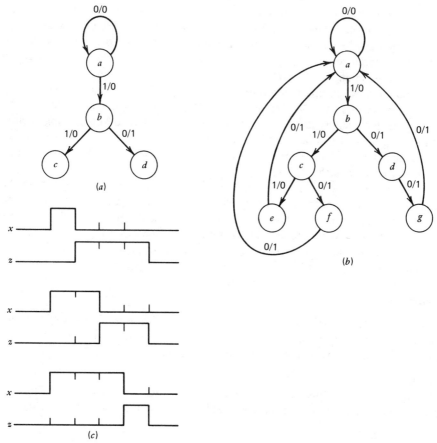

Figure 9.21  Design of pulse-width adjuster.

completes the output signal in all three cases, so the circuit may return to state *a*. You are encouraged to trace each of the three input sequences given in Fig. 9.21*c* through the state diagram to verify that the proper output sequence will occur.  ■

## 9.7  Compatible States

The general technique used in the derivation of the state diagram in Fig. 9.21 was to define new next states for each input arrow until a point was reached where the obvious action was to return to the initial state. As was the case for the pulse width adjuster, this approach results in an inverted tree with an approximate doubling of states at each successive level moving down the tree. Are all these states actually necessary? In many cases the answer is no.

Consider some sequential circuit, M, which might at a particular point in time be in state $q_1$. Beginning from this state, $q_1$, any sequence of allowable inputs, one each clock period, might be applied. For each of these possible input sequences, a clock period by clock period sequence of outputs will result. This collection of input sequences and corresponding output sequences is called the behavior of the sequential circuit beginning in state $q_1$.

Now consider some other state of M, say $q_2$. Suppose that the behavior of circuit M is exactly the same beginning in state $q_2$ as it is beginning in state $q_1$. That is, suppose that every sequence of inputs that can be applied to M beginning in either of the states $q_1$ and $q_2$ will result in exactly the same sequence of outputs regardless of whether the beginning state was $q_1$ or $q_2$. In this case we say that the two states $q_1$ and $q_2$ are *compatible*.

If two or more states of M are *mutually compatible*, it follows naturally that only one of these states is actually necessary. This notion is important, since unneeded states in a sequential circuit will sometimes imply unneeded flip-flops in the realization of that circuit and, therefore, an unnecessarily complicated and expensive realization. The circuit with its state diagram given in Fig. 9.21 is an example.

## Example 9.6

Replace any sets of mutually compatible states in the state diagram of the pulse width adjuster of Example 9.5 with single states to produce an equivalent state table that can be realized with as few flip-flops as possible.

## Solution

The first step is to translate the state diagram of Fig. 9.5*b* to state-table form, as given in Fig. 9.22*a*. Notice in this state table that for input $x = 0$ the next state for

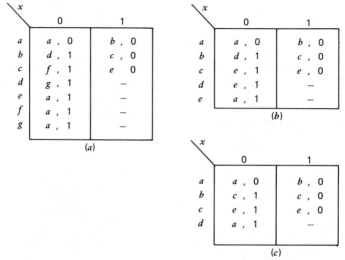

*Figure 9.22   Equivalent state tables for the pulse-width adjuster.*

each of the present states $e$, $f$, and $g$ is $a$. Similarly, for $x = 0$ the outputs for each of these three states is 1. For these three present states an input of $x = 1$ will never occur. It is easy to see that states $e$, $f$, and $g$ are mutually compatible. The first input (always 1) that could be applied beginning in any of these states will result in the same output for each of the states. Since this input will take the circuit to state $a$ from any of these states, any continuation of the input sequence will give the same output sequence for all three of these beginning states. Thus the behaviors are the same beginning in states $e$, $f$, or $g$; and these three states are mutually compatible. We may, therefore, replace states $f$ and $g$ wherever they appear in the state table of Fig. 9.22$a$ with the compatible state $e$. The resulting equivalent state table is given in Fig. 9.22$b$.

An inspection of the table of Fig. 9.22$b$ reveals that the same behavior can be expected if input $x = 0$ is applied to the circuit beginning in either of states $c$ or $d$. The argument is the same as in the previous paragraph, since the next state and the output are the same for both beginning states. Have we succeeded in proving that states $c$ and $d$ are compatible? Almost! A sequence beginning with $x = 1$ will never be applied to the circuit starting in state $d$. Therefore, the output sequences are indeed the same for any input sequences beginning with $x = 1$ that will be applied to the circuit starting in either of states $c$ or $d$. Therefore, the behaviors for these two beginning states are the same, and they are compatible. Replacing state $d$ by state $c$ wherever $d$ occurs in Fig. 9.22$b$ results in the equivalent state table of Fig. 9.22$c$.

Notice that the final state table of Fig. 9.22$c$ includes only four states. Thus it can be realized using only two flip-flops. Three flip-flops would have been required to realize Fig. 9.22$a$. Having saved a flip-flop, we conclude that the process was worthwhile.                                                                                                           ∎

The preceding example of finding and merging mutually compatible states was a simple one. The process used here will not always lead to a state table with the fewest possible states. More complex and more effective procedures have been developed for this task. For a much more complete treatment of this subject, see Ref. [2]. These techniques are very difficult to apply to large systems, and over the years useful applications have been rare. We, therefore, limit our treatment of this subject in this beginning textbook. It is important that you be aware that redundant states can result from a careless approach to the generation of an ASM chart or state diagram. No doubt the best approach is to ascertain that new states are really needed as they are added to the state diagram.

## Problems

**9.1**  Construct an ASM chart for a modulo-4 up–down counter given the transition table of Fig. 8.28.

**9.2**  Construct an ASM chart representing the network given in Fig. P8.3.

**9.3**  A certain clock-mode sequential circuit has an unclocked reset mechanism that will cause circuit operation to begin in a state, $Q0$. An ASM chart is to be developed representing this circuit. The reset input line will not be included in the ASM Chart. The circuit has only one other input and one output. The circuit must generate an output of 1 for one clock period only coinciding with the second 0 input of a sequence consisting of exactly two 1's (no more than two) followed by two 0's. Once the output has been 1 for one clock period, the output will remain 0 until the circuit is externally reset to $Q0$.

**9.4** (a) Derive the ASM chart of a circuit featuring an input line, $X$, in addition to the clock input. The only output line, $z$, is to be 0 unless the input has been 1 for four consecutive clock periods or 0 for four consecutive periods. The output must be 1 at the time of the fourth consecutive identical input. that is, any time the input is the same as it was during the three previous periods.

   (b) The input signal shown in Fig. P9.4 is to be applied to a realization of the ASM chart derived in (a). Plot the resulting output clock period by clock period. All input transitions, state transitions within the circuit, and output transitions are synchronized by falling edges of the clock shown in the figure. Neglect gate delays and memory element response times in the output plot.

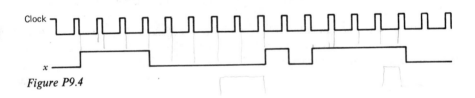

*Figure P9.4*

**9.5** Modify the ASM chart of Fig. 9.7 so that it will respond if a quarter is deposited following a nickel or dime. It will be sufficient for customers to receive merchandise and a nickel change just as if they had deposited a quarter without any other coin.

**9.6** A controller is to be designed for a coin-operated photocopy machine for which the price is $0.10. The machine must accept nickels, dimes, and quarters and has a coin-return button. The inputs may be coded as in the vending-machine example of Section 9.3. In addition to an output, $P$, which will enable the copier, there is an output, $C$, which will cause a nickel and a dime to be refunded. Construct an ASM chart for this controller. You may neglect reset and coin return.

**9.7** A controller similar to the one described in Problem 9.6 is to be designed. There is to be one additional input from a switch, $S$, located inside the machine. If the signal from this switch is 0, the price is to be $0.10. If the signal is 1, the price is to be $0.15. In place of output line $C$, there will be a line $N$ directing $0.05 change. When more than $0.05 change is needed, $N$ should be pulsed two or three times, as required. Obtain an ASM chart for this controller.

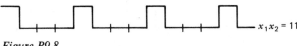

Figure P9.8

**9.8** Obtain an ASM chart for a waveform generator similar to the one described in Fig. 9.11, except that the waveform for inputs $X_1X_2 = 11$ is replaced by the one given in Fig. P9.8.

**9.9** A washing-machine controller is to be constructed. The controller has the following inputs:

> HOT: 1 if hot/cold switch specifies hot-water wash
> STRT: 0 to start washing; 1 to stop, even in midcycle.
> FULL: 1 if water filled to top
> EMPTY: 1 if water completely empty
> TIME: 1 if timer indicates done

The controller must generate the following outputs:

> HOTOUT: 1 to select hot water
> 0 to select cold water
> PUMP: 1 to turn on water pump
> FILL: 1 to direct water into washer
> 0 to direct water out of washer
> AG: 1 to agitate wash
> SPIN: 1 to spin wash
> SETIM: 1 to set the timer to 0

When the controller receives a start signal, it fills the washer with the correct temperature of water and agitates until the timer indicates it is done. It, then, empties the soapy water and fills the washer with cold rinse water and agitates again until the timer indicates it is done. Finally, it spins the clothes dry after emptying the rinse water. There is only one timer, so that the wash, rinse, and spin cycles must take exactly the same amount of time. Draw the ASM chart for this controller. Use conditional outputs.

**9.10** (a) Translate the ASM chart of Fig. P9.10a to state table form, coding the inputs as given in Fig. P9.10b. These are the only combinations of input values that will ever occur. None of these three input combinations will occur twice in two consecutive clock periods. $Q0$ may be regarded as a reset state.

   (b) Obtain a transition table for this chart.

**9.11** Obtain a realization of the transition table determined in Problem 9.10. Use $D$ flip-flops and obtain a minimal sum of products expression for the $D$ input of each memory element. Also determine an expression for the output, $Z$.

**9.12** Obtain a truth-table specification (similar to Fig. 9.16) for a ROM realization of the copy-machine controller described in Problem 9.6.

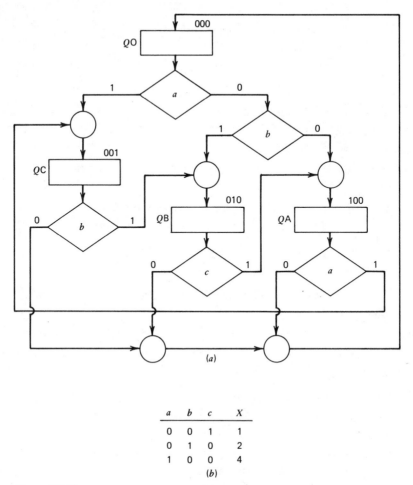

| a | b | c | X |
|---|---|---|---|
| 0 | 0 | 1 | 1 |
| 0 | 1 | 0 | 2 |
| 1 | 0 | 0 | 4 |

(b)

*Figure P9.10*

**9.13**  Develop a transition table from the ASM chart determined for the modified waveform generator in Problem 9.8. Use *D* flip-flops and obtain minimal sum-of-products expressions for the *D* input to each memory element in a realization of this transition table. Also determine a Boolean expression for the output.

**9.14**  Develop a state diagram that will describe the sequential circuit whose ASM chart is given in Fig. P9.10.

**9.15**  Consider the reduced state table of the pulse width adjuster given in Fig. 9.22c. Plot clock period by clock period the output that would be generated in response to application of the input signal given in Fig. P9.15 to a realization of this state table. All transitions on inputs, all state transitions within the circuit, and all transitions on outputs are synchronized by the falling edge of the clock shown in the figure. Assume that the circuit is in state *a* during

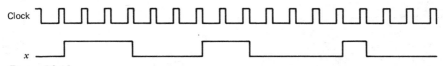

Clock

x

*Figure P9.15*

the first clock period at the left. Neglect gate delays and memory element response times in your output plot.

**9.16** Translate the ASM chart determined in Problem 9.4 to state diagram form.

**9.17** Translate the ASM chart determined in Problem 9.6 to state diagram form.

**9.18** Derive a state diagram and state table for a sequential circuit that is to function similarly to the pulse width adjuster of Example 9.5. In this case input signals on the only input line, $x$, can be 1 only for one or for two consecutive clock periods. Immediately following a 1 on $x$ for just one clock period, the only output, $z$, must be 1 for two clock periods. Immediately following a 1 on $x$ for two clock periods, the output must be 1 for three clock periods. Note that some input present state combinations will be don't-cares.

**9.19** The state table in Fig. P9.19 includes some sets of mutually compatible states. Use the technique illustrated in Section 9.7 to replace these sets of mutually compatible states with single states to obtain an equivalent state table with as few states as possible.

| $x$ | 0 | 1 |
|---|---|---|
| $a$ | $a$ , 0 | $b$ , 0 |
| $b$ | $c$ , 1 | $d$ , 0 |
| $c$ | $a$ , 1 | — |
| $d$ | $e$ , 1 | $f$ , 0 |
| $e$ | $g$ , 1 | — |
| $f$ | $a$ , 1 | — |
| $g$ | $a$ , 1 | — |

*Figure P9.19*

**9.20** Assume that the input signal of Fig. P9.15 is applied to a realization of the reduced state table obtained in Problem 9.19. Obtain a clock period by clock period plot of the resulting single output, $z$.

**9.21** Use the technique of Section 9.7 to obtain a state table with as few states as possible that is also equivalent to the state table obtained in Problem 9.18.

# References

1. Clare, C. H., *Designing Logic Systems Using State Machines*. McGraw-Hill, New York, 1973.

2. Hill, F. J., and G. R. Peterson, *Introduction to Switching Theory and Logical Design,* 3rd ed. Wiley, New York, 1981.

3. Huffman, D. A., "The Syntheses of Sequential Switching Circuits." *J. Franklin Inst.,* 257, No. 3, 161–190; No. 4, 275–303 (March–April, 1954).

4. Kohavi, Z., *Switching and Finite Automata Theory,* 2nd ed. McGraw-Hill, New York, 1978.

5. McCluskey, E. J., and S. H. Unger, "A Note on the Number of Internal Variable Assignments for Sequential Switching Circuits." *IRE Trans. on Electronic Computers,* EC-8, No. 4, 439–40 (December 1959).

6. Mealy, G. H., "A Method for Synthesizing Sequential Circuits." *Bell System Tech. J.,* 34:5, 1045–1080 (September, 1955).

7. Moore, E. F., "Gedanken Experiments on Sequential Machines," in C. E. Shannon and J. McCarthy (eds.), *Automata Studies,* Princeton University Press, Princeton, N.J., 1956.

8. Wiatrowski, C. A., and House, C. H., *Logic Circuits and Microcomputer Systems,* McGraw-Hill, New York, 1980.

# Register Transfer Design

## 10.1 Generalized ASM Output

So far we have used the ASM chart as a means of describing sequential circuits with relatively few flip-flops. It is obvious that there are far more complex digital systems, in particular, digital computers. The ASM chart as defined in Chapter 9 is not sufficient to describe a digital computer or other similarly complex systems. To see why, let us consider the following design example.

---

**Statement of Design Problem**

A sequential circuit is to be designed that has a vector (group) of eight input wires that we shall label **X**, together with two single wire inputs, **a** and **b**. The circuit will have a vector of eight output wires labeled **Z** and a single-wire output labeled **out**. The circuit may be independently reset to an initial state in which it will wait for the input **a** to go to 1. When this occurs, a data vector (i.e., an 8-bit unit of information) will be available on lines **X**. This data vector must be stored by the circuit. Several clock periods after **a** was 1 line **b** will go to 1 to indicate that a second data vector is available on lines **X**. At this time the circuit must consider the two vectors as binary numbers and compare their magnitudes. It must then make the larger-magnitude vector available on output lines **Z** for at least four clock periods. During the first four of these clock periods the output line **out** must be 1. At all other times, **out** = 0. Once **out** goes back to 0 the circuit should return to its initial state. By this time it may be assumed that both **a** and **b** will have returned to 0.

---

This problem description may seem lengthy, but it is so only so that the required timing can be specified precisely. The actual function of the system is quite simple. Two successive input vectors on lines $X$ are compared and the largest in magnitude of the two is made available as an output for four successive clock periods. The first of the two input vectors must be stored by the system until the second appears.

The simplicity of function of this system does not imply that it is easily expressed using only the ASM chart as defined in the previous chapter. The simple fact that an 8-bit vector must be stored alone makes this form of representation impractical. Storage of the 8-bit vector will require eight flip-flops that may assume any of 256 states. Combining these with additional flip-flops to control the sequence of events means some multiple of 256 ASM blocks on the chart. Clearly this is not practical.

Although some other mechanism is required to represent storage of data, we do not wish to give up the ASM chart completely. It will prove very convenient for representation of timing relationships and the relation between the system and control inputs such as $a$ and $b$ in the example just given. To have the "best of both worlds," we propose the model depicted in Fig. 10.1. This model divides the system into two separate sequential circuits. Block A will include the numerous flip-flops required to store vectors of information such as the first input vector in the problem statement just given. Block B will include the much smaller number of flip-flops required to "control" the functioning of the system. Block B will be a sequential circuit describable by an ASM chart. The relationship between the two blocks will become clear as we develop the system with this specification.

Before moving to the example in the next section, let us make a few general remarks about the interpretation of block B of Fig. 10.1 as an ASM chart. Notice that the block has two sets of input lines and two sets of output lines. From a functional point of view, there will be no distinction between the external inputs, labeled *control inputs* in Fig. 10.1 and the *status signals*, which are fed to block B from the data unit of block A. All these will be allowed to influence the sequencing through

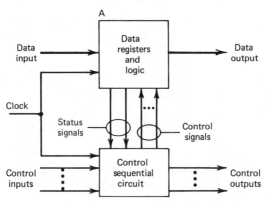

*Figure 10.1  Separation of data and control.*

the states of the *control sequential circuit*. Those outputs labeled *control outputs* in Fig. 10.1 may be expressed in the ASM charts in the manner described in Chapter 9. Those ouputs from block B to block A labeled *control signals* have not been seen before. An active signal on one of these lines will cause some action to take place on the data registers in block B, for example, a transfer of data from one register to another or an increment of a counter. The function of the *control sequential circuit* in block B is to assure that these action-causing control signals reach the data and logic unit of block A in the proper sequence and at the proper time.

## 10.2   ASM Chart Representation of a Control Unit

There is no mechanical way to determine the number and application of the memory elements to be used in block A of a system design. This is always a task requiring engineering judgment. We have already observed that at least one register of eight flip-flops will be required in the "sequential comparator" design example that was specified in the previous section. Since the 8-bit output vector must be stored for four clock periods, a register is also required for this purpose. But why do we need the first input vector anymore after the output has been determined? We don't! The same register can be used for both purposes! Let us call this register $A[8]$, indicating that it consists of eight bits.

What else is needed? The answer is that a counter must be included to count through the four clock periods for which $out = 1$. A 2-bit counter, which we shall call $COUNT[2]$, will be sufficient for this purpose. As a graphic illustration of our progress with the design, thus far, we show these two registers in block A of Fig. 10.2. It may not be evident to you that these are the only two data registers that will

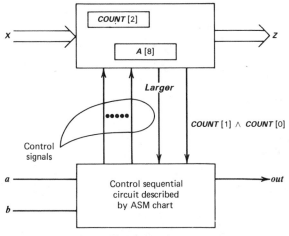

*Figure 10.2   Sketch of sequential comparator.*

be required, but let us proceed to see if we can complete the design with these two registers. In practice it is not uncommon for designers to discover that they have not anticipated all their data storage needs while in the middle of the development of an ASM chart. When this happens, they add to block A and begin again on the ASM chart.

Also established in Fig. 10.2 is the fact that **a** and **b** are control inputs that will be used to determine branching in the ASM chart. Similarly, **out** will be considered a control output, whereas **X** and **Z** are data inputs and outputs, respectively. The interpretation of the control-signal lines is yet to be determined.

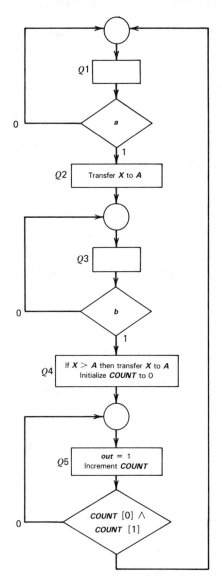

*Figure 10.3   ASM chart for sequential comparator.*

The first block of the ASM chart of Fig. 10.3 represents the initial state in which the system waits for input *a* to go to 1. After *a* goes to 1, the system goes to a state that causes the values of the input vector *X* to be transferred into register *A*. For now we understand the notation "Transfer *X* to *A*" to mean that while the system is in state $Q2$, the control line (or lines) that effect this transfer are active. The system remains in state $Q2$ for only one clock period, after which it goes to $Q3$ to wait for the second control input, *b*, to go to 1. After *b* goes to 1 the circuit spends a clock period in $Q4$ to accomplish two actions. If the second input vector is larger than the one stored in *A* this larger vector must be transferred into *A*. Otherwise *A* is left alone. A signal is always generated by state $Q4$, which causes the contents of the two counter flip-flops to be reset to 0.

The control circuit will remain in state $Q5$ for four clock periods. How this is accomplished should be clear from the ASM chart. When the circuit enters state $Q5$, the values stored in **COUNT** will be 00. During each clock period in which the circuit remains in this state the counter will be incremented. At the same time, the old value of the counter will be tested to determine the next state of the circuit. At the time of the fourth clock period in $Q5$ the counter will contain 11. When this is detected, the circuit will return in state $Q1$. During the four clock periods in which the circuit remains in $Q5$, *out* will be 1 and the contents of *A* will remain unchanged, since no control signal is sent to this register.

With respect to the behavior of the control sequential circuit itself and the control outputs, the ASM chart of Fig. 10.3 conforms to the notation defined in Chapter 9. The notation calling for control signals to be sent to the data unit is informal, since so far nothing has been defined for this purpose. Not only do we want an unambiguous notation for specifying these transfers, but we must also have a precise mechanism through which this notation can be translated to hardware implementation. This will be the subject of the next few sections.

We do intend to obtain a realization of the system of Fig. 10.3. This will be accomplished in Section 10.7 once we have established sufficient background to translate this ASM chart to an unambiguous notation. Please stay with us.

# 10.3  Register Transfer Language (RTL) Notation

In the previous sections we demonstrated that the ASM chart as defined in Chapter 9 was inadequate for the representation of large digital systems. We then proposed to separate such systems into separate control and data sections. We found that the ASM chart will remain a convenient tool for description of the control section by itself, even for fairly large systems. Next we extended the notation available for use within the state box of the ASM chart to provide for informal description of the data section. In the process we destroyed the one-to-one correspondence between the ASM chart and a hardware realization that we carefully cultivated in Chapter 9. This one-to-one correspondence or ability to translate a specification mechanically into its hardware realization is important. We would like a notation that preserves

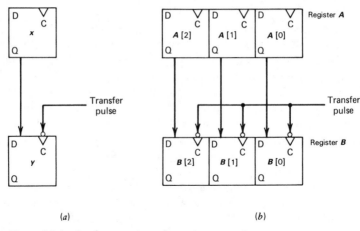

(a)                                                  (b)

*Figure 10.4   Implementation of a register transfer.*

this property for the data section as well as the control section. Such a tool is available in the form of a *register transfer language* or RTL.*

The primary RTL operation is the transfer of information from one place to another. Consider first the circuit shown in Fig. 10.4a. Here we have two $D$ flip-flops, denoted by $x$ and $y$. We shall denote individual flip-flops (not part of registers) by a string of one or more lowercase letters.† We have connected the output of the $x$ flip-flop to the $D$ input of the $y$ flip-flop. When we apply a pulse to the clock input of $y$, the current value of $x$ will be transferred to (copied into) $y$; this operation will be represented by the *register transfer statement,*

$$y \leftarrow x$$

Note carefully that a transfer operation is a copying operation, the source is not altered.

In Figure 10.4b we show two 3-bit registers, $A$ and $B$, connected in a manner similar to the connection of $x$ to $y$ in Fig. 10.4a. Each register is represented by a string of one or more uppercase letters, and the individual positions, or bits, in the register are numbered from right to left as shown. The right-most bit is always bit 0, with the bit numbers increasing from right to left. Thus, if the register contains $n$ bits, the left-most bit is numbered $n - 1$. Like dimensions, bit numbers are given in square brackets. We shall soon see that this is not an ambiguity, since bit numbers and dimensions will never appear in the same part of an RTL description. The 3-bit

---

*With the exception of the reverse ordering of bits in vectors, the RTL to be described in this section is consistent with the language AHPL developed by the authors and treated in detail in Ref. [2]. Not all the features of this language will be presented here.

†The convention of using lowercase letters for single flip-flops or lines and uppercase for vectors is consistent with AHPL and will be observed in this chapter. In later chapters, however, we shall be dealing with standard devices for which the nomenclature does not conform to this convention. In such cases we will use uppercase for single lines where the meaning is obvious.

registers of Fig. 10.4 are an example.

$$A = A[2], A[1], A[0] \quad \text{and} \quad B = B[2], B[1], B[0]$$

If we apply a pulse to the clock inputs of all three $B$ flip-flops, the values of the $A$ flip-flops will be transferred to (copied into) the corresponding $B$ flip-flops, represented by the set of transfer statements

$$B[2] \leftarrow A[2]; B[1] \leftarrow A[1]; B[0] \leftarrow A[0]$$

or, more compactly, by the single statement

$$B \leftarrow A$$

The type of register transfer represented by the last statement, the transfer of the complete contents of one register to another, is the most common, but there are other possibilities. Consider the situation shown in Fig. 10.5a. Here we have two 4-bit registers, $A$ and $B$, and a single flip-flop, $x$. With the connections shown, a pulse applied to the clock inputs of $B$ will result in the transfer

$$B \leftarrow x, A[3:1]$$

which is equivalent to

$$B[3] \leftarrow x; B9[2] \leftarrow A[3]; B[1] \leftarrow A[2]; B[0] \leftarrow A[1]$$

The comma in

$$x, A$$

indicates *catenation,* the joining together of two units of information from different sources into a single unit, or vector, of information. An equivalent representation would be

$$x, A[3], A[2], A[1], A[0]$$

A variant of this transfer is shown in Fig. 10.5b, where $x$ has been replaced by a constant, logical 0, and could be written as follows.

$$B \leftarrow 0, A[3:1]$$

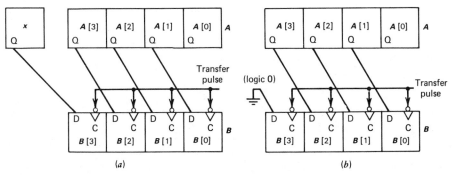

*Figure 10.5   Implementation of a shift.*

Note that this is a shift of **A** one place right into **B**, with a 0 shifted into the left-most bit position. Register transfer statements can also represent rotations of the bits within a single register. Figure 10.6 shows a single register **A** connected for a right rotate. When this register is clocked, the transfer taking place is given by

$$A \leftarrow A[0], A[3{:}1]$$

which is equivalent to

$$A[3] \leftarrow A[0]; A[2] \leftarrow A[3]; A[1] \leftarrow A[2]; A[0] \leftarrow A[1]$$

Note that this transfer within a single register is valid because we assume that the flip-flops are either edge-triggered or master–slave, so that the outputs do not change until after the clock is no longer active. Thus, the present values of the outputs are stable at the register inputs for the required period.

   In illustrating all these transfers we have shown discrete flip-flops in the interests of clarity, but the same notation can be used to describe transfers in MSI parts. For example, if a 74LS295B shift register is connected as shown in Fig. 10.7, the operation taking place when the chip is clocked is given by

$$A \leftarrow A[0], A[3{:}1]$$

   The example of shifting illustrates the very important concept that the simple moving of information is a form of processing. We can also perform specific logical operations on information as it is transferred. If we connect the $Q$ outputs of the **A** register to the inputs of the **B** register, as shown in Fig. 10.8, clocking **B** will result in the transfer

$$B \leftarrow \overline{A}$$

   Figure 10.9 shows a system with three 3-bit registers, **A**, **B**, and **C**. In Figure 10.9*a*, each bit of **A** is ORed with the corresponding bit of **B**, so that clocking **C** will result in the transfer

$$C \leftarrow A \vee B$$

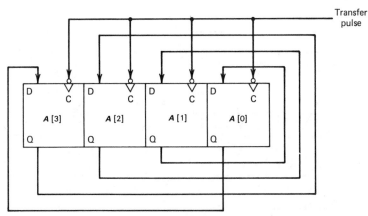

*Figure 10.6   Implementation of a right rotation.*

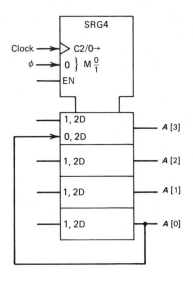

*Figure 10.7    Rotation in an MSI shift register.*

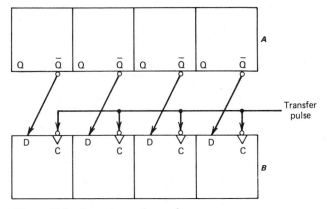

*Figure 10.8    Inversion and transfer.*

In a similar manner, the system of Fig. 10.9*b* will accomplish the transfer

$$C \leftarrow A \wedge B$$

The register transfer notation can be extended to input and output data lines. For example, if data from three input lines, $X[2]$, $X[1]$, $X[0]$, are to be transferred into a 3-bit register $A$, we use the notation

$$A \leftarrow X$$

Connecting the contents of a register to a vector of output lines requires a different notation than the register transfer notation presented thus far. Unlike a transfer, which is executed by a transfer pulse that appears at the end of some clock period, a connection is in effect for one or more entire clock periods. The "$=$" symbol

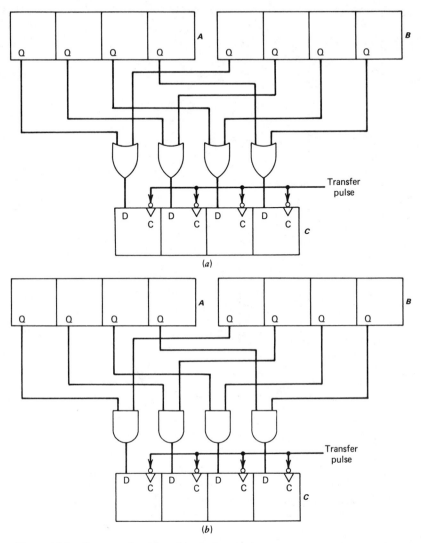

*Figure 10.9   Storage of results of logical operations.*

will be used to denote a connection. For example, the connection of the outputs of register *A* to an output vector *Z* may be expressed as follows.

$$Z = A$$

None of the register transfer implementations shown thus far could have been connected as they were shown if the number of bits on the right side of a transfer arrow were not the same as the number of bits of the target on left side. Similarly, bit-by-bit logical operations between vectors imply that these vectors have the same number of bits. We may state a formal rule that an RTL transfer expression is valid only if the source and target have the same number of bits. With one exception, both arguments of a vector logical operation must have the same number of bits. The exception is that either of the arguments may be a single bit. In this case, the single bit is treated as if it were a vector with the same number of bits as the other argument, but with all bits having the same value. Consider, for example, a 3-bit input vector, $X$, and a 1-bit register, $a$. In this case we may write

$$a \wedge X$$

to express

$$a \wedge X[2], a \wedge X[1], a \wedge X[0].$$

# 10.4  Construction of a Data Unit from an RTL Description

As was discussed in Section 10.1, a complete register transfer description is made up of two distinct sections, the data section and the control section. The data section includes the registers and data paths and logic required to implement a specified set of transfers. The control section issues the necessary signals, in the appropriate sequence, to cause the desired transfers to take place in the specified order. We shall see that the RTL sequence can directly specify the hardware necessary in both the data and control sections. As an example, consider a system with three 2-bit registers, $A$, $B$, and $C$, 2-bit input lines $X$, and 2-bit output lines $Z$, in which the following sequence of operations is to take place. The data on the $X$ lines are to be loaded into register $A$. The data in $A$ are then to be complemented into $C$, with the data in $C$ next shifted into $B$. The data in $A$ and $B$ are then ORed, with the result being stored in $C$. Finally, the data in $C$ are placed on the $Z$ output lines. All this can be compactly described in RTL notation as shown in Fig. 10.10.

First, we simply assign a name to the system, or *module,* for ease of reference. We next specify the principal components of the system in a set of *declaration* statements. These specify the *memory,* that is, those components used to store information, the *inputs,* and the *outputs.* For registers or sets of lines with more than one

```
MODULE: DATA MOVER
   MEMORY: A[2]; B[2]; C[2].
   INPUTS: X[2].
   OUTPUTS: Z[2].

   1   A ← X.
   2   C ← Ā.
   3   B ← C[0],C[1].
   4   C ← A ∨ B.
   5   Z = C.
ENDSEQUENCE.
```

*Figure 10.10   RTL description of DATA MOVER.*

bit, we specify the number of bits, the dimension, in brackets. Next comes the control sequence, a set of numbered steps that specify what is to happen and the order in which it is to happen. The sequence is terminated by the statement END-SEQUENCE. In this sequence we see the two uses of the brackets. In the declaration section, numbers in brackets are always dimensions, for example, **A**[2] indicates that **A** is a 2-bit register. In the control sequence, numbers in brackets are always bit numbers, for example, **C**[0] refers to bit 0 of register **C**.

We shall now show, step-by-step, how the actual hardware design can be determined from this description. First, the right sides of the statements in the sequence specify the signals that must be developed for storage in registers or connection to output lines. From this information we determine the logic and connections required to develop these signals, as shown in Fig. 10.11. For this simple example, the only logic necessary is a pair of OR gates to develop **A** $\vee$ **B**. All other signals are obtained by simple connections. The complement of **A** is obtained from the $Q$ outputs of the **A** register, and the shifting of **C** is accomplished simply by reversing the order of connections to **C**.

The next step is to connect these signals to the destinations specified by the left

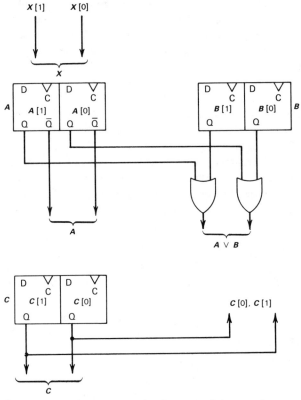

*Figure 10.11   First stage in development of data section.*

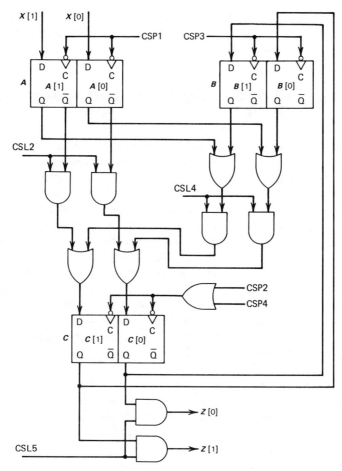

*Figure 10.12    Completion of data section for DATA MOVER.*

sides of the statements in the sequence. These connections are shown in Fig. 10.12. Step 1 calls for loading the **X** lines to the D inputs of register **A** and clocking **A** with *control step pulse* 1 (CSP1), a clock pulse routed to **A** during the time allocated for step 1. Step 2 calls for transferring $\overline{A}$ to **C**. Looking ahead to step 4, we see that a different signal, $A \lor B$, must be transferred to **C**. Thus we cannot connect $\overline{A}$ permanently to the D inputs of **C** but must provide two banks of AND gates to select the inputs to **C** and a bank of OR gates to combine their outputs. The signal $\overline{A}$ is gated through to **C** by *control step level* 2 (CSL2), a level signal that is up during the time allocated for step 2. The signal $A \lor B$ is similarly gated to C during step 4 by CSL4. Step 3 is realized by connecting the shifted output of **C** to the D inputs of **B** and clocking **B** with CSP3. Step 5 is realized by a pair of AND gates that connect the outputs of **C** to the **Z** lines when CSL5 goes high, during the time allocated for step 5. Note carefully the distinction between a transfer into a register (indicated by "←"), and a connection to a set of lines (indicated by "="). When data is loaded into a register, it remains there until something else is loaded. When data is connected to a set of lines, it is on the lines only during the steps for which that connection is specified.

## 10.5   Timing of Connections and Transfers

It is intended that the steps in the DATA MOVER module be executed in order, one each clock period beginning with step 1. In Fig. 10.12 we observed that certain control levels were required in the data unit to accomplish the transfers and connections required in the various steps. If the steps are to be executed in the proper order, the control levels and pulses must be issued in the corresponding order.

The relation between execution timing and the data network can be more clearly understood by considering the timing diagram showing the control signals required by this module, as given in Fig. 10.13. The overall timing of the system is synchronized by a clock signal, and we have assumed that all state changes are triggered by the falling-edge of this clock. Each step in the sequence is allocated one clock period, an interval from the end of one clock pulse to the end of the next. The time in which each step is active is marked by the corresponding CSL signal (control step level). Each CSL signal can be ANDed with the system clock to generate a single clock pulse, the CSP signal for that step. The execution of the sequence starts when CSL1 goes to 1.

In this example CSL1 is not needed in the data section, but it gates a clock pulse onto line CSP1 to load **X** into **A**. The same clock pulse that is gated to form CSP1 must cause the control unit to change states, that is, advance from step 1 to step 2. When CSL2 goes to 1, $\overline{A}$ is gated through the busing network to the data

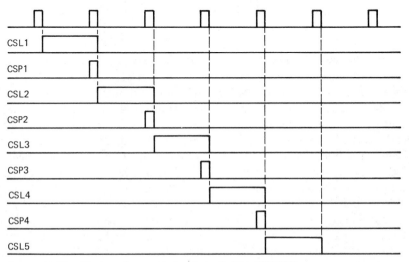

*Figure 10.13   Timing of control signals in DATA MOVER.*

inputs of **C**. At the end of this period, a clock pulse passes through an AND gate to form CSP2 to load **A** into **C**. At step 3, CSL3 gates a clock pulse to CSP3 to transfer the rotated value of **C** into **B**. At step 4, CSL4 gates **A** $\vee$ **B** to the data inputs of **C** and gates a clock pulse to CSP4 to complete the transfer. At step 5, CSL5 gates the outputs of **C** onto the **Z** data lines. No CSP5 is needed, since no registers are loaded at step 5.

A better understanding of execution timing can be obtained by recording the results of execution of the five steps of the RTL sequence. To perform this trace of execution we assume initial values in the registers at the time step 1 is activated and then determine the contents of the registers and the values at the outputs for each succeding clock period. Step 1 is active during the first clock period, and the initial register values are listed in this first column in Fig. 10.14. Notice that the connection of register **C** to the outputs **Z** is effective only in step 5. When any other step is active, **Z** must be at the default values 00. This is the situation during the first clock period as indicated in the first column of Fig. 10.14. For this example the input values were held constant throughout the trace.

In Fig. 10.14 we show the results of a trace of the operation of the DATA MOVER. For ease of interpretation we have circled the contents immediately following the transfer of data into that register. (Recall our basic timing assumption that the contents of a register change only at the end of a clock period for which a transfer into that register is active.) We assume all registers cleared initially and **X** = 10 throughout the sequence. At the end of step 1, **X** is copied into **A**. This value is available at the outputs of register **A** during the second clock period when step 2 is active and will remain there until **A** is again the target of an active transfer. At the end of step 2, $\overline{A}$ is placed in **C**. At step 3, **C** is rotated into **B**. At step 4, **A** $\vee$ **B** is transferred into **C**, and **C** is finally placed on **Z** at step 5. This very simple system does not do much of any interest, but the trace procedure is a very powerful technique in checking digital systems for correct operation. Computer programs called *simulators* are available that can automatically trace the operation of very complex systems, given nothing more than an RTL description.

The function of the control section of the module is to generate the sequence of CSL and CSP signals just described. There are many possible realizations for the

| **Period:** | 1 | 2 | 3 | 4 | 5 |
|---|---|---|---|---|---|
| **Active CSL:** | 1 | 2 | 3 | 4 | 5 |
| **X** | 10 | 10 | 10 | 10 | 10 |
| **A** | 00 | (10) | 10 | 10 | 10 |
| **B** | 00 | 00 | 00 | (10) | 10 |
| **C** | 00 | 00 | (01) | 01 | (10) |
| **Z** | 00 | 00 | 00 | 00 | (10) |

**Register or vector** | **Values**

*Figure 10.14   Trace of DATA MOVER sequence.*

control unit of a particular RTL description, but the often simplest form uses one flip-flop for each control step. A control unit for the DATA MOVER module is shown in Fig. 10.15. We cause execution of the sequence to begin by placing flip-flop 1 in the set state and all the other flip-flops in the reset state, by means that will be discussed later. This places us on the timing chart at the point where CSL1 has just gone up. As noted earlier, this signal is not used in the data section but gates a clock pulse through to form CSP1. At the time of the clock pulse at the end of CSL1, the input to control flip-flop 2 is at the 1-level and the inputs to all the other flip-flops are at the 0-level. On the trailing edge of this clock pulse, the 1 in is transferred to flip-flop 2, driving CSL2 to 1; and 0 is transferred into the other control flip-flops, including control flip-flop 1. The CSL2 signal then gates $\overline{A}$ to the inputs of $C$ and gates a clock pulse to CSP2 to complete the transfer. At the end of CSL2 the process is repeated. Since the input to flip-flop 3 is the only control flip-flop input at the 1-level, the trailing edge of the next clock pulse transfers control activity from flip-flop 2 to flip-flop 3. This process is repeated until all five steps have been executed, at which time all control flip-flops go to the reset state. The alert reader will recognize that this is nothing more than a shift register, into which we initially load one 1 into the first bit position and then shift this 1 through to place 1's successively on each of the five CSL lines.

We noted earlier that we must get the control unit into step 1 to start the sequence. This is basically done in much the same way we reset counters to a desired starting count, through the use of the direct set and reset lines on the flip-flops. However, there is a special problem in that the reset must be done in such a manner as to assure that we get at least a full clock period for the first step. The details of how this is done will not be considered at this time. However, the RTL description should reflect the desired starting condition; for this purpose, after ENDSEQUENCE we add the statements:

CONTROL RESET (1).
END.

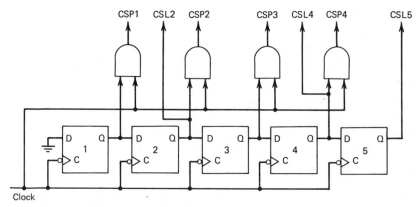

*Figure 10.15   Control unit for DATA MOVER.*

The first statement indicates that circuitry must be included to reset the control unit to step 1, the second signifies the end of the RTL description.

We have now completed the design of the DATA MOVER module. The data section of Fig. 10.12 and the control section of Fig. 10.15 taken together comprise the complete system. It is worthy of note that the process of translating the RTL description into hardware is very straightforward, almost mechanical, so much so that it is quite possible to program a computer to do the translation. This is an important consideration in the design of large systems, in which the sheer labor of just generating the design documentation makes computer assistance imperative.

## 10.6   Sequencing of Control

The sequence discussed in the previous section is a rather uninteresting one in that it executes a single unvarying sequence of transfers and then stops. At the beginning of this chapter we developed the ASM chart for a more interesting circuit, the sequential comparator. The control unit for this example did not merely advance to the next control state each clock period, as was the case for the data mover. In the sequential comparator, control waits at states $Q1$ and $Q3$ for external inputs to go to 1. It also loops in state $Q5$ for four clock periods until the counter is incremented to 11. It is important that we be able to represent this more general sequencing of control in an RTL description.

In programming terms, we need a branching capability. For this purpose, each RTL step may consist of two parts—a *transfer statement* and a *branch statement*. We have already seen examples of transfer statements; they simply specify what is done at a particular step. The branch part specifies the next step to be executed. A numbered step need not include both parts. If the branch statement is omitted, the next sequential statement is executed. If the transfer statement is omitted, a branch is made to another statement, but nothing else is done.

As in programming, we have two types of branches, conditional and unconditional. The *unconditional branch* has the general form

$$\rightarrow (S)$$

where S is a statement number, and simply means, "GO TO S." In the example sequence in the last section, assume that after executing step 5 we wish to go back to step 1 and repeat the sequence. To do this we simply add the unconditional branch

$$\rightarrow (1)$$

to statement 5.

The *conditional branch* has the general form

$$\rightarrow (f_1, f_2, \ldots f_h)/(S_1, S_2, \ldots S_n)$$

The functions $f_1, f_2, \ldots$, are logic functions of the system variables that take on the values 0 or 1. If a given $f_i = 1$, then $S_i$, the statement number in the corresponding position to the right of the slash is the number of the statement to be executed next.

Only one of the $f_i$ may be 1 at a time; if all $f_i = 0$, the next sequential statement is executed. As an example, the conditional branch statement

$$4 \rightarrow (\overline{X[1]}, X[1])/(5,7)$$

specifies that statement 5 is to be executed next if $X[1] = 0$, and statement 7 is to be executed next if $X[1] = 1$.

We will not be ready to write the RTL description for the sequential comparator until still another language feature is discussed in the next section. For now, we illustrate our branching notation by extending the data mover.

### Example 10.1

Write the sequence for a new DATA MOVER module that has an additional register, $D$. Following step 3, we wish to choose between sending $A \vee B$ to $C$ or $D$ and connecting $C$ or $D$ to $Z$, depending on $A[0]$. Control should return to step 1 after connecting $C$ or $D$ to $Z$.

Solution

```
        MODULE; DATAMOVER2
           MEMORY: A [2]; B [2]; C [2]; D [2].
           INPUTS: X [2].
           OUTPUTS: Z [2].
              1 A ← X.
              2 C ← A̅
              3 B ← C [0], C [1];
                 →(A̅ [̅0̅]̅, A [0])/(4,6)
              4 C ← A ∨ B.
              5 Z = C;
                 →(1).
              6 D ← A ∨ B.
              7 Z = D;
                 →(1).
        ENDSEQUENCE
           CONTROL RESET (1).
        END.
```

Following step 3, steps 4 and 5 are executed if $A[0] = 0$; steps 6 and 7 are executed if $A[0] = 1$.  ∎

Where we have a two-way branch with one destination, the next statement, we can simplify the statement by taking advantage of the default option that the next statement is executed if all $f_i = 0$. For example, step 3 in Example 10.1 can be simplified to the form

$$3. \quad B \leftarrow C[0], C[1];$$
$$\rightarrow (A[0])/(6).$$

This form of branch can be interpreted to mean, "IF $A[0] = 1$, THEN go to 6, ELSE go to 4." This IF-THEN-ELSE form is standard in many programming languages. If there are more than two branch destinations, the default option should not be used, because it is too easy to overlook the default destination in generating the hardware.

The presence of branches in the sequence has no effect on the data section. Every transfer or connection in the sequence must be provided for, irrespective of whether the transfers are executed every time the sequence is executed or only sometimes. Branches do somewhat complicate the realization of the control unit, but less than might be expected. Figure 10.16 shows the control unit for DATAMOVER 2. The two unconditional branches at steps 5 and 7 simply require routing the outputs of flip-flops 5 and 7 to the input of flip-flop 1; since there are two steps leading to step 1, we need an OR gate at the input to flip-flop 1. The conditional branch is implemented by the AND gates following flip-flop 3. During the execution of step 3 the signal CSL3 is at the 1-level. If $A[0] = 0$, this 1 is routed to the input of flip-flop 4; if $A[0] = 1$, it is routed to flip-flop 6. When the clock pulse at the end of CSL3 arrives, it turns off FF3 and turns on FF4 or FF6, depending on $A[0]$. Following this, the sequence proceeds unconditionally to step 5 or step 7 and thence back to step 1.

The procedure for implementing branches is always the same as in this example. For conditional branches, a set of AND gates is required to route the CSL from the branch statement flip-flop to the appropriate next-step flip-flop, as determined by the $f_i$ functions. If the $f_i$ functions are more complicated than single bits, separate logic may be necessary to realize them. If a single step is the destination of two or more branches, an OR gate will be required to combine the branch signals at the input of the destination flip-flop.

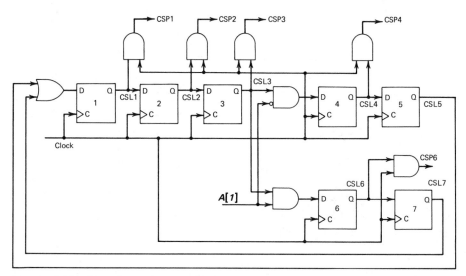

*Figure 10.16  Control unit for DATA MOVER 2.*

## 10.7  Combinational Logic and Conditional Transfers

We are now ready to turn our attention back to the sequential comparator of Fig. 10.3. Let us focus in particular on the activity that takes place when the circuit is in control state $Q4$.

---

If $X > A$, then transfer $X$ to $A$.

---

This would seem to be a simple transfer statement except that it does not always take place when the control unit is in state $Q4$. That is, the transfer is to happen only if the condition "$X > A$" is satisfied. Hence we call this transfer a *"conditional transfer."* If we assume that step 4 of an RTL description corresponds to control state $Q4$ on the ASM chart, we may represent the control and data unit hardware involved in the implementation of this conditional transfer as shown in Fig. 10.17. Of the eight data flip-flops composing the vector $A$, only one, $A[0]$, is actually shown in the figure. The control pulse line will actually fan out to all eight of the data flip-flops.

We now have a hardware implementation of the conditional transfer but have not yet been able to express it in our RTL. To do this we need, in addition to a notation for conditional transfer, a formal way of expressing "$X > A$". Figure 10.17 does not show the origin of the wire labeled "$X > A$" that will be 1 if the binary number represented by the 8-bit vector $X$ is greater than the 8-bit number stored in $A$. This line is the output of a combinational logic network with 16 inputs including all the elements of $X$ and $A$. We would not like to write out the Boolean expression for this network explicitly in our AHPL description.

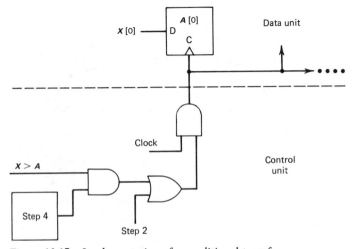

*Figure 10.17   Implementation of a conditional transfer.*

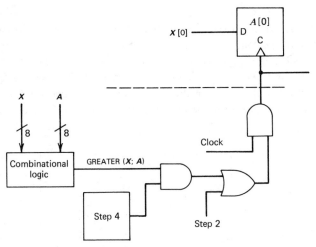

*Figure 10.18   Using a combinational logic unit.*

Let us instead use a function-type notation to refer to the output of networks that we know are strictly combinational logic. For example, GREATER($X$;$A$) will be used to represent a variable that will be 1 whenever "$X > A$" and 0 otherwise. The physical realization of this function is symbolized by a box in Fig. 10.18, and the output is connected into the same AND gate input of the same network given in Fig. 10.17. We will not develop the actual combinational logic network represented by the box at this time, but it can be done easily using repetitive sections much like the adder network. We shall find the following practice helpful throughout the book. Once the function of a combinational logic network is clearly understood, we may introduce its output into an RTL description using functional notation. A combinational logic network realization may be generated at our leisure.

It is clear from the circuit of Fig. 10.17 that a clock pulse will be gated into register $A$ and a transfer actually take place at step 4 only if GREATER($X$;$A$) = 1. The notation for expressing this additional condition in our RTL is as follows.

$A * \text{GREATER}(X;A) \leftarrow X$

The "*" when appearing on the left-hand side of a transfer will uniformly mean that the vector at the right of the "*" will be the *target* of the transfer (get a clock pulse) only if the scalar variable at the right of the "*" is 1. We shall refer to transfers denoted in this manner as "conditional transfers."

We are finally ready to develop the RTL description of the sequential comparator of Fig. 10.3 as an illustration of our notations for combinational logic and the conditional transfer.

### Example 10.2

Translate the ASM Chart of Fig. 10.3 to an RTL Description of the Sequential Comparator.

Solution

It was established earlier in the chapter that each block of the ASM chart corresponds to a state of the control unit. Similarly, it was established in Section 10.5 that each step of an RTL description corresponds to a control unit state. Therefore blocks $Q1$, $Q2$, $Q3$, $Q4$, and $Q5$ in Fig. 10.3 can correspond 1, 2, 3, 4, and 5 of an RTL description. Anticipating a control unit realization using one flip-flop per control state, we may now assign state-variable values to the control states to yield the more

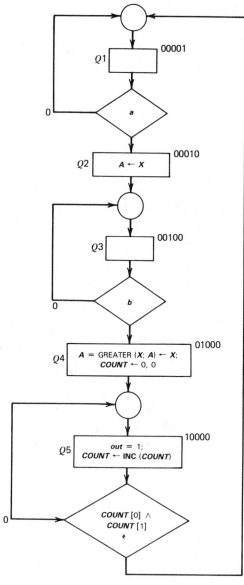

*Figure 10.19  Sequential comparator ASM chart with control-state variables assigned.*

complete ASM chart of Fig. 10.19. We have also replaced the informal representation of transfers in the various boxes with RTL notation.

The first three lines of the following RTL description are declarations of the input and output lines and the memory elements given in the sketch of the sequential comparator of Fig. 10.2. The remainder of the RTL description was obtained by merely copying the ASM chart of Fig. 10.19 into line-by-line program format. The transfer statements are in the boxes, and the branch statements are obtained from the decision diamonds of the same state blocks.

```
MODULE: SEQUENTIAL COMPARATOR
INPUTS: X[8] ; a; b.
MEMORY: A[8]; COUNT[2].
OUTPUTS: Z[8]; out
1.  → (a̅)/(1).
2.  A ← X.
3.  →(b̅)/(3).
4.  A * GREATER(X ; A) ← X; COUNT ← 0,0.
5.  COUNT ← COUNT[0] ⊕ COUNT[1], COUNT[1];
    out = 1;
    →( ∧/COUNT, ∧/COUNT)/(1, 5).
ENDSEQUENCE
CONTROLRESET(1) ; Z = A.
END.
```

The purpose of the first statement in step 5 is to increment the counter. As with the counters of Chapter 9, the least significant bit will change values every clock period. The more significant bit, $COUNT[0]$, will change values if and only if $COUNT[1] = 1$, as expressed using the exclusive-OR gate.  ∎

The 2-bit vector, $COUNT[1] ⊕ COUNT[0]$, $\overline{COUNT[1]}$, in this example could have been expressed as a combinational logic function. The notation

$$COUNT ← INC(COUNT)$$

will be used to express the incrementing of counters of any number of bits in the remainder of the book.

Also new in this example is the statement, $Z = A$, after ENDSEQUENCE. Statements appearing after this keyword are in effect every clock period. Connections, such as $Z = A$, are permanent, and transfers appearing here are executed by every clock pulse.

A special form of branch is the waiting loop, which is used when we wish to wait at a certain point in a sequence for something to happen before proceeding. In the example sequence we read in a value from the $X$ lines at steps 2 and 4. The waiting loops at steps 1 and 3 provide for waiting until external control signals tell us that there is meaningful data available on lines $X$. Step 1 is an example of a branch-only step that executes no transfers, but simply waits at step 1 until $a$ goes

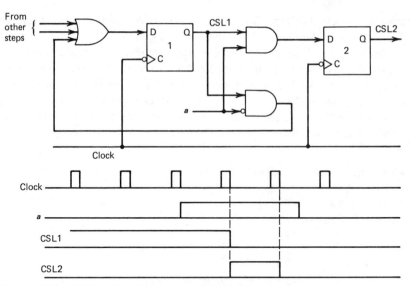

*Figure 10.20   Waiting loop hardware and timing diagram.*

up, at which time control advances to step 2. Figure 10.20 shows the control circuit to implement this waiting loop and the corresponding timing diagram.

At the beginning of the timing diagram we assume that CSL1 is already up, as the result of a CONTROL RESET or a branch from some other step. As long as $a = 0$, the control level is gated back to the input of FF1, which, therefore, remains set each time that it is clocked. When $a$ goes to 1, CSL1 is gated to the input of FF2 so that the next clock pulse turns off CSL1 and turns on CSL2. There are some important restrictions on the $a$ signal. It must be synchronized with the sequential comparator clock, and it should go down before the sequence returns to step 1, unless it is desired to gate in the same data from $X$ again. These restrictions must be taken care of in the design of the unit that loads the $X$ lines and supplies the control signals.

### Example 10.3

Obtain a simulation trace of the functioning of the SEQUENTIAL COMPARA-TOR similar to that obtained for the DATA MOVER in Fig. 10.14. Use the RTL description of the SEQUENTIAL COMPARATOR exactly as given here except let $X$, $A$, and $Z$ be 4-bit data vectors to reduce the space required by the data tabulation. Assume that execution begins in control step 1 following a control reset.

### Solution

The inputs as shown in the first three rows of Fig. 10.21 for all 11 clock periods are assumed to be supplied from outside the comparator. The task is to execute the steps of the control sequence using these inputs in the proper clock periods to determine the entries shown in the last five rows of the chart.                                    ∎

You should give careful attention to the chart of Fig. 10.21 as a way of review-ing the timing relationship between the control and data units. Several features are

## Period

| | 1 | 2 | 3 | 4 | 5 | 6 | 7 | 8 | 9 | 10 | 11 |
|---|---|---|---|---|---|---|---|---|---|---|---|
| a | 0 | 1 | 1 | 0 | 0 | 0 | 0 | 0 | 0 | 0 | 0 |
| b | 0 | 0 | 0 | 0 | 1 | 1 | 0 | 0 | 0 | 0 | 0 |
| X | 0000 | 0001 | 0001 | 0000 | 0010 | 0010 | 0000 | 0000 | 0000 | 0000 | 0000 |

## Control step

| | 1 | 2 | 3 | 4 | 3 | 4 | 5 | 5 | 5 | 5 | 1 |
|---|---|---|---|---|---|---|---|---|---|---|---|
| A | 0000 | 0000 | 0000 | 0001 | 0001 | 0001 | 0010 | 0010 | 0010 | 0010 | 0010 |
| COUNT | 01 | 01 | 01 | 01 | 01 | 01 | 00 | 01 | 10 | 11 | 00 |
| out | 0 | 0 | 0 | 0 | 0 | 0 | 1 | 1 | 1 | 1 | 0 |
| Z | 0000 | 0000 | 0000 | 0001 | 0001 | 0001 | 0010 | 0010 | 0010 | 0010 | 0010 |

Figure 10.21   Simulation of the SEQUENTIAL COMPARATOR.

illustrated here, including conditional branching, conditional transfers, waiting loops, and counting, that were not present in the data mover example. Notice in particular that the system waits in step 1 until *a* goes to 1 during the second clock period. The clock pulse at the end of this period sends the control unit to step 2. The pulse at the end of the third clock period transfers the value of *X*, 0001, into *A*, as called for by step 2, and sends control to step 3. The circuit remains in step 3 until *b* goes to 1 during clock period 5; and the pulse at the end of this period sends control to step 4. Since the value of *X*, 0010, during period six is greater than the contents of *A*, step 4 causes this value to be transferred into *A* while the counter is reset to 00. Notice that control remains in step 5 for four clock periods until the value of *COUNT* reaches 11. *COUNT* = 11 during period 10, and the pulse at the end of this clock period returns control to step 1 as specified by the branch in step 5.

The connection statement, *out* = 1, is effective only for step 5, so this value is 0 except for clock periods 7–10 when control is in step 5. The connection *Z* = *A* appears after ENDSEQUENCE, so *Z* has the same value as *A* during every clock period.

## 10.8   Graphical and RTL Bus Notation

We have seen that we can select among several sources for transfer to a particular target. We can also select among several targets for a transfer. As an example, consider the DATAMOVER 2 module, for which we discussed the modifications to the control unit in Example 10.1. In that module we wish to make provision for transferring either *A* or *A* ∨ *B* to either of two targets, *C* or *D*. The necessary modifications to the data section are shown in Fig. 10.22. The outputs of the OR gates combining the gated sources are connected to both target registers. We select the source, *A* or *A* ∨ *B*, with a CSL signal and the target, *C* or *D*, with a CSP signal. There is one difference between source selection and target selection; we can select only one source at a time, but there is no reason it cannot be gated into both registers at the same time, simply by clocking both of them at the same time.

Just as we needed the register transfer notation to make descriptions of complex systems reasonably compact, so we need more compact block diagram forms. The data section shown in Fig. 10.2 is complicated even for the 2-bit registers implemented; if the registers were 16 bits, the diagram would be very unwieldy. In most cases we shall instead use the simple form of Fig. 10.23, which shows the complete data section for DATA MOVER 2. Registers are represented by single boxes with the number of bits indicated. Sets of identical gates will be indicated by single-gate symbols, again with the number of bits indicated. Multiple lines are indicated in the manner first shown in Fig. 7.32, by lines with a slash through them and the number of bits shown next to the slash.

Whenever we have multiple sources and targets, we need AND–OR networks of the form shown in Fig. 10.22. A network of this form providing a standard communication path between registers, is known as a *bus* and is indicated on a standard block diagram by a single heavy line. Buses differ greatly in details of electronic

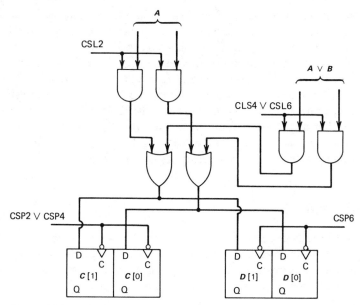

*Figure 10.22   Connection to multiple targets.*

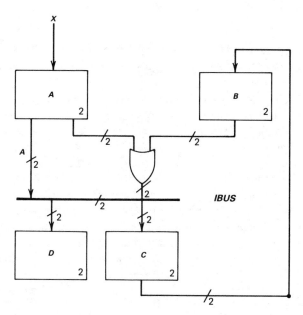

*Figure 10.23   Simplified bus-block diagram.*

realization, but we shall assume they always take the same logical form. An $n$-bit bus consists logically of $n$ OR gates driven by as many banks of $n$ AND gates as there are separate sources. Since the simplified block diagram does not show the CSL and CSP signals, it conveys no information about the sequence of transfers. The RTL sequence and the control-section diagram provide that information.

If a user wishes to employ a bus with contents that can be transferred into more than one target register as shown in Fig. 10.22, it is necessary to declare the existence of that bus. A single-target bus is generated automatically in the process of assembly of the various transfers to that target into network form. In our RTL we may declare a bus with a statement of the following form in the declaration section.

       BUSES: *IBUS* [2]; . . .

Once a bus has been declared, connections to the buses may be specified in connection statements throughout the RTL sequence. Like connections to outputs, data connected to a bus appears at the bus output during the entire clock period of the step in which the connection is active. Therefore, executing

       8. *IBUS* $=$ *A* $\lor$ *B*; *C* $\leftarrow$ *IBUS*.

accomplishes exactly the same results as does

       8. *C* $\leftarrow$ *A* $\lor$ *B*

when the step is active.

## 10.9  Timing Refinements in RTL Systems

Throughout this chapter we have favored the use of a trailing-edge $D$ flip-flop for implementation of RTL descriptions. The reason for this is the convenience of ANDing a control level with the system clock to generate a transfer pulse. If the leading edge of the clock were used as the synchronizing edge, this ANDing could not be accomplished, since the control level would not overlap the clock pulse.

A practical problem often arises when an RTL system is to be physically implemented because many standard TTL chips come with rising-edge triggering. For example, a standard $D$ flip-flop chip is the 7474, which is rising-edge triggered. In dealing with this problem it is most important to keep in mind the goal, which is that all changes of state in the control unit and in the data unit are to be triggered by the trailing edge of clock pulses. Accomplishing this in the control unit requires only that the system clock be passed through an inverter before it reaches the clock input of rising-edge triggered control flip-flops. This is depicted on the second line of the timing chart of Fig. 10.24.

In order that the same trailing edge of the clock synchronize transfers in the

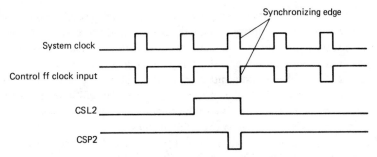

*Figure 10.24   Using rising edge-triggered flip-flops.*

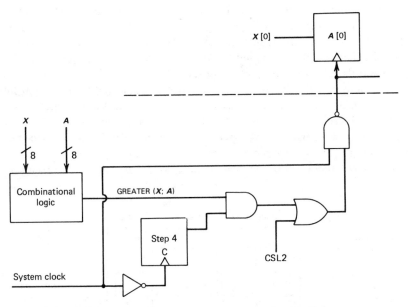

*Figure 10.25   Modification of Fig. 10.18 for rising-edge flip-flops.*

data unit as well as in the control unit, the control step pulses must be effectively inverted. This is easily accomplished by using a NAND gate rather than an AND gate to generate the control step pulses. This approach is illustrated by the modification of Fig. 10.18 to use rising edge-triggered pulses, as shown in Fig. 10.25. The timing relationship of the control-step pulse corresponding to an occurence of a 1 on CSL2 is depicted by the last two lines of Fig. 10.24.

# 10.10  An RTL Design Example

To complete this chapter, let us consider a more complex example. It is desired to design an automatic tester for logic circuits having six inputs and six outputs. The only way to test a logic circuit completely is to apply all possible inputs and check to see that the outputs are correct. With six bits of input there are 64 possible input combinations, so that doing this manually would be quite tedious. We shall store the 64 possible input combinations and the corresponding outputs on tape, one pair of 6-bit characters for each input–output combination. The basic block diagram of the system is shown in Fig. 10.26. When the tester is reset it will issue a *rewind* signal to the tape reader, which will rewind to the start of the tape and issue a *ready* signal when the first input is available on the *X* input lines.

When the operator has loaded a circuit to be tested into the test fixture, he or she will press the *go* button. The tester will then read the first input from *X* into the register *IN* and issue the *step* signal to the tape reader. The reader will drop the *ready* signal, advance the tape, and raise *ready* again when the new character is stable on *X*. When the tester sees *ready* go high again, it will read the next character, containing the correct circuit output, into the register *OUT,* and issue the *step* signal so that the reader will be bringing the next input character into position while the tester proceeds with the actual testing of the circuit.

The output of the *IN* register is permanently connected to the output lines *Z*, which in turn drives the circuit under test. The outputs of the test circuit drive the lines *Y*, which are exclusive-ORed with the desired output stored in *OUT,* the result being stored in *A*. If the outputs agree, the result in *A* will be all 0's. We therefore OR all six bits of *A* together and load the results into flip-flop *bf*. If the circuit output was not correct, there will be at least one in *A*, so that *bf* will be set, raising the line

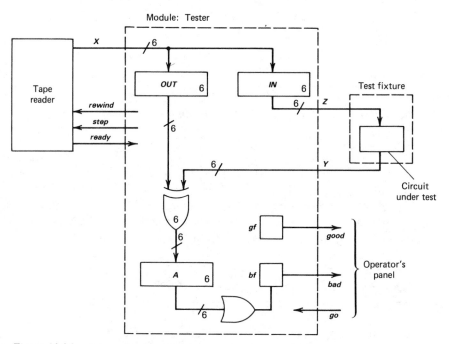

*Figure 10.26   Automatic circuit-tester system.*

*bad,* which will turn on a light to notify the operator that the circuit has failed the test. If the circuit was bad, we then go back to step 1 to rewind the tape and wait for the operator to load another circuit. If the circuit passed this test (*bf* = 0), we must check to see if all inputs have been applied. The inputs will be on the tape in numerical order; that is, the first test input is 000000, the last is 111111. At this point, then, we check the register *IN* to see if it contains all 1's. If it does, the circuit has passed all tests, so we set the flip-flop *gf,* which drives the line *good* to signal the operator, and return to step 1 to rewind and wait for the next circuit. If we are not finished with the test, we return to read the next pair of characters. The RTL sequence for the tester is shown in Fig. 10.27.

Note how compact this description is compared to the long verbal description. A few items of new notation that require explanation. In the declarations, if a dimension is 1, that is, a single flip-flop or line, the dimension is omitted. Two or more transfer statements in the same step, separated by semicolons, indicate separate

MODULE: TESTER
    MEMORY: *IN*[6]; *OUT*[6]; *A*[6]; *gf; bf.*
    INPUTS: *X*[6]; *Y*[6]; *ready; go.*
    OUTPUTS: *Z*[6]; *rewind; step; good; bad.*

    1  *rewind* = 1.

    2  →($\overline{go}$)/(2).

    3  *bf* ← 0; *gf* ← 0;
       →($\overline{ready}$)/(3).

    4  *IN* ← *X*; *step* = 1.

    5  →($\overline{ready}$)/(5).

    6  *OUT* ← *X*; step = 1.

    7  *A* ← *out* ⊕ *Y*

    8  *bf* ← ∨/*A.*

    9  →(*bf*)/(1).

    10  NO DELAY
        →($\overline{∧/IN}$)/(3).

    11  *gf* ← 1;
        →(1).

ENDSEQUENCE
    CONTROL RESET (1);
      *Z* = *IN*; *good* = *gf; bad* = *bf.*
END

*Figure 10.27   RTL sequence for TESTER.*

actions taken in the same step. At step 3, for example, the **bf** and **gf** flip-flops are both cleared. As a matter of fact, **bf** and **gf** will be repetitively cleared as long as **ready** = 0; this does no harm and saves us a step. At steps 8 and 10 we see examples of a new notation. The notation of a logical operator followed by a slash and the name of a register or set of lines means that the operation is successively applied to all bits. The notation

$$bf \leftarrow \vee / A$$

is equivalent to

$$bf \leftarrow A[5] \vee A[4] \vee A[3] \vee A[2] \vee A[1] \vee A[0]$$

In step 10 we call for **AND**ing together all the bits of **IN** and complementing the result. The result of **AND**ing together all the bits in a single bit that is 1 only if all the bits of **IN** are 1. Step 10 thus calls for returning to step 3 unless all the bits of **IN** are 1. Finally, we note connection statements after ENDSEQUENCE. These denote permanent connections of output lines to the outputs of registers. For example, the statement

$$bad = bf$$

after ENDSEQUENCE indicates that the line **bad** is permanently connected to the Q output at the **bf** flip-flop. By contrast, connection statements in the sequence denote connections that are made only during the execution of a specific step.

The data and control sections for this module are shown in Figs. 10.28 and

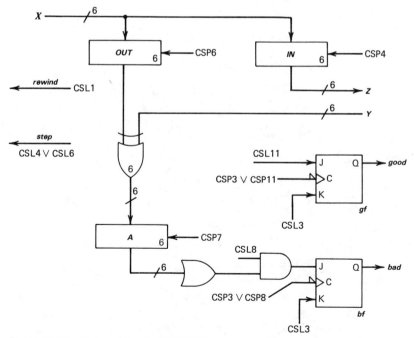

*Figure 10.28   Data section for TESTER.*

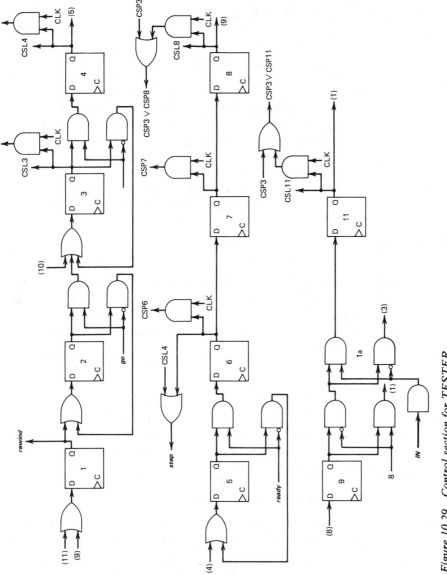

Figure 10.29  Control section for TESTER.

10.29. The derivation of the data section from the RTL sequence should be fairly obvious except for a couple of points. When we have flip-flops that need to be separately set and reset, for example, the *bf* and *gf* flip-flops in this example, it is often convenient to use *J-K* flip-flops, controlling the *J* and *K* inputs with CSL signals and clocking the flip-flops with CSP signals when it is desired to set or reset them. When it is desired to set an output line to 1 for a single step, as in steps 1, 4, and 6 in this sequence, all that is necessary is to connect the output line to the corresponding CSL line.

The derivation of the control unit should also be reasonably evident except for one point. You will note the comment **NO DELAY** after step 10. Steps 9 and 10 call for two branch decisions in a row, and there is no reason why they should be separated by a clock period. The contents of **IN** were established several steps previously and are not affected by the decision at step 9. Indeed, steps 9 and 10 could be combined into a single three-way branch, but writing them as separate steps makes it easier to write the sequence and to design the logic. In terms of the control section hardware, the **NO DELAY** statement simply means that there is no flip-flop associated with step 10; the hardware for step 10 consists of just the branch logic. In terms of timing, both branches are executed in the clock period associated with CSL9.

## Problems

**10.1**   Perform a simulation trace on the DATA MOVER 2 module of Example 10.1. Obtain a chart similar to Fig. 10.14 showing the values of all declared inputs, memory elements, and outputs for as many clock periods as required by parts (a) and (b) that follow. Also indicate the active control step for each clock period.
   (a) Start at step 1, assuming *X* = 01, and continue the trace until the module returns to step 1.
   (b) Repeat part (a), assuming *X* = 10.

**10.2**   Shown in Fig. P10.2 is the RTL description of a simple module.
   (a) Perform a trace on this module, assuming *X* = 0101 at the time of step 1, *X* = 1001 at the time of step 3, and *y* = 0 throughout. Carry the trace through to Step 7.
   (b) Repeat (a), assuming *X* = 1001 at the time of step 1, *X* = 0101 at the time of step 3, and *y* = 1 throughout.

**10.3**   Given the system described by the RTL sequence of Fig. P10.3, draw detailed logic diagrams of the data section and the control section.

**10.4**   (a) Draw the data section for the KLUGE module of Problem 10.2. Use the simplified form similar to Fig. 10.23.
   (b) Draw the control section for the KLUGE module.

**10.5**   In a certain RTL sequence, step is a branch only (no transfer). The drawing to the right shows the The section of the control unit for this step is shown in Fig. P10.5. Determine the RTL statement for step 3.

MODULE: KLUGE
MEMORY: $A[4]$; $B[4]$; $C[4]$.
INPUTS: $X[4]$; $y$.
OUTPUTS: $Z[4]$.

1    $A \leftarrow X$.

2    $B \leftarrow \overline{A}$

3    $A \leftarrow X$.

4    $B \leftarrow B \vee A$;
     $\rightarrow (A[3]\,)/(7)$.

5    $A * y \leftarrow A$.

6    $A \leftarrow A \wedge B$.

7    $C \leftarrow A \vee B$;
     $Z = A$.

ENDSEQUENCE
     CONTROL RESET (1).
END.

*Figure P10.2*

MODULE: SHIFTER
MEMORY: $R[3]$.
INPUTS: $X[3]$; $a$; $b$.
OUTPUTS: $Z[3]$; *ready*.

1    $R \leftarrow X$;
     $\rightarrow ((\overline{a} \wedge \overline{b}), a, (\overline{a} \wedge b))/(1,2,4)$.

2    $R \leftarrow R[0], R[2:1]$;
     $\rightarrow (\overline{b})/(4)$.

3    $R \leftarrow R[0], R[2:1]$.

4    *ready* $= 1$; $Z = R$;
     $\rightarrow (1)$.

ENDSEQUENCE
     CONTROL RESET (1).
END.

*Figure P10.3*

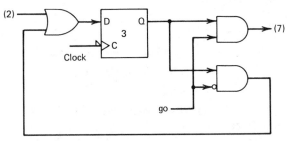

*Figure P10.5*

**10.6**  The RTL description of Fig. P10.6 specifies an alternate realization of the up–down counter of Fig. 8.29. Obtain a network realization of this description. DEC is a combinational logic function representing the decremented value of the argument *CNT*.

(a) Compare the cost of the hardware realization of this circuit with that of Fig. 8.29.

(b) Does this example accurately reflect the typical cost of a realization obtained from an RTL description with that of the most economical possible realization? Why or why not?

**10.7**  Rewrite the RTL description of the up-down counter of Problem 10.6 to provide for two inputs *"up"* and *"down."* If *up* $= 1$, the counter is to count

MODULE:UPDOWNCOUNTER
INPUTS: *x*.
MEMORY: *CNT*[2].
OUTPUTS: *Z*[2].

ENDSEQUENCE
$CNT \leftarrow (INC(CNT) \land \bar{x}) \lor (DEC(CNT) \land x); Z = CNT.$
END.
*Figure P10.6*

up. If **down** = 1, the counter is to count down. During any clock period in which both **up** and **down** are 0, the counter values are to remain the same. **Hint:** Use conditional transfer notation to cause a transfer into the counter to take place only if either up or down is 1.

**10.8**   Obtain a simulation trace for the RTL description given in Fig. P10.8a. The input sequences and initial control step are shown in Fig. P10.8b. The problem is solved by filling in all blank spaces on the chart.

MODULE:   TEST

INPUTS: *a, x*
MEMORY: *CT*[2]; *b*.
OUTPUTS: *Z; result*
1. $b \leftarrow x; \rightarrow (\bar{a})/(1)$
2. $b \leftarrow x \oplus b;$
3. $CT \leftarrow INC (CT); Z = 1;$
   $\rightarrow (\land/CT, \overline{\land/CT})/(1,3).$

ENDSEQUENCE
   CONTROLRESET(1); *result* = *b*,
END

(*a*)

| Clock period | 1 | 2 | 3 | 4 | 5 | 6 | 7 | 8 |
|---|---|---|---|---|---|---|---|---|
| Control step | 1 | | | | | | | |
| *X* | 0 | 0 | 1 | 1 | 1 | 1 | 1 | 1 |
| *a* | 0 | 1 | 1 | 1 | 1 | 1 | 1 | 1 |
| *b* | 0 | | | | | | | |
| *CT* | 0,1 | | | | | | | |
| *result* | 0 | | | | | | | |

(*b*)

*Figure P10.8*

**10.9** Rewrite the RTL description for the sequential comparator so that three consecutive input vectors can be received as signaled by three control inputs *a*, *b*, and *c*. The largest of the three is to be available on the output lines for five clock periods while *out* = 1.

**10.10** Construct a generalized ASM chart of the form of Fig. 10.19 for the SHIFTER module of Problem 10.3. Assign control-state values for each box so that a control unit with one flip-flop per control state will be specified.

**10.11** A digital system has 8-bit data input lines, *X*[8], and a single input line *ready.* It has two output lines, *same* and *ask.* Each time the system places a 1 on line *ask* for one period, an external system will place new data on the *X* lines, the presence of these new data being indicated by *ready* going high for one period. The data on *X* will remain stable until a new *ask* signal is issued. Each time the input data are exactly the same as the data received at either of the two previous requests, the line *same* should go to 1 for four periods before issuing a new *ask* signal. If the data are not the same, the *same* signal should remain 0 and a new *ask* signal should be issued.
  (a) Construct a generalized ASM chart for the system just described.
  (b) Write an RTL description of this system, specifying registers as needed.
  (c) Draw a simplified diagram of the data section and a complete diagram of the control section.

**10.12** A digital system has two input lines, *control* and *data,* 8-bit output lines *Z*[8], and a single output line, *ready.* Data representing 8-bit words will arrive serially (one bit at a time) over the line data. It is to be converted to parallel form to be put out on the *Z* lines. The serial data may arrive most significant bit first, or least significant bit first. In either case, the parallel output should always be in the standard form, most significant bit in the left-most position.
  Synchronization is provided by interaction between the *control* input and the *ready* output. When the system is ready to receive a new word it will raise the *ready* line and start monitoring *control.* Two consecutive 1's (that is, the line is at 1 for two consecutive clock periods) on line *control* will indicate that the next bit on *control* will indicate the direction of the serial data, 0 for MSB first, 1 for LSB first. The eight bits of data will then follow on the data line the next eight clock periods. *Ready* should go down when the two consecutive 1's on *control* are detected and should go back up when all eight bits have been received. The parallel word should be gated onto the *Z* lines anytime *ready* is up.
  (a) Construct an ASM chart for the serial-to-parallel converter described here.
  (b) Write an RTL description of a system to accomplish this function.
  (c) Draw the detailed block diagram of the data section of the system. Use two 74194 shift register chips for the serial-to-parallel conversion.
  (d) Draw the control section of the system.

**10.13**   An RTL description is to be written for a more sophisticated version of the vending-machine controller of Section 9.3. The price of the item is to be $0.85. The controller is to keep track of received coins using three separate counters for nickels, dimes, and quarters. A single combinational logic function with the three counters as inputs and a single output line may be used. This function, whose realization need not be specified, will be 1 when the value of received coins is greater than or equal to $0.85. A second logic function of these three arguments with two output lines indicating the number of nickels change to be given may also be used. In the process of writing the RTL description construct a generalized ASM chart containing four control states: one for initializing the counters, one for receiving the coins, one in which $A = 1$ for one clock period, and one to provide change. The controller will remain in the last state with $C = 1$ for one clock period for each nickel to be returned.

# References

1. Chu, Y., "Why Do We Need Hardware Description Languages?" *Computer,* December 1974, pp. 18–22.
2. Hill, F. J., and G. R. Peterson, *Digital Systems: Hardware Organization and Design,* 2nd ed., Wiley, New York, 1978.
3. Lipovski, G. J., "Hardware Description Languages," *Computer,* June 1977.

# Small Computer Organization and Programming

## 11.1 Introduction

We are now ready to consider the most versatile, useful, and important type of digital system, the computer. We have studied the basic building blocks of digital systems, gates and flip-flops, and have seen how they can be combined to form more complex elements such as counters, registers, and adders. We have then seen how these elements can be connected to form still more complex systems, described in RTL form. These more complex systems perform functions such as comparing numbers, modifying numbers, transferring numbers from point to point, functions of the sort often performed by computers. Thus, it is a relatively small step from the concepts of the simple register transfer systems we have so far considered to the concepts of complete computers.

The types of systems we have considered through the medium of RTL design procedures perform fixed functions determined by the design of the system. By contrast, computers are versatile devices, designed to be capable of performing a wide variety of tasks, to be determined by the user, not the designer. The user controls the function of the system by means of a set of instructions, a *program*. The concept of controlling a register transfer system is not completely new. In the automatic circuit tester studied earlier, we had control signals, *ready* and *go,* by means of which an operator could control the system. If we were to set up a time sequence of changes in the values of these two signals, this sequence could be considered a program for controlling the system. But it is important to consider how this "program" would be supplied to the system. An operator supplying new values of the two signals at appropriate times would be *controlling* the system, not *programming* it. We shall consider a system to be programmed when the information to control it is prepared in advance and stored in some manner, to be accessed by the system when required.

We now wish to consider practical means of achieving this programmed control, in the *stored-program* computer. In so doing we must consider not only the hardware, but also the means of programming it. This approach is something of a departure from that taken in most books on computers. It has been traditional to consider hardware and software as two separate subjects, obviously related, but rarely considered together. In recent years, however, it has become increasingly evident that the truly competent professional in the computer field must thoroughly understand both hardware and software and the way they interact. In the hopes of developing such understanding, we shall parallel our study of the hardware organization of computers with consideration of the means by which this hardware is controlled by programming.

## 11.2 Central Processor and Memory Organization

That part of a computer that does the actual computing is known as the *central processing unit* (CPU). It is a register transfer system of exactly the sort we have already studied, except more complex. The data section consists of a number of registers, combinational logic for performing various arithmetic and logical operations, and necessary interconnecting data paths. The control unit performs the same functions as the control sections of the register transfer systems already studied; it issues the appropriate control levels and pulses to cause the desired sequence of operations to take place. It differs from the control sections previously studied only in being far more complex, capable of developing many different sequences of control signals, as determined by the stored instructions.

The CPU does the actual processing of data; to do so, it must have data to process and instructions as to how to process these data. All this information is obtained from the *memory,* which also provides a place to store the results. Many different types of memories are used in computer systems, and many computer systems have several different types of memories. The memory with which the CPU communicates most directly is always a *random access* memory.

There are many different types of random access memory in terms of physical realization. Whatever the realization, a random access memory is logically equivalent to a large set of registers, each comprising a storage location. Associated with each location is an identification number, or *address,* by means of which it may be accessed, or referenced, for the purpose of storing information or retrieving previously stored information. The number of locations will typically be in the range of about a thousand to several million. The number of locations is normally a power of two, because the addresses are binary numbers. For example, if the address consists of 10 bits, we can address $2^{10} = 1024$ locations, numbered consecutively from 0 (0000000000) to (1111111111). Since 1024 is close to 1000, it is common to express memory size in multiples of 1024 and use the abbreviation "K" to indicate 1024. A memory with 1024 locations would be referred to as a 1K memory, a memory with 4096 locations (12-bit addresses) would be referred to as a 4K memory, and so forth.

The information stored in the memory is always stored in binary form, that is, strings of 1's and 0's. The amount of information that can be stored in an individually addressable location, expressed in terms of the number of bits, is known as the *word length* of the memory. Nomenclature is sometimes inconsistent, you may hear such terms as byte-organized and variable word length applied to memory. Whatever the type of organization, we shall use the term *memory word* to signify the smallest amount of information that can be individually accessed, or addressed, in memory. Memory-word lengths in computers typically range from 4 bits in some microprocessors to 64 bits in some very large computers.

A model for a typical random access memory is shown in Fig. 11.1. When an address is placed on the **ADDRESS** lines and the *read* line is raised, the contents of the addressed location will be read out onto the **DATAOUT** lines, available for use by the CPU. There is, of course, some delay from the time that *read* is raised to the time the requested word appears on the **DATAOUT** lines. In some cases, the CPU will wait a fixed time before using the data; in other cases, the memory may have an output line that will go up when the data is ready. For simplicity, we shall assume the former in our discussions.

In most cases, the operation of reading does not change the contents of the location read, so that the data is still available for later use. When an address is placed on the **ADDRESS** lines and data on the **DATAIN** lines and the **write** line is raised, the data on the **DATAIN** lines are stored in the addressed location, replacing whatever was there before. Again, there may be an output signal to indicate when the write operation is complete, but we shall assume that the CPU simply waits a fixed time before attempting to use the memory again.

The term *random access* signifies that any randomly chosen location can be accessed just as fast as any other, a requirement that usually necessitates electronic access to the storage locations. Memory systems that require physical movement of the storage medium are not random access. For example, magnetic tape is widely used for memory; the length of time required to access a given unit of information clearly depends on its location on the tape.

In addition to the CPU and the memory, there are two other sections in the

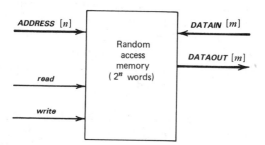

*Figure 11.1   Model of random access memory with n-bit addresses and m-bit words.*

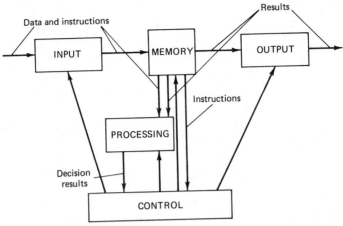

*Figure 11.2   Basic computer organization.*

typical computer system, the *input section* and the *output section*. The input section, consisting of devices such as keyboards and card readers, provides the means for feeding data and instructions to the computer. The output section, consisting of devices such as card punches and printers, provides the means of getting results to the user. The overall block diagram of a typical computer system is shown in Fig. 11.2. It is important to note that the input section puts information into the memory, and the output section gets results from memory. In some computers, the CPU may be involved in input/output operations, but the purpose of input is to get information from the outside world into the memory, and the purpose of output is to get information from the memory to the outside world.

## 11.3   CPU Organization and Instruction Formats

If we wish to run a program on a computer, the first thing we have to do is load the program and the data into memory. Some programs include instructions for loading the data, but we shall assume, for the present, that the programs and the data have already been loaded, by means that need not concern us here. We then enter the address of the first instruction, often by means of a keyboard or switches provided for the purpose. When we press the start button, this address is sent to memory and the first instruction is read out. This information in this instruction will be needed at various times while it is being executed, so we load it into the *instruction register (IR)*.

With the instruction safely preserved, the information contained in that instruction is interpreted by the control unit to determine what is to be done. Let us consider what information is needed to execute an instruction. First, the computer needs to know what operation is to be performed, addition, negation, shifting, or whatever. The *op code* is a numeric code, typically four to eight bits, indicating what operation is to be performed. This part of the instruction is clearly essential; in some cases, such as HALT, it is all the information needed. More commonly, the specified operation is to be performed on some unit of data, and it is necessary to know where the data are located. Some operations, such as the logical complement, involve only one unit of data, or *operand;* an arithmetic operation such as addition involves two operands. Next it is necessary to know what is to be done with the result of the operation and, finally, to know where to go to get the next instruction.

Instructions may be classified by how much information they contain. If an instruction contained all four addresses (two operands, result, next instruction) it would be known as a four-address instruction. However, an instruction containing all that information would be impractically long and unwieldy. Various schemes are used to reduce the amount of information needed in each instruction. In most cases, programs are loaded into memory in the order they are to be executed, one instruction after the other, in sequential locations. To take advantage of this orderly arrangement, we provide a *program counter (PC)*. When we start a program we load the address of the first instruction into *PC* and increment *PC* each time an instruction is executed. As a result, instructions need not include a next-instruction address except for those cases where the next instruction to be executed is not the one in the next sequential location. In computers with program counters (which includes practically all computers) instructions will thus fall into two categories, *transfer instructions* and *branch instructions*. Transfer instructions specify an operation, and branch instructions specify where to go in memory to find the next instruction.

The parallel to branch and transfer steps in RTL descriptions is far from coincidental. It reflects the basic algorithmic nature of digital systems, in which we view all processes as consisting of a series of steps, some specifying particular operations, others specifying the order in which operational steps are to be performed. In RTL systems we combine an operation and a branch into a single step, in programming we usually consider operations and branches as separate steps, but there is no real conceptual difference. This algorithmic view of processes is extremely important because it is one of the main reasons for the power and wide applicability of digital techniques.

Many different formats for transfer instructions are used in various computer systems. Two of the most popular formats are the two-address replacement format and the single-address format. In the two-address replacement format, the two addresses specify locations in memory from which two operands will be obtained, and the result replaces one of the operands. This approach reduces instruction length by eliminating an address, but there are obvious problems with overwriting one operand, thus making it unavailable for further computations. In the single-address format, there is only one memory address, which may specify the source of an operand or the destination of the result, depending on the op code. Registers in the CPU are used to provide the source of the other operand (if any) and/or the destination of the result, with the registers to be used being specified by the op code. If there is a single register, used both as an operand source and destination result, it will typically be called the *accumulator (AC)*. Single-address machines with accumulator are the most common type of computer, and we shall restrict our present discussion to such computers.

The CPU of a single-address computer must contain three registers, the

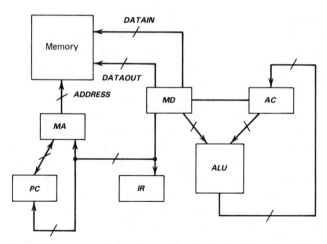

Figure 11.3   *Core organization of a typical single-address computer.*

instruction register *(IR)*, the program counter *(PC)*, and the accumulator *(AC)*. In addition, many computers include two registers used in communicating with memory. The memory address register *(MA)* is the source of addresses for memory; the memory data register *(MD)* is the source of words to be stored in memory and the destination of words read from memory. Depending on the type of memory, these registers may or may not be needed, but we shall assume them in our discussions. Figure 11.3 shows the organization of the CPU and memory of a typical single-address computer. No two computers are organized exactly the same way, but Fig. 11.3 represents the basic structure of most single-address computers with sufficient accuracy for our purposes. The block labeled "ALU" contains the combinational logic necessary to implement the various arithmetic and logical operations on the contents of the registers.

## 11.4  Fundamental Internal Sequence of a Single-Address Computer

Since a computer is clearly a register transfer system, an RTL sequence is the logical way to describe the internal functioning. In the last section, we kept the discussion as general as possible to provide you with an overview of basic features of computer organization. It is not possible to maintain this degree of generality when writing an RTL sequence. We whall have to restrict our discussion to a very specific organization. However, this organization includes features found in many computers, so that the understanding gained from the discussion will be widely applicable. In addition, we shall indicate alternative ways of accomplishing various functions in machines with different organizations. Since our computer will have an 8-bit word length, some of its features will be specific to 8-bit organizations. In most respects, however, its organization is typical of microcomputers and minicomputers of different word lengths.

For ease of reference, we shall name our example computer the *TB6502,* short for textbook version of one of the most commonly used 8-bit microprocessors, the MOS Technology 6502. A few features of the 6502 will not be included in the TB6502, but programs written for the TB6502 will execute on the 6502. Readers with other access to other microprocessors should nonetheless find the concepts of the next few sections useful and easily translatable to the framework of their own laboratory computers. Some other microprocessors, such as the 6809, are also very close to direct extensions of the TB6502. We shall start with a simplified version of the TB6502 and gradually add features as we proceed.

We shall start with an internal organization exactly as shown in Fig. 11.3. The memory word length will be eight bits, **AC, MD,** and **IR** will be eight bits, and **PC** and **MA** will be 16 bits. The use of 16 bits in **PC** and **MA** provides the TB6502 CPU with an address space of 64K words. The *address space* is simply the number of memory locations a CPU is capable of addressing. The fact that a CPU has a certain address space does not mean that a given system need implement that many locations; a user will physically implement only as much of the address space as needed for his other application. An 8-bit word is often referred to as a *byte.* Since the TB6502 is an 8-bit machine, we shall use the terms *word* and *byte* interchangeably in discussing the TB6502, but you should remember that a byte is not equivalent to a word in discussing systems with other word lengths. We shall assume 8-bit opcodes, stored in a single byte in memory. Addresses, if required, will be stored in succeeding bytes.

The RTL sequence that we shall develop will implement the execution of a single instruction. The execution of a complete program will be accomplished by the passage of control through this RTL sequence repeatedly, once for the execution of each instruction. At any time during the running of a program the location in memory of the current instruction can be determined from the contents of the program counter. Just prior to the start of the execution of an instruction, the contents of **PC, MA, MD,** and **IR** will be as shown in Fig. 11.4a. The program counter will contain the 16-bit number $\alpha$, which is the address of the opcode of the instruction to be executed. The remaining three registers will contain information left over from the last instruction, which will no longer be of interest.

The first step in executing an instruction is to fetch the opcode; in the TB6502 this is accomplished by the following three steps. Step 1 transfers the address of the next instruction from **PC** to **MA**. In step 2, we introduce a new notation: **M⟨MA⟩** specifies the location in memory for which the address is in **MA**. Step 2 transfers the contents of this location to **MD,** that is, it reads the instruction from memory and increments the program counter in preparation for fetching the next byte when required. Step 3 transfers the opcode to **IR.**

1. **MA ← PC**
2. **MD ← M⟨MA⟩; PC ← INC(PC)**
3. **IR ← MD**

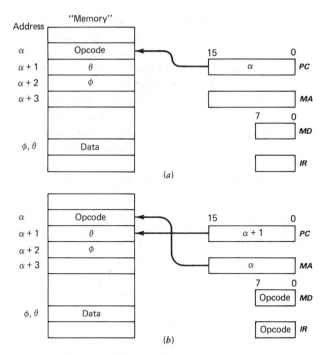

*Figure 11.4   Execution: (a) Begin, (b) after step 3.*

The contents of the four registers of interest are shown again in Fig. 11.4*b* after the completion of step 3. **PC** now contains the number $\alpha + 1$, and **MA** contains $\alpha$. For emphasis, arrows from each of these registers point to the location in memory whose address is stored in that register.

At this point the opcode is in **IR** and the bits of the opcode will be referred to as **IR**[7] to **IR**[0]. The next step is to decode the opcode to determine what is to be done. The usual practice is to divide the instruction set into groups having certain features in common and to use the opcode as a basis for making branches into sequences for various groups. Instructions may be categorized on the basis of the type of instruction, that is, data manipulation, memory reference, transfer of control, and so on, or on the number of bytes in each instruction. The 6502 actually uses a combination of these two approaches. For the TB6502, in the interests of simplicity, we shall decode on the basis of instruction type only. The resultant sequence will differ from that of the 6502 in details of timing, but the end results for all instructions will be identical. In the TB6502 bits **IR**[3:0] will distinguish between general categories of instructions. The remaining bits will specify particular instructions and addressing modes (to be discussed later) within these general categories. Figure 11.5 shows the four major categories of instructions with which we shall be concerned: (1) branch instructions, (2) register only instructions, (3) instructions that reference memory but do not use **AC**, and (4) **AC** and memory instructions.

A memory reference instruction (MRI) is an operation that uses a byte from

| *IR*[0] | *IR*[2] | | |
|---|---|---|---|
| 0 | 0 | *IR*[3] = 0   Transfer of control<br>*IR*[3] = 1   Register only | Non-MRI |
| | 1 | *IR*[1] = 0   Other MRIs<br>*IR*[1] = 1   Memory to memory<br><div align="center">Memory ↔ *X*</div> | MRI |
| 1 | | *AC* and memory | |

*Figure 11.5   Categories of instructions.*

memory as an operand or stores the result of the operation in memory. In Fig. 11.5 we see the instruction will be an MRI unless *IR*[0] = *IR*[2] = 0. Step 4 enforces a branch to another step, if the instruction is not an MRI. Hardware control will pass to step 5 for all instructions that require a data byte to be fetched from memory. For now we shall discuss only those instructions that use the two bytes of memory following the opcode to form the 16-bit address of the desired data. The low-order (least significant) eight bits of the address immediately follows the opcode and is followed by the high-order eight bits. Steps 5 through 8 fetch the 16-bit address and transfer it to *MA*.

    4. → ($\overline{IR[2]}$ ∧ $\overline{IR[0]}$)/(Seq. for non-MRI)
    5. *MA* ← *PC; PC* ← INC*(PC)*
    6. *MD* ← *M⟨MA⟩; MA* ← *PC; PC* ← INC*(PC)*
    7. *MD* ← *M⟨MA⟩; MA*[7:0] ← *MD*
    8. *MA*[15:8] ← *MD*

Step 5 is the first of three transfers in which a single register is the source for one transfer and the destination for another. This is valid because the timing of the registers is the same as in the RTL systems discussed in Chapter 10; that is, the transfer of new data into a register does not cause a change at the output of the register until the end of the time period allocated for that transfer step. In step 5, we transfer the address of the next byte from *PC* to *MA* and, at the same time, increment *PC*. The new value of *PC* is not seen at the outputs of that register until the transfers are complete, so that *MA* receives the original value of *PC*. Such simultaneous transfers are used simply because they save time. Notice from Fig. 11.6*a* that, following step 5, *MA* points to (contains the address of) the low-order address byte and *PC* points at the high-order address byte. At step 6, the first address byte is read out of memory into *MD;* at the same time the address of the next byte is transferred to *MA* and *PC* is again incremented in preparation for fetching the next byte when required. The register contents after completion of step 6 are shown in Fig. 11.6*b*. The program counter now points at the opcode for the next instruction and will remain unchanged until execution of the current instruction has been completed.

    At step 7, the low-order address byte is transferred to the lower half of *MA* at

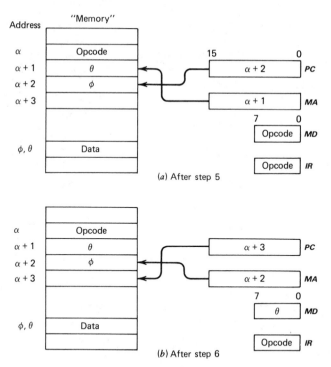

*Figure 11.6   Execution continued: (a) After step 5, (b) after step 6.*

the same time the high-order byte is read out of memory into **MD**. Finally, at step 8, the high-order address byte is transferred to the upper half of **MA**. In Fig. 11.7*b* we see that the complete 16-bit address vector $\phi$, $\theta$ is now in **MA**. This address will be used—most frequently to obtain an operand from memory—during the execution of the instruction.

In the preceding sequence, the separate fetching of the opcode and the address is made necessary by the short word length. In machines with longer words, the opcode and address will normally be stored in a single word in memory. In that event, steps 5 to 8 will not be required, the address being transferred directly to **MA** when the instruction word is fetched.

With the address in **MA,** it is now necessary for the control unit to determine precisely which memory reference instruction is to be executed. The first to be considered are the accumulator and memory instructions, for which **IR**[0] = 1. The TB6502 computer will have seven instructions in the **AC**-and-memory category, as tabulated in Fig. 11.8. Bits **IR**[7:5] specify which of these operations is to be performed, as shown in Fig. 11.8. The "**C**" term in the ADC and SBC operations is the *carry flag,* a single flip-flop that serves a variety of purposes, to be explained later.

The sequence for carrying out these intructions is shown following. At step 9, **IR**[0] is tested to determine if we have an **AC**-and-memory instruction or some other type of MRI instruction. All the **AC**-and-memory commands except STA require reading an operand from memory, so step 10 makes a branch based on the opcode

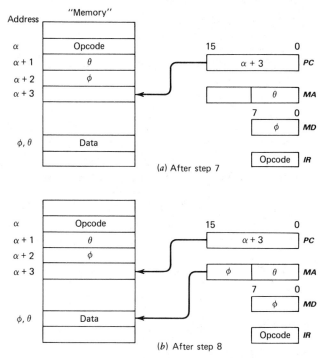

Figure 11.7   *More execution: (a) After step 7, (b) after step 8.*

for STA. If the command is not STA, step 11 fetches the operand from memory. Step 12 executes one of the six commands that modify the contents of the **AC** and returns to step 1 for the next instruction fetch. This step is a complex operation requiring notation we have not developed, so we have simply indicated the purpose of this step. Basically, the contents of **AC** and **MD** are applied to a combinational logic unit (the ALU), which carries out the desired operation (AND, OR, etc.) as determined by the opcode, and the result is stored in **AC**. For STA, steps 13 and 14 store the contents of **AC** in memory and return to step 1 for the next instruction fetch.

| Command | Mnemonic | Opcode/$R[7:5]$ | Operation |
|---|---|---|---|
| Or | ORA | 0 0 0 | $AC \leftarrow AC \vee M\langle MA \rangle$ |
| And | AND | 0 0 1 | $AC \leftarrow AC \wedge M\langle MA \rangle$ |
| Exclusive Or | EOR | 0 1 0 | $AC \leftarrow AC \oplus M\langle MA \rangle$ |
| Add with carry | ADC | 0 1 1 | $AC \leftarrow AC + M\langle MA \rangle + C$ |
| Subtract w/C | SBC | 1 1 1 | $AC \leftarrow AC - M\langle MA \rangle - \overline{C}$ |
| Store accumulator | STA | 1 0 0 | $M\langle MA \rangle \leftarrow AC$ |
| Load accumulator | LDA | 1 0 1 | $AC \leftarrow M\langle MA \rangle$ |

Figure 11.8   *Accumulator-and-memory instructions in the TB6502.*

9. → ($\overline{IR\,[0]}$)/(Seq. for other MRI types)
10. → ($IR\,[7] \wedge \overline{IR\,[6]} \wedge \overline{IR\,[5]}$)/(13)
11. **MD ← M⟨MA⟩**
12. Execute ORA, AND, EOR, ADC, SBC, or LDA based on **IR** [7:5]
    → (1)
13. **MD ← AC**
14. **M⟨MA⟩ ← MD**
    *→ (1)*

At this point we have all the instructions needed to write a simple program. Assume we want to add three numbers that have been previously stored in locations 40, 41, and 42, and store the sum in 43. The following simple program will do the job, assuming the carry flag is initially cleared and the addition does not produce a sum that overflows the capacity of the 8-bit accumulator. We load the program starting at location 200. The first column shows the address of the first byte of each instruction, the second shows the mnemonic for the opcode, and the third shows the operand address. In this listing as in all to follow, 16-bit addresses are represented by four hex digits.

```
0200  LDA   0040
0203  ADC   0041
0206  ADC   0042
0209  STA   0043
020C  next  instruction
```

The first instruction loads the first operand from 40 into **AC**. The second adds the operand from 41, the third adds the operand from 42, and the fourth stores the sum in 43.

The **AC**-and-memory type of instruction is the most common type of MRI instruction, and the only type in many machines. In the TB6502 we have a second type, instructions that modify the contents of a memory location directly, without using **AC**. For these instructions, **IR**[2:0] = 110. There are six instructions, as listed in Fig. 11.9. In the shift and rotate commands, **C** is again the carry flag, which acts as a 1-bit extension, that is, the ninth bit, in these operations.

| Command | Mnemonic | Opcode**IR**[7:5] | Operation |
|---|---|---|---|
| Shift left | ASL | 0 0 0 | $C,M\langle MA\rangle \leftarrow M\langle MA\rangle,0$ |
| Rotate left | ROL | 0 0 1 | $C,M\langle MA\rangle \leftarrow M\langle MA\rangle,C$ |
| Shift right | LSR | 0 1 0 | $M\langle MA\rangle,C \leftarrow 0,M\langle MA\rangle$ |
| Rotate right | ROR | 0 1 1 | $M\langle MA\rangle,C \leftarrow C,M\langle MA\rangle$ |
| Decrement | DEC | 1 1 0 | $M\langle MA\rangle \leftarrow M\langle MA\rangle - 1$ |
| Increment | INC | 1 1 1 | $M\langle MA\rangle \leftarrow M\langle MA\rangle + 1$ |

*Figure 11.9   Memory-only instructions in the TB6502.*

The contents of a memory location cannot be manipulated directly in the memory. The execution of these instructions will involve reading the operand out of memory, processing it in the CPU without using **AC,** and writing the result back into memory. Because of this procedure, such instructions are sometimes referred to as memory-to-memory instructions. At step 9 we branched out of the **AC**-and-memory sequence following fetch of the address. The sequence for other MRI instructions will start at step 15.

9.  → $(\overline{IR[0]})$/(15)

There are other categories of MRI instructions, involving memory and other registers not yet defined, so we specify a branch at step 15 to separate memory-to-memory instructions from other MRI's.

15.  → $(\overline{IR[1]})$/(other MRI's)

Step 16 fetches the operand, step 17 carries out the operation specified by the opcode, and step 18 completes the operation by storing the result in the addressed location. The operand fetch could have been shared with the **AC**-and-memory sequence but that would have complicated things considerably.

16. **MD ← M⟨MA⟩**
17. Execute ASL, ROL, LSR, ROR,
    DEC, INC with result going to **MD.**
18. **M⟨MA⟩ ← MD**
    → **(1)**

At this point the sequence has become a bit complicated, so it may be helpful to look at a flowchart of the sequence, as shown in Fig. 11.10. The first three boxes represent the fetch of the instruction. The branch at step 4 represents the decision-choosing between sequences for instructions that access memory and those that do not. The remainder of the flowchart can be similarly interpreted with reference to the preceding sequence.

# 11.5   Transfer-of-Control and Register-Only Instructions

The simplest transfer-of-control instruction is the unconditional jump, JMP. The JMP instruction is exactly analogous to the unconditional GOTO of FORTRAN. For example, if we have a jump instruction at location 0220, this simply tells the computer to go to location 0240 for the next instruction rather than going on to the next sequential instruction.

0220 JMP 0240

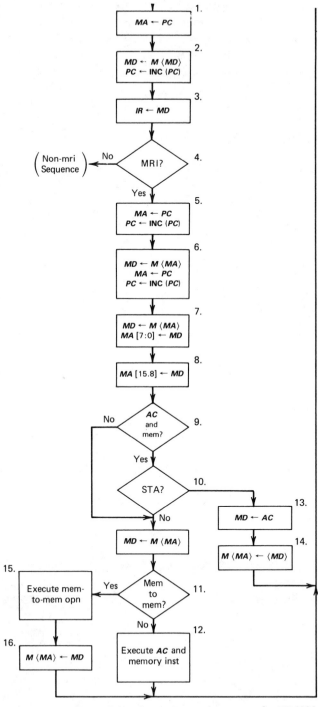

*Figure 11.10 Partial flowchart of control sequence for TB6502 computer.*

The JMP instruction is a three-byte instruction, with the opcode byte followed by the jump address in the next two locations. This instruction is logically a non-MRI instruction, but it does require a two-byte address. For this reason, it is included in the other-MRI category, with opcode 4C (0100 1100), in order to share the fetch of the two-byte address. We shall assume that the branch to other MRI's at step 15 goes to step 30.

15.  $\rightarrow \overline{(IR[1])}/(30)$

At step 30 we separate the JMP command from the remaining MRI's, which will start at step 32.

30.  $\rightarrow (IR[7] \wedge \overline{IR[6]})/(32)$

When control branched at step 15, the two-byte address was assembled in **MA**. For JMP this is the address of the next instruction, so execution of the jump is accomplished by transferring this address to **PC** and returning to step 1 to fetch the next instruction.

31.  **PC** $\leftarrow$ **MA**
   $\rightarrow$ **(1)**

The JMP instruction provides an unconditional transfer of control; we also need instructions that provide conditional transfer of control, that is, that choose between alternative courses of action based on tests of the results of previous operations. We could use conditional jumps of the same format as the unconditional jump except that different opcodes would specify that some quantity should be tested to decide whether to use the jump address or continue on to the next sequential location. This would work, but before concluding that this is a desirable approach, we should look at a factor not previously considered, the number of bytes in each instruction.

All the instructions considered so far have three bytes. This will work, but we pay a cost penalty in the amount of memory required to accommodate three bytes per instruction and a time penalty in the time required to fetch three bytes per instruction. Reducing the number of bytes per instruction can potentially reduce cost and increase speed. The 16-bit address in the JMP instruction gives us the ability to jump from any location in the address space to any other. However, this capability is not often needed; in typical programs most jumps are over a fairly narrow range of locations. For example, a common use of the conditional jump is in the termination of a loop. A *loop* is simply a set of instructions that are executed over and over until some specified condition is met. At the end of the sequence of instructions comprising the loop, we place a conditional jump that tests for the terminating condition. As long as this condition is not met, we jump back to the beginning of the loop; only when the condition is met do we continue on to the next part of the program. A typical loop might consist of 30 or 40 instructions, so we need jump back only 30 or 40 locations.

To meet such requirements in the TB6502, we use the branch instruction, which is a two-byte conditional jump using relative addressing. In relative addressing, the second byte is not an address at all, but a displacement, a signed binary number (positive or negative) that is added to the program counter to determine the address of the next instruction. For example, the instruction

**BEQ 08**

will cause the computer to jump forward eight locations if the results of the last computation were zero, to go to the next sequential location if they were not. (Note that the displacement is only two hex digits since it is only one byte.) With numbers in 2's complement form, eight bits gives us a range of $-128$ to $+127$, so we can jump backward up to 128 locations, forward up to 127 locations. Where this range is not sufficient, we can branch to a location containing an unconditional jump, from which we can jump anywhere in memory.

We also need to consider what information will be tested to determine whether or not a branch should be executed. In many computers, branches test the condition of some specific register, often the accumulator. In the TB6502, we use a more flexible scheme, in which branches are based on the condition of *flags,* flip-flops that are set or cleared on the basis of the results of a variety of data transfers involving registers or memory. There are four flags in the TB6502, the zero flag, **Z,** the negative flag, **N,** the carry flag, **C,** and the overflow flag, **V.** The **Z** and **N** flags are set or cleared by any of the MRI operations discussed earlier except STA. The **Z** flag is set if the result stored in the destination register or memory location is zero and is cleared if the result is not zero. The **N** flag is set if bit 0 of the result, the sign bit, is 1 and is cleared if it is 0. The functions of the **C** and **V** flags for ADC and SBC will be discussed in Section 11.6. For shift and rotate operations, **C** serves as the ninth bit, as shown in Fig. 11.9, and is set or cleared according to what is shifted into it. Since there are four flags, there are eight branch instructions to provide branching on either condition of any flag. These eight instructions are shown in Fig. 11.11.

Branch instructions are in the non-MRI category, which we shall start at step 50, making 50 the destination of the branch at step 4.

4.  $\rightarrow (\overline{IR[2]} \land \overline{IR[0]})/(50)$

The non-MRI category is divided into two basic types, branch instructions and register-only instructions, so we branch on **IR[ 3 ]** at step 50, to the branch sequence at step 51 or the register-only sequence at step 60.

50.  $\rightarrow (IR[3])/(60)$

At step 51, we transfer the address of the next byte (the displacement) to **MA** and increment **PC** in preparation for fetching the next instruction if the branch condition is not satisfied. The branch portion of step 51 tests **f,** a function of the opcode and the flags, which will be 1 if the branch condition is satisfied. If it is not, control returns to step 1 to fetch the next sequential instruction. If the branch condition is satisfied, step 52 fetches the displacement and step 53 adds it to **PC** and returns to step 1 for the next instructions fetch.

| Opcode | | Command | Mnemonic | Operation |
|---|---|---|---|---|
| $IR[3:0]$ | $IR[7:4]$ | | | |
| 0000 | 0001 | Branch on plus | BPL | Branch if $N = 0$ |
| | 0011 | Branch on minus | BMI | Branch if $N = 1$ |
| | 0101 | Branch if no overflow | BVC | Branch if $V = 0$ |
| | 0111 | Branch if overflow | BVS | Branch if $V = 1$ |
| | 1001 | Branch if carry clear | BCC | Branch if $C = 0$ |
| | 1011 | Branch if carry set | BCS | Branch if $C = 1$ |
| | 1101 | Branch on not equal to zero | BNE | Branch if $Z = 0$ |
| | 1111 | Branch if equal to zero | BEQ | Branch if $Z = 1$ |

Figure 11.11   Branch and flag control instructions in TB6502.

51. $\mathbf{MA} \leftarrow \mathbf{PC}; \mathbf{PC} \leftarrow \text{INC}(\mathbf{PC})$
    $\rightarrow (\overline{(\bar{R}|\mathbf{IR}[7:4]}, \mathbf{N,Z,C,V}))/(1)$
52. $\mathbf{MD} \leftarrow \mathbf{M}\langle \mathbf{MA}\rangle$
53. $\mathbf{PC} \leftarrow \mathbf{PC} + \mathbf{MD}; \rightarrow (1)$

The flowchart of Fig. 11.12 summarizes the execution of the instructions discussed to this point.

To avoid unnecessary detail, we will not provide the control sequence for the register-only instructions, which would begin at step 60. To provide for programming convenience, the register-only instructions listed in Fig. 11.13 are included in the TB6502. The first is the NOP or no operation instruction that will cause absolutely no change in the TB6502 registers or TB6502 memory. NOP instructions can be inserted in a program to provide for possible program changes or to control the timing of output signals. The remaining four entries in Fig. 11.13 call for shift or rotate

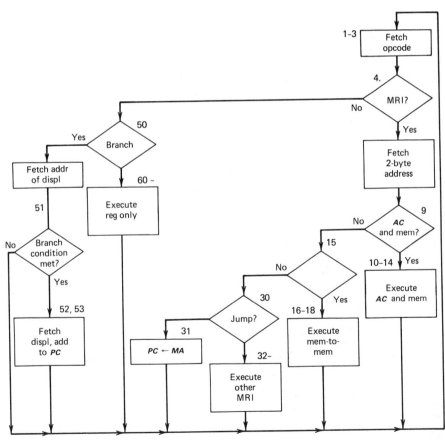

*Figure 11.12　Summary flowchart of TB6502.*

| Opcode | | | | |
|---|---|---|---|---|
| $IR[3:0]$ | $IR[7:4]$ | **Command** | **Mnemonic** | **Operation** |
| 1010 | 1100 | No operation | NOP | |
| | 0000 | Shift *AC* left | ASL A | $C, AC \leftarrow AC,0$ |
| | 0010 | Rotate *AC* left | ROL A | $C, AC \leftarrow AC,C$ |
| | 0100 | Shift *AC* right | LSR A | $AC,C \leftarrow 0, AC$ |
| | 0110 | Rotate *AC* right | ROR A | $AC,C \leftarrow C, AC$ |

*Figure 11.13   Some TB6502 register-only operations.*

in the accumulator. These function in the same manner as the memory shift and rotate commands shown in Fig. 11.9, except that no memory fetch is made. Whatever is in the accumulator is shifted or rotated in conjunction with the *C* flag as shown in Fig. 11.9. The mnemonics for these commands are ASL A, ROL A, LSR A, and ROR A. Although this form of mnemonic might suggest that A is an address, it is not. It is an extension of the opcode mnemonic, and these instructions are single-byte instructions, consisting of an opcode only.

Let us consider now a simple program to see how some of these instructions can be used. Suppose we have a competition in which eight judges vote, Pass or Fail. The votes are sent into a computer over eight lines, 1 for Pass, 0 for Fail. By means that we cannot consider now since we have not looked at input/output, assume the information on these eight lines has been stored in location 0040 in memory. The number of 1's in location 0040 is thus equal to the number of Pass votes, and we wish to write a program to count the votes. Also assume that location 0041 has been set to 00 (hex) and location 0042 has been set to 08, also by means that we will not worry about for now. With these initial settings, the program shown here will count the 1's in location 0040, leaving the count in location 0041.

```
0211    ASL    0040
0214    BCC    03
0216    INC    0041
0219    DEC    0042
021C    BNE    F3
0213    Continue
```

The basic idea of this program is that we shall use the ASL command repeatedly to shift the bits representing the votes from 0040 into *C*. Each time we do so we shall test *C* and increment the count of Pass votes in 0041 if *C* = 1. We shift a bit into *C* and test it with BCC. If the bit is 1, we continue on to 0216 and increment the count of Pass votes in 0041. If the bit is 0, we skip the increment by branching forward three locations to 0219. Note very carefully that the branch is taken relative to the next instruction, not the branch instruction itself. The reason for this can be seen in the control sequence. At the start of the fetch cycle for the BCC instruction, *PC* = 0214. When the opcode is fetched (step 2), we increment *PC* to 0215. When the displacement is fetched (step 36), we increment **PC** again so that the displacement is added to 0216, the address of the next instruction. Since we want to branch to 0219, the desired displacement is 03.

At 0219 we decrement the bit count in 0042. Recall that this was initially set to 08, the number of bits to be checked. Each time we check a bit we decrement this count and then check the count to see if it has reached zero. If it has not, the BNE at 001C branches back to 0211 to repeat the process. When the count reaches zero, after all bits have been checked, the branch is not satisfied and the program continues on to whatever comes next. The branch is from 021E back to 0211, which is $-13(10)$ locations; F3 is the 2's complement form of $-13(10)$, in hex notation.

# 11.6 Addition and Subtraction with Carry

Addition and subtraction are simple enough operations, but the manner in which they are implemented in a microprocessor warrants some special consideration, especially in regard to the role of the carry flag in the operations. Adders will vary in details of logic from one microprocessor to another, but all are logically equivalent to the cascade of full-adders shown in Fig. 6.14. As discussed in Section 6.7, the **C** flag stores the carry-out of the adder, $C_n$ in Fig. 6.14. For the TB6502, one input to the adder will come from the accumulator, the other from memory via **MD,** and the sum will be returned to **AC,** as shown in Fig. 11.14. In some computers, this carry-in is permanently set to 0 so that it has no effect on the addition. In most microprocessors, however, it is driven by the output of the carry flag, as shown in Fig. 11.14. The carry-in from the **C** flag makes it possible to cascade addition of data spread across two or more bytes.

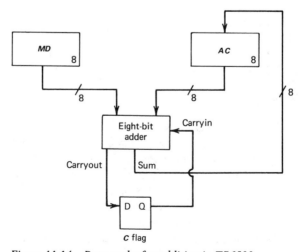

*Figure 11.14   Data paths for addition in TB6502.*

A problem with an 8-bit microprocessor is that you cannot store a large number in a single byte. When large numbers are needed, they will be stored in several adjacent words in memory. For example, assume we wish to perform the addition shown here.

$$4\ 3\ 9\ 6$$

$$+2\ 1\ 8\ 3$$

$$6\ 5\ 1\ 9$$

Assume the first number is stored with 96 in location 0050 and 43 in 0051, and the second with 83 in 0060 and 21 in 0061. The following sequence of instructions will add them, placing the first byte of the sum in 0070, the second in 0071.

```
CLC
LDA     0050
ADC     0060
STA     0070
LDA     0051
ADC     0061
STA     0071
```

We start with a new instruction, CLC (Clear Carry, to be discussed in Section 11.7), which sets $C = 0$. The next step adds $96 + 83$ to produce the sum (hex) 19 with a carryout, which sets the $C$ flag. The next addition adds $43 + 21$ plus the carry-in to produce the correct second byte of the sum.

It is important to note that this cascade addition over several bytes works only for unsigned integers. As discussed in Section 6.7, the $C$ flag does not correspond to a numeric overflow when the leading bits are sign bits. Addition of signed numbers spread over several bytes will require use of the ADC or SBC commands to combine the numeric magnitudes represented in unsigned integer form, with the signs represented and processed separately. On the other hand, the use of the $C$ flag for carryin does not invalidate 2's complement addition of 1-byte numbers. With $C = 0$ before execution of ADC, addition of 2's complement signed numbers will be carried out exactly as described in Chapter 6, and the $V$ flag will correctly indicate signed overflow. You should clearly understand that the operation of these two flags is invariant, as described in Section 6.7. Their significance in a specific program will depend on the meaning that the programmer assigns to the operands.

For subtraction the situation is somewhat more complicated.* Note (Fig. 11.8) that the complement of the carry is subtracted for SBC. The reasons for this relate to the manner in which subtraction is accomplished. Rather than use special subtraction circuitry, the TB6502 does subtraction by using the 2's complement of the number in *MD*. This is done by taking the logical complement of *MD*, flip-flops and

---

*The material on subtraction in the remainder of this section is not essential to the continuity of development at this point and may be skipped and referred to later as needed.

adding $+1$ in the form of a carryin. Suppose we want to carry out the subtraction $78 - 43 = 35$. The situation in the registers and adder is shown here.

| Registers | | Adder | |
|---|---|---|---|
| | | 1 | ← Carryin |
| **AC =** | 01111000 | ⟶ | 01111000 |
| | | Logical | |
| **MD =** | 01000011 | ⟶ | 10111100 |
| | | Complement | |
| | Carryout | = 1 ← | 00110101 |
| | | | 3    5 |

So, it seems to work, but why do we specify subtracting $\overline{C}$ in Fig. 11.8? With $C = 1$ we got the correct answer, that is, $78 - 43 - C = 78 - 43 - 0 = 35$. If $\overline{C}$ were equal to 0, we would add only the logical complement, which is one less than the 2's complement, and obtain $78 - 43 - \overline{C} = 78 - 43 - 1 = 34$. Note carefully, this process of subtraction requires that the $C$ flag be set prior to a subtraction, by means of the SEC (Set Carry) command, which will also be discussed in Section 11.7.

The preceding provides part of the reason for using $C$ flag as an input for subtraction, but what about cascading? Just as we must provide for carry from one byte to the next for addition, we must provide for borrow from one byte to the other for subtraction. Consider the subtraction

$$
\begin{array}{r}
5\,6\,7\,8 \\
-4\,1\,4\,3 \\
\hline
1\,5\,3\,5
\end{array}
$$

Clearly, the subtraction $78 - 43$ in the first byte requires no borrow from the second byte. From this subtraction, however, we note that a carryout is produced, and that is just what we want. This 1, stored in the $C$ flag, will provide the carryin to the second byte necessary to take the 2's complement of the subtrahend byte.

Next, consider a case with borrow.

$$
\begin{array}{r}
5\,6\,4\,3 \\
-4\,1\,7\,8 \\
\hline
1\,4\,C\,B
\end{array}
$$

Here we clearly need a borrow from the second byte into the first. Let us see how this works out in binary, starting with the subtraction in the first byte.

|     | **Registers** |              | **Adder**  |              |
| --- | ------------- | ------------ | ---------- | ------------ |
|     |               |              |            | ← Carryin    |
| *AC* = | 01000011   | ⟶            | 01000011   |              |
|     |               | Logical      |            |              |
| *MD* = | 01111000   | ⟶            | 10000111   |              |
|     |               | complement   |            |              |
|     | Carryout      | = 0 ←        | 11001011   |              |
|     |               |              | C   B      |              |

The first byte is correct, and we have $C = 0$ into the second byte.

|     | **Registers** |              | **Adder**  |              |
| --- | ------------- | ------------ | ---------- | ------------ |
|     |               |              |            | ← Carryin    |
| *AC* = | 01010110   | ⟶            | 01010110   |              |
|     |               | Logical      |            |              |
| *MD* = | 01000001   | ⟶            | 10111110   |              |
|     |               | complement   |            |              |
|     | Carryout      | = 1 ←        | 00010100   |              |
|     |               |              | 1   4      |              |

The second byte of the result is correct, with a carryout, that is, no borrow. From all this we conclude that the *C* flag represents the complement of the borrow when using the SBC command.

Next, let us consider the result of the last subtraction more carefully. The hex number 14CB is the correct positive difference for (5643 − 4178). On the other hand, let us view the subtraction of the first byte (43 − 78) in terms of an addition of unlike-signed numbers. Recalling the procedures of 2's complement arithmetic from Chapter 6, we would expect the result to be −35 in 2's complement form. Note that

$$35 \text{ (hex)} = 00110101$$

so that the 2's complement is

$$11001011 = CB \text{ (hex)}$$

Thus, we have a rather surprising outcome. The results of subtraction with the subtrahend larger than the minuend may be regarded as the positive difference with a borrow out, or as the negative result in 2's complement form. This dual identity of the difference is no accident but is a consequence of the basic definition of the 2's complement.

Finally, consider the action of the *C* and *V* flags. As we have seen, the *C* flag copies the carryout from the adder and drives the carry-in, whether the operation is ADC or SBC. The operation of this flag is invariant with respect to the operations of the adder, although its significance will depend on what command is being executed and on the meaning of the operands. Similarly, the operation of the *V* flag is a fixed function of the operation of the adder. This flag is driven by logic that checks bit 7 of each operand applied to the adder and bit 7 of the sum. If the two operand bits are the same, but differ from bit 7 of the sum, the *V* flag will set; otherwise, the *V* flag will clear. This means that the meaning of the *V* flag is consistent for subtraction. Overflow for subtraction can only occur if the two operands have opposite signs. The complementing of the subtrahend reverses the sign, so like signs seen at the adder means unlike signs on the operands.

This discussion illustrates in a very clear manner why we believe it is important that engineers who will program at this level understand the internal functioning of the processor. You could reduce all the above, including the action of the flags, to a set of rules that could be followed without understanding the hardware. But these rules would be so complicated that they would be very difficult to understand and almost impossible to apply with any consistency. If you understand what is going on internally in terms of basic adder functions, you can always figure out the results for any situation.

## 11.7  Commands That Affect Only the Flags

A few commands in the TB6502 affect only the flags. The simplest of these are those that directly set or clear a flag. In Section 11.5 we saw the need to be able to clear or set the *C* flag prior to executing ADC or SBC. For this purpose we have the CLC (Clear Carry) and SEC (Set Carry) commands. There are no clear or set commands for the *N, Z,* or *V* flags. These are set or cleared only as a result of memory or register operations, as discussed earlier. There are two more flags in the TB6502, in addition to those already discussed, the *D* (Decimal) and *I* (Interrupt) flags. These two flags are very different from the other four, in that they do not indicate the results of operations, but rather control operations. The *D* flag controls the manner in which ADC and SBC are executed. When $D = 0$, addition and subtraction are carried out in binary as discussed in Section 11.5. When $D = 1$, these operations are carried out in BCD. Decimal arithmetic will be discussed in the next chapter. The *I* flag controls the interrupt system of the TB6502, giving the programmer the option of allowing or blocking interrupts, as will be discussed in Chapter 15. The SED and CLD commands control the *D* flag, and the SEI and CLI commands control the *I* flag. These set and clear commands are in the register-only category and are listed in Fig. 11.15.

The other two commands, CMP and BIT, provide a means of testing a memory location without actually modifying the value of any data except the flags. The compare command (CMP) performs the subtraction $AC - M\langle MA \rangle$ and sets the *C, N,* and *Z* flags accordingly but does not change the value in *AC.* Referring to Fig. 11.15,

| Opcode | | Command | Mnemonic | Operation |
|---|---|---|---|---|
| $IR[3:0]$ | $IR[7:4]$ | | | |
| 1000 | 0001 | Clear carry | CLC | $C \leftarrow 0$ |
| | 0011 | Set carry | SEC | $C \leftarrow 1$ |
| | 0101 | Clear interrupt | CLI | $I \leftarrow 0$ |
| | 0111 | Set interrupt | SEI | $I \leftarrow 1$ |
| | 1101 | Clear decimal | CLD | $D \leftarrow 0$ |
| | 1111 | Set decimal | SED | $D \leftarrow 1$ |
| 1101 | 1100 | Compare memory and AC | CMP | Set flags on $AC - M\langle MA \rangle$ |
| 1100 | 0010 | Test memory bits on AC | BIT | Set $Z$ flag on $AC \land M\langle MA \rangle$ |

*Figure 11.15   Commands that affect only flags.*

the correct difference of $AC - MD$ is formed at the output of the adder and the three flags are set or cleared accordingly, but the difference is not clocked into $AC$.

This command is useful when we wish to test a whole series of memory locations for the presence of some value. For example, suppose locations 0050, 0051, 0052 contain ASCII codes and we wish to check to determine if one of them is a carriage return (OD hex), branching to 0240 if a carriage return is found. We assume that location 0040 has previously been loaded with a carriage return.

```
0200    LDA    0040
0200    LDA    0040
0203    CMP    0050
0206    BEQ    38
0208    CMP    0051
020B    BEQ    33
020D    CMP    0052
0210    BEQ    2E
```

The first command loads $AC$ with a carriage return character. We then repeatedly compare the memory location to $AC$. If a memory location contains a carriage return, the subtraction will produce 00, the $Z$ flag will be set, and the BEQ command will branch to 0240. IF we did not have the CMP command, we would have to perform the actual subtraction, and $AC$ would have to be reloaded after each comparison. This way, we load $AC$ once and test as many times as necessary. Note that CMP does not affect the $V$ flag. Also, it is not necessary to set the $C$ flag prior to execution of CMP, as the carryin is automatically set to 1 to obtain the correct subtraction.

The BIT command functions in the same manner except that it sets or clears only the $Z$ flag, on the basic of $AC \land M\langle MA \rangle$. The name BIT comes from the fact that the basic purpose of this command is to test individual bits in memory words. To check a particular bit, set the same bit in *AC* to 1, all other bits in *AC* to 0. The BIT command will produce all 0's, that is, $Z = 1$, if the tested bit is 0, and $Z = 0$ if the tested bit is 1. As with CMP, the main advantage of BIT is that a test can be performed repeatedly without resetting the value in *AC*. In terms of logical function, both CMP and BIT fall into the *AC*-and-memory category, but BIT turns out to be in the other-MRI category in terms of opcode classification.

   This sort of apparent inconsistency is not unusual. A designer will attempt to group opcodes in a logical manner in order to simplify decoding. However, there are usually a few instructions that just do not fit neatly into any particular category. In this case, the arrangement of the opcodes is such that there is room for only eight *AC*-and-memory instructions. Seven of these are listed in Fig. 11.8 and CMP is the eighth, leaving BIT to be assigned codes in another group. This complicates the decoding design somewhat but is of no particular importance to the programmer, who simply works from a list of apparently arbitrary codes.

# 11.8   Programming Procedures

We have now seen a variety of computer instructions and have studied the way these instructions are executed in a typical computer. We have also seen how these instructions can be used to write programs to accomplish simple computational tasks. We now wish to look at the process of programming *per se* in more detail.

   The writing of a program involves a number of steps, which can be summarized as follows.

> 1. Defining the problem
> 2. Outlining the solution
> 3. Coding
> 4. Debugging
> 5. Validation

The importance of the first step is obvious, and yet it is often overlooked. The initial specifications of problems are usually vague, often ambiguous. Do not move on to the next phase until you have determined exactly what the program is to do. Among other things, you need to determine just exactly what the inputs are and in what form they are presented and what outputs are desired in what form.

   Once you are clear just what is to be accomplished, the next step is to determine your general method of solution. This step involves determining the basic procedures to be followed, not in terms of specific instructions but in terms of functional procedures. This phase may include specifying algorithms for certain parts of the program. For example, we might use existing routines for sorting or code conversion, tasks of the sort that appear in many different programs. Flow-charting is often useful in this phase of programming because it provides a convenient way to display the relationships between the various parts of the program.

*Coding,* the writing of the actual sequence of instructions, is what most people mean by "programming," but it is actually only a small part of the process. If the first two parts of the process have been done properly, the writing of the code should be straightforward, almost mechanical. The biggest mistake made by most programmers is to start right off writing code without properly defining the problem and outlining the method of solution.

When you try to run a program for the first time, it almost never works. *Debugging* is the process of trying to find out why and correcting the code accordingly. If the first two phases were carried out properly, most of the errors at this stage will be mechanical, that is, typing errors, missing commas, incorrect statement numbers, or such. If you have skipped the first steps and started to code directly, you will probably have errors in the basic logic of the program such that it will not do what you want even though it is mechanically correct and will run. At this point you get to the last phase, *validation.* Even though the program runs, it may not be producing the correct results. To validate the program you must devise a set of test data that will exercise the program adequately to demonstrate that it does what it is supposed to do.

The amount of time required for these various phases will depend on the complexity of the program, but, in any case, an essential part of programming at every stage is adequate documentation. This cannot be overemphasized. More time is wasted in the programming and use of computers because of inadequate documentation than for any other reason. Documentation is important for several reasons. First, as you program, having to explain what you are doing as you do it will clarify your thoughts and reduce the likelihood of error. Second, when you get to the debugging phase, it will be very difficult to figure out what is wrong if you have not explained what the program should be doing. Finally, documentation is essential to keep a program in use. We are not talking now about programs being written and used only once and then discarded. Writing programs is a lot of work, work that should not be undertaken unless the program is to be used repeatedly, probably by many users. Clearly, no one else can use your program unless they know what it does. Even if you are the only user, sooner or later the day will come when it does not work properly and you have forgotten just what it is supposed to do. It is frightening to contemplate the number of perfectly good programs that have been discarded because inadequate documentation made it faster to write a new program than figure out the old one.

The exact form the documentation takes will depend on the complexity of the program, the level at which it is written, and the language used. We will be working at the assembly level, for which comments are the primary form of documentation. Every program or subroutine should start with identifying comments, explaining what the program or routine is supposed to do. Additional comments may be inserted at various points to explain the function of various segments of the program. Finally, every line of code, or at least every three or four lines, must include a comment to explain the function of that line or set of lines.

Consider the following TB6502 program as an example of a program that is adequately documented through the use of comments.

```
*   PROGRAM TO COMPUTE SIGNUM FUNCTION OF
*     SUM OF TWO NUMBERS, (M+N)
*     UPON ENTRY
*       0040 = INITIAL VALUE OF M
*       0041 =    "     "   "  N
*       0050 = +1 (CONSTANT)
*       0051 =  0       "
*       0052 = -1       "
*     UPON EXIT
*       0043 = +1 IF (M+N) POSITIVE
*       0043 =  0 IF (M+N) ZERO
*       0043 = -1 IF (M+N) NEGATIVE
*
  0210    CLC                /CLEAR CARRY FLAG
  0211    LDA    0040         /M TO AC
  0214    ADC    0041         /(M+N) TO AC
  0217    BMI    09           /IF (M+N) NEG,
*                             /   THEN 222, ELSE 219
  0219    BEQ    0E           /IF (M+N) = 0,
*                             /   THEN 229, ELSE 21B
  021B    LDA    0050         /(M+N) POS, +1 TO AC
  021E    STA    0043         / STORE +1 IN 43
  0221    HLT                 /FINISHED
  0222    LDA    0052         /(M+N) NEG, -1 TO AC
  0225    STA    0043         /STORE -1 IN 43
  0228    HLT                 /FINISHED
  0229    LDA    0051         /(M+N) = 0, 0 TO AC
  022C    STA    0043         /STORE 0 IN 43
  022F    HLT                 /FINISHED
```

Now you have written your well-documented program and you try to run it. No matter how careful you have been, it probably will not work the first time. So you now enter what most find the most frustrating part of the programming process, the debugging phase, in which you try to figure out what went wrong and what to do about it. There are many debugging techniques, but the basic strategy is usually to break the program up into small parts and check the functioning of each part. Just how this is done depends in part on what sort of access you have to the computer. If your only access is input through a card reader and output through a printer, the usual method is to insert a lot of print statements to print out intermediate results. This is the usual approach when running batch mode on a large system. If you are running on a small system, such as a minicomputer or microcomputer, and have direct access to the computer via a console or keyboard, one of the best methods is to "single-step" through the program, one instruction at a time, checking after each instruction to see if the computer did what you expected. This is where good documentation is so important. If you have documented your program properly, as in the preceding example, it is very easy to determine what should be in the various registers and memory locations after every instruction. Without good documentation you will waste endless hours trying to figure out if what did happen was what was supposed to happen.

You will recognize that this process of going through a program one step at a time and determining the resultant contents of pertinent registers and memory locations is basically the same tracing process we saw in Chapter 10. To trace a program most effectively, you need some knowledge of the internal organization and sequenc-

ing of the computer. Let us assume the initial contents of registers and memory locations are as shown here.

$AC$ = 08     $PC$ = 0220     $M\langle 0040\rangle$ = 0A

and the instruction at locations 0220, 0221, 0222 is

### ADC 0040

(Remember that the TB6502 $AC$-and-memory instructions require three bytes.)

If this single instruction is executed, what will be the resultant contents of the
same locations? First we recall that the $PC$ will be incremented three times, so the
final contents of $PC$ will be 0223. The instruction calls for adding the contents of
location 0040 to $AC$. So the final contents of $AC$ (assuming $C$ = 0) will be

$AC$ = 08 + $0A$ = 12

recalling that all numbers are in hex. Finally, since reading from memory is nondestructive, the contents of 0040 are not altered.

Assume we have the initial values,

$C$ = 0     $PC$ = 0230

and the instruction at locations 0230, 0231 is

### BCC 08

Determine the final values. During the fetch of the instruction, $PC$ is incremented
twice, to 0232. The instruction BCC calls for a branch if $C$ = 0; since it does, we
add the displacement to $PC$ to give the final value

$$PC = 0232 + 08 = 023A$$

The value of $C$ is not affected.

In this general manner we can trace the operation of a program on an instruc-
tion-by-instruction basis. However, checking the contents of every register and relevant memory location after every instruction will be very time-consuming for long
programs. When we trace a complete program we usually take a segment of code,
say, 10 steps, that carries out some readily identifiable process, execute that segment
and then check to see if the results are correct. If they are not, we then come through
the segment one step at a time to find out where it went wrong.

As an example, let us trace the TB6502 program for computing SIGNUM (M
+ N) given earlier. Assume the following initial values in registers and memory.

$PC$ = 0210     $X$ = 00     $M\langle 0040\rangle$ = 08     $M\langle 0041\rangle$ = F3

Trace the program to determine the final contents of $PC$, $AC$, and $M[0043]$.

We assume 2's complement representation, for which a 1 in the leading bit position indicates a negative number. In hex terms this means a leading digit $\geq 8$, so we see that the number in 41 is negative. By converting it to binary and recomplementing, we find the number is $-13$ (decimal) so that the sum $(M + N)$ will be negative. For this case, $-1$ is stored in 43 at step 0225, and the program halts at step 0228. The final contents of the locations will be as follows.

$$PC = 0229 \qquad AC = FF \qquad M(0043) = FF$$

Note that *PC* is incremented during the fetch of the HLT instruction, and that FF is the hex representation of $-1$ in 2's complement.

Note how simple the tracing process is with the comments. We have seen in the previous examples that we can trace a single instruction without knowing what function it serves in the program. If we did not have the comments we would have to go through the program step-by-step to find out what each instruction was supposed to do. With the comments we can see immediately where the program will go for a given set of inputs and go directly to the final values without considering intermediate steps. Furthermore, if the program does not work and you have to go through it step-by-step, the comments will make it much easier to figure out what should be happening. For example, suppose we have values of *M* and *N* that produce a zero sum and we step through to the branch at 0219, and then execute that branch. The comment tells us that *PC* should now be at 0229. Without the comment we would have to do the hex addition to the incremented *PC* to figure out where the program should be.

We said it before, we shall say it again, *the importance of documentation cannot be overemphasized.* When you start out to write a program it is quite natural to want to get it written as fast as possible and into the computer. Documentation takes time, and you will often be reluctant to spend that time. We hope you will accept the wisdom based on the accumulated experience of programmers beyond number and believe that *time spent* on documentation is *time saved.*

## Problems

**11.1**  Add complete documentation to the final program of Section 11.6, the program to count votes.

**11.2**  For each part of this problem, assume the following initial values for the locations indicated.

$$PC = 0210 \qquad AC = 08 \qquad C = 0 \qquad M\langle 0040 \rangle = 0A$$

$$M\langle 0041 \rangle = F4$$

For each of the following instructions, assumed to be located at 0210, determine the values in the preceding locations after execution of the instruction.

**11.3** Shown here are several short sequences of instructions. Prior to execution of each sequence, assume various memory locations contain the hex values shown. Also assume all flages = 0.

$$M\langle 0040 \rangle = 6D \quad M\langle 0041 \rangle = 35 \quad M\langle 0042 \rangle = 94$$

$$M\langle 0043 \rangle = C3 \quad M\langle 0044 \rangle = CB \quad M\langle 0045 \rangle = 34$$

Show the values of th e *N, Z, C,* and *V* flags after execution of each sequence.

| (a) | LDA 0040 | (b) | ADC 0043 | (c) | SEC |
|-----|----------|-----|----------|-----|-----|
|     | ADC 0041 |     | ADC 0040 |     | LDA 0041 |
|     |          |     |          |     | ADC 0041 |

| (a) | AND 0040 | (b) | STA 0041 |
|-----|----------|-----|----------|
| (c) | ADC 0041 | (d) | ROL 0041 |
| (e) | DEC 0040 | (f) | SEC |
| (g) | BCC 0A   | (h) | BCS 0A |

| (d) | LDA 0041 | (e) | LDA 0041 | (f) | SEC |
|-----|----------|-----|----------|-----|-----|
|     | ADC 0044 |     | SBC 0044 |     | LDA 0041 |
|     |          |     |          |     | SBC 0044 |

| (g) | SEC      | (h) | LDA 0040 |
|-----|----------|-----|----------|
|     | LDA 0040 |     | SBC 0043 |
|     | SBC 0042 |     |          |

**11.4** Given the following TB6502 program

```
*    PROGRAM TO COMPARE TWO NUMBERS, M AND N
*      UPON ENTRY
*        0040 = INITIAL VALUE OF M
*        0041 =    "        "    " N
*        0050 = +1 (CONSTANT)
*        0051 = 0  (CONSTANT)
*      UPON EXIT
*        0042 = +1 IF M = N, 0 IF M / N
*
     0210    LDA  0040    /M TO AC
     0213    EOR  0041    /M EXOR N TO AC
     0216    BEQ  07      /IF M = N,
*                         / THEN 1F, ELSE 18
     0218    LDA  0051    /M = N, 0 TO AC
     021B    STA  0042    /STORE 0 IN 42
     021E    HLT          /FINISHED
     021F    LDA  0050    /M = N, +1 TO AC
     0222    STA  0042    /STORE +1 IN 42
     0225    HLT          /FINISHED
```

Assume initial values as follows:

$$PC = 0210 \qquad AC = 00 \qquad M\langle 0040 \rangle = 0A \qquad M\langle 0041 \rangle = 03$$

$$M\langle 0042 \rangle = FF$$

Determine final contents of these locations after execution of the program.

**11.5** Shown here is a short program segment. Also shown are the initial contents of certain registers and memory locations. Perform a program trace, showing the contents of *PC, AC, C, V, N,* and *M⟨0050⟩* after execution of each instruction.

$$PC = 0200 \qquad AC = 00 \qquad C = V = N = 0$$

$$M\langle 0050 \rangle = 40 \qquad M\langle 0051 \rangle = 64 \qquad M\langle 0052 \rangle = 72$$

$$M\langle 0053 \rangle = 21 \qquad M\langle 0054 \rangle = 35$$

```
0200    LDA    0051
        ADC    0052
        BVC    04
        SEC
        ROL    0050
        CLC
        LDA    0053
        ADC    0054
        BVC    04
        SEC
        ROL    0050
        LDA    0050
```

**11.6** Shown here are several short sequences of instructions. Prior to execution of each sequence, assume various memory locations contain the hex values shown. Also assume all flags = 0.

$$M\langle 0040 \rangle = 6D \qquad M\langle 0041 \rangle = 38 \qquad M\langle 0042 \rangle = 91$$

$$M\langle 0043 \rangle = 08 \qquad M\langle 0044 \rangle = 80$$

Show the values of the *N, Z, C,* and *V* flags after execution of each sequence. Also show the contents of *AC* and any memory location referenced in the sequence.

| | | |
|---|---|---|
| (a) LDA 0040 | (b) LDA 0041 | (c) LDA 0042 |
|     CMP 0041 |     ADC 0042 |     CMP 0040 |
| (d) LDA 0043 | (e) LDA 0044 | (f) LDA 0040 |
|     BIT 0040 |     BIT 0040 |     BIT 0043 |

**11.7** Given an integer $N$, $1 < N < 6$, initially stored at location 40. Write a TB6502 program to compute two times the sum of the integers from 1 to $N$, that is,

$$X = 2 \, x \sum (1 + 2 + \ldots + N)$$

and store the result at 41. Start the program at 0200.

**11.8** Three numbers, $K$, $L$, $M$ are stored in locations 40, 41, 42. All three are 8-bit positive integers (no sign bit). Write a program that will arrange them in order, the largest in 40, the next largest in 41, the smallest in 42. Start the program at location 0200.

**11.9** Routines to test a word entered from a keyboard are important in interactive programming. Assume an operator has typed in either YES or NO in response to a question, and the first letter (Y or N) is stored in location 40, in ASCII code (Fig. 6.3). Write a program that will store 01 in location 41 if the letter is Y, and store FF in 41 if the letter is N. If it is neither, location 41 should be set to 00. The letters may be entered in either uppercase or lowercase. Start the program at 0200.

# Addressing and Assembly Language

## 12.1 Microprocessor Addressing

It is expected that most users of Chapters 11 through 15 of this book will at the same time be engaged in laboratory experience with microprocessors. Although 16-bit microprocessors have been available for some time, 8-bit microprocessors remain most appropriate for many design applications. At most institutions at least part of the stations in the microcomputer laboratory will be based on an 8-bit microprocessor. Chapters 11 through 15 are intended to support the programming instruction for any 8-bit microprocessor.

Clearly there are differences between the various microprocessors used in the commonly available microcomputers. Therefore, the treatment of the TB6502 must be extended and adapted somewhat by the instructor to match the available microcomputer. This extension* will always be straightforward and much easier than starting over with the real microcomputer assuming no background.

Up to now we have formed the 16-bit address of an operand in only one way, by catenating the next two bytes in memory following the opcode. There are other ways to form addresses. Each approach to forming a 16-bit address will be called an *addressing mode*. The addressing mode to be used will, in most cases, be specified

---

*The TB6502 is closest to MOS Technology 6502 and Motorola 6809 microprocessors. In the former case it is only necessary to add the preindexing feature. In the latter, it is only necessary to extend the index registers to 16 bits. A discussion of differences between the TB6502 and the 6502 will be found in Appendix D.

| Memory location | Contents | Comment |
|:---:|:---|:---|
| $n$ | Opcode | Always present |
| $n + 1$ | Postbyte | Sometimes used |
| $n + 2$ | Low-order address byte | If required by opcode |
| $n + 3$ | High-order address byte | If required by opcode |

*Figure 12.1    Typical instruction arrangement with postbyte.*

by the opcode. The actual 16-bit address to be used in referencing memory (the *effective address*) will be formed in a variety of ways, as described in the following.

Some microprocessors, such as the Z80 and the Motorola 6809, have insufficient bits in the opcode to completely specify all the addressing modes that are available. These devices require a supplementary opcode byte, or postbyte, between the opcode and address bytes, as illustrated in Fig. 12.1. If a postbyte is allowed, it can be used to specify bit manipulations within registers as well as a richer choice of operands and addressing modes. A postbyte will never be required by a TB6502 opcode. Nonetheless, most of the important addressing modes will be found in the TB6502, as discussed in the next section.

## 12.2  Addressing Modes

The instruction set developed for the TB6502 computer to this point is, except for input/output, entirely adequate for most any sort of program. Indeed, some computers have even simpler sets of instructions. Most computers, however, have a number of other features that, although not absolutely essential, can make programs simpler and yet more powerful. In this section we wish to investigate the manner in which variations in the way the address is formed can add to our programming power.

As we shall see, an address to be used in accessing memory can be obtained or formed in a variety of ways. In all cases the address that actually goes to memory— via **MA** in the TB6502—is known as the *effective address*. When the complete effective address is provided in the instruction, we have *absolute addressing*. This is the only mode of addressing we have used so far for the **AC**-and-memory and the memory-only instructions in the TB6502. In the branch instructions in the TB6502 we use *relative addressing,* in which the "address" portion of the instruction is not an address at all, but a *displacement,* which is added to the program counter to obtain a branch destination relative to the current location in the program.

In discussing relative addressing we noted that one advantage is that it saves one byte per instruction, compared to absolute addressing. If we could similarly reduce the various register and memory instructions by one byte, we could achieve further saving in memory space and execution time. This is accomplished in the

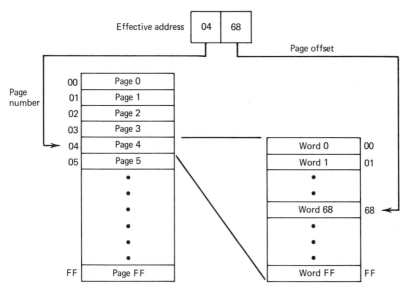

*Figure 12.2   Memory pages in the TB6502 computer.*

TB6502 through the use of *zero-page addressing*. The address space of 65K words is considered to consist of 256 pages of 256 words each. The most significant byte of the effective address specifies the *page number;* the least significant byte specifies the *page offset*. For example, address 0035 refers to location 35 on page 00; address 0468 refers to location 68 on page 04. The concept of paging is illustrated pictorially in Fig. 12.2.

You will recall from Section 11.4 that the first byte after the opcode in absolute addressing is the lower byte, that is, the page offset. In zero-page addressing we fetch the page offset from the byte immediately following the opcode and then assume that the page number is 00, thus eliminating the fetch of the second address byte. This form of addressing can reference only the first 256 words in memory, but a surprising number of programs can be accommodated within this limitation.

In the sequence of Section 11.4, bits $IR[3:0]$ indicated the basic type of instruction and bits $IR[7:5]$ indicated the specific opcode. The remaining bits are used to indicate the type of addressing. We shall delay writing the detailed sequence until we have seen some other types of addressing, but we shall modify the flowchart to show what needs to be done. In Fig. 12.3 we see a portion of the flowchart of Fig. 11.10, modified to include zero-page addressing. At step 5 we move the address of the page-address byte to $MA$ and increment $PC$, as before. We add a branch, based on $IR[4:2]$, to choose between absolute and zero-page addressing. Absolute addressing proceeds as before. For zero-page, we fetch the page-offset byte and then transfer it to the lower half of $MA$ along with 00 to the upper half.

When writing programs with zero-page addressing in mnemonic form, an asterisk (*) is inserted between the opcode and the address to indicate this address

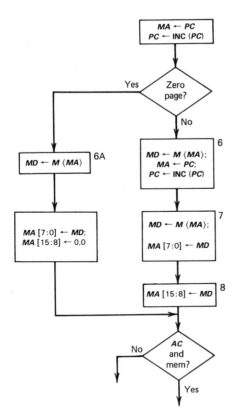

*Figure 12.3   Modification of sequence to include zero-page addressing.*

mode. For example, the program to add three numbers would appear as follows with zero-page addressing:

```
0110    LDA     *40
0112    ADC     *41
0114    ADC     *42
0116    STA     *43
0118    Next instruction.
```

The use of zero-page addressing does not make any difference in the way we program. If we compare this program with the corresponding program using absolute addressing, written in Section 11.4, we see exactly the same sequence of instructions, but a reduction in size of the program from 12 bytes to 8 bytes by using zero-page addressing. Also note that we have moved the program itself off page 0. This is not necessary, but it is good programming practice. Since the program counter is a full 16 bits, there is no advantage to locating instructions on page 0, so we normally restrict this page to data to get the optimum use of the 256 locations available.

The next form of addressing is *immediate addressing,* which provides a convenient way to specify constants directly in the body of the program. Immediate addressing is really not addressing at all, since it does not generate an address to be used in accessing memory. With immediate addressing there are two bytes in the instruction, and the second byte is the operand. The symbol for immediate addressing in mnemonic form is "#", placed before the "address" byte. For example,

> LDA #00

means, "load the accumulator with the quantity 00," which is equivalent to the CLA (clear the accumulator) instruction found in many computers. Because this form is available, the TB6502 does not have a CLA instruction. Next consider the form,

> EOR #FF

Since FF is hex for all 1's, the net result is to ex-OR each bit of *AC* with 1. Since $1 \oplus 1 = 0$ and $1 \oplus 0 = 1$, execution of this command forms the logical complement of the contents of the accumulator. Computers without immediate addressing usually include a command CMA, complement the accumulator. Another useful immediate form is

> ADC #01

which simply adds 1 to the accumulator, that is, increments *AC* (the IAC command in some computers).

In the examples here we have used immediate addressing to effect common operations on *AC*. This form can also be used to introduce constants in a program. Consider the program in Section 11.5 to compute the signum function, which calls for storing $+1$, 0, or $-1$ in a result location. In that example we had to store these three constants in memory before executing the program. Immediate addressing can eliminate these stored constants. For example, step 1B of that program,

> 001B    LDA  0050    /(M+N) POS, +1 TO AC

can be replaced by

> 001B    LDA #01    /(M+N) POS, +1 TO AC

Making this change does not save any steps in the program, but it does save memoi˙ locations by eliminating stored constants. Immediate addressing is available in all the *AC*-and-memory instructions except STA.

The next type of addressing, *indirect addressing,* has a feature in common with paging in allowing us to reduce the amount of address information in the instruction and also offers some additional programming flexibility. In indirect addressing the address portion of the instruction is not the address of the operand but is the *address of the address* of the operand. In the case of the TB6502, indirect addressing requires two-byte instructions with the second byte used as a zero-page address pointing to the first of two locations containing the effective address. In the mnemonic form we

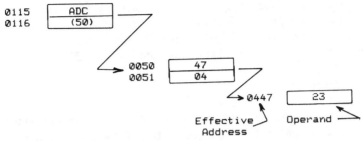

*Figure 12.4   Indirect addressing in the TB6502.*

indicate indirect addressing by placing parentheses around the address, indicating that it is the address of an address. Assume we have the command

**0115   ADC (50)**

Figure 12.4 illustrates how the operand is obtained. The indirect address (50) points to location 0050, which contains the lower byte of the effective address, with the upper byte following in location 0051. In this case we assume 0050 and 0051 contain 47 and 04, respectively, so the effective address of the location from which the operand will be obtained is 0447. Note that indirect addressing enables us to access any location in the memory with a two-byte instruction, but at the price of using up two locations on page 00.

One possible use of indirect addressing can be illustrated by considering the problem of adding a string of numbers. We have already written a program for adding three numbers in which we basically just repeated the add instruction three times. This approach could be extended to any number of operands, but to do so would be very inefficient for a long string of numbers. Assume we have a long string of numbers to be added that are stored in consecutive locations starting at 0410. The following approach can be used to add these numbers without writing a program with as many ADC commands as there are numbers in the string.

```
*    PARTIAL PROGRAM TO ADD A STRING OF NUMBERS
*       0060 = LOWER BYTE OF POINTER
*       0061 = UPPER BYTE OF POINTER
*       UPON ENTRY, STRING OF NUMBERS STORED IN
*          CONSECUTIVE LOCATIONS STARTING AT 0410
*
 0100       LDA   #10       /10 TO AC
 0102       STA   *60       /INIT PNTR LOW BYTE
 0104       LDA   #04       /04 TO AC
 0106       STA   *61       /INIT PNTR HIGH BYTE
 0108       CLC             /CLEAR CARRY FLAG
 0109       LDA   #00       /CLEAR AC
 010B       ADC   (60)      /ADD NUMBER
 010D       INC   *60       /INCREMENT POINTER
 010F       JMP   010B      /JUMP BACK FOR NEXT
```

In this program we use a pointer, a location (a pair of locations in the TB6502) that is initially loaded with the address of the first element in the array to be processed. Each time a number is added, the pointer is incremented, so that it continuously "points" to the number to be added next. The first four steps in the program initialize the pointer to 0410, and the next two clear **C** and **AC** in preparation for the addition. We then add indirect through the pointer at 0060. Since the pointer contains 0410, the first number in the string is added. We then increment the pointer to 0411 and jump back to add again. We then add the second number in the string, increment the pointer to 0412, jump back to add again, and so on. This program is incomplete and would loop endlessly as written. To complete the program, we would have to add a means to determine when the complete string has been added, but this has nothing to do with indirect addressing and has not been shown here.

It is important to note in this program that only the page-offset portion of the indirect pointer is incremented, so that the string of numbers cannot cross the page boundary. If the string were long enough that page offset reached FF, the next increment would take it to 00, and the pointer would be 0400; that is, it would wrap around to the beginning of the page. If the string does cross the page boundary, the pointer increment steps should be as follows.

```
010D    INC   *60      /INCR. PTR PAGE OFFSET
010F    BNE   02       /PAGE OFFSET = 0?
0111    INC   *61      /YES, INCR. PAGE NMBR
0113    JMP   010B     /GO BACK, ADD NXT NMBR
```

If the increment does not bring the page offset to zero, we branch to the jump and continue. If the page offset is zero, we increment the page number to cross the page boundary, and then continue.

This program shows that indirect addressing provides an effective way to step through an array of data, but it does have the disadvantage of requiring extra memory accesses to obtain data. Another popular approach to array handling, which does not have this disadvantage, is *indexed addressing*. The TB6502 has two index registers, **X** and **Y**, which can be loaded, incremented or decremented, and manipulated

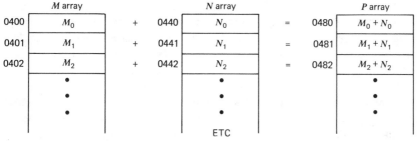

*Figure 12.5   Addition of array.*

in other ways. In index addressing, we fetch a two-byte address just as in absolute addressing and then add the contents of the index register to form the effective address. To see how this can be used, assume we have two arrays of numbers, *M* and *N,* with *M* starting at 0400 and *N* starting at 0440. We wish to add each element of *M* to the corresponding element of *N* and store the sum in the corresponding element of a third array, *P,* starting at 0480. Figure 12.5 illustrates pictorially what is to be done, and the program that follows will do the job.

```
*     PROGRAM TO ADD TWO ARRAYS AND STORE SUM
*       AS THIRD ARRAY.
*       UPON ENTRY:
*         ARRAY M STARTS AT 0400
*         ARRAY N STARTS AT 0440
*       UPON EXIT:
*         ARRAY P TO START AT 0480
*
0120    LDX    #00        /SET INDEX TO 00
0122    CLC               /CLEAR CARRY FLAG
0123    LDA    0400,X     /LOAD ELEMENT OF M
0126    ADC    0440,X     /ADD ELEMENT OF N
0129    STA    0480,X     /STORE SUM IN P
012C    INX               /INCR. INDEX REGISTER
012D    JMP    0122       /JUMP, ADD NEXT PAIR
```

In this program we have two new commands, LD*X*, which loads the index register ***X***, and IN*X*, which increments ***X***. The first step sets ***X*** to 00 by an immediate load, and the second step clears ***C*** in preparation for addition. The third step is an example of the mnemonic notation for index addressing, with the address portion,

**0400, X**

indicating that the address is 0400 modified by ***X***. At this point ***X*** contains 00, so the effective address is 0400 and the first element of *M* is loaded. In the same manner the fourth step adds the first element of *N* and the fifth step stores the result in the first element of *P*. We then increment the index register, to 01, and return to 0122. Now the indexed load adds 01 to 0400 to obtain the effective address 0401, loading the second element of *M*. In the same manner the next two steps now add the second element of *N* and store the sum in the second element of *P*. The program continues in this manner, stepping through all three arrays. Again the program is incomplete because it includes no provision for ending the loop. In this program we could just as well have used index register ***Y***, simply by replacing ***X*** with ***Y*** at every point in the program.

Now consider the advantages of this approach relative to indirect addressing. First, as noted earlier, index addressing is faster than indirect because it requires fewer memory accesses. Second, memory space is saved by eliminating the pointers. Third, the program is simplified. If we used indirect addressing we would have to have three pointers, one for each array, and each would have to be incremented on each pass through the loop. With this approach we have only a single increment of the index register. You should not conclude from this, however, that indexing is always preferable to indirect addressing. Indexing works in this example because we are stepping through all three arrays at the same rate. If we needed to step through all three arrays at different rates, we would have to use indirect pointers unless we had three index registers. Some computers do have more index registers, but there will always be some programs for which we do not have enough index registers and indirect pointers will have to be used.

Implementation of the variety of addressing modes just discussed will require a considerable modification of the original control sequence set forth in Section 11.4. In any microprocessor the addressing mode will be specified by the values of certain bits in the opcode. In Section 11.4 we illustrated branching on opcode bits to distinguish between opcode types. The remaining bits are used in a similar manner to select the addressing mode. The flowchart of Fig. 12.6 shows the sequence of branches necessary to implement the various addressing modes. Assuming that programming will be done in mnemonic form, so that the programmer need not know exactly which bits indicate which modes, the branches in the flowchart simply indicate the nature of the decisions being made, without reference to specific bits.

Immediate, absolute, zero-page, and indirect addressing are all provided for in this flowchart as well as indexing on $Y$. Notice that steps 6 and 7 from the sequence of Section 11.4 have been expanded and rearranged to form steps 6, 6a, b, c and 7, 7a, b. In addition to immediate addressing there are six combinations of addressing modes depending on the steps between steps 6 and 9 in Fig. 12.6. For example, absolute addressing without indexing is accomplished by a path through steps 6, 6b, 7a, and 8. This is the only form of addressing described in Section 11.4.

Notice that to keep the illustration as simple as possible only index register $Y$ is shown in Fig. 12.6. As shown, indexing using $Y$ may be applied following indirect addressing. The use of this mode will be illustrated in section 12.5. If indexing is employed without indirect addressing, either $X$ or $Y$ may be used; and both forms will be used at various points in the book. The use of $X$ together with indirect addressing does not conform to Fig. 12.6. The resulting combined addressing mode is of less interest and is treated only in the appendix. Notice that a carry may be created in the process of adding the least significant address byte to the contents of the index register. This is stored in the carry flag flip-flop, $C$, until the most significant address byte is fetched from memory by step 7a. If the carry was 1, the most significant byte is incremented at step 8.

The use of indexing requires some additional instructions to manipulate the index register. The instructions provided in the TB6502 for $Y$ are listed in Fig. 12.7. The addressing modes available are immediate, zero page, and absolute for LDY and CPY, zero page and absolute for STY; the other four instructions are one-byte instructions with no address. An identical set of instructions is available for $X$. Just substitute $X$ for $Y$ in all cases. The control sequence we have developed so far could be extended to implement the instructions shown in Fig. 12.7.

Our purpose in presenting the RTL implementation of the various instructions has been to assist you in understanding what is accomplished by each instruction. At this point the meaning of the instructions of Fig. 12.7 should be almost self-explanatory, and you should be able to develop the RTL implementation if so desired. The

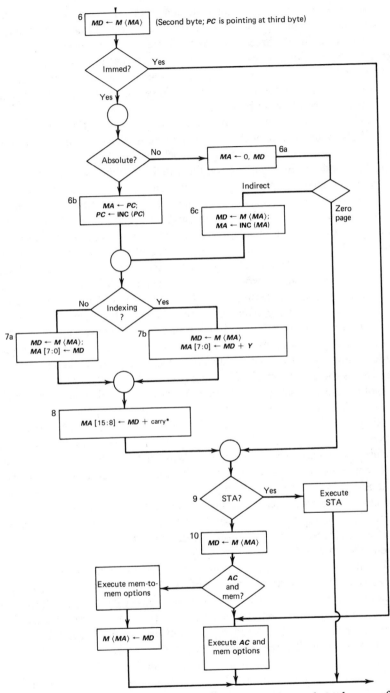

*Figure 12.6   TB6502 addressing options. *Carry will be 1 only in the case of indexing across a page boundry.*

| Command | Mnemonic | Operation |
|---------|----------|-----------|
| Load index $Y$ | LDY | $Y \leftarrow M\langle MA \rangle$ |
| Store index $Y$ | STY | $M\langle MA \rangle \leftarrow Y$ |
| Compare $Y$ | CPY | Set flags on $Y - M\langle MA \rangle$ |
| Increment $Y$ | INY | $Y \leftarrow Y + 1$ |
| Decrement $Y$ | DEY | $Y \leftarrow Y - 1$ |
| Trans $AC$ to $Y$ | TAY | $Y \leftarrow AC$ |
| Trans $Y$ to $AC$ | TYA | $AC \leftarrow Y$ |

*Figure 12.7   Index instructions in the TB6502.*

CPY instruction can be used for testing the contents of the index register. The contents of the addressed location are subtracted from the index register; the index-register contents are not altered, but the flags are set according to the results of the subtraction. For example, if the contents of the addressed location are equal to the index-register contents, the $Z$ flag will be set.

# 12.3 Assembly Language Programming

In the preceding programs we have used a form that we have called *symbolic machine language*. We have written all addresses in the hex form, but have used symbolic representations (mnemonics) for the opcodes. The use of the mnemonics has the advantage of relieving the user of the need to remember the binary codes for the various operations. This is easier in writing the program and is also simpler when reading a program. It is obviously easier to remember that the mnemonic "ORA" stands for logical OR than to remember the corresponding hex or binary code. Before a program can be run on the computer the mnemonics must be converted to the hex codes. This can be done by the programmer or can be done by a computer program. We can further simplify the programmer's task by using symbolic names for variables and for destinations of branches and jumps.

As an example consider the program in Section 11.5 to compute the signum function of the sum of two numbers. Using symbolic addresses this program would appear as follows.

```
*    PROGRAM TO COMPUTE SIGNUM FUNCTION OF
*      THE SUM OF TWO NUMBERS (M+N)
*        UPON EXIT:
*          SIGNUM = +1 IF (M+N) IS POSITIVE
*          SIGNUM =  0 IF (M+N) IS ZERO
*          SIGNUM = -1 IF (M+N) IS NEGATIVE
*
START    CLC
         LDA   M          /M TO AC
```

```
       ADC   N          /(M+N) TO AC
       BMI   NEG        /IF (M+N) NEGATIVE,
  *                     /   GO TO NEG
       BEQ   EQ         /IF (M+N)=0, GO TO EQ
       LDA   #+1        /(M+N) POS, +1 TO AC
       STA   SIGNUM     /STORE RESULT
       HLT              /FINISHED
NEG    LDA   #-1        /(M+N) NEG, -1 TO AC
       STA   SIGNUM     /STORE RESULT
       HLT              /FINISHED
EQ     LDA   #0         /(M+N) = 0, 0 TO AC
       STA   SIGNUM     /STORE RESULT
       HLT              /FINISHED
```

In this program, wherever a variable is referred to, we simply use its name rather than the actual address in memory where it will be stored. Wherever a branch is made, we simply label the instruction that is the destination of the branch, rather than specifying the displacement from the current location. This form of program offers several advantages. The programmer can concentrate on the logic of the program without worrying about where in memory the various data terms and instructions will be stored. All the information necessary to determine what the program will do is present, and in a form that is much easier to interpret than the form showing actual addresses. This form of programming, using symbolic names throughout, is known as *assembly language programming*. Note that the assembly language program parallels the equivalent machine language program exactly. Each assembly language step corresponds to a machine language instruction.

When the program is to be run it must be *assembled*. This process can be carried out by the programmer, a process sometimes called "hand assembly," or it can be done by a computer, by a program known as an *assembler*. In either case, addresses must be assigned to all the variables and labeled locations in the program and the mnemonic opcodes must be converted to their hex equivalents. The addresses of the variables must replace the variable names wherever they appear in the program, and the displacements must be computed for the branch instructions. If the assembly is done by the computer, the assembler will often generate a *symbol table,* a list of the addresses assigned to the various named locations. For the SIGNUM program, if we assume the same absolute locations used in the original program in Section 11.5, the symbol table might appear as shown here.

```
      *********   SYMBOL TABLE   **********

  *    SYMBOL              ADDRESS
       ------              -------
  *    START               0210
  *    NEG                 0222
  *    EQ                  0229
  *    M                   0040
  *    N                   0041
  *    SIGNUM              0043
```

If the program is to be hand-assembled, it is not essential to generate a symbol table, but it is a convenient way to summarize information that will be needed when the program is run. Note that some of the information in the symbol table is information that would otherwise have to be listed in the initial comments section.

If the program is to be hand-assembled, you can use any forms in the assembly program that you find convenient. If you are going to use an assembler, however, you will have to follow the rules of the assembler. First, you must use the established mnemonics for the opcodes, so that the assembler will recognize them. There will usually be limits on the size of symbolic names. We shall allow names of up to 10 alphanumeric characters, with the first character always alphabetic. For example, X2, SIGNUM, FOO45, and M are all legal names, whereas 24XY (starts with a number) and INVENTORYTAX (too many letters) are not. The single letter "A" may not be used as a variable name because it is reserved for indicating the accumulator in commands such as ASL A. Numbers used in an assembler are assumed to be decimal unless preceded by a % (for binary) or a $ (for hex). For example,

**LDA  #10**

will load the binary equivalent of decimal 10 (00001010) into *AC*, whereas

**LDA  # % 10**

will load (00000010) into *AC*, and

**LDA  #$10**

will load (00010000) into *AC*.

This convention of assuming numbers are decimal unless specifically indicated otherwise is quite general in assemblers. We have gotten so used to talking about hex numbers at this point that the assumption of decimal form may seem awkward. However, this convention is followed because one purpose of assemblers is to make life simpler for people, and people are used to working with decimal numbers. We will follow this convention in the rest of this book, except where it is obvious from context that we are referring to addresses (in hex) or to binary numbers.

Indirect addressing and index addressing can be indicated in an assembly language program in the same manner as in symbolic machine language. For example, suppose we are adding a string of numbers in an array *M*, by indirect addressing through a pointer MPTR. In assembly language the add command would be written

**ADC  (MPTR)**

meaning to add the number with its address found in MPTR. If we wish to step through the array *M* by use of the index register *Y*, we would write the command in form

**ADC  M, Y**

meaning to add element *Y* of array *M*.

   The notation for zero-page addressing will also be the same as given in the previous section. For example,

   **INC *PNTR**

would call for incrementing the contents of a location PNTR on page 0. Although this notation does indicate that the variable is on page 0, it does not give the exact address. This observation leads to the more general question regarding how the assembler knows where to locate the program and data. The assembler will have *assembler directives,* or pseudo-ops, special instructions that the programmer issues to the assembler, telling it where to start the program, what area in memory to set aside for data, how large the arrays are, and so forth. These directives will vary from one assembler to another, but it will be useful to consider a few of the standard directives, which are found in virtually every assembler.

   The EQUATE, or Define Internal, directive is used to assign addresses to variables. Typical notations for this directive are shown here.

```
SUM     EQU     $480
CNT     .DI     $40
PNTR    =       $61
```

   The exact notation depends on the assembler, but all three statements perform the same function, to assign an address to the named variable. Thus, these statements will assign variable SUM to location 0480 in memory, CNT to 0040, PNTR to 0061. Note that this directive does not assign a value to the variable, just an address.

   The ORIGIN, or Base Address, directive, tells the assembler where the program is to start. For example, suppose we wanted the signum program considered earlier to start at memory location 0240. At the beginning of the program we would insert one of the following statements, depending on the assembler.

```
ORG     $240
.BA     $240
*=      $240
```

   The result of this directive will be that the opcode for the first instruction in the program (CLC in this example) will be placed at 0240, with the remaining code following in the usual manner. The END, or .EN, directive is placed at the end of any program to notify the assembler that it has reached the end of the program. The signum program is repeated again following, with the necessary directives added.

```
*    PROGRAM TO COMPUTE SIGNUM FUNCTION OF
*      THE SUM OF TWO NUMBERS (M+N)
*      UPON EXIT:
*         SIGNUM = +1 IF (M+N) IS POSITIVE
*         SIGNUM =  0 IF (M+N) IS ZERO
*         SIGNUM = -1 IF (M+N) IS NEGATIVE
*
          .BA  $210
```

```
M           .DI   $40
N           .DI   $41
SIGNUM      .DI   $43
START       CLC
            LDA   M          /M TO AC
            ADC   N          /(M+N) TO AC
            BMI   NEG        /IF (M+N) NEGATIVE,
 *                          /  GO TO NEG
            BEQ   EQ         /IF (M+N) = 0,
 *                          /GO TO EQ
            LDA   #+1        /(M+N) POS, +1 TO AC
            STA   SIGNUM     /STORE RESULT
            HLT              /FINISHED
NEG         LDA   #-1        /(M+N) NEG, -1 TO AC
            STA   SIGNUM     /STORE RESULT
            HLT              /FINISHED
EQ          LDA   #0         /(M+N) = 0, 0 TO AC
            STA   SIGNUM     /STORE RESULT
            HLT              /FINISHED
            .EN
```

With the information provided by these directives, the assembler could then produce the symbol table shown earlier. The directives really are not complicated. Their basic function is to provide the assembler with the information you would need to put the program in machine language form manually. In a sense, the directives are heading comments placed in a form accessible to the assembler. In the original version of the signum program in Section 11.5, we listed the addresses of the variables in the heading, for reference as we generated the hex code. We need the information to write the code; the assembler needs the same information. Note, however, that we do not need to provide the addresses of the labeled statements, such as START, NEG, and EQ, since the assembler can figure them out. Since START is the label of the first instruction, START corresponds to the base address, 0210. The assembler then puts one byte (for CLC) at 0210, three bytes (for LDA M) at 0211, 0212 and 0213, and so forth. In this manner, when it gets to the statement labeled NEG, it can determine the address of that statement, 0222. We could use page 0 addressing for steps involving M, N, and signum, for example, LDA *M. However, absolute addressing was the only type of addressing available in Chapter 11, and we have retained it here for consistency.

A feature provided in many assemblers is the ability to do simple arithmetic to compute addresses. Typically, the assembler will accept $(+)$ and $(-)$ as arithmetic symbols when they appear in operand labels. For example, in the program using indirect addressing in the previous section, we set aside locations 0060 and 0061 for the two bytes of the indirect pointer. In assembly language form we could assign separate names to these two bytes, such as PNTRA and PNTRN, and write the first four steps in the program in the form

```
LDA  #$10
STA  PNTRA
LDA  #$04
STA  PNTRN
```

This will work but will require that we use .DI directives to assign addresses to both PNTRA and PNTRN. A simpler way is to name the first byte PNTR and assign it to location 0060 by

```
PNTR     .DI     $60
```

and write the first four steps in the form

```
LDA     #$10
STA     PNTR
LDA     #$04
STA     PNTR+1
```

This notation indicates that the address associated with the second STA command is the address associated with PNTR plus 1, that is, 0061. Not only is this simpler in requiring the defining of fewer labels, but it is easier to interpret, since it makes it clear that we are referring to two consecutive locations.

Other common directives are BLOCK, or Define Storage, which allocate blocks of memory for storage of arrays, and WORD or BYTE, which assign values (not addresses) to variables. However, these work in different ways in different assemblers, and we will not consider them further since they are not needed unless the program is actually going to be run using a specific assembler. From this point on all programs will be written in assembly language as defined in this section.

## 12.4   More Programming

As you gain experience in programming, you will note that there are a number of common structures in programming, situations that occur again and again in all sorts of problems. Efficient programming requires that you develop standard procedures for dealing with these situations, so that you do not have to "reinvent the wheel" for every program. One thing we note is that programming a computer can be a lot of work, and that effort can rarely be justified unless there are tasks to be performed repetitively, over and over. *Looping,* the repetitive use of a segment of code, is the basic technique for doing repetitive operations; the programmer must be thoroughly competent at handling loops.

Several common features are found in many loops. One is that loops often process arrays of data, in which event we need ways to step through arrays. In Section 12.2 we saw two ways of doing this—through the use of indirect addressing and the use of index addressing. Indirect addressing is a specific example of *pointer addressing,* which is the most common technique for stepping through arrays. The general notation for pointer addressing is of the form

```
LDA  (XX)
```

where **XX** represents the address, the name, the label, the identification, of a storage location that contains the address of the data element that is to be processed (in this case, loaded into the accumulator). In the case of the TB6502, **XX** is the page address of the first of two locations on page 0 containing the data address. In another computer, **XX** might be the name of a register containing the data address. To use pointer addressing to step through an array, we first load the pointer with the starting address of the array. Each time we access the array, we execute two instructions, for example,

```
LDA  (XX)
INC  XX
```

In this manner, each successive execution of **LDA (XX)** accesses the succeeding element in the array.

Whether or not a loop accesses data arrays, an obvious basic problem is that of terminating the loop, of deciding when it has been executed enough times. There are two basic approaches. One is to test for some specified condition in the data. This method is used when it is not known in advance how many times the loop is to be executed.

The more common approach, used when it is known in advance how many passes through the loop are required, is to set up a *counter,* to count the number of passes through the loop. The counter may be implemented in memory or in a register. It can initially be set to the number of passes and decremented each time the loop is executed until it reaches zero. It can be set to minus the number of executions and incremented until it reaches zero, or it can be set to zero and incremented until it reaches the desired count. The choice between these methods usually depends on what instructions are available for manipulating and testing the counter. In the TB6502, the most flexible approach for most purposes is to decrement a counter in page 0. For example, assume a certain loop is to be executed eight times. The following illustrates the standard structure.

```
              LDA  #8
              STA  *CNTR     /INIT. COUNTER TO 8
     LOOP      .
               .
              body of loop
               .
               .
              DEC  *CNTR     /DECREMENT COUNTER
              BNE  LOOP      /NOT DONE, REPEAT LOOP
               .
              continuation
               .
               .
```

With these two standard structures in mind, let us consider again the program for adding a string of numbers using indirect pointers and assume that there are 10 numbers in the string. Shown here is the modified program with provision added to end the loop after 10 numbers have been added.

```
*    PROGRAM TO ADD A STRING OF TEN NUMBERS
*      UPON ENTRY:
*        STRING OF NUMBERS STORED IN CONSECUTIVE
*        LOCATIONS STARTING AT 0410
*      UPON EXIT:
*        SUM IN ACCUMULATOR
*
START     LDA    #$10        /INIT. PAGE OFFSET
          STA    *PNTR       /   OF POINTER
          LDA    #$04        /INIT. PAGE NUMBER
          STA    *PNTR+1     /   OF POINTER
          LDA    #10         /
          STA    *CNTR       /INITIALIZE COUNTER
          CLC                /CLEAR C
          LDA    #00         /CLEAR AC
LOOP      ADC    (PNTR)      /ADD NUMBER
          INC    *PNTR       /INCREMENT POINTER
          DEC    *CNTR       /DECREMENT COUNTER
          BNE    LOOP        /DONE?  IF NOT,
*                            /   GO TO LOOP
          HLT                /YES, DONE.

**********  SYMBOL TABLE  **********

*   SYMBOL          ADDRESS
    ------          -------
    PNTR            0060        /LOW BYTE OF POINTER
    CNTR            0062        /COUNTER
```

We have provided a counter at 0062 and added two steps to the program to initialize the counter to 10. The first six steps of the program are used to initialize the counter and pointers. It is very important that this initialization be included in the body of the program. Some novice programmers assume that pointers and counters can be initialized at the same time the data are loaded. This is poor practice; anything that has to be done whenever the program is run should be included in the program if possible. After each addition, we decrement the counter and test to see if it has reached zero. If it has not, we return to add again; if it has, we halt.

The pointer approach is quite general, but it has its limitations. If we want to step through several arrays at the same time, we need several pointers and each must be incremented in each pass through the loop. As we saw in Section 12.2, the use of index addressing may provide a superior way of handling such problems. Suppose we simply wanted to copy all the elements of an array ABLE into the corresponding elements of an array BAKER. The basic structure would be as shown here.

```
LDA    ABLE, X
STA    BAKER, X
INX
```

ABLE and BAKER represent the beginning addresses of the two arrays. Prior to entering the loop, **X** is set to 00. On the first pass through the loop the program will transfer from ABLE to BAKER, on the second pass from ABLE + 1 to BAKER + 1, and so on, with only one increment of **X** being required on each pass since **X** modifies the address of both instructions.

We still have the problem of terminating the loop. Assuming we know how many elements are in the arrays, we could use the same counter method just discussed. However, since we are using the index register anyway, we can let it serve a dual purpose, serving to step through the arrays and also serving as counter. After each increment, the index register indicates how many elements have been transferred. Thus, all we need to do is compare that number to the number of elements to be transferred. This is done through the use of the CPX command (Compare X) as shown in the following structure.

```
          LOOP        .
                      .
                   body of loop
                      .
                      .
                   INX
                   CPX    #NN
                   BNE    LOOP
                      .
                      .
                   continuation
                      .
```

Here NN represents the number of times we want to pass through the loop. The CPX command compares the contents of **X** to NN; if they are not equal, it clears the **Z** flag; if they are equal, it sets the **Z** flag. The BNE instruction will thus return to LOOP until the desired number of passes has been executed. This approach is attractive only because the TB6502 has a CPX instruction. In computers lacking this instruction, the use of index registers as counters is not as satisfactory. In any event, the counter approach discussed earlier is usually preferable unless the index register is being used anyway.

The program to add two arrays discussed in Section 12.2 is shown here in assembly form, with provision for terminating the loop included.

```
*    PROGRAM TO ADD TWO 16-ELEMENT ARRAYS AND
*      STORE SUM AS THIRD ARRAY, P = (M + N)
*      UPON ENTRY:
*        ARRAY M STARTS AT 0400
*        ARRAY N STARTS AT 0440
*      UPON EXIT:
*        ARRAY P STARTS AT 0480
*
START    LDX    #0          /CLEAR INDEX REGISTER
LOOP     CLC                /CLEAR C
         LDA    M,X         /LOAD ELEMENT OF M
         ADC    N,X         /ADD ELEMENT OF N
         STA    P,X         /STORE ELEMENT OF SUM
         INX                /INCREMENT INDEX
         CPX    #16         /COMPARE INDEX TO 16
         BNE    LOOP        /DONE?  IF NO, REPEAT
         HLT                /YES, DONE
```

```
********* SYMBOL TABLE *********

*  SYMBOL        ADDRESS
   ------        -------
*  M             0400      /START OF M ARRAY
*  N             0440      /START OF N ARRAY
*  P             0480      /START OF P ARRAY
```

# 12.5  Data Conversion, Decimal Arithmetic, and Subroutines

In the last section we saw examples of loops terminated by counters. As noted, another way to terminate a loop is to look for some special condition in the data. One type of program in which such situations are often encountered is the data conversion program. Data formats convenient for input of data are rarely suitable for internal calculation, so that the programmer is often faced with the need to convert data from one form to another.

One common form of input is from a keyboard. Each time a key is struck a code is sent to the computer. When alphanumeric keyboards are used, the most common code is the ASCII code, which was tabulated in Fig. 6.3. In many situations the operator will be permitted to enter a message of arbitrary length, terminated by a *delimiter* character, such as a comma, period, or carriage return. When using an 8-bit computer such as the TB6502, the loading routine will load each successive character into a successive location in memory until the delimiter is sensed. When the complete message has been entered and stored, it may then be converted to some other format.

Let us assume that a string of decimal digits representing a positive integer has been entered, starting at location 0050 and terminating with a carriage return. One convenient way of representing decimal data is binary-coded decimal (BCD) in which each decimal digit is represented by its 4-bit binary equivalent. Since each digit requires only four-bits, it is customary to pack two BCD digits into each byte. The conversion of ASCII to BCD is very simple. As seen in Fig. 6.3, the low-order four-bits in the ASCII codes for the decimal digits are BCD codes, so that all we have to do is strip off the upper four bits and then pack two 4-bit codes per byte. Let us assume we wish to convert the string of ASCII digits indicated by ASCPTR to a string of packed BCD digits, indicated by BCDPTR. The program shown here will accomplish this task.

```
*  PROGRAM TO CONVERT ASCII STRING TO PACKED
*  BCD STRING, ASCII STRING TERMINATED
*  WITH CARRIAGE RETURN.  POINTERS TO ASCII
*  AND BCD STRINGS SET BY ROUTINE THAT
*  LOADS ASCII STRING.
*  UPON ENTRY:
*     ASCII STRING LOADED WITH MOST
*     SIGNIFICANT DIGIT (M.S.D.) AT LOCATION
*     INDICATED BY ASCII POINTER (ASCPTR)
```

```
*     UPON EXIT:
*        BCD STRING STARTS WITH M.S.D. AT
*        LOCATION INDICATED BY INITIAL VALUE
*        OF BCD POINTER (BCDPTR)
*
START      LDA   (ASCPTR)   /GET ASCII CHARACTER
           CMP   #$0D       /COMPARE TO CARR. RET.
           BEQ   END        /IF CR, GO TO HALT
           ASL   A          /
           ASL   A          /SHIFT BCD DIGIT TO
           ASL   A          /UPPER HALF OF BYTE
           ASL   A          /
           STA   (BCDPTR)   /STORE UPPER BCD DIG.
           INC   *ASCPTR    /INCR. ASCII POINTER
           LDA   (ASCPTR)   /GET ASCII CHAR.
           CMP   #$0D       /COMPARE TO CR
           BEQ   END        /IF CR, GO TO HALT
           AND   #$0F       /MASK OFF UPPER BYTE
           ORA   (BCDPTR)   /ADD 2ND DIG. TO 1ST
           STA   (BCDPTR)   /STORE PACKED DIGIT
           INC   *ASCPTR    /INCR. ASCII POINTER
           INC   *BCDPTR    /INCR. BCD POINTER
           JMP   START      /GO BACK FOR NEXT
END        HLT              /DONE, HALT
```

The end of the string is marked by a carriage return. On a typewriter or CRT keyboard, striking the CR key will start a new line and send the corresponding ASCII character, having no special meaning other than that assigned to it by the programmer. To terminate this conversion, we test each ASCII character as it is brought from memory to see if it is a carriage return, using the CMP command. For this program we use CMP #0D to subtract the carriage return character from each ASCII character as it is brought from memory. If the character is a carriage return, the subtraction will produce all 0's, setting the *Z* flag. The BEQ command then tests the *Z* flag and branches to HLT if it is set. Otherwise we continue, *with the character intact for further processing.*

When we process strings of decimal digits with BCD arithmetic, it will generally be necessary to know how many digits there are. For this reason, when we set up a conversion routine such as the one just shown, we will often include provision for counting the digits as we convert them to packed BCD form.

Another complication arises with regard to the order in which the digits are read in and converted. When data are entered from a keyboard, they will normally be entered in the same order in which they are written, starting with the most significant digit. On the other hand, arithmetic processing normally starts with the least significant digit. If we know the location of the most significant digit and the number of digits, we can easily determine the location of the least significant digit. But, if the number of digits is odd, this routine will leave the least significant digit in the upper half of a byte, even though BCD arithmetic is most convenient if it starts with the lower half of a byte. For this reason, many conversion routines will start with a loop that scans the string for a CR and counts characters and then does the conversion to packed BCD in the reverse order from that used here, starting with the least significant digit.

Once strings of decimal digits have been entered and converted to packed BCD form, it may be desired to perform arithmetic on them, for example, adding two decimal strings. Many computers can do arithmetic only in the binary form, in which event BCD arithmetic will require special programming. The TB6502 computer includes hardware provisions for BCD arithmetic. If we execute the one-byte instruction SED (set decimal), the ADC command will function as a BCD add command until we execute CLD (clear decimal). Only the ADC and SBC commands are affected by the switch to decimal mode.

The decimal operation of ADC differs from the binary in that the 8-bit byte is treated as two positive BCD digits and the addition is according to the rules of BCD arithmetic with appropriate carry from the lower digit to the upper digit and setting of the carry flag if there is a decimal carry out of the upper digit position. Since the carry flag also serves as an input to the addition, it provides carry from one pair of digits to the next.

Assume we wish to add the two decimal numbers 1476 and 2351, stored as packed BCD strings in two bytes each. After clearing the carry flag, we execute ADC on the low-order bytes, giving the following result.

$$0 \leftarrow C$$

| | |
|---:|:---|
| 76 | 0111 0110 |
| 51 | 0101 0001 |
| 127 | 1 0010 0111 |

C     AC

The partial sum 27 is stored, and the operation is completed by executing ADC on the high-order bytes.

$$1 \leftarrow C \qquad\qquad 1 \leftarrow C$$

| | |
|---:|:---|
| 14 | 0001 0100 |
| 23 | 0010 0011 |
| 38 | 0011 1000 |

We store the partial sum 38 in the next byte in the sum area, giving the complete sum 3827 in two bytes.

Assume we have two strings of BCD digits stored in the arrays OPND1 and OPND2, least significant digits first. The number of digits, the same for each string, is stored at NODIG. The sum is to be stored in a string SUM, least significant digit first. A program for carrying out this addition is shown here.

```
*    PROGRAM TO ADD TWO STRINGS OF BCD DIGITS
*      UPON ENTRY:
*        NUMBER OF DIGITS LOADED IN NODIG
*        INPUT STRING IN ARRAYS OPND1 AND
*        OPND2, LEAST SIG. DIGIT FIRST.
*      UPON EXIT:
*        SUM STRING TO BE IN ARRAY SUM,
*        LEAST SIG. DIGIT FIRST.
*
START    LDA    NODIG      /GET DIGIT COUNT
         STA    *CNTR      /LOAD COUNTER
```

```
           CLC                /CLEAR CARRY
           SED                /SET DECIMAL MODE
           LDX    #0          /CLEAR INDEX
LOOP       LDA    OPND1,X     /LOAD BYTE, 1ST OPND
           ADC    OPND2,X     /ADD BYTE, 2ND OPND
           DEC    *CNTR       /DECR. DIGIT COUNTER
           BEQ    ODD         /DONE?
           DEC    *CNTR       /NO, DECR. DIG. CNTR
           BEQ    EVEN        /DONE?
           STA    SUM,X       /NO, STORE SUM BYTE
           INX                /INCR. INDEX
           JMP    LOOP        /BACK FOR NEXT BYTE
ODD        AND    #$0F        /SET UPPER DIG TO 0
EVEN       STA    SUM,X       /STORE FINAL BYTE
           HLT                /DONE, HALT
```

If the number of digits is odd, the upper half of the final sum byte should be 0. Depending on the conversion routine, the upper half of the final operand bytes may not be 0, in which case the final sum byte will not be correct. After forming a sum byte, we decrement the digit counter to see if we have reached the final digit. If so, we branch to ODD to set the upper half of the byte to 0 and then store it. If the first digit in the byte is not the final digit, we decrement the digit counter again. If this is the final digit, we branch to EVEN to store the final sum digit. If neither digit in the current byte is the final digit, we store the sum digit, increment the pointers, and repeat the loop.

This program assumes that there will be no carryout; that is, the number of digits in the sum will be the same as the number of digits in the operands. If this were not the case, we would have to check the carry flag after adding the final digits and generate an additional 1-digit as the leading digit of the sum.

You may note that we used indirect addressing to step through the arrays in the ASCII–BCD program and indexing in the BCD add program. The indexing is clearly simpler and faster, so why did we not use it in both routines? The reason is that the ASCII–BCD routine steps through the two arrays at different rates. Two ASCII characters are needed for each packed BCD word. We could have used indexing for one array and indirect for the other, but it is generally better to be consistent. Later in this section we shall see a possible advantage to using indirect addressing in the BCD add program.

If data have been processed internally in the form of packed BCD, they must be converted back to ASCII form for output. Each packed byte must be unpacked and converted to two ASCII bytes, with the end of each line signified by the sending of a CR character. Let us assume that the program just presented has just been executed to form a BCD sum and we now wish to transmit the sum to a printer.

At the conclusion of this program the pointer SUMPTR is pointing to the byte containing the most significant digit. We shall again use ASCPTR to designate the location of the ASCII string and assume it has been initialized to an appropriate starting value prior to the execution of this program. Recall that the most significant digit may not be in the upper half of a byte at the end of the previous program. Assuming that we do not want to print leading zeros, the lines immediately preceding LOOP check for all zeros in the upper half of the leading byte. If a leading zero is found, we branch to LOWER to start conversion with the lower half of the first byte, otherwise conversion of a full byte is started at LOOP. Note that we check the digit counter for zero only after every other decrement, since we know that the last digit will come on a byte boundary.

```
*    PROGRAM TO UNPACK BCD STRING AND
*      CONVERT TO ASCII STRING
*        UPON ENTRY:
*          BCD STRING AND POINTERS LOADED
*          COUNT OF DIGITS IN NODIG
*        UPON EXIT:
*          ASCII STRING AT LOCATION
*          SPECIFIED BY POINTER (ASCPTR)
*
START     LDA   NODIG        /GET DIGIT COUNT
          STA   *CNTR        /LOAD COUNTER
          LDA   (SUMPTR)     /GET FIRST BYTE
          AND   #$F0         /TEST UPPER DIGIT
          BEQ   LOWER        /IF UPPER DIGIT 0,
*                            /   GO TO LOWER
LOOP      LDA   (SUMPTR)     /LOAD FULL BYTE
          LSR   A            /
          LSR   A            /SHIFT UPPER DIGIT TO
          LSR   A            /LOWER HALF OF BYTE
          LSR   A            /
          ORA   #$30         /FORM ASCII CHAR.
          STA   (ASCPTR)     /STORE ASCII CHAR.
          INC   *ASCPTR      /INCR. ASCII POINTER
          DEC   *CNTR        /DECR. COUNTER
LOWER     LDA   (SUMPTR)     /GET BYTE
          AND   #$0F         /MASK OFF UPPER DIG.
          ORA   #$30         /FORM ASCII CHAR.
          STA   (ASCPTR)     /STORE ASCII CHAR.
          INC   *ASCPTR      /INCR. ASCII POINTER
          DEC   *SUMPTR      /DECR. SUM POINTER
          DEC   *CNTR        /DECR. DIGIT COUNT
          BNE   LOOP         /NOT DONE, GET NEXT
          LDA   #$0D         /DONE, FORM CR
          STA   (ASCPTR)     /STORE CR
          HLT                /DONE, HALT
```

We noted earlier that efficient programming requires that segments of code be used many times. Looping provides one way of doing this. Another important technique is the use of subroutines. A *loop* is a segment of code that is used repeatedly at one particular point in the program. A *subroutine* is a segment of code, often a complete program in itself, that can be used at many different points in a larger program. As an example, suppose we have a program that requires BCD addition at three different points. Rather than repeat the code just developed three times in the program, we set the BCD addition program up as a subroutine and call it, or invoke it, whenever we need it. The basic relationship between the main program and the subroutine is illustrated in Figure 12.8, where we have assumed that the BCD routine has been loaded at 0200. Here we see that the subroutine is called at three points in the main program, at locations 0320, 0350, and 0375. In each case, following execution of the subroutine, control must return to the main program immediately following the instruction that called the subroutine.

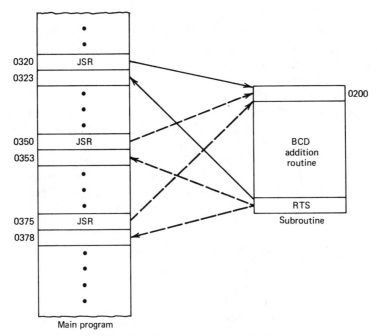

*Figure 12.8   Relationship of main program and subroutine.*

The procedure for getting to the subroutine and back to the main program at the right place is often referred to as the *linkage.* Getting to the subroutine is no problem. If we know where the subroutine is located, we can jump to it. Getting back is a more difficult problem. When the subroutine is written, it is not known where in the main program it will be called. The information required to return cannot be included in the subroutine; rather, it must somehow be preserved when the subroutine is called. The usual procedure is to store the program counter contents at the time the subroutine is called in some location from which they can be retrieved when the subroutine has been executed. In the TB6502 this is accomplished with two commands, JSR (jump to subroutine) and RTS (return from subroutine). JRS is a 3-byte absolute-address command. When this command is executed, the current program counter contents are stored in a reserved location in memory and the address portion of the instruction, the address of the subroutine, is transferred to the program counter. When RTS is executed, the stored address of the return point in the main program is retrieved from the reserved location and placed in the program counter.

As an example, refer again to Fig. 12.8. The subroutine, starting at location 0200, is first called at location 0320 with the command

**0320 JSR 0200**

In the fetch of this instruction the program counter is incremented to 0323, the address of the next instruction in the main program. This address is stored in the reserved location in memory, 0200 is transferred to the program counter, and control returns to step 1 of the CPU sequence to fetch the first instruction of the subroutine. When the subroutine is complete, the RTS instruction retrieves the address 0323 from the reserved location and loads it into the program counter and returns to step 1 to fetch the next instruction in the main program.

In addition to getting to and from the subroutine, we have to get the operands to the subroutine and the results back. This process is sometimes referred to as *parameter passing,* and the set of instructions for accomplishing it is known as the *calling sequence.* Many different kinds of calling sequences are used in various computers. The simplest method is to use the processor registers, such as **AC** and **Y** in the case of the TB6502. As an example, suppose we want to take the absolute value of a number. All that is necessary is to get the number into **AC** and execute JSR ABS. This procedure is probably too simple to justify setting up a subroutine, but it will serve to illustrate parameter passing in its simplest form.

```
*   SUBROUTINE TO TAKE ABS. VALUE OF INTEGER
*     UPON ENTRY:
*       NUMBER IN AC
*     UPON EXIT:
*       ABSOLUTE VALUE OF NUMBER IN AC
*
ABS       ORA   #$00      /SET FLAGS
          BPL   DONE      /IF POSITIVE, RETURN
          EOR   #$FF      /LOGICAL COMPLEMENT
          CLC
          ADC   #1        /ADD 1 TO FORM
*                         /  TWO'S COMPLEMENT
DONE      RTS             /DONE, RETURN TO MAIN
```

First, note the heading comments, which specify what the subroutine does and where the input and output parameters will be found. Obviously, you cannot use the subroutine unless you know these things. The first step is done to set the flags, on the assumption that we do not know if the current setting of the flags corresponds to the contents of **AC**. Perhaps the main routine loaded **AC** and then executed some other commands that affect the flags before jumping to the subroutine. ORing with 00 does not change the contents but sets the flags. We then test the sign of the number and take the 2's-complement if it is negative, leaving the result in **AC** for return to the main program.

This procedure is very simple but limited to just a few operands, depending on how many registers the processor has. A more general approach is to use locations in memory instead of registers, setting aside a *parameter area* in memory through which parameters are passed between main program and subroutine. In the TB6502, parameters are usually passed through page 0 because that page can be used for indirect and zero-page addressing. As an example, assume we wish to compare two numbers.

```
*   SUBROUTINE CMPR,
*     TO COMPARE TWO POSITIVE INTEGERS
*     UPON ENTRY:
*       ONE NUMBER IN 0060, ONE IN 0061
```

```
*      UPON EXIT:
*         LARGER INTEGER IN AC
*
CMPR      LDA    *60        /GET NUMBER
          CMP    *61        /COMPARE SEC. NUMBER
          BCS    DONE       /BRANCH IF FIRST
*                           /  NUMBER LARGER
          LDA    *61        /SECOND NUMBER LARGER
DONE      RTS
```

The subroutine is simple enough. It simply subtracts one number from the other, using 2's-complements, tests the results to see which is larger, and leaves the larger in **AC**. In this example we have used a combination of the two techniques, using a parameter area for the operands passed to the subroutine and **AC** for the result passed back to the main program. In the previous example there was no calling sequence *per se;* when JSR ABS is executed, the absolute value of whatever is in **AC** will be taken. In this case we have to add steps (a calling sequence) to the main program to get the operands into the parameter area. Assume the operands to be compared are ABLE and BAKER.

```
                        Calling Sequence

MAIN          .
              .
              .
          LDA    ABLE       /GET FIRST NUMBER
          STA    *60        /MOVE TO SUBROUTINE
*                           /    PARAMETER AREA
          LDA    BAKER      /GET SECOND NUMBER
          STA    *61        /MOVE TO SUBROUTINE
*                           /    PARAMETER AREA
          JSR    CMPR       /JUMP TO SUBROUTINE
          continue
              .
              .
```

In these two examples we have passed the actual values of the parameters. When dealing with arrays of data it is usually more practical to pass the addresses of the arrays. As an example, let us assume that the decimal addition routine is to be used as a subroutine. As a subroutine it needs to be provided with the addresses of the operands and the sum and the number of digits. The number of digits is no problem; we just let the main program load the CNTR location. But how do we get the addresses into the subroutine? When we wrote the original BCDADD program, we assumed fixed locations for the arrays OPND1, OPND2, and SUM, which permitted us to use index addressing. When writing a subroutine, we do not know where the arrays will be; that depends on the main program. The only really satisfactory way to get addresses into a subroutine is to use indirect addressing. The subroutine identifies the arrays by means of pointers, and the main program loads the pointers with the actual addresses of the arrays. For this example we shall specify the locations for the digit count and the various pointers as described in the heading comments for the subroutine.

```
*   SUBROUTINE TO ADD TWO STRINGS OF BCD DIGITS
*     UPON ENTRY
*       00E0 = CNTR, SET TO NUMBER OF DIGITS
*       00E1,00E2 = OP1PTR, OP1PTR+1,
*                   POINTER TO FIRST STRING
*       00E3,00E4 = OP2PTR, OP2PTR+1,
*                   POINTER TO SECOND STRING
*       00E5,00E6 = SUMPTR, SUMPTR+1,
*                   POINTER TO SUM STRING
```

In the body of the program, we can now use indirect addressing to access the three arrays, but this would seem to complicate the process of stepping through the arrays, which is done in the original program by indexing. We can increment each pointer separately, but this adds steps to the program. It would be very convenient to be able to fetch the indirect address from the pointer and then index it. The TB6502 has an additional addressing mode, *indirect indexed addressing,* which does exactly that. An example of this type of addressing is

**0115    ADC    (50), Y**

Figure 12.9 illustrates how the operand is obtained. The indirect address (50) points to an address stored at 0050 and 0051. In this case we assume that 0050 and 0051 contain 47 and 04 respectively, giving an address 0047. This address is then indexed by adding the contents of **Y** (assumed to be 05 in this example) to produce the effective address 004C. Note that the name of the mode indicates the order in which the steps are taken; we first fetch an address indirectly and then index it.

With this addressing mode available, the complete BCDADD subroutine will appear as shown here.

```
BCDADD  CLC                  /CLEAR CARRY
        SED                  /SET DECIMAL MODE
        LDY  #0              /CLEAR INDEX Y
LOOP    LDA  (OP1PTR),Y      /LOAD BYTE 1ST OPND
        ADC  (OP2PTR),Y      /ADD BYTE 2ND OPND
        DEC  *CNTR           /DECR. DIG. COUNTER
        BEQ  ODD             /DONE?
        DEC  *CNTR           /NO, DECR. DIG. CNT
        BEQ  EVEN            /DONE?
        STA  (SUMPTR),Y      /NO, STORE SUM BYTE
        INY                  /INCR. INDEX Y
        JMP  LOOP            /REPEAT, NEXT BYTE
ODD     AND  #$0F            /ZERO UPPER DIGIT
EVEN    STA  (SUMPTR),Y      /STORE FINAL BYTE
        RTS                  /DONE, RETURN
```

We skip the first two instructions of the original routine since the main program can load the count of digits directly into CNTR. All references to the arrays are changed from indexed to indirect indexed addressing. In other respects there is no change in the addition routine except that the HLT instruction is replaced with RTS.

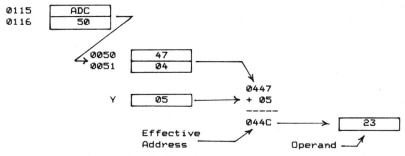

*Figure 12.9   Indirect indexed addressing.*

Let us assume that the main program is to call the subroutine to add two strings of numbers that have been loaded starting at 0300 and 0360, with the sum to be returned starting at 03C0. Also assume that the number of digits has already been loaded into the accumulator. The following calling sequence will do the job.

```
030A    STA     *E0        /LOAD DIGIT COUNT
030C    LDA     #$00       /LOAD PAGE ADDRESS
030E    STA     *E1        / OF FIRST OPERAND
0310    LDA     #$60       /LOAD PAGE ADDRESS
0310    STA     *E3        / OF SECOND OPERAND
0314    LDA     #$C0       /LOAD PAGE ADDRESS
0316    STA     *E5        /OF SUM
0318    LDA     #$03       /
031A    STA     *E2        /LOAD PAGE NUMBERS
031C    STA     *E4        /OF ALL POINTERS
031E    STA     *E6        /
0320    JSR     BCDADD     /JUMP TO SUBROUTINE
0323    xxx     xxxx       /NEXT INSTRUCTION
```

We have used absolute addressing for the various pointer locations to clarify the process. Symbolic addresses can be used as long as care is taken to see that both the main program and the subroutine assign the same addresses to the various symbols.

One important feature of this subroutine is that its functioning is independent of where it is located in memory. It can be loaded in any convenient area of memory, and the only change necessary is the address portion of the JSR instruction. This is particularly important if we want to work in assembly language. We can assign a name (symbol) to the subroutine and the assembler can locate the subroutine wherever convenient and assign the starting location address to the symbolic name.

In the above discussion we indicated that the JSR instruction placed the *PC* in a "reserved location" in memory. This reserved memory location is actually located at top of what will be called a *stack*. A special register called a *stack pointer* will point at this location. Our discussion of stack operations will be deferred until Chapter 15.

## Problems

**12.1**  Convert the program of Problem 11.3 to assembly language form.

**12.2**  (a) Convert the program of Problem 11.4 to assembly language form.
(b) Repeat for Problem 11.5.
(c) Repeat for Problem 11.6.

**12.3**  Assume the following values (in hex) are intially stored in the locations indicated.

$AC$ = 04      $PC$= 0112      $Y$ = 03      $M\langle 0347\rangle$ = 0A
$M\langle 0348\rangle$ = 41      $M\langle 0349\rangle$ = 26      $M\langle 034A\rangle$ = 17
$M\langle 0060\rangle$ = 47      $M\langle 0061\rangle$ = 03      $C$ = 0

For each of the following instructions, determine the effective address.
(a) **ADC  ∗60**    (b) **ADC  (60)**    (c) **ADC  0347,Y**
(d) **ADC  #%01**    (e) **ADC  0348**

**12.4**  For the initial values in Problem 12.3 and for each of the instructions there, assuming each instruction is located at 0112, determine the contents of $AC$, $PC$, and $C$ after execution of the instruction.

**12.5**  Assume two arrays, $M$ and $N$, of 10 elements each, with $M$ starting at 0210 and $N$ starting at 0220. A program is to be written to compare the two arrays on an element-by-element basis. For each pair of elements, $+1$, 0, or $-1$ is to be stored in the corresponding element of an array $C$ starting at 0230, $+1$ if the element of $M$ is the larger, 0 if they are equal, and $-1$ if the element of $N$ is the larger. Write a TB6502 program in assembly language to accomplish this function.

**12.6**  Given a 16-by-16 array of numbers, with the first row stored in locations 0200, 0201, 0202, and so forth, the second row in locations 0210, 0211, 0212, and so forth. We wish to sum the elements of each row, with the row sums being stored in a 16-element array starting at 0300. Write a TB6502 program in assembly language form to accomplish this function.

**12.7**  Modify the input data conversion routine so that it counts the number of characters and packs starting with least significant digit. Write the program as a subroutine, passing parameters through page 0, including returning the count of characters to the main routine via page 0. Write a calling sequence to call this subroutine, assuming the ASCII string starts at 0200 and the packed BCD string starts at 0200. You may assume the maximum number of digits is 120. If the number of digits is odd, the final byte stored should have 0 in the upper four bits. For example, if the number being stored is 45932, the final byte stored would be 04.

**12.8**  Assume that two equal length BCD strings have been packed by the routine in Problem 12.7, starting at locations 0280 and 0380. Modify the BCD add routine as required to add these two strings, allowing for the fact that the sum may have one more digit than the operands and providing a count of the number of digits in the sum.

**12.9** Write a subroutine to compute the absolute value of an array of not more than 40 numbers and to store the absolute values in another array. The main routine will provide the addresses of the two arrays and the number of elements. Set aside locations in the subroutine as needed for receiving these parameters, and be sure the subroutine is fully documented. Then write a typical main program calling sequence to use this subroutine.

**12.10** An array contains positive integers. Write a subroutine to find the maximum element in the array. The subroutine should receive the starting address of the array and the number of elements from the main program. It should return the largest element in the accumulator. Write a typical calling sequence for this subroutine, assuming the array starts at 0340 and has 25 (decimal) elements.

# Memory and Input/Output

## 13.1 Memory Mapping

In our consideration of computers to this point we have deferred consideration of how information gets in and out of the computer. In all our programs we have assumed that the necessary data and instructions were already in the computer. Obviously this is not a realistic assumption; in the practical use of computers we must be able to communicate with them. From the standpoint of both hardware and software, input/output poses some of the most difficult problems in designing and using computer systems. Many of the difficulties derive from the fact that input/output requires communications between devices and systems of radically different characteristics.

Many input/output devices are at least partly mechanical and orders of magnitude slower than typical electronic devices. If the devices interact directly with humans, as in the case of a keyboard, the differences in speed are even more extreme. We also frequently have differences in format. Computers generally work in terms of units of information (words) of a fixed size. Human users do not find it convenient to organize data this way in many cases. Taking a keyboard as an example again, data are transmitted one character at a time, with no necessary limit on the number of characters, and with no necessary correspondence between the size of a character and the size of a computer word. The general process of making the signals of one device compatible with the signals of another is often referred to as *interfacing*. Interfacing may also involve purely electronic considerations, as when the voltage levels in one device are different from those in another device. However, we shall not consider such problems in this chapter, assuming any devices to be interconnected are electronically compatible.

In this chapter we wish to deal with some of the hardware aspects of interfacing. We shall assume that all input/output operations are basically controlled by the central processor. Input/output (I/O) operations may take place in response to requests from external devices, but the overall control always is exercised by the CPU. To control an I/O operation the CPU must be capable of issuing two basic types of signals, identification signals and control signals. As the name implies, identification signals are used to identify the device that is to communicate with the CPU. Control signals are used to control the actual transfer of information. There are many different I/O organizations found in computer systems, but most fall into one of two basic categories. In an *independent I/O* system, the CPU issues special signals that differentiate between I/O operations and memory operations and may have special signal lines dedicated to I/O operations. In such systems, the CPU is considered to have a certain number of *I/O ports,* with each port commonly dedicated to communications with a specific device. When this approach is used, the CPU will have a number of special I/O instructions by means of which a programmer can select a port and control signaling operations at that port.

The second basic technique is known as *memory-mapped I/O.* There are no special lines or signals provided for handling I/O. Instead, I/O devices simply implement specific locations in the address space of the CPU and communications take place over the same lines and in the same manner as communications with memory. Each I/O device is regarded as a memory location, and no special instructions are needed. The same instructions used to manipulate memory can be used for I/O.

Both approaches have advantages and disadvantages. The independent I/O method is usually the faster and more efficient of the two and is more likely to be used in larger, more expensive systems. The memory-mapped approach is simpler, cheaper, and more flexible but is usually slower. It is the most popular approach in microprocessors and other small systems. Since our primary interest is in the smaller systems, this is the approach we shall concentrate on in this chapter.

At this point you should refer to Fig. 11.3, showing the general organization of a single-address computer. In the case of a microprocessor, everything in that figure except the memory is on the microprocessor chip itself. The memory will usually be realized on separate chips. As suggested by that figure, there will be two categories of lines from the microprocessor chip to the memory chips, the address lines and the data lines. Rather than having separate lines for **DATAIN** and **DATAOUT,** most microprocessor systems have just one set of bidirectional data lines, used for transmitting the data in both directions. For the TB6502 these two sets of lines will be known as **ADBUS** (16 bits) and **DBUS** (8 bits), as shown in Fig. 13.1. There are also a number of control lines interconnecting the TB6502 and the memory and I/O interface packages. These are not shown in Fig. 13.1 but will be introduced as needed in succeeding sections.

In the memory-mapped I/O system there is no logical distinction between a location in memory and an I/O device. The identification of the memory location or the I/O device is provided by the address sent out over the address lines. A given address, say, 6423, might refer to a location in memory or to a printer or to a key-

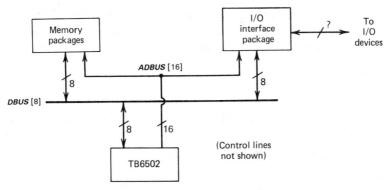

*Figure 13.1* CPU–memory–I/O interconnection.

board, or to any other kind of device. Recall that the address space is simply the number of locations that can be referred to (addressed) by the CPU. In the case of the TB6502 there are 16 bits of address, so the address space is $2^{16}$ = 65536 locations, or 64K. We say that a location in the address space is *implemented* if there is something "out there" that can accept information or provide information or both.

Let us consider how we use this address physically to select a location in the address space. One possibility would be to apply the address to a 16-line-to-64K-line decoder, producing an output on one of 64K lines, corresponding to the selected location. This is seldom practical, and the decoding is generally broken down into several steps. Another consideration is the fact that it is not necessary to decode all 16 bits unless the entire address space is actually implemented. The size of the address space is fixed by the address length, but we need physically implement only as many locations in that space as are required for a given application.

Consider a very simple application for which only 256 memory locations are required. We can locate the 256 locations anywhere in the address space that might be convenient. Wherever we put them we need decode only eight bits of the address to select one location out of 256. If we use these locations to implement addresses 0000 to 00FF—the most obvious choice—we need decode only the lower eight bits (the lower byte) of the address. It would not make any difference whether the CPU issued the address 0063 or the address 4963; it would still access location 63 in the group of 256, since the upper byte is ignored.

A user of a memory-mapped system need not be concerned with hardware difficulties involved in allocating the address space. All that is needed is a listing of the addresses that are implemented and the form of implementation, that is, RAM, ROM, I/O, and so forth. This information is often provided in the form of a *memory map*. If we are designing a system, the problem is more complex because we must specify the memory map. This first of all requires a thorough understanding of the required functioning of the proposed system, on the basis of which decisions can be made regarding how much RAM is needed, how much ROM is needed, how many I/O devices must be controlled, and so on. With these decisions made, the next problem is to decide where to place the locations in the address space. In resolving this question we must consider several factors, including possible effects of address choices on programming and the hardware needed for decoding the addresses. The following example will illustrate some of the problems associated with setting up a memory map.

## Example 13.1

Develop a memory map for an application for the TB6502 for which 2K of ROM for storing permanent programs, 256 words of RAM for data and temporary storage, and 16 I/O locations are required.

## Solution

The first decision is to use the 256 words of RAM to implement page 00. This use of RAM to implement page 00 is virtually standard in computers with zero-page addressing. Only by implementing page 00 in RAM can we make full use of the addressing capabilities of the CPU. This is an important example of the way address choices can affect programming.

With this decision made, it might seem reasonable to place the 16 I/O locations in the next 16 addresses following the RAM, followed by the 2K of ROM, but we should first consider in more detail how the address decoding is to be accomplished. When we check for availability of devices to realize the ROM we find that 1K ROMs are a standard item available from a variety of sources, so we decide to use two of these to realize the desired 2K of ROM. Figure 13.2a shows the pin diagram of a typical 1K ROM. We have 10 address pins, A9 to A0, to which we must connect the 10 address lines required to select one location out of 1024. These pins are connected to an internal decoder that accesses the desired location in the ROM, so we do not have to worry about this part of the decoding. We could connect

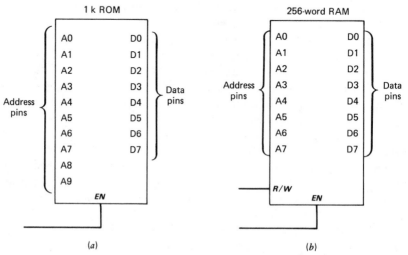

*Figure 13.2   Pin arrangement of typical ROM and RAM chips.*

any 10 of the 16 address lines to these 10 pins, but the obvious thing to do is to connect the 10 low-order bits, so that the 1024 locations will occupy 1024 contiguous addresses.

Connecting the 10 low-order address bits to the address pins on the two ROM chips enables us to select a specific location on each chip, but we also need a means a selecting one chip or the other. For this purpose we have an *EN* (enable) input on the ROM, which must be at the 1-level to gate the contents of the desired location onto the data pins, D7 to D0.

If these two ROM chips were the only devices in the address space, a single additional address bit would be sufficient to choose between them. But we also have the RAM and I/O locations to be selected. The RAM will be organized in much the same way as the ROMs except that there will be only eight address pins, A7 to A0, and there will be an additional pin, typically labeled *R/W*, to control whether the RAM reads or writes when enabled by a 1 on the *EN* line (Fig. 13.2b).

Since we have assigned the RAM to locations 0000 to 00FF, we must connect the eight low-order address bits to its address pins and use a signal derived from one or more additional bits to enable the RAM when a page 00 address is generated by the CPU. The simplest way to satisfy all these requirements is to use address bits 10 and 11 to select among four segments of 1024 addresses each. The first segment will contain the RAM in its first 256 locations, the second will contain the I/O locations in its first 16 addresses, and the ROMs will occupy the third and fourth segments. This is illustrated in Fig. 13.3, in which we show the memory map and the way the address bits are decoded to select specific locations. The rectangle in the center of the figure is the memory map itself, containing the information a programmer needs.

Address bus bits 11 and 10 are connected to the inputs of a two-to-four-line decoder, producing a signal on one of four lines, K0 to K3, to select one of four 1K segments. Within each segment we decode as many low-order bits as required to address the number of locations implemented in that segment. You should note carefully that we have decoded no more bits than necessary to address the implemented locations. It might seem that we have in some sense "wasted" address space since we use only 256 locations in the first 1K and only 16 locations in the second 1K. But address space costs nothing unless it is implemented, and gaps in the memory map will be of concern only if all the address space is needed. Since not all the address bits are used, each location will correspond to several 16-bit addresses. For example, the memory map indicates that the address of the last location in ROM is 0FFF, but any address of the form $x$FFF, where "$x$" can be any digit, will access this same location, since the upper four bits are ignored. Nevertheless, a programmer should use only the addresses shown on the memory map, since he or she will not generally know exactly how the addresses are decoded. We shall defer consideration of the way the four low-order bits are used to select a specific I/O location. ∎

You will recognize Example 13.1 as similar to the realization of the 8K ROM considered in Section 7.4. In the earlier case the entire $2^{13}$ bytes of memory addressable by the system were used. Example 13.1 was more difficult in that it required the allocation of small regions of memory within a much larger address space.

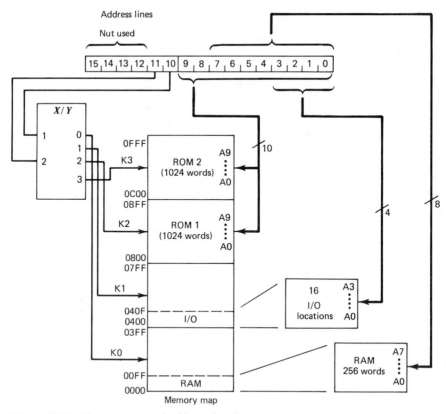

*Figure 13.3*   *Memory map and address decoding.*

## 13.2   Timing of Memory Operations

We noted earlier in our discussion of Fig. 13.1 that the CPU must issue two types of signals to I/O devices, address identification lines and control signals. We have now seen how the address bus provides identification of words in memory and I/O devices in a memory-mapped system. The control signals that we shall begin to discuss in this section tell the selected device what to do, and when to do it. Again, many different methods are used in various systems. The method used in the TB6502 is about as simple as possible and is typical of many other microprocessor systems. There will be just two control lines, the internal CPU clock, which will also be used as the external clock, and **R/W** (read/write), which will carry control signals from the TB6502 to memory and I/O interface chips. As the name implies, the **R/W** signal indicates whether the selected device is to read, that is, send information to the CPU (**R/W** = 1) or to write, that is, receive information from the CPU (**R/W** = 0). Devices that can do only one or the other, such as ROMs, will ignore this signal.

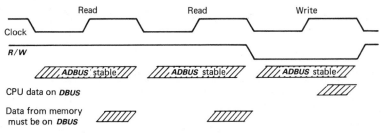

Figure 13.4   READ and WRITE timing in the TB6502.

During every clock period some address will appear on the address bus. If $R/W = 1$ and the vector on the address bus is a valid memory address of some memory or I/O interface package, the addressed device will cause a data word to be placed on the data bus. This word will be accepted by the TB6502 at the end of the positive cycle of the clock. If the address on the address bus falls outside the implemented address space or the data are for any reason not desired by the TB6502, the CPU merely ignores the data. During any clock period in which the operation is to be a WRITE the line $R/W$ will go to 0 and the addressed device will accept the data at the end of the clock period. The timing relationships among the control signals and the data bus and address bus are shown in Fig. 13.4 for three clock periods (two read cycles followed by a write cycle).

Any read or write operation starts with the CPU placing a new address on the address lines. The $R/W$ line is at the same time set to the desired value. For a read operation, the selected device must place the requested word on the data bus and hold it there for a specified minimum length of time prior to the end of the clock pulse. At the end of this clock pulse the word is gated into the CPU registers. For a write operation, the word is placed on the data bus at the same time as the address and is held there until the end of the clock period. The clock pulse enables the acceptance of this word by the addressed peripheral.

This system is an example of *noninteractive signaling*. The CPU expects the memory or I/O interface packages to respond properly to one data transfer operation each clock period. It does not interact with the peripheral in the sense of waiting for some confirmation from the peripheral that the signal has been received or the desired action completed. When controlling memory chips, the clock signal is often combined with address signals to drive the enable input. For example the *EN* of ROM 1 in Fig. 13.3 will be driven by the AND of K2 and the clock, as shown in Fig. 13.5. When the address is such as to select this 1K segment of the address space, K2 will go up. When the clock goes up, the chip will then be enabled, causing the contents of the selected location to be gated to the data pins, after a delay determined by the

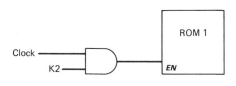

Figure 13.5   Use of the clock to enable memory.

technology of the ROM chip, and to hold on the data pins until the clock, and thus
*EN,* goes back down.

## 13.3  Parallel Input/Output Interfacing

The information transmission discussed in the previous section required that memory
and interface packages respond to commands from the CPU within one clock period.
In some cases the user may have some control over the clock rate, but the range of
such control is limited by the internal operation of the microprocessor. It will rarely
be practical to slow the clock and, therefore, the overall speed of operation of the
system to accommodate a single slow peripheral. When dealing with memory there
is seldom any problem in satisfying the timing requirements. The manufacturer of
the CPU chip will either provide compatible memory chips or else make the CPU
timing compatible with standard memory chips. With I/O devices, the situation is

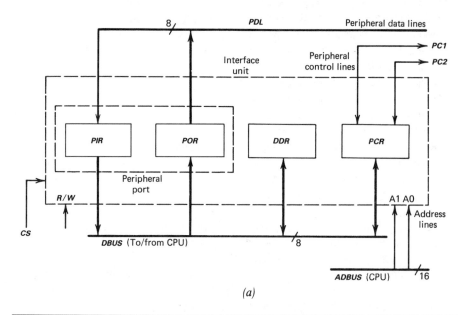

*(a)*

| A1 | A0 | Selected register | Internal selection signal |
|----|----|-----|----|
| 0 | 0 | Peripheral port | *portselect = CS $\wedge$ $\overline{A1}$ $\wedge$ $\overline{A0}$ $\wedge$ clock* |
| 0 | 1 | DDR | *ddrselect = CS $\wedge$ $\overline{A1}$ $\wedge$ A0 $\wedge$ clock* |
| 1 | 0 | PCR | *pcrselect = CS $\wedge$ A1 $\wedge$ $\overline{A0}$ $\wedge$ clock* |
| 1 | 1 | Unused | |

*(b)*

*Figure 13.6   TBPIA, a parallel interface unit for the TB6502.*

much more difficult. There are all manner of devices, with all manner of timing characteristics. Most I/O devices are slower than the CPU, and it is definitely not safe to expect them to respond in one clock period.

Dealing with this timing mismatch generally requires a *data buffer,* a register that can hold the data temporarily. In some cases an *address buffer* may be required as well. In addition, we generally need some sort of *control buffer,* by means of which the CPU can inform a peripheral device that signaling action is required, or a peripheral can inform the CPU that it is ready for some signaling action. These buffering activities can be provided by custom electronics, designed by the user for the purpose and constructed of standard chips, such as flip-flops and gates. It is more usual to use special interface packages, provided by the CPU manufacturer to minimize the amount of custom electronics that must be designed. These are the interface packages anticipated in Fig. 13.1.

Every microprocessor manufacturer provides several types of interface chips, each different in some way from every other. We cannot hope to cover them all; but the unit shown in Fig. 13.6 will serve to illustrate features common to most interface chips. The configuration of Fig. 13.6a is a simplified version of the 6522 Peripheral Interface Adapter, a support package for the 6502 microprocessor. We shall refer to this unit as the TBPIA, for TextBook Peripheral Interface Adapter. It consists of four-8-bit registers, the timing of which is compatible with the TB6502 timing. Although there are physically four registers, they are seen by the CPU as three locations in the address space. A specific register is selected by two address lines, A1 and A0, as indicated in Fig. 13.6b. These two lines are usually connected to the least significant two bits of the **ADBUS.** The chip also has an **R/W** line, serving the same purpose as in a RAM chip, and a **CS** (chipselect) line. The chipselect line will be driven by an ANDing of some subset of the address bits or their complements to select this chip.*

To simplify the discussion as well as the illustrations, the signals that actually select the internal registers have been separately named. Each of these select signals, together with its logical realization, is also shown in Fig. 13.6b. When the chip is selected and clocked, it will read (if **R/W** = 1) or write (if **R/W** = 0), to the location specified by the lines A1 and A0, in exactly the same manner as a RAM. The peripheral control register (**PCR**) and the data direction register (**DDR**) are conventional bidirectional registers, capable of reading or writing. The peripheral port, consisting of two separate registers for reasons to be discussed later, functions as a single location. A write to the peripheral port will load the peripheral output register (**POR**) from **DBUS.** A read from the peripheral port will connect the contents of the peripheral input register (**PIR**) onto the **DBUS.**

Connections to a peripheral device are provided through the 8-bit peripheral

---

*Chipselect (**CS**) and enable (**EN**) inputs serve similar purposes, but there are important differences. An enable input is normally found on a chip that has no clock input and is driven by a signal derived from both the clock and address lines, as shown in Fig. 13.5, to provide both timing and selection. Chipselect inputs are normally found on chips that have a separate clock input and are driven by address lines only, solely for the purpose of selecting the chip. Even though the chip may be selected, nothing will happen until the chip is clocked. The clock is not shown in Fig. 13.6, because it is taken for granted as in the CPU.

MODULE: TBPIA
BUSES: *PDL*[8]; *DBUS*[8].
MEMORY: *POR*[8]; *PIR*[8]; *DDR*[8]; *PCR*[8].
INPUTS *A1*; *A0*; *CS*; *R/W*; *PC1*; *PC2*.
ENDSEQUENCE

*POR* ∗ ($\overline{R/W}$ ∧ *portselect*) ← *DBUS*;
*DDR* ∗ ($\overline{R/W}$ ∧ *ddrselect*) ← *DBUS*;
*PCR*[5:0] ∗ ($\overline{R/W}$ ∧ *pcrselect*) ← *DBUS*[5:0];
*DBUS* = (*PIR* ! *pcr* ! *DDR*) ∗ ((*portselect, pcrselect, ddrselect*) ∧ *R/W*);
*PDL*[7] = *POR*[7] ∗ *DDR*[7]; *PDL*[6] = *POR*[6] ∗ *DDR*[6];
*PDL*[5] = *POR*[5] ∗ *DDR*[5]; *PDL*[4] = *POR*[4] ∗ *DDR*[4];
*PDL*[3] = *POR*[3] ∗ *DDR*[3]; *PDL*[2] = *POR*[2] ∗ *DDR*[2];
*PDL*[1] = *POR*[1] ∗ *DDR*[1]; *PDL*[0] = *POR*[0] ∗ *DDR*[0].
*PIR* ∗ ($\overline{PCR[7] \lor PCR[2]}$) ← *PDL*;
END.

*Figure 13.7   Partial RTL description of TBPIA.*

data lines (*PDL*), which are connected to *PIR* and *POR*. Each of the eight *PDL* lines can be specified as an input line (from a peripheral) or an output line (to a peripheral) under the control of *DDR*. If a bit of *DDR*, *DDR*[*i*], is set to 1, the corresponding line, *PDL*[*i*], functions as an output line; if the bit is set to 0, the *PDL*[*i*] line functions as an input line. A partial RTL description of the TBPIA describing the connections between the *DBUS* and *PDL* and the data registers is shown in Fig. 13.7.

Notice that all the description of the TBPIA module provided in Fig. 13.7 appears after ENDSEQUENCE. Thus all the connections or transfers given are active every clock period in which the respective condition expressions are satisfied. The first three statements implement the WRITE operation by taking data from the *DBUS* and clocking it into whichever of the three registers is addressed. Recall that Fig. 13.6*b*, defines the logic expressions that must be satisfied to select internal registers. Only the lower six bits of *PCR* can be established by a WRITE operation. The setting, resetting, and function of the bits *PCR*[7] and *PCR*[6] will be discussed in the next section.

A new notation of the form

$$BUS = (REG1 \ ! \ REG2) * (a,b) \tag{13.1}$$

in which the condition "∗" appears for the first time on the right side of a connection statement is introduced in the fourth statement in the description of Fig. 13.7. This notation provides for the conditional connection of registers through three-state switches to a bus. In Eq. 13.1, if the condition *a* is 1, register *REG1* is connected to *BUS*. If condition *b* is 1, *REG2* is connected to *BUS*. If *a* = *b* = 0, the bus lines are in the high impedance state unless a connection is established by another statement in the same or another module.

Any one of the three registers *PIR*, *PCR*, or *DDR* is placed on the *DBUS* by the fourth statement of the description of Fig. 13.7, if it is addressed and *R/W* remains 1. Eight separate statements are included to specify the conditional connection of the bits of output register, *POR,* through three-state switches to the I/O bus lines, *PDL.* This is because the placing of each separate bit of the *POR* on *PDL* is controlled by the respective bit of *DDR.* This can be very advantageous and is the whole purpose of including the data direction register, *DDR.* Now some of the *PDL* lines may be used as inputs at the same time others are used as outputs. If *DDR*[$i$] = 1, the output of *POR*[$i$] is gated onto *PDL*[$i$]. If *DDR*[$i$] = 0, then *POR*[$i$] is not connected to *PDL* and that line can function as an input controlled by some external device. The direction of these lines can be "programmed" by writing a byte in *DDR.*

The last statement of the partial description of the TBPIA provides for the acceptance of input data by *PIR.* Notice that the effect of this statement is that *PIR* continuously samples *PDL* each clock period as long as either *PCR*[7] or *PCR*[2] is 0. When both these bits become 1, the data vector is effectively latched in *PIR,* usually to remain there until read by the CPU. The use of the bits *PCR*[7] and *PCR*[2] for interactive control signaling will be discussed in the next section.

### Example 13.2

Construct a logic block diagram showing all interconnections between an arbitrary bit I of each of the registers *PIR, POR,* and *DDR* and the same bit of *DBUS* and the I/O bus, *PDL.* Show the application of all relevant control input lines together with the register select lines as listed in Fig. 13.6b.

Solution
The desired logic network is given in Fig. 13.8. Note from Fig. 13.7 that the clock has already been included in the AND expression generating *portselect.* This signal may, therefore, be connected directly to the clock input of *POR*[$i$]. ∎

Interface chips such as the TBPIA will implement locations in the address space of the computer. Recall that in the memory map of Fig. 13.3 we allocated 16 locations in the address space to I/O. The interface of Fig. 13.6 will occupy four locations in the address space even though only three are physically implemented. (In theory you could separate out the unused code 11 and use it for other devices, but this would not usually be worth the trouble.) Thus, we could use four of these units to occupy the 16 I/O locations in Fig. 13.3, with only 12 of the locations actually implemented. The connection of the address lines for this purpose is shown in Fig. 13.9.

Note from Fig. 13.3 that K1 is used to select the 1K segment of the address space set aside for I/O, and that bits A3–A0 select 1 out of 16 locations. In Fig. 13.9, bits A3 and A2 are applied to a two- to four-line decoder to select one of the interface units. The segment select signal, K1, is applied to the *EN* input of the decoder so that there will be an active decoder output only when an interface chip is to be selected. Address bits A1 and A0 are applied to all four chips, to select *PCR,* *DDR,* or the peripheral port in whichever chip is selected. With this connection of the address lines, a table of addresses of the I/O locations would be as shown in Fig. 13.10.

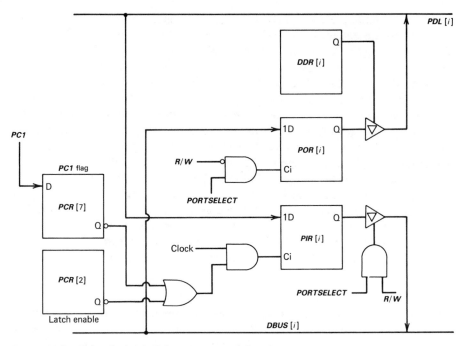

*Figure 13.8   Control of signal direction in peripheral port.*

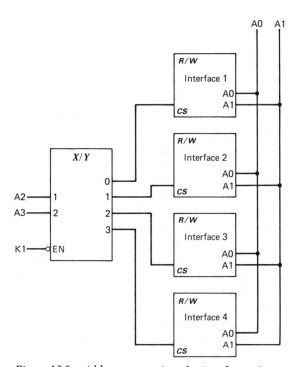

*Figure 13.9   Address connections for interface units.*

| Address | I/O register |
|---------|--------------|
| 0400 | Peripheral port 1 |
| 0401 | DDR 1 |
| 0402 | PCR 1 |
| 0403 | Not used |
| 0404 | Peripheral port 2 |
| 0405 | DDR 2 |
| 0406 | PCR 2 |
| 0407 | Not used |
| etc. | |

*Figure 13.10   Tabulation of I/O addresses.*

Many interface chips have two complete interface units similar to Fig. 13.6 on a single chip, with the two sections being known as the A side and the B side. In some cases there are slight differences between the two sides, and the user should check the specification sheets on this point. For most purposes, however, the two sides are identical and it makes no practical difference to the user that there are two units on one chip rather than on two separate chips.

You may wonder why the *DDR* is included, providing individual control over each peripheral line. Since the CPU is an 8-bit device, would it not make better sense to set the entire interface unit for input or output of 8-bit words? It was pointed out at the beginning of this chapter that peripheral devices often work with different data formats than the CPU. The use of the *DDR* provides a means of dealing with such differences. As an example, suppose we have a simple I/O unit that exchanges numeric information with the CPU in BCD (4-bit) form. This unit displays numeric data received from the CPU on seven-segment indicators and provides for input of numeric data by means of a decimal keyboard that generates BCD codes. If the interface units had to be used entirely for input or output, two of them would be needed to interface this I/O unit. An approach to accomplishing this task with a single port is discussed in the following example.

### Example 13.3

Construct a block diagram of one digit of a BCD keyboard display system controlled by a microprocessor interfaced by a TBPIA, using *PDL*[7:4] for input and *PDL*[3:0] for output. Write the program segments necessary to establish the appropriate values in the *DDR* register, to establish a desired BCD digit on the display, and to receive a BCD digit from the keyboard.

#### Solution

A block diagram of the proposed system is given in Fig. 13.11.

Assume this interface unit is Unit 1 in Fig. 13.9, with addresses 0400, 0401, and 0402. We would initialize the interface by executing the following two instructions.

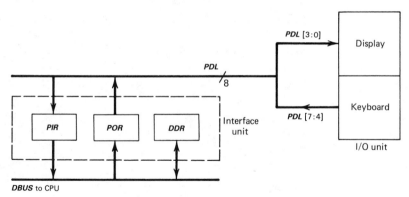

*Figure 13.11   Use of the TBPIA four-bit input and output.*

```
LDA #0F
STA 0401
```

This sets the **DDR** to 00001111; that is, it sets the high-order four bits for input, the low-order four bits for output. With the interface thus initialized, we can send out a BCD code by loading it into the low-order four bits of **AC** and executing

```
STA 0400
```

which loads the digit into the low-order four-bits of **POR,** which in turn drive the low-order bits of **PDL.** This command also loads the upper four bits of **POR** with whatever happens to be in the upper four bits of **AC,** but this has no effect on **PDL** since those four bits are not programmed for output. Similarly, executing

```
LDA 0400
```

will read a BCD code from the keyboard into the upper four bits of **AC.** This will also read the lower four bits of **POR** back into **AC,** and these will have to be masked off or shifted out to isolate the BCD code. This is not a deficiency of the interface but is simply a result of the fact that one cannot do a partial load of **AC.**    ■

## 13.4   Interactive Input/Output

The interface unit described in the last section only partly solves the problems of interfacing the CPU to devices with different timing characteristics. On output, we can write noninteractively to **POR** in the same manner as any memory location and these data will be available to the the **PDL** lines until we change the data. In addition we need some way to signal the peripheral device that data are available and some way to determine when the peripheral device has accepted these data. On input, we can read **PIR** in the same manner as any memory location, but we need some way for the peripheral to signal that data are available on the **PDL** lines and some way to signal the peripheral that the data have been accepted. There are many systems of interactive signaling used; the one we shall use is the simplest and most common. We shall first describe the general technique and then describe how it is implemented through use of the interface unit.

Figure 13.12 illustrates a situation in which two TBPIAs might be used to enable the TB6502 to control the transfer of information between a keyboard and a CRT. Connecting each TBPIA and its respective peripheral device are an 8-bit data vector and two signaling lines, **dataready,** controlled by the device that is sending the data, and **datataken,** controlled by the device that is receiving the data. For the case of input to the TBPIA, the **dataready** line is a control input to this device and the **datataken** line is a control output. A typical timing diagram for relating these control lines for the input case is shown in Fig. 13.13. First, the peripheral device places the data on the data lines (**PDL** in the interface unit described earlier) and then places a signal on the **dataready** line. This signal tells the CPU that there are data available on the input lines and that the data will be held there until the CPU accepts them. When the CPU senses the **dataready** signal, it reads the data from the lines and places a signal on the **datataken** line. When the peripheral senses the **data-taken** signal, it removes the data from the data lines and removes the **dataready** signal, which in turn causes the CPU to remove the **datataken** signal, thus terminating the input operation.

For output the process is similar but in the opposite direction. The CPU places the data on the lines (by loading **POR** in the interface unit) and sends the **dataready** signal. When the peripheral senses the **dataready** signal, it accepts the data (copies them into a register or takes whatever action is appropriate) and sends the **datataken** signal. This signal tells the CPU that the data have been taken and the CPU can proceed as appropriate. Clearly, some provision must be made in the TBPIA for connecting to control signals such as **dataready** and **datataken** and for generating signals on these lines that will satisfy timing diagrams such as Fig. 13.13. The

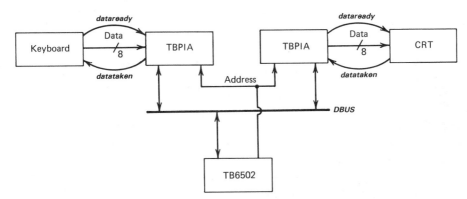

*Figure 13.12 Control signals for parallel data transmission.*

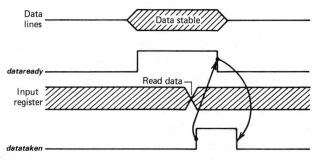

*Figure 13.13    Timning of interactive input signaling.*

approach must be sufficiently flexible to allow for sensing the *dataready* line and driving the *datataken* line for input and for sensing the *datataken* line and driving the *dataready* line for output.

In the TBPIA the peripheral control register (*PCR*) is the primary means of providing interactive signaling. The assignment of bits in *PCR* is shown in Fig. 13.14. In addition to the register, there are two peripheral control lines, *PC1* and *PC2*, which are connected to *PCR*[7] and *PCR*[6]. *PC1* is always an input line, and *PCR*[7], the *PC1* flag, will be set by an active transition on *PC1*. *PC2* may be either an input line or an output line but is usually an output line, in which event it is controlled as described in the following. Recall that the RTL description of Fig. 13.7 indicated that the CPU can write to all but the left-most two bits of *PCR* just as to any other memory location. Only *PC1* can set the flag, *PCR*[7], and it can be reset only indirectly by other operations, as will be explained later in this section. *PCR*[6] is used only in those cases when *PC2* is also considered an input. In those cases it bears the same relation to *PC2* as *PCR*[7] bears to *PC1.* In the more likely case that *PC2* is used for output, this line may be used together with *PC1* to implement the *datataken* and *dataready* signals described earlier. For input, *PC1* is the *dataready* line, by means of which the peripheral can notify the CPU that data are ready, and

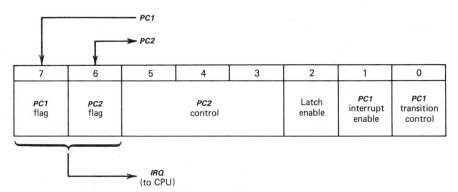

*Figure 13.14    Bit assignments in the PCR register.*

*PC2* is the *datataken* line, used by the CPU to indicate acceptance of the data. For output the assignments are the opposite, *PC2* is the *dataready* line, *PC1* is the *datataken* line.

In the preceding we referred to the *PC1* flag being set by an active transition on *PC1*. We have seen in previous chapters that various devices use active-low or active-high signaling in a more or less random manner. This is also true of peripheral devices. Most all will have lines corresponding to *dataready* and *datataken* in purpose, but they may use active-low or active-high signaling. To allow for this, we can control the active transition on *PC1* by means of *PCR*[0]. If the peripheral uses active-low signaling, we set *PCR*[0] = 0, so that *PCR*[7] will be set by a negative-going (high-to-low) transition on *PC1*. For active-high signaling, we set *PCR*[0] = 1, so that *PCR*[7] will be set by a positive-going (low-to-high) transition on *PC1*.

We also referred to the CPU sensing the presence of *dataready* and responding with *datataken*. But how does the CPU "sense" *dataready*? There are two distinct methods used in most systems. The first is *programmed flag-checking*. In the case of input via the interface unit, we assume that *PC1* is used as the *dataready* line and a transition sets the *PC1* flag, which thus serves as the *dataready* flag. At such intervals as may be appropriate, the CPU inputs the contents of *PCR* and then checks bit 7 to see if the *dataready* flag has been set. When this approach is used, the programmer will often put the CPU into a loop, repeatedly checking the flag until it is set, as in the following.

```
FLGCHK    LDA 0402
          BPL FLGCHK
          LDA 0400
```

This program specifically waits for the flag in interface unit 1 in Fig. 13.10. The first instruction reads *PCR* into *AC*. The *PC1* flag is bit 7, so it is effectively the "sign bit," and the contents of *AC* will appear as a negative number if the flag is set. The BPL command will thus loop back to read *PCR* again until the flag is set, at which time the peripheral port is read.

A drawback to this approach is that it ties the CPU up waiting for the peripheral, which may be a very slow device. The alternative is to let the CPU go on with other tasks and provide a means for the peripheral to "interrupt" the CPU when it requires attention. Figure 13.14 indicates that an *IRQ* line is associated with the two flag bits. This is the interrupt request line. We shall treat interrupts fully in a later section. For now we shall just note that an active signal on *IRQ* will interrupt the program currently being executed and force the CPU into a special program to deal with the interrupt request. Clearly, the programmer should have the capability of deciding whether or not a peripheral will be allowed to interrupt a program. This is the function of bit *PCR*[1], the *PC1* interrupt enable. If *PCR*[1] = 0, the setting of the *PC1* flag will have no effect on the *IRQ* line. If *PCR*[1] = 1, the *IRQ* line will go active when the *PC1* flag is set.

The line *dataready* going active on input indicates that the peripheral has valid data and it was inferred previously that the data will be held on the data lines until the CPU accepts them. But this may not always be possible. Some peripherals may

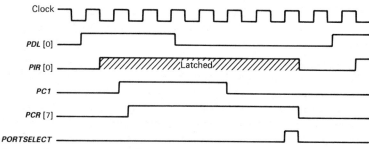

*Figure 13.15   Latching data lines and signaling.*

have data available for too short a time for the CPU to react and read them into **AC**. This problem is handled by treating **PIR** as a register composed of clocked latches. When the latch enable (**PCR**[2]) is 0, **PIR** is continuously clocked, so that the latch is effectively transparent, in the sense of being updated every clock period (Fig. 13.7). When *latching* is enabled (**PCR**[2] = 1), **PIR** is controlled by the **PC1** flag. As long as this flag is not set, **PIR** will follow **PDL** each clock period. When the **PC1** flag is set by a **dataready** signal, the clock to **PIR** will be gated off, latching the current contents of **PIR** until the CPU can respond and read the contents of **PIR**. The read of **PIR** will also reset the **PC1** flag, returning the **PIR** register to the condition of following the **PDL** lines and readying the **PC1** flag to receive a new **dataready** signal. This is the indirect reset of the flag referred to earlier. Either a read of **PIR** or a write of **POR** will reset the **PC1** flag; that is, each occurrence of **portselect** resets this flag. This mechanism of setting **PCR**[7] on a transition of **PC1** and resetting it on each occurrence of **portselect** (a port read or write) is the primary means of signaling the CPU for both input and output operations.

Figure 13.15 illustrates the timing of the latching operation. We assume **PCR**[2] = 1, so latching is enabled, and **PCR**[0] = 1, so a rising transition on **PC1** will set the **PC1** flag. Prior to **PC1** going high, the latch is transparent and **PIR** follows **PDL**. (Figure 13.15 shows only a single bit of data, but the mode of operation is the same for all eight bits.) When **PC1** goes high, **PCR**[7] sets,* gating off the clock to **PIR** and latching the current value from **PDL**. Note that **PIR** no longer follows changes on **PDL**. When **portselect** goes high, **PIR** is gated onto **DBUS**. At the completion of the read operation, when **portselect** goes low, **PCR**[7] is reset, restoring the clock to **PIR**, and **PIR** again follows **PDL**.

---

*As has been the case throughout the book, the design approach in Example 13.4 assumes that the clock-mode requirements are satisfied. This may not be the case. If the I/O device supplying the signal **PC1** is not driven by the CPU clock, a transition on **PC1** could occur at the same time as the active clock transition. The simple design of Example 13.4 does not adequately provide for this case. Some form of synchronization is necessary. Unfortunately, this subject would divert our attention from the matter at hand and must be deferred until Chapter 17. Suffice it to say that recent versions of most real interface packages will function properly in the presence of unsynchronized inputs.

**Example 13.4**

Write the RTL statements necessary to provide for the setting and resetting of
*PCR*[7] as described previously. Neglect declaration statements and use statements
that can appear in the section of the RTL description following ENDSEQUENCE.

Solution*

Since we must sense a transition on *PC1,* we include a flip-flop labeled *old,* which
during each clock period will store the former value of *PC1.* This will be continuously
compared with the current value of *PC1* to determine if a transition in the direction
demanded by *PCR*[0] has taken place. The required description is as follows.

> ENDSEQUENCE
>     *old* ← *PC1*;
> *PCR*[7] * (*old* ∧ *PC1* ∧ $\overline{PCR[0]}$ ∨ $\overline{old}$ ∧ *PC1* ∧ *PCR*[0]) ← 1;
> *PCR*[7] * *portselect* ← 0. ∎

Finally, let us consider the *PC2* signal, which can function as an input or an
output. This line is controlled by bits *PCR*[5:3]. If *PCR*[5] = 0, *PC2* functions as
an input, with *PCR*[6] as the *PC2* flag, in the same manner as *PC1* and the *PC1* flag.
*PCR*[3] controls the active transition in the same manner that *PCR*[0] controls transitions on *PC1.* *PCR*[4] enables interrupts on the *PC2* flag in the same manner as
*PCR*[1] for the *PC1* flag. This is summarized in Fig. 13.16.

When the *PC2* flag, *PCR*[6], is set by an active transition on *PC2,* it will be
reset by a read of *PIR* or a write to *POR.* Unlike *PCR*[7], *PCR*[6] will have no effect
on the loading of *PIR,* as described in Fig. 13.17. Thus, when *PC2* is set for input, it
is used only to pass some form of control information to the CPU.

When *PCR*[5] = 1, *PC2* functions as an output controlled by *PCR*[4] and
*PCR*[3], as summarized in Fig. 13.17. The *manual mode,* selected by *PCR*[4] = 1,
allows direct control of *PC2,* so that the programmer can use this line as a general
purpose output line, in any manner that may be useful. In this mode the *PC2* line
effectively follows *PCR*[3]. The *pulse mode* allows *PC2* to be used as a pulsed *datataken* or *dataready* signal. For example, on input this could serve as the *datataken*
line, sending out a negative pulse when *PIR* is read, to signal the peripheral that the
data have been accepted.

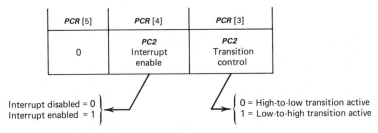

*Figure 13.16   PC2 control as input.*

| PCR[5] | PCR[4] | PCR[3] | Operating mode |
|--------|--------|--------|----------------|
| 1 | 0 | 0 | Handshake mode: *PC2* goes low on a read or write of the peripheral port (*PIR* or *POR*); *PC2* goes high on an active transition of *PC1* |
| | | 1 | Pulse mode: *PC2* goes low for one clock period following a read or write of the peripheral port |
| | 1 | 0 | Manual mode: *PC2* = 0 |
| | | 1 | Manual mode: *PC2* = 1 |

*Figure 13.17   Functioning of PC2 as output.*

## Example 13.5

Assume that *PC2* can function only as an output. Write the part of an RTL description necessary to specify the value on this line.

Solution

A two-step sequence is needed to provide for the one-period 0 on the *PC2* line. In the manual mode (*PCR*[4] = 1), control will remain in step 1 with *PC2* following *PCR*[3]. In either the handshake or pulse mode (*PCR*[4] = 0), step 1 remains active with *PC2* = 1 until a *portselect* signal arrives, advancing control to step 2 and driving *PC2* low. For the pulse mode (*PCR*[3] = 1), control returns to step 1 on the next negative clock transition, thus generating the one-period low signal. In the handshake mode (*PCR*[4:3] = 0,0), step 2 remains active with *PC2* low until the PC1 flag, *PCR*[7], is set. The *PCR*[4] term is necessary in the branch portion of step 2 to return control to step 1 if the programmer should switch to manual mode while *PC2* is low.

1. $PC2 = \overline{PCR[4]} \vee PCR[3]$;
$\rightarrow (\overline{portselect} \vee PCR[4])/(1).$

2. $PC2 = 0$;
$\rightarrow (PCR[7] \vee PCR[3] \vee PCR[4], \overline{PCR[7]} \wedge \overline{PCR[3]} \wedge \overline{PCR[4]})/(1,2).$  ■

The *handshake mode* makes it possible to implement the *dataready*–*datataken* interactive signaling, or handshaking, with minimum special programming. For an input operation in the handshake mode, *PC1* serves as *dataready* and *PC2* as an active-low *datataken*. The timing of a typical read handshake sequence (handshake mode *PCR*[3] = *PCR*[4] = 0) is shown in Fig. 13.18. We assume that *PCR*[0] = 1 so that *PCR*[7] will set on positive transitions of *dataready*. The sequence starts when the peripheral places data on *PDL* and raises the *dataready* signal. This sets the *PC1* flag, *PCR*[7], and the interface unit raises *PC2*. Note that this transition is

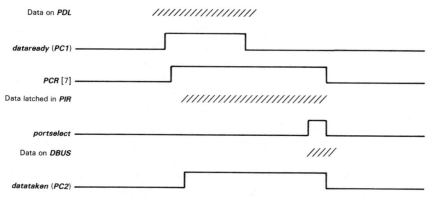

*Figure 13.18   Timing of input handshake.*

not the **datataken** response. This raising of **PC2** is necessary to put **datataken** in the inactive state in readiness for going active when the data are read. Following the setting of **PCR**[7], the data are shown as latched into **PIR**. The **dataready** signal need stay up only long enough to set **PCR**[7], which is now available for interrogation by the CPU. Nothing further will happen until the CPU determines, either as the result of an interrupt or a programmed flag check, that a **dataready** signal has arrived. The CPU then initiates a read of the peripheral port, and **portselect** goes to 1 for one clock period. The interface unit then gates **PIR** onto **DBUS**. When **portselect** goes down, signaling the end of the read operation, the interface unit drops **datataken** (**PC2**) and clears the **PC1** flag to prepare for the next read operation. The negative transition on **datataken** is the signal to the peripheral that the data have been taken and the whole cycle can be repeated at any time.

For an output operation, **PC2** serves as **dataready** and **PC1** as **datataken,** and we again assume that **PCR**[7] sets on positive transitions of **PC1**. The timing of a typical output sequence is shown in Fig. 13.19. The sequence starts when the CPU initiates a write operation to **POR.** When **portselect** goes down, signaling the end of the write operation, the interface drops the **dataready** (**PC2**) signal and clears the

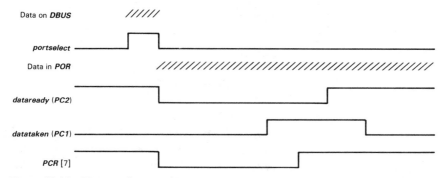

*Figure 13.19   Timing of output handshake.*

*PC1* flag (*PCR*[7]). The negative transition on *dataready* is the signal to the peripheral that the data are available. When the peripheral has accepted the data, it raises the *datataken* signal, which sets *PCR*[7], and the interface raises *dataready* in preparation for the next output operation. The *datataken* flag is now available for testing by the CPU to determine that the data have been taken and new data can be sent.

Once the handshake mode has been established by setting *PCR* to the proper values, the changes on *PC2* occur automatically in response to changes on *PC1* or read or write of the peripheral port, as described earlier, without further intervention by the programmer. The interface unit thus extends the capability of the CPU to include interactive signaling with minimal complications in the programming process. In the output sequence there is a minor complication in initialization, since *PC2* has to be gotten to 1 in the first place. This can be done by using the manual mode to do a "dummy write" to the peripheral to get everything into the proper starting mode. The exact details of this will depend on the characteristics of the peripheral device.

### Example 13.6

To illustrate the use of this interface unit, design a CRT output display. Assume an implementation of the memory map of Fig. 13.2, with four interface units connected as shown in Fig. 13.7. Use interface package 1 to interface to a CRT display to be used for output.

### Solution

We shall assume a very simple CRT unit, with 8-bit data lines for receiving ASCII characters to be displayed, and the two signal lines, *dataready* and *datataken*. The CRT will constantly monitor the *dataready* line. When this line goes low, the CRT will read the data into a register and issue a positive pulse on the *datataken* line. Note that we use active-low signaling on *dataready* because that is the way *PC2* functions in the handshake mode. If the device being interfaced required active-high control signals, it would be necessary to invert *PC2*. The connections between the interface unit and the CRT are shown in Fig. 13.20.

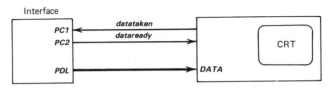

*Figure 13.20   Connections of interface to CRT.*

We assume that a string of ASCII characters has been loaded into memory by a program similar to those discussed in Chapter 12. A pointer (APTR) is intitialized to point to the first location in the string, and the string terminates with a carriage return. The program shown here will output this string of characters to the CRT.

```
            CRTOUT   LDA   #$FF
                     STA   DDR1        /SET DDR FOR PDL OUT
                     LDA   #$34        /USE MANUAL MODE TO
                     STA   PCR1        /   INITIALIZE PC2
                     LDA   #$21        /SET PCR FOR
            *                          / HANDSHAKE, POSITIVE
                     STA   PCR1        / TRANSITION ON PC1
            LOOP     LDA   (APTR)      /GET ASCII CHARACTER
                     STA   PP1         /WRITE PERIPH. PORT
                     CMP   #$0D        /TEST FOR CR
                     BEQ   END         /IF CR, DONE
            FLGCHK   LDA   PCR1        /GET PCR, CHECK  FLAG
                     BPL   FLGCHK      /NOT SET, GET AGAIN
                     INC   APTR        /DATA TAKEN, INC PNTR
                     BNE   LOOP        /GO BACK FOR NEXT
            END      HLT               /DONE


            *********   SYMBOL  TABLE   *********

            *   SYMBOL              ADDRESS
                -------             -------
            *   PP1                 0400
            *   DDR1                0401
            *   PCR1                0402
```

The first two steps initialize **DDR** to all ones to establish the **PDL** lines as output lines. The next two lines use the manual mode to initialize **dataready** to 1, and **PCR** is then set for handshake mode with positive transitions on **PC1**. The main loop starts with a fetch of a character, which is then sent to **POR**. Since the interface is in the handshake mode, this will result in **dataready** dropping, to signal the CRT that a character is available. The character is also checked to see if it is a carriage return, in which case the complete message has been sent and the program halts. The program now goes into a waiting loop, in which it repeatedly reads in **PCR** and checks for the **datataken** flag (**PCR**[7]) being set. When the flag is set, the program leaves the waiting loop, increments the pointer, and goes back for the next character. ∎

It can be seen from Example 13.6 that the interface unit makes the programming of interactive transfers quite simple. The use of a loop to wait for the **datataken** signal does have the possible disadvantage of slowing the CPU down to the speed of the CRT. This can be avoided through the use of interrupt programming, to be discussed later. However, for the sort of simple applications for which microprocessors are often used, this "slowing down" may not be a problem. When a point is reached where an output operation must be carried out before anything else can be done, the CPU will have to wait for the output device in any case.

# 13.5 Programming Time Delays

A frequent situation in applications of microprocessors is the need to delay for a specified length of time while waiting for some event to occur. For this purpose we use a time-delay subroutine. The basic idea is to use a location in memory as a counter, setting it to some initial value, depending on the delay desired, and then decrementing it down to zero in a loop, as shown here.

```
         LDA  #NN
         STA  *CNTR
LOOP     DEC  *CNTR
         BNE  LOOP
```

To determine how long it will take to bring CNTR to zero, it is necessary to know how long the computer takes to execute each instruction. In the TB6502, execution times are multiples of the primary period, which we shall define to be 1.0 $\mu$sec. A number of microprocessors have this same clock period, but some are faster. When writing a delay subroutine for a particular microprocessor, the designer must know the clock rate for that microprocessor and the number of clock periods it requires to execute each instruction within the delay loop. The execution times for the various instructions in the TB6502 are given in Appendix C. A few of the instructions of interest are summarized in Fig. 13.21.

Applying these times to the loop just presented, DEC *CNTR takes five cycles, BNE LOOP takes three cycles, so that each pass through the loop takes eight cycles. Often delays of the order of a few seconds are required to provide for interaction

| | Addressing mode (entries in clock cycles) | | | | |
| --- | --- | --- | --- | --- | --- |
| **Instruction** | **Immed.** | **Page 0** | **Absolute** | **Index** | **Indirect** |
| ORA, AND, EOR, ADC, SBC, LDA, CMP | 2 | 3 | 4 | $4^a$ | 5 |
| STA | | 3 | 4 | 5 | 6 |
| JMP | | | 3 | | 5 |
| JSR | | | 6 | | |
| INC, DEC | | 5 | 6 | 7 | |
| INY, DEY, NOP | | | 2 | | |
| RTS | | | 6 | | |
| BRANCH RELATIVE | 2 if branch not taken; 3 if branch taken on page; 4 if branch taken across page boundary | | | | |

$^a$Add one cycle if indexing goes across page boundary.

*Figure 13.21   Some TB6502 execution times.*

with humans. Since the constant, NN, cannot be greater than 255, delays this long cannot be accomplished using this simple loop alone. It is important that it be as easy as possible for programmers to include a delay of arbitrary length wherever they may wish in a program. This can be done by implementing the delay mechanism as a subroutine, so that the main program can supply a parameter specifying the exact delay time and then call the subroutine. A convenient delay time for the inner loop within the subroutine would be 1.0 msec, or 1000 μsec. If as we have assumed for the TB6502, the clock period is 1 μsec, each pass through the loop will consume 8 μsec. Thus, 125 passes through the loop are required for a 1.0 msec delay, so we set NN = 125.

## Example 13.7

Write a time-delay subroutine. The main routine is to pass the desired delay in milliseconds to the subroutine as a 16-bit word in locations 0050 and 0051, less significant bits in 0050. Thus, the range of delay is from 1 to 65,535 msec, that is, from 1 msec to a little over 1 min. The actual overall delay need be only ±10% of the nominal delay.

Solution

The 16-bit delay time must be loaded into some form of 16-bit counter. Since there are no 16-bit registers in the TB602, a 16-bit counter must be implemented in two cascaded 8-bit words. The lower byte will be decremented each time the counter is to be decremented, but the upper byte will be decremented only when the lower byte reaches zero. The subroutine will consist of two nested loops. The inner loop will be similar to that shown previously, providing 1 msec delay. Each time this loop "times-out," the 16-bit counter will be decremented in the outer loop, until it reaches zero, at which time control returns to the main routine. The following is one possible version of the subroutine.

```
DELAY   LDA   #125
        STA   *CNTR       /INIT. 1 MSEC LOOP
LOOP    DEC   *CNTR
        BNE   LOOP        /125 TIMES THRU LOOP?
        DEC   *50         /DECR LEAST SIG. BYTE
        BNE   DELAY       /REPEAT 1 MSEC LOOP
*                         / IF LEAST BYTE = 0
        LDA   *51         /GET MOST SIG. BYTE
        BEQ   DONE        /BOTH BYTES 0, RTRN
        DEC   *51         /DECR. MOST SIG. BYTE
        JMP   DELAY       /NOT DONE, LOOP AGAIN
DONE    RTS
```

This routine executes the 1-msec loop as soon as it is entered and then decrements the 16-bit counter. Thus, it cannot be used for a delay of 0 msec. The most significant byte is decremented as soon as the least significant byte is found to be 0, so that the most significant byte will be one too small the first time through the loop after JMP DELAY is executed.  ∎

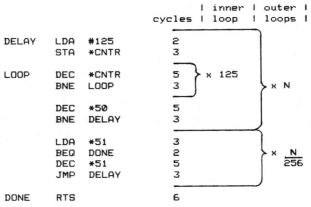

*Figure 13.22   Timing computation for delay subroutine.*

The subroutine presented in Example 13.7 will provide a delay of approximately $N$ times 1 msec, where $N$ is the two-byte parameter passed to the subroutine in locations 0050 and 0051. If the time delay is to be precise, it is necessary to take into account the time required to execute the other instructions in the two outer loops and the time required for the JSR and RTS. The subroutine is repeated in Fig. 13.22 with the comments omitted and with the number of cycles required by each instruction and the number of times each instruction is executed for each call of the subroutine with $N$ stored in locations 0050 and 0051.

Note that there are actually three loops rather than two since the least significant byte of $N$ is decremented $N$ times, whereas the most significant byte is decremented only approximately $N/256$ times. This leads to Eq. 13.2, where one cycle is subtracted associated with leaving each of the two inner loops because BNE requires one less cycle, if the branch is not satisfied. The cycles lost in branching out of the outermost loop and the JSR and RTS cycles are neglected.

$$\text{Delay} = 125N \cdot (8) + N \cdot (-1 + 13) + \frac{N}{256} \cdot (-1 + 13)$$

$$= N\left(1000 + 12 + \frac{12}{256}\right) \tag{13.2}$$

$$\cong N \cdot 1012$$

**Example 13.8**

Modify the subroutine of Example 13.7 so that the delay of the subroutine is as close as possible to $N$ times 1 msec rather than $N$ times 1.012 msec. Compute the percentage error in the nominal delay values after this modification is made.

Solution

It is not written on stone that the inner loop must be executed exactly 125 times. Instead, let us choose the integer, $p$, that will satisfy Eq. 13.3 which disregards the 12 cycles in the lower loop in Fig. 13.22.

$$1000 \geq 8 \cdot p + 12 \tag{13.3}$$

$$988 \geq 8 \cdot p$$

Therefore, $p = 123$.

We can further refine the delay by adding two-cycle NOP (no operation) instructions between DEC *50 and BNE DELAY so that more than 12 cycles are required in the loop that decrements the least-significant byte of $N$. The required number of NOPs, $m$, is found by solving Eq. 13.4.

$$1000 = (8 \cdot 123) + 12 + 2 \cdot m \tag{13.4}$$

$$m = \frac{1000 - 996}{2} = 2$$

The routine is thus modified by changing the first step to

```
DELAY  LDA  #123
```

and adding two NOP-instructions at the point indicated previously. For typical valves of $N$ greater than 256 msec the error in the delay will be less than 0.005%.
■

# 13.6  Device Drivers

A typical microcomputer system will have a variety of peripheral devices, each requiring its own special interface. In the last section we considered a general-purpose interface unit that, by appropriate programming, can be used to interface to many different devices. At the end of that section we saw an example of a program segment to interface to a simple CRT. Suppose those data are to be sent to this CRT at several points in the program. Clearly we do not want to duplicate the code each time, so it seems obvious that this segment of code should be a subroutine, to be called whenever data are to be sent to the CRT for display. Subroutines of this sort, designed to interface to peripherals are important enough to merit a special name. Such routines are known as *device drivers,* or *device handlers,* or *I/O drivers,* or *I/O handlers,* depending on the reference.

Microcomputer systems often run under control of a system program, or *monitor,* a program that takes care of the "nasty details" and allows the user to view the computer simply as a programmable system, with known response to known commands. Among the "nasty details" are the signaling characteristics and requirements of the various peripherals. Programmers do not want to have to worry about **data-taken** and **dataready** signals when they want to print a character. They would prefer to put the character in **AC** and issue a command to PRINT. A device driver for the printer will make this possible. A monitor program for a computer system will include a number of device drivers, subroutines designed to interface to the system peripherals. Writing these drivers in the first place requires detailed understanding of the peripherals and interface units such as introduced in the last section. But once the drivers are written, the programmer views the interface to the peripherals in terms of a programming interface. For example, placing an ASCII character in **AC** and issuing the instruction

**JSR PRINT**

will cause that character to be printed. That is all the programmer needs to know. In this section we shall look at some typical device drivers.

As a first example, let us consider a simple printer, which uses the *dataready–datataken* signaling system discussed in the last section, accepting one ASCII character at a time in parallel over 8-bit lines. We will assume that this printer does not generate its own line feeds. When a carriage return (CR) is received, the print head returns to the beginning of the line, but the paper does not advance. To start a new line, a line feed (LF) character must be sent to the program. We shall include this feature in the driver; whenever a CR is sensed in the ASCII stream, an LF will also be sent.

If you will consider the CRT driver written in the last section, you will note that it consists of two sections, one to initialize the interface, the other to carry out the transfer of data. Normally the initialization need be done only once, when the system is first turned on.

A typical monitor program will start with a whole collection of initialization routines that will be executed whenever the system is powered up. For this printer the initialization will be the same as for the CRT, since the signaling convention and the data format are the same.

```
PINIT    LDA    #$FF
         STA    DDR1          /SET DDR FOR OUTPUT
         LDA    #$38          /USE MANUAL MODE TO
         STA    PCR1          /TO INITIALIZE PC2
         LDA    #$21          /SET PCR FOR HNDSHKE,
         STA    PCR1          /POS. TRANS. ON PC1
```

The PRINT driver will be similar to the CRT driver except for sending an LF when a CR is detected.

```
*    SUBROUTINE PRINT
*      SENDS CHAR TO PRINTER, ADDS LF IF CR SENT
*      UPON ENTRY:
*        ASCII CHAR TO BE PRINTED IN AC
*
PRINT    STA    PP1           /SEND CHAR TO POR
         CMP    #$0D          /CR?
         BNE    FCHK          /NO, SKIP LF STEPS
FCHK1    LDA    PCR1          /GET PCR TO CHECK FLAG
         BPL    FCHK1         /NO FLAG, CHECK AGAIN
         LDA    #$0A          /CR TAKEN, GET LF
*                             / (LF = 0A IN ASCII)
         STA    PP1           /SEND IT
FCHK     LDA    PCR1          /GET PCR TO CHECK FLAG
         BPL    FCHK          /NO FLAG, CHECK AGAIN
END      RTS                  /DONE, RETURN
```

The first two lines send out the character and check for a carriage return. If no CR, skip to FCHK to wait for the character to be accepted. If a CR was sent, wait at FCHK1 for it to be accepted, then send an LF and wait for it to be accepted.

The *dataready–datataken* signaling system, although certainly simple enough, has some potential problems. For example, when we send a *dataready* signal, how do we know the receiving device is ready for it? That may depend on the meaning of *datataken.* Does *datataken* mean only that the data are taken, or does it also mean that the device is ready for more data? Whatever *datataken* means, is the receiving device sensitive to the transition on *dataready* or the level value of *dataready*? In the former case *dataready* must not be sent until it is certain that the device is ready to receive it. In the latter case it does not matter if the device is ready, as long as *dataready* is held active long enough to be seen. There are no uniform answers to these questions; one must read the device manuals carefully, being sure to understand exactly what signaling conventions are being used.

Fortunately, interface devices such as the TBPIA are flexible enough to be adapted for almost any signaling convention. Let us consider a printer again, but this time one using the strobe/acknowledge convention. Again there are two signaling lines, DATA STROBE (*DSTR*), issued by the sending device, and ACKNOWL-EDGE (*ACK*), issued by the receiving device. Both are active-low lines. An exchange starts when the receiving device (the printer) raises *ACK* to indicate that it is ready to receive data. The sending device (the CPU) puts data on the lines and pulses *DSTR* to indicate that data are available. The printer acknowledges receipt of the data by dropping *ACK,* which is held low until the printer is ready to receive new data. Figure 13.23 illustrates how this convention can be implemented using the interface unit in the pulse handshake mode. With *ACK* high, indicating the printer is ready, the CPU loads a character into *POR.* When *portselect* goes low, the interface issues a negative pulse on *DSTR* (*PC2*). The printer acknowledges by first dropping *ACK* and then raising it again when it is ready for a new character. This convention is similar in some respects to the *dataready–datataken* convention, but more information is exchanged since *ACK* conveys more information than *datataken,*

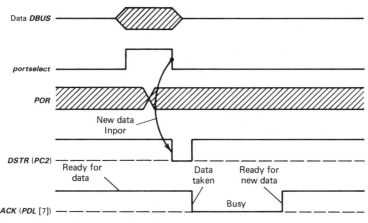

*Figure 13.23  Strobe/acknowledge signaling convention.*

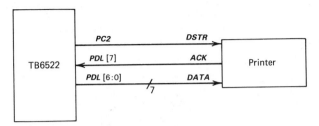

*Figure 13.24   Connecting* **PD**[7] *as a control input.*

using its transitions to indicate first acceptance of the data and then readiness for new data.

Since both transitions of **ACK** are significant, it would be inconvenient to use **PC1** for **ACK,** since **PC1** responds to transitions in only one direction, unless we change **PCR** between transitions. Fortunately, the logic levels on line **ACK** now have meaning. When **ACK** = 1, the printer will be ready to receive data. When **ACK** = 0, the printer is busy printing a character and is, therefore, unable to receive data. To test a level we can ignore the transition sensing **PC1** flag and connect **ACK** as a level input, if such a line is available. Fortunately, most printers expect only seven data bits, since the eighth bit of the ASCII code is commonly arbitrary anyway. Therefore, we shall connect **PDL**[6:0] for data output and connect **ACK** to **PDL**[7], which may be separately treated as an input. This configuration is illustrated in Fig. 13.24.

Latching will not be enabled, so **PIR**[7] will simply follow **ACK,** and its value can be tested at any time. With these assignments of lines, the initialization of the interface will require the following steps.

```
        PINIT1   LDA  #$7F      /SET DDR1, PDL[7]
      *                         / FOR IN, PDL[6:0]
                 STA  DDR1      /  FOR OUT
                 LDA  #$28      /SET PCR FOR PULSE
                 STA  PCR1      / MODE HANDSHAKE
```

Note that initialization of **PC2** to 1 is not needed in the pulse mode; the interface does this automatically. The data transfer portion will be as shown in the following. This is a simple character transfer driver, without the line feed insertion.

```
      *  SUBROUTINE PRINT1
      *     UPON ENTRY:
      *        ASCII CHAR TO BE PRINTED IN AC
      *
      PRINT1   STA  CHAR      /AVOID LOSING CHAR.
      LOOP     LDA  PP1       /TEST PIR FOR ACK
               BPL  LOOP      /IF PIR[7] = 0,
      *                       /  PRINTER NOT READY
               LDA  CHAR      /RETRIEVE CHARACTER
               STA  PP1       /SEND CHAR TO POR
```

```
TEST    LDA  PP1      /MAKE SURE ACK
*                     /  INDICATES BUSY
        BMI  TEST     /WAIT FOR ACK = 0
END     RTS           /DONE, RETURN
```

We have a special problem here. The character to be sent out is in **AC,** but we cannot send it until we check to see if the printer is ready. How do we test **ACK** (**PIR**[7]) while still preserving the character to be sent? In the first step it is stored, temporarily, in location CHAR to be retrieved after it is determined that the printer is ready for a data character.

The second step of the PRINT1 routine loads all eight **PDL** lines in **AC.** The line **ACK** is the sign bit that is tested by BPL. If this bit is 0 (the printer is not ready), the BPL branch repeats the test. When **ACK** goes to 1, the character is sent to **POR,** and a second test loop on **ACK** is executed to make sure the printer has accepted the character. Once the printer is found to be busy, control returns to the calling routine.

Next let us consider the problem of input from a keyboard. A keyboard usually consists of a matrix of switches. In an unencoded keyboard, the position of the switch in the matrix must be determined to identify the key pressed. In a coded keyboard, logic is provided to generate a unique code for each key. We shall first consider a coded keyboard, as shown in Fig. 13.25. This is a full alphanumeric keyboard, producing a 7-bit ASCII code, plus an eighth line, **KEY,** that will go high anytime any key is pressed. This keyboard will also include rollover protection against multiple key closures. If the operator attempts to press a second key before releasing the first, nothing will happen until the first key is released. A problem with mechanical switches is that the contacts often "bounce," making contact and then oscillating between open and closed for a short time. There are electronic techniques for "debouncing" switches, but these add to the cost of a keyboard. If a microprocessor is being used to process the keyboard input, another method is to simply wait a short

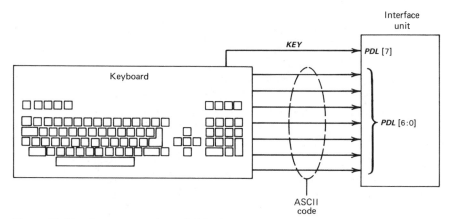

*Figure 13.25   Connections of encoded keyboard to interface unit.*

time after the initial closure for the switches to stop bouncing and then to sample the code. This is the method we shall use.

As shown in Fig. 13.25, we shall connect the ASCII input from the keyboard to **PDL**[6:0] of the interface unit and **KEY** to **PDL**[7]. The control lines, **PC1** and **PC2** will not be used for this application. A computer is so fast compared to a human at a keyboard that we shall assume that it can keep up without the need for any interactive signaling. Checking the status of **KEY** about once every 5000 to 10,000 operations will provide about 100 checks per second, amply fast to keep up with even the fastest typist. The initialization portion of the driver need only set the port for input.

```
KBINIT    LDA   #$00
          STA   DDR1    /SET DDR TO
*                       / ALL 0'S FOR INPUT
```

The main body of the driver first checks **KEY.** If it is not high, **C** is cleared to indicate no character and control returns to the calling routine immediately. Note that the computer is not tied up waiting for a human being. If a key has not been pressed, the computer can do something else and then check again. If a key has been pressed, the driver calls a delay subroutine of the form discussed in the previous section to wait for possible bounce to subside. The amount of delay depends on the type of switch, but 1 to 2 msec is usually sufficient.

Following the bounce delay, the character is temporarily stored in CHAR. The subroutine then waits for **KEY** to go down, indicating that the key has been released. This last is vital. The computer is so fast it could easily read in a character, process it in several thousand steps, and then come back to this routine and find **KEY** still high because the operator has not yet released the key. One of the more challenging aspects of input/output programming is keeping in mind the enormous difference in speed between a computer and most of the devices connected to it. Before exiting the subroutine, the character is returned to **AC,** and **C** is set to 1 to indicate the presence of a character.

```
*   SUBROUTINE KYBD
*      UPON EXIT:
*         C = 0 IF NO KEY CLOSURE
*         C = 1 AND CHAR IN AC IF KEY CLOSURE
*
KYBD      LDA   PP1       /CHECK FOR KEY CLOSED
          BMI   TAKE      /BRANCH IF KEY = 1
          CLC             /NO KEY, CLEAR C
          RTS             /RETURN
TAKE      JSR   DELAY     /WAIT FOR BOUNCE
          STA   CHAR      /STORE CHARACTER
KYCHK     LDA   PP1       /CHECK KEY
          BMI   KYCHK     /WAIT FOR KEY = 0
          LDA   CHAR      /LEAVE CHAR. IN AC
          SEC             /SET C = 1
          RTS             /RETURN
```

Note the use of the **C** flag to carry status information back to the calling routine. This is a fairly common situation, one in which a subroutine can produce several different results and the main routine needs to know which occurred. Where there are only two possibilities, as in this case, the **C** flag is convenient. If there are more possibilities, a register or memory location may be needed. As an example of the use of such information, let us assume a situation in which the computer is controlling some process by means of a complex program called PROCESS. An operator can control the process by typing in commands on a keyboard, with a carriage return indicating that a new command has been entered. Approximately 100 times a second, PROCESS will call KEYBD to see if the operator has pressed a key. If not, the normal operation of PROCESS continues. If a key has been pressed, the character will be checked to see if it is a CR. If not, the character (part of the command string) will be stored in a command buffer and PROCESS continues. This process repeats until a CR is entered, at which time the program interprets the new command and continues as appropriate. Figure 13.26 illustrates the overall process, and the program segment to interact with KYBD and assemble a new command follows.

```
*    PROG.SEGMENT TO ASSEMBLE NEW COMMAND FOR
*    PROCESS PROGRAM.   COMMAND WILL BE
*    ASSEMBLED IN BUFFER POINTED TO BY COMPTR.
*    AFTER COMMAND IS ASSEMBLED,
*    NUMBER OF CHARACTERS IN COMMAND
*    WILL BE FOUND IN COMCNT
*
COMSEG   JSR   KYBD      /CHECK FOR KEY CLOSED
         BCC   PROCESS   /NO KEY, RTRN TO PROC
         AND   #$7F      /MASK UPPER BIT TO 0
         CMP   #$0D      /CARRIAGE RETURN?
         BEQ   COMM      /BRANCH TO INTERPRET
*                        /  NEW COMMAND
         STA   (CMPTR)   /NO, STORE CHARACTER
         INC   CMPTR     /ADVANCE POINTER
         INC   COMCNT    /INCR. CNT OF CHAR.
         BNE   PROCESS   /RETURN TO PROCESS
```

It is assumed that COMPTR and COMCNT will be properly initialized in the body of PROCESS and that their values will then be retained as a new command is assembled. It is interesting to note how this relatively simple routine prevents the computer from being monopolized waiting for an operator command. As long as no key is pressed, each pass through this segment involves only the first two instructions of COMSEG and the first four instructions of KYBD, requiring about 25 $\mu$sec to execute at a 1-MHz clock rate. If the check is to be made 100 times a second, this amounts to 0.25% of the computer's time required to monitor for new commands. When a new command is being entered, things will be slowed down due to the need to wait for the operator to release the key. This delay could be reduced at the expense of some additional program complexity, by using the same periodic sampling technique to check for the key being released.

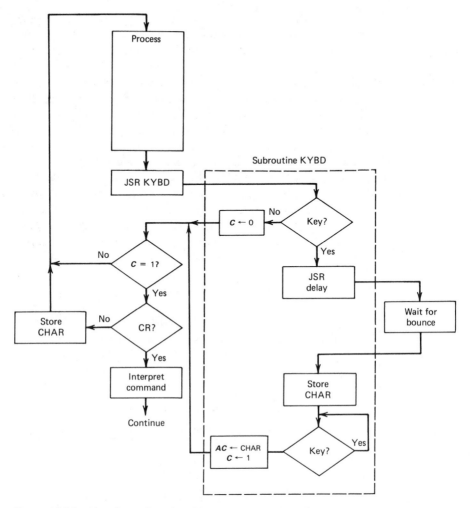

*Figure 13.26   Flowchart of PROCESS interaction with KYBD.*

## 13.7   Substituting Input Scanning for Combinational Logic

The concept of device driver subroutines was introduced in the previous section. In this and the next section two more device driver subroutines will be developed; each one will illustrate another powerful and generally applicable technique from the microprocessor "bag of tricks." In this section we shall look into the use of a microprocessor to replace a network of combinational logic. That this could be done was cited as an advantage of the microprocessor when it first appeared on the scene.

As noted earlier, keyboards come with or without encoding logic. An encoded keyboard is certainly simpler to use than one that is unencoded, but the encoding

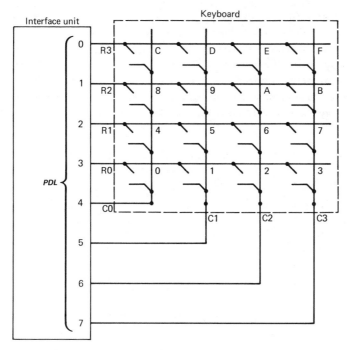

*Figure 13.27   Matrix keyboard connected to interface unit.*

logic costs money. Thus, it is natural to consider using the microprocessor to do the encoding. It would certainly not be economical to use a microprocessor solely for this purpose, but if you have the microprocessor in the system anyway, why not? Figure 13.27 shows a 16-key hex keyboard, with the keys arranged in a $4 \times 4$ matrix. When a key is pressed it connects a row line to a column line. The advantage of a matrix arrangement over separate keys is that it reduces the number of wires. In this case we have 8 wires compared to 16 for separate keys. The advantage is more evident for larger keyboards. For example, a 64-key keyboard in an $8 \times 8$ matrix requires only 16 wires. The keys are numbered in hex as shown, and the rows are numbered R0 to R3, the columns, C0 to C3. The goal is to input the number of the key pressed to the computer.

The problem of encoding a keyboard is a "classic" microprocessor problem; there are many possible solutions. Most approaches involve using one or more interface units to drive the column lines and interrogate the row lines, or vice versa. Assume the keyboard is connected as shown in Fig. 13.27, with the row lines connected to *PDL*[7:4] and the columns to *PDL*[3:0]. We start by setting the *DDR* so that rows are inputs and the columns are outputs, and we set all the columns to 0. The interface is TTL compatible, which means that open inputs appear as logic 1. Thus, if no switches are pressed, all the row lines will appear as logic 1. Now assume that key 9 is pressed. Then row R2 will be connected to column C1, so R2 will go to 0. We can then read in the status of the row lines and determine the row corresponding to the depressed key. The various techniques differ mostly in the way row and column interrogations are combined to isolate a single switch.

Figure 13.28 shows a flowchart for the keyboard scan routine we shall use. The scan starts by setting the columns to 0 and checking the rows. If all the rows are 1, there are no key closures, so *C* is set to 0 and control returns to the calling routine, as in the previous example. Again, computer time must not be wasted waiting for someone to press a key. If there is a key closed, we start by setting SWNO to 0 and

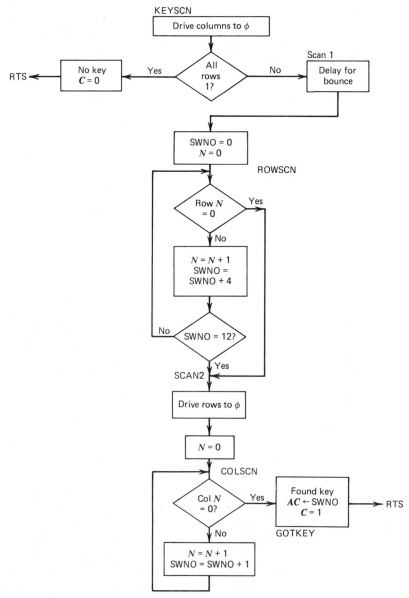

*Figure 13.28  Flowchart of keyboard scan subroutine.*

N to 0. SWNO is a counter that will receive the number of the depressed key and N is the number of the row being checked. We check RN for a 0. If there is none, N is incremented and 4 is added to SWNO, to bring it to the number of the first switch in the next row. If this brings SWNO to 12, the first three rows have been checked without finding a closure, so the closure is in R3, and the processor goes on to check the columns. Otherwise it goes back and checks the next row.

When a closure is found in a row, we then must check the columns. Again assume that key 9 has been pressed. Thus we find a 0 in R2, at which point SWNO has been advanced to 8. We now reverse directions on the interface, connecting the columns as inputs and the rows as outputs, and drive the rows to 0. Thus a column will appear as 1 unless the depressed switch is in that column. Now N is set to 0 and C0 is checked. It will be 1, so N is incremented to 1 and SWNO to 9. Going back to check C1, we find it equal to 0 since key 9 is pressed, so **C** is set to 1; and the switch number moves from SWNO to **AC** followed by an exit. The complete subroutine is shown here.

```
*    SUBROUTINE KEYSCN
*     SCANS A 4 X 4 HEX KEYBOARD
*     UPON EXIT:
*      C = 0 IF NO KEY DEPRESSED
*      C = 1 AND KEY NMBR IN AC IF KEY PUSHED
*
KEYSCN   LDA   #$0F        /SET ROWS FOR INPUT,
         STA   DDR1        / COLS FOR OUTPUT
         LDA   #$F0        /DRIVE ALL COLUMNS
         STA   PP1         / TO ZERO
         CMP   PP1         /CHK ROW FOR ALL 1'S
         BNE   SCAN1       /NOT ALL 1'S, SCAN
         CLC               /NO KEY, C = 0
         RTS               /RETURN
SCAN1    JSR   DELAY       /WAIT FOR BOUNCE
         LDA   #0
         STA   SWNO        /INIT. SWITCH NUMBER
         LDA   #$10        /SET N = 0
         STA   N           /SAVE N
ROWSCN   LDA   N           /GET N
         BIT   PP1         /TEST ROW N
         BEQ   SCAN2       /IF 0, FOUND ROW
         ASL   N           /N = N + 1
         LDA   SWNO        /GET SWITCH NUMBER
         CLC
         ADC   #4          / INCREMENT BY 4
         STA   SWNO        /SAVE IT
         CMP   #$0C        /3 ROWS CHECKED?
         BNE   ROWSCN      /NO, CHECK NEXT ROW
SCAN2    LDA   #$F0        /SET ROWS FOR OUT,
         STA   DDR1        / COLS FOR IN
         LDA   #$0F
         STA   PP1         /SET ROWS TO 0
         LDA   #$08        /SET N = 0
COLSCN   BIT   PP1         /TEST COL N
         BEQ   GOTKEY      /IF 0, FOUND KEY
         LSR   A           /N = N + 1
         INC   SWNO        /SWNO = SWNO + 1
         BNE   COLSCN      /CHECK NEXT COLUMN
GOTKEY   LDA   SWNO        /SW. NUMBER TO AC
         SEC               /C = 1
END      RTS               /RETURN
```

Most of this program should be easy enough to follow with the help of the flowchart, but there are a few points that warrant comment. Note the use of the BIT command for scanning the rows and columns. For example, R0 is connected to **PDL**[4]. So we start the row scan by setting **AC** to hex 10 (0001000). Then the command BIT PP1 will produce a 1 if **PDL**[4] = 1, all 0's otherwise. To check the next row we then shift that 1 in $N$ one position left, to give (0010000), to test **PDL**[5]. Thus, comments such as "$N = 1$" and "$N = N + 1$" are accurate in terms of the basic logical structure of the scan algorithm, but they do not accurately describe what is going on in the computer. $N$ is not truly a number, but is actually a mask word for bit testing. Also note that the storing of $N$ in memory in the row scan segment is necessary because **AC** has to be used for incrementing SWNO by 4. Finally, this is a fairly complicated program, one of the longest we have seen so far. Is it possible that this might take so long that the operator might release the key before the scan routine can identify it? In the worst case, the F key being pressed, which requires a scan of all rows and columns, this routine executes 68 instructions, requiring approximately 0.2 msec. Again we see the power of the computer in being able to do things that seem very complex in practically no time at all in terms of a human time frame.

## 13.8   Time-Sharing of Interface Ports

In switch debouncing and keyboard scanning we have seen two examples of operations that are frequently implemented with custom electronics but that can be done economically with a microprocessor if one is available in the system for other purposes. As another example of this sort of microprocessor application, consider the problem of driving a display of seven-segment indicators. Assume we have decimal information in our system in BCD form that is to be displayed on seven-segment indicators. Then we need to convert from the BCD code to the seven-segment code as listed in Fig. 7.18. Clearly this can be done with logic, and a circuit for such a code converter is shown in Fig. 7.19. But, as we argued before, if the microprocessor that is already in the system can do the job, why pay for additional electronics?

When data conversion is done by programming there are two basic categories of conversions: procedural conversions and table-driven conversions. The conversion from matrix coordinates to key number in the keyboard scan routine is an example of a procedural conversion. This approach is applicable because there is a logical relationship between the matrix coordinates and the key numbers. For the conversion of BCD code to seven-segment code, there is no procedural technique because there is no logical relationship between the two codes. The seven-segment code is an entirely arbitrary code based on the geometry of the arabic numerals. Therefore, we must use a table-driven conversion. The idea is really very simple. We simply store the seven-segment codes in a table, as shown in Fig. 13.29. Since our memory words are eight bits, we arbitrarily set the first bit in each entry to 0, the remaining seven bits being the code as listed in Fig. 7.18. The name of the table (SEVSEG in this case) will refer to the first location in the table. The value of the number to be put in seven-segment code will be the offset into the table. Thus, if we want the code for

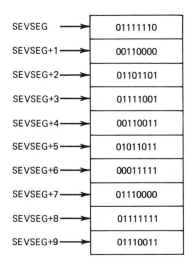

| SEVSEG | → | 01111110 |
| SEVSEG+1 | → | 00110000 |
| SEVSEG+2 | → | 01101101 |
| SEVSEG+3 | → | 01111001 |
| SEVSEG+4 | → | 00110011 |
| SEVSEG+5 | → | 01011011 |
| SEVSEG+6 | → | 00011111 |
| SEVSEG+7 | → | 01110000 |
| SEVSEG+8 | → | 01111111 |
| SEVSEG+9 | → | 01110011 |

*Figure 13.29   Table for seven-segment code.*

5, we go into the table with an offset of 5, that is, at location SEVSEG + 5. Of course, this code must be stored in the microprocessor's memory in some manner, but that is usually no problem. Note that there is nothing new here. The converter of Fig. 7.19 is a ROM with the seven-segment code stored in it. We are using a portion of the memory of our microprocessor system instead of a special ROM dedicated to this one task.

With the seven-segment code stored in memory, the routine to output a decimal digit to a seven-segment indicator is quite simple. This routine, which we shall call DISP7, is given here.

```
*   SUBROUTINE DISP7
*     SENDS 7-SEGMENT CODE TO PERIPHERAL PORT
*       UPON ENTRY:
*         7-SEG CODES IN TABLE STRTNG AT SEVSEG
*         ASCII CODE FOR DECIMAL DIGIT IN AC
*
DISP7     AND    #$0F        /MASK OFF UPPER 4
*                            /  BITS OF ASCII CODE
          TAX                /TRANSFER BCD CODE
*                            /  TO INDEX REGISTER
          LDA    SEVSEG, X   /GET CODE FROM TABLE
          STA    PP1         /SEND TO INTERFACE
          RTS                /DONE, RETURN
```

We assume that the seven-segment indicator is being driven by **PDL**[6:0] of interface 1 and that it has been initialized for output. In the ASCII codes for the decimal digits, the low-order four bits are the BCD code, so the first step converts ASCII to BCD. This is then sent to the index register. The indexed load then reads the table with the correct offset, obtaining the seven-segment code, which is then sent to the interface. That is all there is to it. The simplicity of this approach is charac-teristic of table-driven conversions. Basically we are trading memory space for the

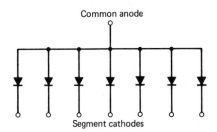

*Figure 13.30  Electrical equivalent for seven-segment indicator.*

table for program complexity. In this case we had no choice because there is no procedure for computing a seven-segment code. However, even when there are procedures, we often use table-driven methods because they are so simple. Sometimes in the literature you will find discussions of table-driven routines that make it sound all very complicated, but it really is not. It just means that, instead of computing an answer, we look it up.

Suppose it is desired to drive a six-digit display of seven-segment indicators. It would be rather expensive to have a separate interface unit for each indicator, so we look for a different approach. First we note that most seven-segment indicators are of the "common anode" configuration, as shown in Fig. 13.30. Each segment is electrically equivalent to a diode, and the anodes of all seven are tied together.

To display a digit, a voltage must be applied to the common anode and the cathodes grounded for the segments that are to be turned on. This arrangement is useful because it allows a digit to be turned on or off simply by controlling the anode. We shall make use of this feature to set up a time-multiplexed display.

A possible arrangement is shown in Fig. 13.31. The cathodes of all six digits are tied together and driven by the *PDL* lines of interface 1 (PP1). The anodes of the

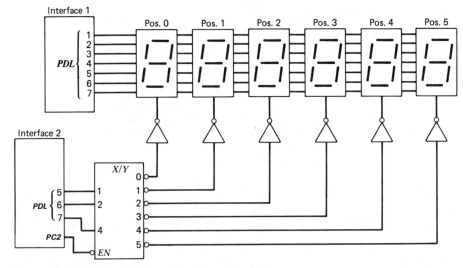

*Figure 13.31  Time-multiplexed display of six digits.*

six digits are driven through a second interface unit and a decoder. We load the first digit to be displayed into PP1 and then turn on the first indicator. Then we turn off the first indicator, move the second digit into PP1 and then turn on the second indicator. We just continue in this manner, until all six digits have been displayed and then repeat the whole thing. If the set of six indicators are scanned at least 10 times a second, the display will appear continuous due to the persistence of vision.

The scanning of the anodes of the six indicators will be accomplished by counting modulo-6 (that is, 0 to 5 and start over) in **POR**$[2:0]$ of interface 2 (PP2). This count will be used to drive a three- to eight-line decoder, successively driving lines 0 to 5 active to turn on the indicators one after the other in a repeating pattern. (The two unused outputs of the decoder have been omitted for clarity.) The **PC2** output of interface 2 is used to drive the EN input of the decoder, so that all indicators can be turned off while the digit in PP1 is being changed.

Why the decoder? Clearly we could use six of the **PDL** lines from PP2 and just shift a 1 along to turn on the indicators in order. The advantage of using the decoder is that the other five **PDL** lines are then available for other purposes, which may be a significant economy, since interface adapters are far more expensive than decoders. This may seem to contradict the argument of the previous section with regard to using programming to eliminate hardware, as in the case of keyboard encoding. But there the situation was different. Whether or not the keyboard was encoded, the interface adapter was still needed. The only cost of eliminating the separate encoding logic was extra programming. There was no new cost in hardware. In the current example, eliminating the decoder will incur extra hardware cost in terms of using up more of the capacity of the interface adapter, and the programming cost will be about the same either way.

We noted that the complete six-digit display must be scanned at least 10 times a second to avoid flicker. This means that about once every 10 to 15 msec, one digit must be turned off and the next one turned on. That rate of speed is very slow to the computer, and we clearly cannot afford to tie up the computer doing nothing but that. We need a subroutine that can be called by the main routine at the appropriate intervals to update the display. This should be a "background" routine, requiring nothing more of the programmer than inserting

**JSR SCAN**

at appropriate points in the main program. Each time this routine is called, it will turn off one digit and turn on the next. It will keep track of which digit is to be displayed each time; no parameters need be passed by the main program, and nothing will be done that will disturb the main program.

We will assume a six-word buffer area starting at SCNBUF, into which the main program has inserted the six digits to be displayed, in seven-segment code. The digit to be displayed at position 0 (see Fig. 13.27) will be in SCNBUF, the digit for position 1 in SCNBUF + 1, and so on. We shall use the index register **X** to keep track of the digits, starting it at 0 and incrementing until it reaches 5 and then setting it back to 0. Each time the routine is called, it will turn off the current digit, bring the next code out of the buffer area of PP1, and then turn on the next position.

```
*    SUBROUTINE SCAN
*      UPON ENTRY:
*        DIGITS TO BE DISPLAYED STORED IN SIX
*        WORDS STARTING AT SCNBUF.  CURRENT
*        DIGIT COUNT (IN RANGE 0-5) IN DIGCNT.
*        DIGCNT INITIALIZED TO 00 WHEN
*        SCNBUF LOADED.
*
SCAN      STX   TEMP1       /SAVE CURRENT VALUES
          STA   TEMP2       / OF X AND AC
          LDX   DIGCNT      /MOVE DIG. COUNT TO X
          LDA   #$38        /SET PC2 = 1,
          STA   PCR2        /TURN OFF INDICATORS
          STX   PP2         /DIGIT COUNT TO POR2
          LDA   SCNBUF,X    /GET NEXT DIGIT
          STA   PP1         /SEND CODE TO POR1
          LDA   #$30        /PC2 = 0, TURN ON
*                          /  INDICATORS
          STA   PCR2
          INX               /INCREMENT DIGIT COUNT
          CPX   #6          /COMPARE TO 6
          BNE   SAVE        /IF NOT 6, SAVE
          LDX   #0          /COUNT REACHED 6,
*                          /  RESET TO 0
SAVE      STX   DIGCNT      /SAVE DIGIT COUNT
          LDX   TEMP1       /RESTORE CONTENTS OF
          LDA   TEMP2       / X AND AC
          RTS               /DONE
```

We assume that both interface units have been properly initialized by the main routine. The subroutine starts by saving the contents of **X** and **AC**. This is very important. The programmer wants to be able to call this routine at any time without worrying about the current status of the main program. Since the subroutine uses both **X** and **AC**, their current contents must be saved so they can be returned to the main return intact. (In the next chapter a section on stack handling will provide a better way to do this.) We then move the digit count to **X** and turn off all the indicators by using the manual mode to set **PC2** = 1, which drives the ENable of the decoder inactive. We then move the digit count, which points to the next digit to be displayed, to **POR2** in preparation for turning on the next digit. In the interests of simplicity, we assume that **PDL**[7:3] are programmed for input so that the setting of **POR**[7:3] does not matter. If this were not the case, we would have to use masking so that only the low-order three bits of **POR** would be changed. We then use index addressing to generate the proper offset into SCNBUF, to bring out the next digit to be displayed. This is sent to PP1 to drive the cathode lines of the indicators, and **PC2** is then set to 0 to ENable the decoder and turn on the proper indicator. The next steps increment the digit count and store it in DIGCNT in preparation for displaying the next digit the next time SCAN is called. Finally, the contents of **X** and **AC** are restored to their original values. Execution time for this subroutine is 45 $\mu$sec. If called 60 times per second, for 10 complete scans of the display per second, this amounts to about 0.25% of the computer's time, the same as required to monitor the encoded keyboard.

## Problems

**13.1** Modify Fig. 13.3 as required to include 1K of RAM, implementing addresses 0000 to 03FF, assuming 256-word RAM chips are to be used.

**13.2** A system is to be designed with the following memory configuration.

> RAM: Two 1K chips, addresses 0000–07FF
> ROM: Four 1K chips, addresses F000–FFFF
> I/O: 128 locations, addresses 1000–107F

Show the memory map and address decoding for this system in a diagram similar to Fig. 13.3

**13.3** Repeat Problem 13.2 for the following configuration.

> RAM: Four 1K chips, addresses 0000–0FFF
> ROM: Four 2K chips, addresses 1000–27FF and F800–FFFF
> I/O: 32 locations, addresses 3000–301F

**13.4** Assume the 32 I/O locations in Problem 13.3 are to be implemented with eight interface chips of the type shown in Fig. 13.6. Show, by a diagram similar to Fig. 13.9, how the address decoding will be done for these chips. Also list the addresses associated with each chip, for example, PP1 = 3000, DR1 = 3001.

**13.5** During a particular clock the following listed memory registers and input lines to a TBPIA have the following listed values.
$CS$ = 1, A1 = 1, A0 = 0, $PCR$ = 80(hex), $DDR$ = F0, $PIR$ = 39, $POR$ = 4A, and $R/W$ = 1
  (a) Use the partial RTL description of Fig. 13.7 to determine the values of $DBUS$ during that clock period.
  (b) Determine the values on lines $PDL$ during the same clock period. Indicate those lines left in the high impedance state by the TBPIA having value X.

**13.6** Repeat Problem 13.5 with the same values except that now $DDR$ is 0F and A1,A0 is 01.

**13.7** In Fig. 13.7, what must be the values of the TBPIA inputs and buses during a particular clock period to cause the hex number 1F to be written into $POR$ at the end of that clock period?

**13.8** For the input/output system of Fig. 13.11, assume the keyboard issues the signal 1111 when no key is being pressed. Write a subroutine that reads the keyboard and compares the input with the output. If no key is being pressed, it does nothing. If a key is being pressed and the digit coming in is the same as that currently being displayed, it does nothing. If the digit is different, it should be sent to the display, that is, moved to $POR$.

**13.9**   In the system of Fig. 13.11, BCD digits are being sent out by the computer
to be displayed on seven-segment indicators, so that the output device will
have to include code converters from BCD to seven-segment. An alternative
approach would be to use seven output lines and send out the seven-segment
code from the computer. Assume the following configuration: We shall use
interfaces 1 and 2 in Fig. 13.10. Interface 1 will be used to input 8-bit ASCII
characters with odd parity from a keyboard. Interface 2 will be used to send
out seven-segment codes on **PDL**[6:0], and **PDL**[7] will be used for input
from a push button.

First show the commands necessary to initialize the **DDR**s correctly.
Then write a program to function as follows: The program should wait in a
loop for the push-button input to go to 1. When this happens, it should read
in an ASCII character from the keyboard. The program should check this
character for parity. If the parity is wrong, it should send out the code
1100111, which will display E (for Error) on a seven-segment display. If the
parity is correct, the program will check the ASCII character to determine
if it is a decimal digit. If it is not, it should send the code 0100111, which
will display C (for character). If it is a decimal digit, the BCD code should
be loaded into the lower four bits of **AC,** and the program should jump to a
subroutine SEVSEG that will return the seven-segment code to **AC,** to be
sent out to the display. Whatever is sent out, the program should then wait
for the push button to go back to 0 and then go back to the beginning to
wait for the push button to be pressed again.

Write the program in assembly-language form, using symbolic names
such as PP1 and DDR1, together with a symbol table to identify the corresponding addresses. Start the program at 0200, and assume SEVSEG is at
0400. You are not responsible for writing SEVSEG.

**13.10**   Assume a system configured as in Problem 13.9 except that there is no push
button. Instead, the keyboard will issue a **dataready** signal that will go high
anytime a key is pressed. This line will be connected to the **PC1** line of interface 1 and latching will be enabled. Thus, anytime a key is pressed, the flag
will be set and the ASCII code latched into **PIR**. This approach has the
advantage that the computer does not have to be continuously checking for
a key closure to avoid missing data.

Rewrite the program as a subroutine that will check the **PC1** flag
whenever called. If the flag is not set, the subroutine will return without
doing anything. If it is set, the program will proceed as before. Use handshaking with the **PC2** signal serving as a **datataken** signal to the keyboard
to prevent sending a new character until the previous one has been read. You
may assume that the line formerly used for the push button is now an output
used for other purposes. Also write a subroutine INIT consisting of the steps
necessary for initialization of the interfaces for this application.

**13.11**   Write a partial RTL description similar to the one given at the end of Example 13.4 that will provide for setting and resetting flip-flop **PCR**[6]. It must
only be possible to set this device to 1, if **PC2** is programmed as an input.

| | **Period** | | | | | | | | | | | | | | |
|---|---|---|---|---|---|---|---|---|---|---|---|---|---|---|---|
| | **1** | **2** | **3** | **4** | **5** | **6** | **7** | **8** | **9** | **10** | **11** | **12** | **13** | **14** | **15** |
| *PC1* | 1 | 1 | 0 | 0 | 0 | 1 | 1 | 1 | 1 | 1 | 1 | 1 | 1 | 1 | 1 |
| *portselect* | 0 | 0 | 0 | 0 | 0 | 0 | 0 | 0 | 0 | 0 | 0 | 1 | 0 | 0 | 0 |

*Figure P13.12*

**13.12** Consider the RTL sequence of Example 13.5 in conjunction with that of Example 13.4. Let $PCR[4] = PCR[0] = 0$, and $PCR[3] = 1$. Suppose $PCR[7]$ is initially 0 and the two-step sequence of Example 13.5 is initially in step 1. Let *PC1* and *portselect* be as given by Fig. P13.12 for 15 clock periods. Determine the values of $PCR[7]$ and *PC2* for these same 15 clock periods.

**13.13** Write a Boolean expression for the interrupt output line, *IRQ,* of the TBPIA. Consider all relevant flip-flops in *PCR*.

**13.14** A common device in microprocessor systems is the analog-to-digital converter, which converts an analog (continuous) voltage within a specified range into a binary number. We shall assume a converter with a single control input, *CONV*. When a pulse is placed on *CONV,* the converter will then sample the analog voltage and convert it to digital form. When the conversion is complete, the signal *RDY* will go high and the binary value will be held on the 16-bit *DATA* lines until a new *CONV* pulse is received. We shall use two interface units to input data from this device. The upper eight bits will go to interface 1, the lower eight bits to interface 2. The *RDY* line will go to *PC1* of interface 1 and *CONV* will be driven by *PC2* of interface 1, with interface 1 operating in the pulse handshake mode. Thus, each time a word is read in, a *CONV* pulse will be issued to start the next conversion.

First write a subroutine INIT to initialize the interfaces properly. Note that a "dummy" read will be needed to issue the first *CONV* pulse. Then write a subroutine GET that will input data from the converter. Each time GET is called, it will first check to see if the *PC1* flag is set. If not, it will return without doing anything. If the flag is set, the word should be read in. Since it is 16 bits, this will have to be done in two steps. The word should be returned to the main routine through two reserved locations on page 0, DATAU and DATAL for the upper and lower bytes, respectively. Write in assembly-language form, including symbol table.

**13.15** Modify the driver KYBD assuming that *KEY* is connected to *PC1* as a *data-ready* line to set a flag when a character is ready. This will eliminate the need to wait for *KEY* to go back down because the flag will be cleared by the READ of the character and the flag responds only to transitions in one direction.

**13.16**   Modify subroutine SCAN of Section 13.8 so that only *POR*[2:0] of interface 2 will be altered when the digit count is updated.

**13.17**   Write a program that will input a string of decimal digits from a keyboard in ASCII form and then convert to a packed BCD string. Use the subroutine KYBD from Section 13.5 and the program for ASCII to BCD conversion in Section 12.5 as subroutines in this program. Include provision for ignoring ASCII characters that are not decimal digits.

**13.18**   Write a program to unpack a packed BCD string, convert it to ASCII, and send it out to a printer. Use subroutine PRINT from Section 13.5 and the BCD to ASCII conversion routine from Section 12.5 as subroutines. Assume the printer is limited to 72 characters on one line, and include provision for inserting a carriage return and line feed if the string has more than 72 digits.

**13.19**   Write the subroutine SEVSEG mentioned in Prob. 13.9. This routine will receive a BCD code in the lower four bits of *AC* and return the seven-segment code in the lower seven bits of *AC*. Use a table-driven approach.

## References

1. Andrews, M. *Programming Microprocessor Interfaces for Control and Instrumentation,* Prentice-Hall, Englewood Cliffs, N.J., 1982.
2. Artwick, B. A., *Microcomputer Interfacing,* Prentice-Hall, Englewood Cliffs, N.J., 1980.
3. Sargent, M., III, and R. L. Shoemaker, *Interfacing Microcomputers to the Real World,* Addison-Wesley, Reading, Mass., 1981.
4. Taub, H., *Digital Circuits and Microprocessors,* McGraw-Hill, New York, 1982.
5. Tocci, R. J., and L. P. Laskowski, *Microprocessors and Microcomputers,* 2nd ed., Prentice-Hall, Englewood Cliffs, N.J., 1982.
6. Wakerly, J. F., *Microcomputer Architecture and Programming,* Wiley, New York, 1981.

# Serial Input/ Output

## 14.1  Serial to Parallel Conversion

In all the input/output situations considered to this point, data have been transmitted in parallel, over several lines at a time, as in the case of transmitting an ASCII character from an encoded keyboard over eight *PDL* lines. Of course, the transmission in such a case is serial in the sense that the characters are sent one at a time, and such data transfer is sometimes referred to as serial-by-character, parallel-by-bit. However, when people talk about serial data transmission, they usually mean one bit at a time, over a single line. Actually there will be two lines if the necessary ground return path is included, but that is usually taken for granted and not included in the count of data lines.

A variety of sophisticated modulation techniques make it possible to send continuous streams of serial data over a single line at very high data rates. The desireability of using the already existing single-voice-line telephone system for data transmission over long distances has strongly influenced the adoption of a serial approach. Parallel communication on eight lines seldom exists outside of a single computer system. Whether between devices in different rooms in the same building or between cities connected by telephone lines, communications will ordinarily be over a single serial data path.

Since data are ordinarily represented in the computer in parallel form, as words, or characters, or bytes, serial transmission will require conversion from parallel to serial, or vice versa. As we saw in Chapter 8, the basic technique for such conversions is the use of shift registers. Consider a shift register with a certain byte

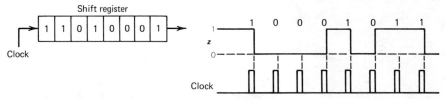

*Figure 14.1   Parallel-to-serial conversion in a shift register.*

initially stored, as shown in Fig. 14.1. Now clock the register eight times, and look at the serial output on line *z*, the output of the right-most flip-flop. Would the signal shown on line *z* be a suitable serial output to send to another device? That all depends. Suppose you saw only the signal on *z* without the clock. How would you know that it represents the sequence of bits 11010001? It is easily seen that you would not know without some information about the timing. If you had the clock signal, it would be easy; the clock tells you when to sample the information. So the answer seems obvious, send the clock signal along. That is indeed what we do when the transmission is within a single system, on a common clock. How can we send the clock when we have only a single line? In terms of modern communications technology, there are a variety of methods for transmitting timing information along with the data. However, the most common signaling techniques go back to a time long before computers and modern electronics, to the early days of telegraphy.

We are so accustomed to thinking of binary techniques in connection with computers, which are very modern, that it is something of a shock to realize that the first practical application of electricity, the telegraph, was a serial binary communications system. We are all familiar with the idea of an operator tapping on a key to switch voltage on and off a line, in short and long pulses (dots and dashes) making up the Morse code. This is certainly binary (voltage or no voltage), but it is not serial in the same way as the output of a shift register, since it is not the values of 0 and 1 that are significant, but the duration, or width, of the pulses.

In modern terms, Morse code is an example of pulse-width modulation. Since one of the signal levels in a telegraph system is no voltage, a problem is raised that we have mentioned before, the distinction between 0 as a unit of information and nothing at all, that is, the simple absence of any signal. When the voltage goes to 0 in a telegraph system, how do you know whether that means that information is being sent or simply that the line broke? To eliminate this sort of confusion, the convention was established that the normal, or "resting," condition of a telegraph line would be with voltage applied, with the line in the "MARK" condition. When the operator presses the key, it opens the line, to place it in the "SPACE" condition. Even though this is a very old-fashioned idea, the terms MARK and SPACE are still widely used in referring to serial data systems. In logical terms, MARK is equivalent to 1 and SPACE to 0.

In Morse code the timing, or synchronization, problem is solved by using time itself, in terms of pulse duration, to convey the information, rather than values of voltages. It was soon recognized that the need for operators trained in Morse code was a limiting factor in the use of telegraphy. Well over 100 years ago devices were being developed to use keyboards to send data and printers to receive data. As such devices were developed, it was found that Morse code was ill-suited to such purposes, so different types of codes were developed, based on using the values of the signals (MARK or SENSE) to convey the information.

One such code was developed in 1874 by Emile Baudot. Although rarely used in the United States today, the Baudot code is still widely used in other parts of the world. Also, the name Baudot has come down to us in the term "baud rate," the rate of information transfer in units of information per second. In a binary system the unit of information is the bit, so baud rate is the same as bits per second. Thus, a 300-baud line transmits information at a rate of 300 bits/sec.

With codes based on the values on the signal lines, we are back to the problem discussed in connection with Fig. 14.1, we need to know when to sample the data. Knowing when to sample involves two pieces of information—how often to sample, that is, the bit rate or clock rate, and when to start sampling. The "when" is important because the data are not just a continuous stream of bits but are divided into data units, such as 8-bit ASCII characters. The designers of early teleprinters (later teletypewriters) solved the synchronization problem in a simple but ingenious manner. The sampling rate was determined from the line frequency of the local power source. This was long before electronics, and the sampling was done by mechanical switches driven by synchronous motors, that is, ac motors with speeds determined by the power-line frequency. Assuming the power-line frequencies are the same—a pretty good assumption in the United States—we can thus be assured that two devices at widely separated locations are sampling at the same rate.

The information as to when to start sampling is provided by the coding scheme illustrated in Fig. 14.2. Every character is preceeded by a 0-bit called the START bit. Between characters the line is in the resting state, that is, at MARK. A transition to SPACE is the signal to the receiving device that a START bit has been sent. However, to prevent response to a spurious transition due to noise, the receiver then waits one-half bit time and then samples the line. If the line is still 0, it is assumed that this is a start bit and the receiver can now start sampling the data. If it is not

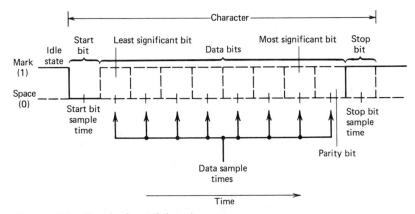

*Figure 14.2  Standard serial data format.*

0, the transition was an error, and the receiver goes back to waiting for a START bit. On the receipt of a START bit, assuming an 8-bit character, the receiver samples the line at the next eight bit times, shifting the values at these times into a shift register. To provide further check on the synchronization, each character is followed by a 1-bit known as the STOP bit (two stop bits in the case of 10 character/sec teletypewriter transmission). At the ninth bit time following the START bit, the receiver looks for a STOP bit. If the line is not 1 (MARK) at this time, a *framing error* has occurred, and some sort of error flag will be set. Assuming everything is all right, the receiver will then go back to waiting for a MARK-to-SPACE transition indicating the start of a new character.

One very desirable feature of this scheme is that it resynchronizes on every character, so that small differences in the sampling rate between the two devices will not result in a cumulative loss of synchronization. Figure 14.3 illustrates a situation in which the clock rate at the receiving device is slightly slower than that at the sending device. Since we start in sync at the beginning of the character, we can tolerate about a 10% difference in clock rate without losing sync within the 10-bit character. If we kept on for more bits, we would soon lose sync, that is, we would be sampling at the wrong time.

Resynchronizing on each character cannot prevent loss of synchronization if the difference in clock rates is large, but it does allow a considerable margin for error, depending on the clock rate. At 300 baud, the rate most commonly used for signaling on standard telephone lines, the bit time is 3.3 msec. Assuming a 10% deviation in bit time is acceptable, this means we need to maintain the bit time within a margin of about $\pm 300$ $\mu$sec, which is simple with modern technology. At 19,200 baud, the fastest rate used for this technique, the margin for error is about $\pm 5$ $\mu$sec, not impossible to maintain, but much more of a challenge. Modern systems do not rely on the power-line frequency for synchronizing. Instead, crystal-controlled oscillators are used to generate the local clock signals. Crystals with the accuracy required up to 19,200 baud are readily available. This mode of signaling is known as *unclocked signaling* because no clock (timing) signal is sent with the data. As noted, 19,200

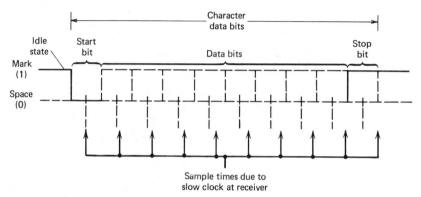

*Figure 14.3   Effect of differences in sampling rate.*

baud is the limit for unclocked signaling. For higher data rates a clock signal must be transmitted with the data.

### Example 14.1

A receiver for serial transmission is to be designed that will accept 10-bit characters at a maximum data rate of 300 bits/sec. The only crystal-controlled clock available has a clock frequency of 10 MHz with a square wave output at standard TTL logic levels. A counter is to be used to generate a local clock frequency of approximately 300 Hz to synchronize the receipt of characters. Determine the form and number of bits for the required counter. Estimate the potential timing error over the duration of a single character.

Solution:
Let us first consider the simplest form of counter, one that will merely divide the available clock frequency by a power of two. Should no such counter be satisfactory, a more complex counter involving a modulo-3 or modulo-5 section might be tried.

If we divide $10^7$ by $2^{15}$ or 32,768 we obtain the following result.

$$10,000,000/32768 = 305.17 \qquad (14.1)$$

This tells us that a 15-bit counter driven at a clock rate of 10 MHz will have an output frequency of 305.17 Hz. The corresponding clock period is given by

$$1/(305.17) = 3.27 \text{ msec} \qquad (14.2)$$

For 300 bits/sec the period is 3.33 msec, and the duration of a 10-bit character is 33.3 msec. Ten periods at 305.17 Hz require 32.7 msec. To find the cumulative error over one 10-bit character we merely subtract

$$33.3 - 32.7 = 0.6 \text{ msec} \qquad (14.3)$$

This difference of 0.6 msec is much less than one half the bit time, so we may safely use the output of a 15-bit counter to sample at the center of each data bit if some form of resynchronization is employed at the beginning of each new character. ∎

In succeeding sections the more commonly used devices for interfacing serial communications lines to microcomputers will be introduced. To illustrate how these might be designed, we choose a simplified example based on the clock-rate determination in Example 14.1.

### Example 14.2

Write an RTL description of a device that will accept serial 10-bit characters at a maximum rate of 300 bits/sec. Only the data line is available as an input. The system clock is 10 MHz. Once a data character is received, it is to be placed in a special buffer register, so that the device is immediately ready to accept a new character. The outputs of the buffer register are to be device outputs. Each time a new character is placed in the buffer register a positive transition should occur on line *dataready*, and this line should remain 1 for approximately the duration of one data bit.

INPUTS: *data*
MEMORY: *COUNT*[15]; *BITCOUNT*[3] *SR*[8]; *BUFREG*[8].
OUTPUTS: *Z*[8]; *a.*
1. *BITCOUNT* ← 0,0,0;
    → (*data*)/(1).
2. *a* = 1.
3. → ($\overline{COUNT[14]}$)/(3).
4. *a* = 1.
5. → $\overline{\wedge / COUNT}$/(5).
6. *SR* ← *SR*[6:0], *data*; *BITCOUNT* ← INC(*BITCOUNT*);
    → ($\wedge$/*BITCOUNT*, $\overline{\wedge/BITCOUNT}$)/(7,4).
7. *BUFREG* ← *SR*; *a* = 1.
8. *dataready* = 1;
    → ($\wedge$/*COUNT*, $\overline{\wedge/COUNT}$)/(1,8).
ENDSEQUENCE
CONTROLRESET(1);
*COUNT* * $\overline{a}$ ← INC(*COUNT*);
*COUNT* * *a* ← 15 T 0;
*Z* = *BUFREG*
END.

*Figure 14.4    RTL description of serial receiver.*

Solution

An RTL description of the device complete with declarations is given in Fig. 14.4. A 15-bit counter, *COUNT*, is used to derive a 305.17-Hz sampling signal from the 10-mHz system clock. Also included are a 3-bit couter, *BITCOUNT*, to count the eight data bits in each character; an 8-bit shift register, *SR*; and a buffer register, *BUFREG*.

   Notice from the portion of the description after ENDSEQUENCE that the 16-bit counter is incremented every clock period unless line *a* happens to be 1, in which case it is reset. (The notation 15*T* 0 signifies a 15-bit encoding of zero, that is, a string of 15 zeros.) We assume that the system waits in step 1 until a SPACE or START bit appears on the input line *data*.* When this occurs *COUNT* is reset at step 2, and the system waits in step 3 for one half of a 3.27-msec period, that is, until the most significant bit of *COUNT* is 1. Control will pass to step 4 at precisely the middle of the start bit. In step 4 the 16-bit counter is again reset. The system remains idle in step 5 until *COUNT* contains all ones, indicating the center of the first data bit. This data bit is then shifted into *SR* while *BITCOUNT* is incremented in step 6. Control will circulate in steps 4, 5, and 6 until the eighth data bit is shifted into *SR*. At this time *BITCOUNT* will contain all ones, and control will pass to step 7 where the received character is passed on to the buffer register and the 15-bit counter is

---

*The branch in step 1 of Fig. 14.4 involves an input that is not synchronized to the system clock. This could very well cause problems; but, as stated in Section 13.4, these considerations must await Chapter 17.

once again reset. Control remains in step 8 until half way through the start bit. Thus the next SPACE on line **DATA** encountered by step 1 will be a legitimate start bit of another character. The control line **DATAREADY** is held at 1 for 3.27 msec by step 8, as required by the problem specification. ∎

You may well wonder why we are going into all this "ancient history." This standard is old, but it is a good one, and it is still in use. If we were to start over today to create a standard for asynchronous serial data communications, we would very likely do it differently. But this standard has been in use for so long that there is too much equipment installed to permit changing the standard without great economic upheaval.

We have mentioned transmission of data by this standard over telephone lines. But how does this work? We have been talking in terms of telegraphy, in which voltages on a line are just switched on and off, and that is fine for computers, which also switch voltages on and off. But telephones do not work that way. To bridge the gap, we use devices called *modems* (modulator/demodulator) that convert voltages into audio signals, and vice versa. The system is simple in principle; at the transmit end, a MARK level is converted into an audio tone at one frequency, a SPACE level to a tone at a different frequency. This is known as *frequency-shift keying* (FSK). At the receiver end, one frequency is converted into MARK level, the other into SPACE level.

Suppose a computer is to send data to a printer at a remote site over a telephone line. The complete system would be as shown in Fig. 14.5. We assume both the computer and printer handle 8-bit data bytes in parallel. The computer sends the data to a parallel-to-serial converter that puts the data in the standard format discussed previously and sends this to the modem. The modem converts the voltage levels to audio frequencies. At the other end another modem converts the frequencies back to levels and sends the serial signal in the standard format to a serial-to-parallel

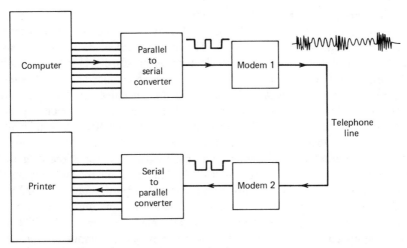

*Figure 14.5   Typical computer communication system.*

converter, which strips off the START and STOP bits and sends the parallel characters to the printer.

Since a printer can only receive data, the system is a one-way system that operates in *half-duplex* mode. In many cases we wish to set up communications between devices that can both send and receive, for example, a computer at one end and a CRT terminal at the other. In that event, modems are required that can both transmit and receive: and two converters, parallel-to-serial and serial-to-parallel, are required at each ends. Such systems can still run half-duplex, with the two devices taking turns, each alternating between transmitting and receiving. More often such systems run in *full duplex,* both modems being capable of transmitting and receiving at the same time. This is done by using two pairs of frequencies. One modem is known as the originate modem, the other as the answer modem. The originate modem transmits at (typically) 1070 Hz (SPACE) and 1270 Hz (MARK), and the answer modem receives at these frequencies. The answer modem transmits at 2025 Hz (SPACE) and 2225 Hz (MARK), and the originate modem receives at these frequencies. This may seem strange, but it really isn't; it is no different than two people on a phone line talking at the same time. This process of sharing a single line for two different messages at different frequencies is known as *frequency division multiplexing.*

Figure 14.5 shows only one line, the data line, between the converters and the modems. Some control lines are required just as was the case for parallel transmission. The fact that the serial-to-parallel converter is looking for a START bit to start sampling does not solve all problems of controlling the interaction between the devices. For example, before the computer starts sending characters, it needs to know that the modems are in communication, with modem 1 in the sending state, sending the MARK frequency, and modem 2 in the receiving state. The necessary control lines are discussed in the following paragraphs. Note that the existence of these lines does not make the system parallel. There is still only a single line between modems. The control lines exist only between the terminals and modems that are physically adjacent.

There are a number of standards, or protocols, for serial data communications involving modems and telephone lines, which may be divided into two categories, timing protocols and electrical protocols. The timing protocol discussed previously is virtually universal, but there are several electrical protocols, of which the most widely used in the United States is RS-232. RS-232 is a recommended standard published by the Electronic Industries Association (EIA) that specifies the identity and function of control signals, signaling protocols, voltage levels, even connector types and pin assignments. Ideally, if you have two devices following RS-232, you should be able to plug them together and have them work properly. (Provided, of course, you know what you are doing.) In the design of most data communications devices, it is necessary to deal with only a small part of the RS-232 standard and only a subset of the available control lines. Figure 14.6*a* depicts the portion of the RS-232 standard that we shall consider here. There are two data lines and four control lines that connect the DCE (data communications equipment) or modem to the DTE (data terminal equipment). A DTE may be a CRT terminal, a computer, a printer, or almost any piece of digital equipment. Names are assigned to these lines that identify their functions as tabulated in Fig. 14.6*b*.

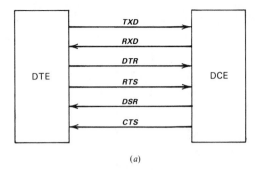

(a)

| Name | Meaning | Source |
|------|---------|--------|
| *TXD* | Transmitted data | DTE |
| *RXD* | Received data | DCE |
| *DTR* | Data terminal ready to function | DTE |
| *RTS* | Request to send data | DTE |
| *DSR* | Data set (DCE) ready to function | DCE |
| *CTS* | Clear to send: If the modem is idle waiting for a request to transmit, this signal will be 0; it will go to 1 in response to an *RTS* | DCE |

(b)

Figure 14.6   Simplified RS-232 standard.

The flowchart given in Fig. 14.7 describes the operation of a half-duplex device operating in accordance with RS-232. The *DIT* and *RTS* signals are output lines from the DTE and DCE. The box labeled "Receive character" represents an implementation of a control sequence similar to that given in Fig. 14.4. An RTL sequence of about the same complexity and using the same counters could be written to represent the "Transmit character" box as well. The DTE indicates that it is ready to start operations by driving *DTR* active, and then waiting for the DCE to respond with *DSR*.

While no transmission is in progress, the *RXD* line is continuously checked for a start bit. If a start bit does not appear, the terminal checks internally to see if there is a character to be sent out. If so, it makes sure the *CTS* from the modem is 1 (inactive). It then issues an *RTS* and waits for *CTS* to go to 0 (active) in response. After a character is transmitted, *RTS* is returned to 1 (inactive). Note that all four signal lines, *DTR*, *DSR*, *RTS*, and *CTS* are active low.

In descriptive literature on such devices, these signal names will often be given with an overbar (e.g., *RTS*) to indicate that they are active low. We shall not use these overbars, because they are not consistent with the RTL language we are using, in which an overbar always denotes a negated signal and has nothing to do with active levels. A full-duplex device will be more complicated than that suggested by Fig. 14.7, since it must be capable of sending and receiving simultaneously.

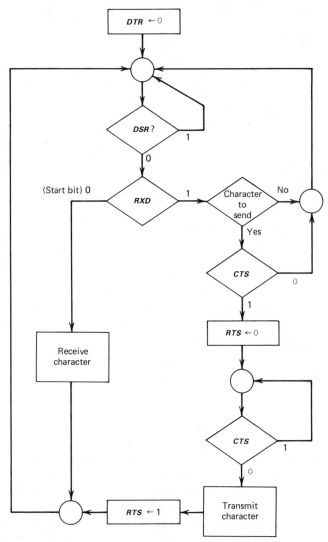

*Figure 14.7   Flowchart for half-duplex DTE.*

## 14.2   A Serial Communications Interface Adapter TBSIA

In Example 14.2 a separate device was designed to accomplish the serial-to-parallel conversion of data to be accepted from a communications system by a CPU. Is a separate device always needed, or can the computer be programmed to accept or output serial data through a standard interface adapter such as the TBPIA discussed in Chapter 13? Parallel-to-serial or serial-to-parallel conversion requires shifting, which the computer can certainly do. If the signaling lines perform functions similar to *dataready* and *datataken*, presumably the interface adapter can handle them.

   In some cases, where economy is at a real premium, RS-232 communications are handled by programming and standard interface units. However, it's a complicated job and tends to tie up the computer. Most often, a more general version of the serial interface adapter unit designed in Example 14.2 will be used. Such an interface unit may be known as an ACIA (asynchronous communications interface adapter), or a UART (universal asynchronous receiver/transmitter), or a USART (universal synchronous/asynchronous receiver/transmitter). The various units differ in details, but all serve the same function—to interface between a parallel device and a serial data path functioning under RS-232.

   Unlike the serial receiver of Fig. 14.4 or a UART, a serial interface adapter is typically designed for convenient direct connection to the buses of a particular microprocessor. It will include several addressable registers as did the TBPIA parallel interface adapter package. In this section we shall lay out the organization of a TBSIA serial interface adapter for use with the TB6502. In the next section we shall write a TB6502 program to transmit and receive streams of data characters through this device. This hypothetical TBSIA will include features of several of the popular units now available.

   The block diagram of the TBSIA is shown in Fig. 14.8. The lines to the left of the figure are those connecting to the TB6502, the 8-bit *DBUS*, *CS*, *R/W*, and one address line. Additional off-chip decoding will be needed to locate this chip in the address space. The lines to the right of the figure are those for connecting to a standard RS-232 full-duplex modem. The registers for handling data are the four at the top of the figure. To transmit data, the computer loads a word into *TDR* (transmit data register). The word is automatically moved to *TSR* (transmit shift register) and shifted out in serial on line *TXD*, with START and STOP bits inserted automatically. For receiving data, *RSR* (receive shift register) monitors *RXD* for a START bit. When a START is sensed, the data word is shifted in, the START and STOP bits stripped off, and the data word is transferred to *RDR* (receive data register), from which it can be read into the computer.

   The other three registers, *AMR* (asynchronous mode register), *ACR* (asynchronous control register), and *ASR* (asynchronous status register), are used for establishing data formats and controlling interactions between the computer and the modem. All registers are read-only or write-only, and all communicate with the TB6502 through the *DBUS*. The addressing of the registers is summarized in Fig. 14.9. The A0 bit selects the data side or the control side. If A0 = 0, *R/W* specifies a read of *RDR* or a write to *TDR*. For A0 = 1, *R/W* specifies a read of *ASR* or a write to *AMR* or *ACR*.

   A write to the control side writes *AMR* immediately after a RESET. All subsequent writes to the control side will write *ACR*. The *AMR* establishes the format for serial data, and it is assumed that this would rarely be changed once the system is initialized. In discussing the serial format in connection with Fig. 14.2 we assumed the most common arrangement of eight data bits, including parity, with START and STOP bits. However, most serial interface adapters allow a number of variations to fit different requirements.

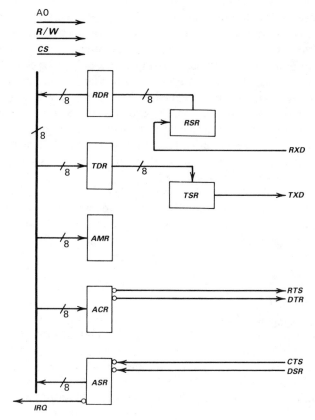

*Figure 14.8   TBSIA for the TB6502.*

In the TBSIA the data word may be five, six, seven, or eight bits, not including the parity bit; and there may be one or two stop bits and a parity bit, odd or even. If parity is enabled, the transmit section will insert the appropriate parity bit following the data bits as the word is shifted out, and the receiver section will check for parity as a word is received, setting a parity error flag in **ASR** if it is not correct. All these choices are specified by the word loaded into **AMR** following reset, as shown in Fig. 14.10. The user can also specify one of eight possible baud rates. (There are other standard baud rates under RS-232, but these are the most common.)

The significance of the bits in the command word loaded into **ACR** is indicated

|     | **R/W** |     |
| --- | --- | --- |
| A0  | 0   | 1   |
| 0   | *TDR* | *RDR* |
| 1   | *AMR* or *ACR* | *ASR* |

*Figure 14.9.   Addressing of registers in the TBSIA.*

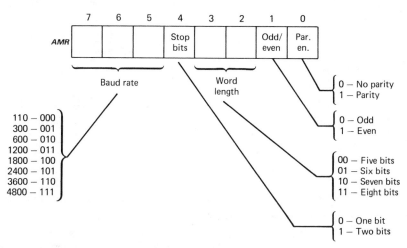

*Figure 14.10 Mode word in the TBSIA.*

in Fig. 14.11. There are two active-low control lines, **DTR** (data terminal ready) and **RTS** (request to send) going to the modem. Setting the **DTRF** or **RSTF** bits in **ACR** will drive these lines active. The next three bits are used to enable interrupts. There is an **IRQ** (interrupt request line) that follows **ASR**[7] and will interrupt the computer when it goes low. For this to happen, the **IRQ** bit in **ASR** must be enabled. If **ACR**[2] is set, **IRQ** will go low when a word is transferred from **TDR** to **TSR**, leaving room for the TB6502 to send another word to **TDR**.

If **ACR**[3] is set, **IRQ** will go low when a word is loaded into **RDR** from **RSR**, available to be read by the TB6502. Three kinds of errors can be detected by the receiving circuits. If **ACR**[4] is set, any of these errors will cause an interrupt. These errors also set flags in **ASR**. Setting **ACR**[5] will clear these flags. A complete reset of the entire chip is provided by setting **ACR**[6]. This will make it possible to change the contents of **AMR**. Finally, **ACR**[7] is the BREAK bit; setting this bit will drive **TXD** to SPACE (0), regardless of what else may be going on. Note that this literally

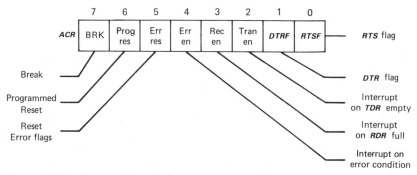

*Figure 14.11 Command word in TBSIA.*

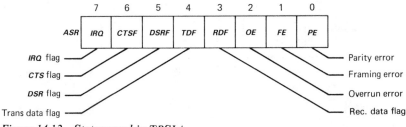

*Figure 14.12   Status word in TBSIA.*

breaks off communications. It can, for example, be used to enable a "simple-minded" device such as a printer to tell a computer to stop sending data.

The meaning of the bits in the status register is shown in Fig. 14.12. The bits **ASR**[2:0] are error flags, for parity error, framing error, and overrun error. The parity flag will be set if parity is enabled and the received word does not have the correct parity. *Framing error,* which was mentioned earlier, occurs when a STOP bit is not found at the proper time. *Overrun error* occurs when the receiving section has assembled a word in **RSR** and moves it to **RDR** and finds that the previous word sent to **RDR** has not been picked up by the computer, and is therefore lost (overrun). **ASR**[3] is RDF (received data flag) which is set whenever a word is loaded into **RDR** and cleared when **RDR** is read by the computer. (Thus, overrun occurs if RDF is still set when a new word is sent to **RDR**.) **ASR**[4] is the transmit data flag, which is set when a word is moved from **TDR** to **TSR** and cleared when a new word is loaded into **TDR** by the TB6502. Transfer of a word from **TDR** to **TSR** is not under program control but takes place as soon as the previous word has been shifted out of **TSR** on **TXD**. Therefore, **ACR**[4] = 1 may be treated as the indication to the CPU that a new word may be transmitted to **TDR**.

There are two active-low input control lines from the modem, **DSR** (data set ready) and **CTS** (clear to send), which drive the **DSRF** and **CTSF** flags. **ASR**[5] is set if **DSR** is active and cleared if it is inactive; **ASR**[6] is set if **CTS** is active and cleared if it is inactive. Note very carefully that **CTS** and **DSR** are level signals, and the flags simply follow them. This is in contrast to the other flags in the TBSIA and the flag in TBPIA, which are set by the occurence of some event, such as an active transition or a register becoming available, and can be reset only by some action of the CPU. Finally **ASR**[7] is the IRQ-flag. It will go low on the appropriate condition, if enabled as discussed in connection with **ACR**. If it goes low as a result of **TDR** being empty, it will be set when **TDR** is loaded. If it goes low as a result of a word in **RDR**, it will be set when **RDR** is read. If it goes low as a result of one of the three error conditions, it will be set when the error flags are cleared.

This is a very powerful unit that can be used in many ways. When used with a standard modem, a typical sequence of operations would be as follows. When reset, either at initial turn-on, or by a programmed reset, all registers are cleared, except that **TDF** in **ACR** is set since **TDR** is empty, and the unit is neither transmitting nor receiving. The first step is to load **AMR** to establish the data format and baud rate. The unit will remain inactive until **ACR**[1], the DTR flag, is set to 1. At this time the receiver will first look for a MARK as an indication that the other modem is online and in the resting state and then look for a START bit. When a word is assembled, it will be moved to **RDR** and the receiver will continue to look for data. This is the only control over the receiving section; any time **DTR** is active, the receiver will be looking for data.

To initiate transmission, the computer will first load a word into **TDR** and then set the RTS flag, which will drive **RTS** active. If the modem has established communications with the other modem, it will drive **CTS** active and the character will automatically be moved from **TDR** to **TSR** and transmission will start. As soon as the word in **TSR** has been sent, the next word will be moved from **TDR** and transmitted. If **TDR** has not been reloaded, the transmitter section will wait until it is loaded and then continue.

Operations will continue in this manner as long as both **RTS** and **CTS** are active. If either goes inactive, the transmitter will complete sending the word currently in **TSR** and then stop. Although **CTS** does set a flag in **ASR**, this does not have to be checked by the computer to start transmission. Transmission will start as soon as three conditions are met—there is a word in **TDR** and both **RTS** and **CTS** are active—regardless of the order in which these events occur.

The four control lines to the modem (**DTR**, **DSR**, **RTS**, and **CTS**) are those specified by RS-232, but they do not all have to be used. For example, the DTR flag (**ACR**[1]) must be set for the TBSIA to do anything, but the **DTR** line may or may not be needed, depending on the characteristics of the modem or whatever else happens to be "out there." With some modems this line is used to signal the modem that it should establish communications with the other modem; with other systems this may be automatic and the **DTR** line may not be needed.

We did not mention the **DSR** line. In some cases it is used to indicate that the modem is powered up and ready to establish communications when **DTR** goes active; in other cases, it may not be used at all. All it does in the interface is to set a flag that is available for testing by the CPU, and it may be regarded as a general purpose input status line, available for whatever purpose may be desired. It is important for users to keep in mind that, while serial interface adapters (and UARTs and USARTs) are designed with the characteristics of RS-232 modems in mind, they are general-purpose serial interfaces that can be used with a great variety of devices besides modems.

## 14.3   A Subroutine for Serial I/O through a TBSIA

As a first example of a use of the TBSIA, assume we have a TB6502-based microcomputer equipped with an ASCII keyboard and a CRT display. We shall write a subroutine to enable this computer to function as a terminal, communicating by telephone with a host computer system. The operator will establish communications with the host computer by telephone, acoustic coupler, and modem. The modem will be connected to the microcomputer through a TBSIA. The interfaces to the keyboard and the CRT were already discussed in Chapter 13. We assume the keyboard to be interfaced to the TB6502 exactly as given in Fig. 13.25 and the subroutine KYBD,

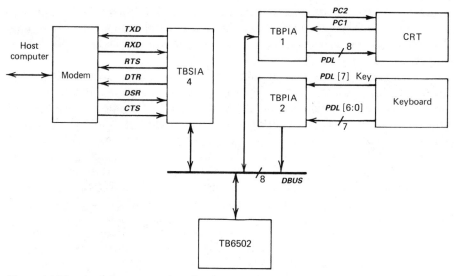

*Figure 14.13*    *A microcomputer-based terminal.*

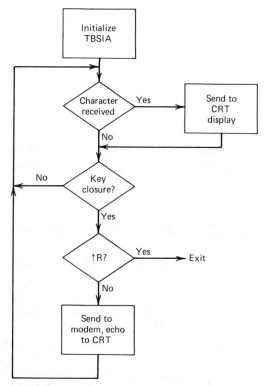

*Figure 14.14*    *Flowchart of terminal subroutine.*

as discussed in Section 13.6, to be available. Similarly, the CRT is connected to the TB6502 through a second TBPIA as given in Fig. 13.20. The complete terminal is shown in Fig. 14.13. The TBSIA and modem have been added to connect the configuration to the host computer.

The flowchart of the subroutine is shown in Fig. 14.14. Following initialization of the TBSIA, the routine simply goes into a loop, alternately checking the modem for a character received from the host computer and the keyboard for a key closure. If a character is received from the modem, it will be displayed on the CRT. When a key is pressed, the character will be sent to the modem for transmission to the host computer and displayed on the CRT. The operator can leave this routine by typing "Control-R" (↑R), that is, by pressing R and the CTRL key at the same time.

We assume the host computer communicates at 300 baud, seven data bits, one stop bit, and odd parity. The TBSIA will be connected as interface 4 in Fig. 13.10, placing the data side at 040C and the control side at 040D. We assign the symbolic names MDATA and MCONT to these two locations. The subroutine is shown here.

```
*    SUBROUTINE TERM
*       A SUBROUTINE TO PERMIT A SMALL COMPUTER
*       TO FUNCTION AS A TERMINAL FOR A HOST
*       SYSTEM, COMMUNICATING AT 300 BAUD VIA
*       A TELEPHONE LINE.
*
TERM        LDA   #0
            STA   MCONT      /DUMMY LOAD
            LDA   #$40
            STA   MCONT      /RESET
            LDA   #$29       /INITIALIZE AMR FOR
*                           / 300 BAUD, 7 BITS, 1
            STA   MCONT      / STOP BIT, ODD PARITY
            LDA   #$03
            STA   MCONT      /ACTIVATE DTR, RTS
RECTST      LDA   #$08
            BIT   MCONT      /CHAR IN RDR?
            BNE   KEYCHK     /NO, LOOK FOR KEY
            LDA   MDATA      /YES, CHAR FROM RDR
            AND   #$7F       /MASK OFF PARITY BIT
            JSR   CRT        /SEND TO CRT DISPLAY
KEYCHK      JSR   KYBD       /LOOK FOR KEY
            BCC   RECTST     /NO KEY, CHECK MODEM
            AND   #$7F       /MASK OFF UPPER BIT
            CMP   #$12       /^R?
            BEQ   END        /YES, EXIT
            STA   TEMP       /NO, SAVE CHAR
            LDA   #$40
            BIT   MCONT      /TEST CTS
            BEQ   ERROR      /MODEM INACTIVE, ERROR
            LDA   #$10
TRTST       BIT   MCONT      /TDR EMPTY?
            BEQ   TRTST      /NO, WAIT UNTIL IT IS
            LDA   TEMP       /YES, RETRIEVE CHAR
            STA   MDATA      /SEND TO MODEM
            JSR   CRT        /ECHO CHAR TO CRT
            JMP   RECTST     /CHECK FOR REC. CHAR.
CRT         BIT   PCR1       /GET DT FLAG OF CRT
            BEQ   CRT        /WAIT FOR DT = 1
            STA   PP1        /SEND CHAR TO CRT
END         RTS              /EXIT CRT
```

In many cases the TBSIA might be initialized at initial power-on for the system. To be on the safe side, this routine assumes we do not know if the TBSIA has been initialized or not, so that we do not know if a write to MCONT will access **AMR** or **ACR**. The first two steps do a "dummy" write to MCONT. If the unit has not been initialized, this will do nothing, but it will give us access to **ACR**. If it has been initialized, this will clear **ACR**. In either event, it is known that the next command will access **ACR**, so a programmed reset is executed, which provides access to **AMR**. **AMR** is then loaded with the appropriate combination of bits for 300 baud, seven bits, and so on, as expected by the host computer. The final write to MCONT loads **ACR** to turn on **DTR** and **RTS**, turning on the TBSIA and enabling transmission.

Starting at RECTST, **RDF** is tested to see if a character has been received from the modem. If so, it is read and sent to the CRT display. At KEYCHK the KYBD subroutine is called to check for a key closure. This is the same routine described in Section 13.6, which returns **C** = 0 if no key has been pressed, and **C** = 1 and the character in **AC** if a key has been pressed. We check **C**; if there is no character, we go back to check for a received character again. If there is a character, a check is made to see if the character is a control-R (12 is the hex code for control-R). If so, the routine is exited. If the character is not a control-R, it is saved; and then **CTS** is checked to see if the modem is ready for transmission. If it is not, we jump to ERROR, a subroutine not shown here, which will display a message such as "MODEM INACTIVE" to warn the operator that he or she is not in communication with the host. If the modem is OK, we wait for **TDF** to set, indicating that **TDR** is available to accept a character for transmission, and then retrieve the character and send it to the modem. Finally, the key is echoed to the CRT and we go back to look for another received character.

We assume the CRT is connected through interface 1 as shown in Fig. 13.20 and that this unit has been properly initialized for output handshaking. The CRT section of the routine first checks for **datataken** = 1, indicating the the CRT is ready to accept a character. When this condition is met, the character is sent. Finally, what about speed? We already know the KYBD routine is amply fast to keep up with the keyboard. At 300 baud, the host, when transmitting, will be sending 30 characters a second, that is, a character every 33 msec. As long as no key is pressed, this routine could easily keep up with 30,000 characters per second. When a key is pressed, the delay is unknown since KYBD waits for the operator to release the key. Thus, data might be lost if the operator pressed a key while the host was transmitting. However, about the only reason to try to transmit to a host while it is transmitting is to tell it to stop, so there should be no problem here.

## 14.4   Interconnecting RS-232 Equipments

Although the standards of RS-232 are derived from the requirements of signaling by telephone or telegraph lines, serial signaling in accordance with RS-232 is often used even when devices are close together and no telephones or modems are involved.

The usual reason for this is that we have two devices that we want to connect and RS-232 is the only "language" they have in common. For example, this book was written on a personal computer that has only RS-232 ports for input/output, so that it was necessary to use a printer equipped for RS-232 signaling. At first glance, it may seem ridiculous to resort to serial communications between a computer and a printer on the desk right next to it, but that is the simplest thing to do when both units come so equipped from the factory.

As we noted earlier, the TBSIA is a very versatile device, and there are no real technical problems in using it to communicate with other RS-232 units without modems in the middle, but there are some nomenclature problems that often cause considerable confusion. In RS-232, all equipment is divided into two categories, data terminal equipment (DTE) and data communications equipment (DCE). The DCE category includes all devices, such as modems, that serve to interface between communications lines, for example, telephone lines, and all other devices. The DTE category includes all the "other" devices, for example, computers, terminals, printers—all the devices that "talk" to modems or other DCEs. DCE units are also referred to as *datasets*. These categories are important because the functions of the signaling lines are rigidly defined with reference to them.

The signal **DTR** (data terminal ready) is always an output from a DTE and an input to a DCE, informing the DCE that the DTE is ready to do something or other. Conversely, the signal **DSR** (data set ready) is always an output from a DCE and an input to a DTE, informing the DTE that DCE is ready to so something or other. The signal **RTS** (request to send) is always an output from a DTE to a DCE, requesting permission to send data, and **CTS** (clear to send) is always an output from a DCE to a DTE, giving permission to send data. When manufacturers state that a device conforms to RS-232, they indicate that these lines will conform to these definitions.

As long as we have a full RS-232 system including DTEs and DCEs, as shown in Fig. 14.15a, there are no problems with these signal definitions. The outputs of the DTEs connect to the identically named inputs of the DCEs and vice versa. However, if we want to connect two DTEs without intervening DCEs, as shown in Fig. 14.15b, we seem to have a problem. The inputs and outputs of the two devices have the same names, but certainly cannot be connected outputs-to-outputs or inputs-to-inputs. The solution is not to worry about the names of the lines but to connect them together in accordance with their functions.

There is still a possible source of confusion here. If you are trying to connect two devices together and they both have lines labeled **DTR**, the natural thing to do is to connect them; but the system will not work if they are both DTEs. A further problem is found in the hardware specifications of RS-232. The standard specifies a 25-pin connector and specifies the pin number for each line. For example, **DTR** is assigned to pin 20 for both DTE and DCE, and RS-232 cables are "pin-to-pin" cables, connecting a pin on one device to the same pin on the other. Thus, if we connect two DTEs together with a standard cable, we shall be connecting inputs to inputs and outputs to outputs.

There is even a problem with the data lines. The term, **TXD**, denotes transmitted data and **RXD** denotes received data. For a DTE these names mean exactly what they say, but the situation is more complicated for DCE. For a modem, trans-

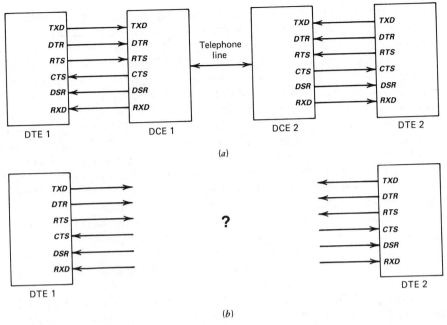

*(a)*

*(b)*

*Figure 14.15   DTE and DCE interconnections.*

mitted data are data received from the DTE, to be transmitted over the phone line, and received data are data received over the phone line, to be transmitted to the DTE. Again, no problem in a complete system, the **TXD** line of one DTE cannot be connected to the **TXD** of another.

As a specific example of how these problems can be handled, consider the connection of a TB6502 to drive a printer via the TBSIA. We shall assume a printer with a line buffer. As characters are received, they are stored in a buffer memory until a carriage return is received or the buffer is full, at which time a full line is printed. Since the buffer is electronic, it can accept data at any RS-232 baud rate. Printing, being a mechanical operation, is slow by electronic standards, so the printer needs a way to tell the computer to stop sending data while a line is being printed. Since the printer is a DTE under RS-232 rules, it has only two possible control outputs, **DTR** and **RTS** (request to send); but **RTS** is inappropriate since a printer is a receive-only device and cannot ask for permission to send. For a printer conforming to RS-232 this problem is usually solved by providing the printer with only one output line, **DTR**, which will be used as a BUSY signal, to tell the computer when it is printing. When the printer is ready to receive data (load its buffer) **DTR** will be active (low). When a carriage return is received or the buffer is full, **DTR** will go high until the line is printed.

On the computer end we shall connect the **DTR** line from the printer to the **CTS** input of the TBSIA. (Note carefully, do not connect it to the **DTR** line of the TBSIA, that is an output!) Recall from Section 14.2 that the **CTS** input to

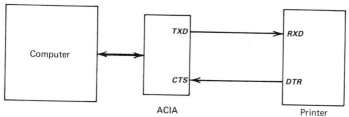

*Figure 14.16   Connections to RS-232 printer.*

the TBSIA stops transmission when it goes high. We could also connect the **DTR** from the printer to the **DSR** input of the TBSIA and let the computer check the **DSR** flag to see if the printer is busy, but it seems simpler to use the **CTS** input to stop transmission automatically when the printer is busy. Then the computer need only check **TDF** to see that **TDR** is available to receive a character. The connections are shown in Fig. 14.16.

There is one complication to the solution depicted in Fig. 14.16, however. The **TDF** flag sets as soon as a word is transferred from **TDR** to **TSR**. As long as transmission is enabled (**CTS** active), a word is transferred from **TDR** to **TSR** as soon as **TSR** is emptied. If **CTS** goes inactive while a character is being transmitted, transmission of that character will be completed. At 300 baud, it takes 33 msec to transmit a 10-bit character. Assume a carriage return has just been transferred from **TDR** to **TSR** for transmission. Long before 33 msec is up, the computer may check **TDF**, find it set, and send the first character of the next line to TDR. When the printer recognizes the carriage return, it will raise **DTR** (the BUSY signal) to stop transmission. Sounds reasonable, but suppose it takes the printer longer to recognize the CR and raise **DTR** than it takes the TBSIA to recognize the empty **TSR** and move the next character in from **TDR**? In that event the TBSIA will go ahead and transmit that next character, which will be lost since the printer is busy. It is a question of relative speed; and the information as to how long such peripheral operations as carriage returns require may not be readily available. Since the computer is always the "smartest" component in the system, it is often called upon to compensate when control signals from external equipments are insufficient. Example 14.3, which develops a subroutine to drive the printer of Fig. 14.16, is an illustration.

## Example 14.3

Write a TB6502 subroutine that will send text to a printer connected as shown in Fig. 14.16. The routine must assure that no characters in the text are lost. Assume that the only characters to be printed and carriage returns will be sent to the printer. The printer can accept characters at the rate of 300 baud and will not check parity.

Solution:

To avoid possible problems with the carriage return, the subroutine will wait 40 msec after sending this character to allow the printer ample time to recognize it and raise **DTR** to stop transmission of further characters.

```
*   SUBROUTINE PRINT
*     SUBROUTINE TO SEND ONE CHARACTER TO
*     AN RS-232 PRINTER
*       UPON ENTRY:
*         7-BIT CHARACTER TO BE PRINTED IN AC
*
PRINT     STA   TEMP        /SAVE CHARACTER
          LDA   #$10
TDCK      BIT   MCONT       /CHECK TDF
          BEQ   TDCK        /WAIT FOR TDR EMPTY
          LDA   TEMP        /RETRIEVE CHARACTER
          STA   MDATA       /SEND TO TDR
          CMP   #$0D        /CARRIAGE RETURN?
          BNE   END         /NO, RETURN
          JSR   DELAY       /YES, WAIT 40 MSEC.
END       RTS               /DONE, RETURN
```

We assume the TBSIA is interface 4 of Fig. 13.10, as we did for the terminal subroutine of Section 14.3, and assume it has been properly initialized. Since the printer operates at 300 baud, seven bits, one stop bit, no parity, the proper word to be loaded into **AMR** would be 00101000, and **ACR** should be loaded with 00000011 to enable the TBSIA. With this done, each time the main routine wants to print a character it loads the seven bit ASCII character into **AC** and calls the print subroutine given in this example. The subroutine saves the character and then waits for **TDF** to set, indicating that **TDR** is empty. When that occurs, the character is sent to **TDR** and the routine checks to see if it is a carriage return. If not, it returns; if so, it delays 40 msec before returning to allow the printer time to respond to the carriage return. The routine is almost trivial, because the TBSIA does most of the work automatically.                                                                              ■

## 14.5   Computer Networks

From the early days of computers up through about 1980, organizations that used computers typically had a central computer that was available for batch processing of programs or was accessible in a *time-sharing* mode from noncomputing terminals. Except for industrial process control and research, computers were not dedicated to single tasks or single users. Since 1980 the state of affairs has been changing very rapidly. Computers have become sufficiently inexpensive that almost anyone with a need can have a computer for his or her use alone or for the use of only a small group of co-workers.

Computer equipment has not and will never become so cheap that all potential users will have control of enough computing power and enough I/O equipment to meet their most sophisticated needs. A way around this dilemma lies in providing communications channels between computers, so that users can do routine tasks on their own computers and send data to and receive results from bigger machines as the need arises. Once a communications network among a set of computers is in place, the ability to exchange messages between pairs of these machines may be exploited in a great variety of applications.

Communications networks may link computers located in different parts of the world or may interconnect only computers at a single site, such as a university campus or industrial plant. Although the goals are the same in both cases, the technology employed may be different for the two cases. Networks of the latter type are called *local area networks* or LANs. In local area networks it is both desirable and possible to pass messages of the order of complete program listings between computers in relatively short times. Data rates of up to 10 megabits per second must be used. This is in contrast to the typical data rate of 300 baud and the maximum of 19,200 baud for the RS-232 standard discussed in the last four sections. Clearly, a new standard for configuring messages is in order.

There are a variety of potential organizations for computer networks. These include the star (Fig. 14.17a), the loop (Fig. 14.17b), the bus (Fig. 14.17c), and combinations of the three. For any network configuration a set of hardware and software conventions called *protocols* must be developed to provide for the exchange of messages on that network. Hardware designers and system programmers must then conform to these protocols as they design computers and operating system programs to function on a network set up according to a particular standard. A system of protocols may consist of up to seven layers. The lowest two layers are the hardware protocol, the provinces of the circuit designer and logic designer. The highest layer of protocol is the network operating instructions, which must be followed by the most casual user of the network. In between are protocols of interest to programmers of various levels, including the designer of the operating system. An IEEE standard 802, which will specify LAN protocols for both loop and bus networks, is under development at the time of this writing.

The smallest message unit in a computer network is usually referred to as a *frame*. A frame is a serial string of binary bits that includes both control and infor-

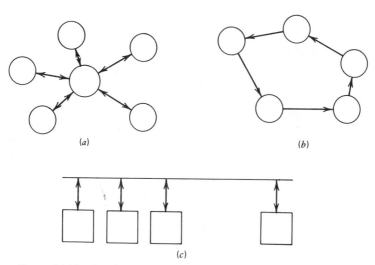

Figure 14.17   Local area network (LAN) configurations: (a) Star, (b) loop, (c) bus.

| Flag | Address | Control | Information | FCS | Flag |
|---|---|---|---|---|---|
| 01111110 | One or more octets | One or two octets | Zero or more bits | Two or four octets | 01111110 |

$(a)$

**Idle link**

1111111111111111111111111111111111111111111111111111111111111111

$(b)$

**Active link**

0111111001111110011111100111111001111110011111100111111001111110

$(c)$

*Figure 14.18    Typical data link protocol.*

mation. A typical frame format is indicated in Fig. 14.18a. The definition of the bits in the various sections of this frame may be thought of as constituting the next to lowest level of protocol, or *data link protocol*. The onset of a message is signaled by an 8-bit flag that is easily distinguished from the all logical 1 condition of the inactive line as illustrated in Fig. 14.18b. The *start flag* is followed by an *address field* that will indicate the destination and possibly the source of the message. The *control field* provides for passing commands between computer nodes on the network and in some cases indicates the length of the information field. The *information block* is typically optional. If a message is only a command to facilitate further communications, no information block will be present. The coding of information in the information block is usually not specified in the data link protocol, but ASCII and EBSIDIC are often used. The *frame check sequence*, FCS, block is provided for error control. Errors may arise in many ways, including lack of synchronization, when transmitting information on a single wire at 10 megabits per second, so some means of detecting and correcting errors is essential.

Mathematically elaborate schemes have been developed for generating error check bits for the FCS block as a function of the bits in the address, control, and information blocks. Errors may be corrected or retransmission requested if the FCS bits regenerated by the receiving system do not match the transmitted FCS bits.

Every frame is terminated by a *flag*. If a given computer node has control of the link but is not currently transmitting a message, it may transmit a sequence of flags as shown in Fig. 14.18c to distingish the condition of the link from idle. In the bus-configured LAN of Fig. 14.17 there will always be potential problems when two computer nodes attempt to simultaneously or almost simultaneously take control of the bus for transmission of messages. This state of affairs is detectable but will always result in wasted time and ultimately in retransmission of the messages in question. Proposed IEEE standard 802.3 provides for carrier sense multiple access/ collision detection (CSMA/CD) on bus networks. The frame configuration for this standard, which is similar to the already implemented ETHERNET, is similar to Fig. 14.18a except that an access control field is included immediately following the start flag.

The problem of simultaneous access collisions is avoided in loop or ring networks by a scheme called *token passing*. Here a single token message continuously circulates around the loop. When a particular computer node senses the start of transmission flag on the link from the predecessor node, it receives the message. If this node is not the addressee of the message, it simply retransmits it to the next node in the loop. Eventually the message will reach its destination. A message token will always circulate even though no node may be currently interested in transmitting a message. A node desiring to transmit must wait until it receives the token without a message. It then adds an information block and alters the control octet to indicate the presence of a message and passes the frame to the next node. The proposed IEEE standard for token-passing loops is 802.5. The notion of token passing may also be used to avoid collisions in bus networks for which the proposed standard is 802.4.

## Problems

**14.1**  Shown in Fig. P14.1 are the address connections to the memory and I/O chips in a certain system. There are two ROM chips, one RAM chip, a parallel interface unit, and TBSIA serial interface. Assuming all unused bits in any address are 0, determine the addresses of the TBSIA registers.

Now write an initialization routine to initialize the TBSIA for 300 baud, seven data bits, one parity bit, odd parity, and one stop bit; to turn on *RTS*; and to enable interrupt on *TDR* empty. Assume you do not know whether or not the unit has just been reset.

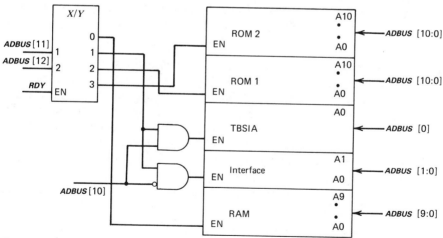

*Figure P14.1*

**14.2**  A small single-board computer can connect to the outside world only via RS-232. The computer includes a TBSIA as interface 4 in Fig. 13.7. We wish to connect this computer to a terminal consisting of a CRT display and an

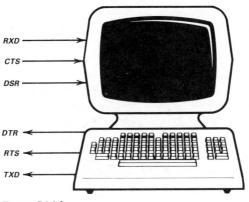

*Figure P14.2*

ASCII keyboard. This terminal is also an RS-232 device, set up to communicate with seven bits, one stop bit, no parity, at 1200 baud. It has the standard signal lines for a DTE, as shown in Fig. P14.2. Assume this is to be connected directly to the computer, without intervening modems or telephone lines. First, show the connections to the TBSIA on the computer. Then write two driver routines, KYBD32 and CRT32. KYBD32 should be so designed that it can directly replace the routine KYBD in any program requiring input from a keyboard. CRT32 should accept a character in the accumulator and send it to the CRT for display. Include the routine to initialize the TBSIA. Then write a program to continuously monitor the terminal for a character from the keyboard. When a character is received, it should be echoed to the CRT for display. The program should insert a carriage return if the operator types more than 72 characters without a carriage return.

**14.3** Some very simple printers, for example, teletypewriters, receive one character at a time and print it immediately. Assume such a printer with an input data line, *datain*, receives serial data at 110 baud, eight data bits without parity, and two stop bits. The only control line is an active low line labeled *busy*. The *busy* line will be inactive when the printer is ready to receive a character, will go active when a start bit is received, and will remain active until the character has been printed. Write a TB6502 program to drive this printer via a TBSIA. Specify the connections of the lines, *datain* and *busy* (which bit in *ASR*?), to the TBSIA; and write an appropriate sequence of program steps to accomplish initialization.

**14.4** The serial receiver in the RS-232 connection shown in Fig. P14.4 is driven by a 10-MHz clock and is described exactly by the RTL description of Fig. 14.4. No parity check takes place. The serial receiver is capable of receiving data continuously at the prescribed baud rate. The two address locations assigned to the TBSIA are 0FFE and 0FFF. The CPU must be interrupted if TDR is empty.

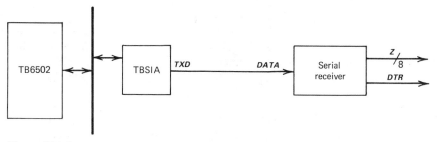

*Figure P14.4*

(a) What addresses are assigned to **AMR**, **ASR**, **ACR**, **RDR**, and **TDR**?

(b) An initialization routine must be written to load **AMR** and **ACR**. List in binary the bytes that must be loaded into these two registers. If a bit does not matter, indicate it as 0.

(c) Write a short sequence of instructions that will accomplish this initialization. Assume that it is not known whether or not **AMR** or **ACR** have been addressed since the last reset.

**14.5**  Assume that TBSIA in Fig. P14.4 has been initialized as specified in Problem 14.4. Write a short TB6502 routine to transfer a block of data to the serial receiver. Let the first data byte in the block be located at 04C0 and the last byte at 04FF.

**14.6**  As discussed in Section 14.4., printers with line buffers will normally have a busy signal that will go active while a line is being printed. When the printer is connected directly to the computer, as in Fig. 14.16, this busy signal can be used as a **CTS** signal to tell the computer to stop sending characters. However, if the printer is connected to the computer over a telephone line, as in Fig.14.5, this simple approach will not work. Control signals sent by the printer through its serial adapter to its modem will not get any information back to the computer. For that purpose, the communication channel must be full-duplex, and the printer must have enough "intelligence" to generate appropriate messages for the computer. Many printers use a microprocessor to control printer operation, and it might be possible to use this same processor to control interactions with the computer.

Let us assume we have a printer with a TB6502, enough RAM to serve as the line buffer, and a TBSIA to handle communication with the computer. The computer will receive characters and store them in the line buffer. It will print a line whenever it receives a carriage return character, or when there are 72 characters in the line buffer. When it prints, it will notify the computer to stop sending characters by sending a control-S character (13 hex) to the computer. When it is ready to start receiving characters again, it will send a control-Q character (11 hex) to the computer. Since the control-S character is sent serially, there will be a delay before the computer can respond, and some additional characters may be sent.

Include provision for receiving one or two more characters after sending a control-S. These should be held in a temporary buffer and moved to the line buffer after printing is complete. Write a program to accomplish these functions, including the dumping of the line buffer to the printer. Assume a character is printing by loading it in *AC* and calling subroutine PRINT. Assume that the TBSIA has been properly initialized.

**14.7** Write a subroutine SEROUT to enable the TB6502 with a TBPIA to carry out the functions of the TBSIA in the transmit mode only. Assume a fixed format of 1200 baud, seven data bits plus an odd parity bit, and one stop bit. It should implement the *TXD*, *RTS*, and *CTS* lines, using whichever TBPIA lines you think appropriate. Each time the subroutine is called, the word should be transmitted in serial, assuming *RTS* and *CTS* are in the appropriate condition.

**14.8** Repeat Problem 14.7 to implement the receiving function of the TBSIA. Implement and assume that detection of a start bit will cause an interrupt that will jump to subroutine SERIN to receive the character. Detection of a parity error should set the *C* flag.

# References

1. Andrews, Michael, *Programming Microprocessor Interfaces for Control and Instrumentation,* Prentice-Hall, Englewood Cliffs, N.J., 1982.
2. Artwick, B. A., *Microcomputer Interfacing,* Prentice-Hall, Englewood Cliffs, N.J., 1980.
3. McNamara, J. E., *Technical Aspects of Data Communications,* Digital Press, Maynard, Mass., 1977.
4. Parker, R., and S. Shapiro "Untangling Local Area Networks," *Computer Design,* March 1983, pp. 159–172.
5. Sargent, M., III, and R. L. Shoemaker, *Interfacing Micocomputers to the Real World,* Addison-Wesley, Reading, Mass., 1981.
6. Stone, H. S., *Microcomputer Interfacing,* Addison-Wesley, Reading, Mass., 1982.
7. Tanenbaum, A. S., *Computer Networks,* Prentice-Hall, Englewood Cliffs, N.J., 1981.
8. Weissberger, A. J., "Bit-Oriented Data Link Controls" *Computer Design,* March 1983, pp. 195–206.

# Additional Programming Topics

## 15.1 Pointer Addressing and Stacks

Another addressing mode found in many computers is *pointer addressing,* or *register indirect addressing.* This functions in the same general manner as indirect addressing except that the address of the data is in a register instead of a memory location. This mode offers two possible advantages relative to regular indirect addressing. First, if the hardware is such that register references are faster than memory references, there is a speed advantage. Second, the instructions will be shorter.

If there is only one register used as a pointer, its use will be implied in the opcode, so there will be no address. If there are several registers, only a very short address—two or three bits—will be needed. Compared to other addressing modes, it has the same advantages and disadvantages as regular indirect addressing. It is very useful for handling arrays, but it imposes programming overhead for initializing and maintaining the pointers. Because of this, it is very inconvenient for accessing isolated data elements; the pointer first has to be loaded with the address. Some early microprocessors used only this mode for accessing memory, making them rather awkward to program. However, modern microprocessors that include this mode also include all the other modes we have discussed, so this just adds another dimension of programming flexibility.

A special form of pointer addressing that is included in virtually all microprocessors is *stack addressing.* The term *stack* refers to a type of data structure in which data elements are loaded into successive memory locations one at a time and then removed in the opposite order. Such structures are known as *last-in, first-out (LIFO)* stacks. (There is another type of stack, the FIFO stack (first-in, first-out), which is described in problem 15.17, but the term stack normally refers to the LIFO form unless otherwise indicated.) Stacks can be implemented in software using the addressing modes already discussed; this approach will be treated later in the chapter. However, most microprocessors incorporate a hardware stack, using a *stack pointer*. The operation of such a stack is shown in Fig. 15.1. The stack itself is simply a set of locations in memory; it is nothing physically separate from the rest of memory. The stack pointer (**SP**) is a register that is initialized to the address of the first location in the stack, which we shall assume to be 01FF (Fig. 15.1a).

Assume there is a data element XX in the accumulator that we wish to put on the stack. This is done by executing a *push accumulator (PHA)* command, which moves the element from the accumulator to the location pointed at by the stack pointer (01FF) and then decrements **SP**. The situation after pushing the first element into the stack is shown in Fig. 15.1b. Now assume there is another element, YY, to be put on the stack. Again it is copied into the location pointed at by **SP** (01FE) and **SP** is then decremented (Fig. 15.1c). Thus, the stack pointer always points at the first empty location on the stack.

Next assume we are in the situation shown in Fig. 15.1d, with some arbitrary data in the accumulator, and we wish to read the top item off the stack to the accumulator. For this purpose we execute a *pull accumulator* command (PLA), which first increments **SP** to point at the top element on the stack (at 01FE) and then reads that element to **AC** (Fig. 15.1e). Since memory reads are nondestructive, this read does not actually "empty" that location, but we show it empty since the location pointed to by **SP** between operations is considered to be the first empty location on

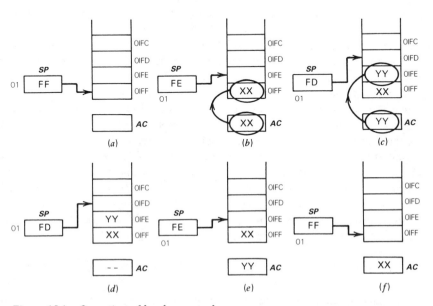

*Figure 15.1  Operation of hardware stack.*

the stack. Another PLA command will repeat this operation, incrementing *SP* and then reading the top element off the stack to the accumulator (Fig. 15.1*f*).

Stacks can also be implemented by incrementing on push and decrementing on pull; this just reverses the direction in which the stack moves through memory. Another variation is to have the stack pointer pointing at the top word on the stack between instructions. In that event *SP* will be decremented (or incremented) *before* writing for PHA and incremented (or decremented) *after* reading for PLA. Although designers debate the matter, there is not really much practical difference between these approaches as far as the programmer is concerned. One advantage of having *SP* point at the top word on the stack is that it is then possible to have a command to read the top word on the stack without pulling it.

For the TB6502 we shall assume the approach shown in Fig. 15.1. In the TB6502 the *SP* register will be eight bits and the upper byte of the 16-bit address sent to memory will be set by the hardware to 01 for PHA and PLA. This means that the stack in the TB6502 is restricted to page 1, and the maximum depth of the stack is 256 words, from 01FF to 0100. Some would consider that a major disadvantage, and the stack pointers in some processors are full length, permitting the programmer to put the stack wherever desired. Again, this is a debatable matter. In microprocessors, stacks are primarily used for handling subroutines and interrupts, and 256 words will usually be ample for this purpose. Also, it will be easier for the programmer to avoid inadvertent overwriting of the stack area if it is always in the same place. Finally, note that the PLA command is often referred to as the POP command, and programmers will talk of "popping" the stack.

We have mentioned initialization of the stack pointer; and sometimes it is desirable to save the setting of the stack pointer for later use. Thus, we need commands to read and write *SP*. There are two such commands in the TB6502, TSX (transfer *SP* to *X*) and TXS (transfer *X* to *SP*). It may seem strange that there are so few commands for manipulating the stack pointer, but this is a common characteristic of computers. The stack pointer is used in so many ways by the computer itself, for example, in the JSR command, that great care must be exercised in working with it. As a result, designers have often made it inconvenient to manipulate the stack pointer to serve as a "gentle reminder" to the programmer that the stack pointer should be handled with care.

Two other commands are associated with the stack. Recall that we have defined four flags, *C*, *Z*, *N*, and *V*, which are set as a results of various operations. In addition there is a *D* flag, enabling decimal arithmetic, which is set or cleared by the SED and CLD commands discussed in Chapter 12. There is also an *I* flag, associated with the interrput system, to be discussed in a later section. Although these flags are logically independent, for programming convenience they are considered to make up six of the eight bits (the other two bits are not used) of the processor status register (*PS*). The assignment of these bits in *PS* is shown in Fig. 15.2.

| *N* | *V* | | | *D* | *I* | *Z* | *C* |
|---|---|---|---|---|---|---|---|
| 7 | 6 | 5 | 4 | 3 | 2 | 1 | 0 |

*Figure 15.2  Processor status register (PS).*

| Command | Mnemonic | Operation |
|---------|----------|-----------|
| Push accumulator on stack | PHA | $M\langle SP \rangle \leftarrow AC$;<br>$SP \leftarrow SP - 1$ |
| Pull stack to accumulator | PLA | $SP \leftarrow SP + 1$;<br>$AC \leftarrow M\langle SP \rangle$ |
| Push processor status on stack | PHP | $M\langle SP \rangle \leftarrow PS$;<br>$SP \leftarrow SP - 1$ |
| Pull stack to processor status | PLP | $SP \leftarrow SP + 1$;<br>$PS \leftarrow M\langle SP \rangle$ |
| Transfer *SP* to *X* | TSX | $X \leftarrow SP$ |
| Transfer *X* to *SP* | TXS | $SP \leftarrow X$ |

*Figure 15.3   Stack commands in the TB6502.*

When going to subroutines or dealing with interrupts, it is often desirable to preserve the status of the main program. For example, suppose the main program is in decmal mode when a subroutine is called, but the subroutine is in binary mode. When the main program is resumed, it should be back in the decimal mode. For this purpose, the TB6502 has the PHP (push processor status) and PLP (pull processor status) commands, which make it possible to save the processor status on the stack and then retrieve it. The various stack commands are summarized in Fig. 15.3.

These six commands are the only commands that deal explicitly with the stack or stack pointer, but there are several commands that use the stack in execution. As mentioned earlier, the stack is used in subroutine handling and in interrupt procedures. Both JSR and RTS use the stack, as does RTI (return from interrupt). Details of these operations will be presented in subsequent sections, but it is important to realize that the stack is used in many ways and it is important for the programmer to protect the stack area from accidental misuse.

Since the stack in the TB6502 is restricted to page 1, we recommend that programmers never use that page in the TB6502 programs, even though 256 words of stack will rarely be needed. If this restriction is followed, there is really no need to initialize the stack pointer because the stack is circular in page 1. Since the 01 byte of the address is automatically generated in hardware, a decrement from 0100 on PHA will wrap-around to 01FF, and an increment on PLA will go from 01FF to 0100. Thus, there really is no "first" location in the stack area. If the system implements only a small portion of the address space, say, 1K or less, it may not be possible to set aside 256 locations for the stack. If the stack is to be used only for simple subroutine and interrupt procedures, 30 to 40 words of stack will be ample. In that event, SP should be initialized to FF with the commands:

```
LDX #$FF
TXS
```

Locations 0100 to 01CF could then be used for general data storage, leaving 48 words for the stack.

Wherever the stack is located, and whatever its extent, it is important that the programmer avoid using addresses in that area. Even though a program may never explicitly address the stack area, two common mistakes can result in inadvertent use of the stack area. One is erroneous displacements in branch statements. Suppose your program starts at 0200 and you have a branch back to the start of the program. A mistake in the displacement could branch you into the stack. If you are using an assembler, which will compute the displacements for you, this will not happen. Another common error, which the assembler will not catch, is going too far when stepping through an array. Suppose you have an array of 10 elements from 0200 to 0209 and are decrementing through this array. A mistake in setting the loop parameters could result in stepping right into the stack. The best way to prevent this is to increment or decrement away from the stack. If the array is above the stack, increment; if it is below the stack, decrement.

## 15.2  Subroutines

In our initial discussion of subroutines in Chapter 12, we observed that JSR stores the return address, that is, the contents of *PC* after fetching the JSR instruction, in a reserved location, and the RTS instruction retrieves this address and puts it back in *PC*. We noted at that time that the reserved location was the stack, but did not discuss the matter further, since we had not yet discussed the operation of the stack. JSR is executed, the computer pushes the contents of *PC* onto the stack. When RTS is executed, the computer pulls the return address off the stack and puts it in *PC*. This is done automatically, no PHA or PLA instructions are needed. The advantage of using a stack for storing return addresses is that it permits nesting of subroutines, that is, subroutines can call other subroutines. Consider the situation shown in Fig. 15.4 in which the main program executes a JSR to SUB1, which in turn calls SUB2.

The status of the stack and PC at various points in this subroutine calling process is shown in Fig. 15.5. Starting at Fig. 15.5*a,* we see the situation just before fetching JSR SUB1 and at Fig. 15.5*b* we see the situation just after executing JSR SUB1. The *PC* is incremented to 0223 during the fetch of JSR, this is pushed onto

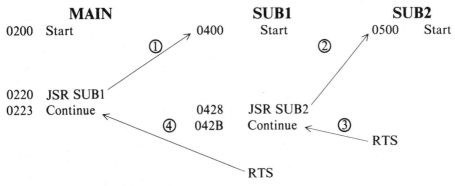

*Figure 15.4   Nested subroutines.*

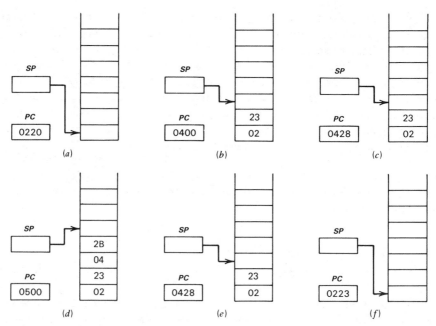

*Figure 15.5   Stack status during calling of nested subroutines.*

the stack, high byte first, and the address of SUB1 is moved to **PC**. At Figs. 15.5*c* and *d* we see the situation immediately before and after executing JSR SUB2. We now have both return addresses stacked. At Fig. 15.5*e* we see the situation after executing the RTS from SUB2, during which the return address to SUB1, 042B, is retrieved from the stack and sent to **PC**. Finally, at Fig. 15.5*f,* we see the situation after executing the RTS from SUB1.

It can be seen that this nesting can be carried to any number of levels, subject only to the capacity of the stack. With a 256-word stack, subroutines could be nested 128 deep, which is hardly practical. Even though the computer can keep track of return addresses to an arbitrary depth, it is very difficult for programmers to keep track of the interactions between subroutines when nested to a depth of more than three or four. Also, all this will work in the neat orderly fashion shown here only if the return addresses on the stack do not get messed up. This means that programmers must be careful when using the stack in subroutines. They must be careful not to pull return addresses off the stacks and must do an equal number of pushes and pulls to be sure the stack pointer is pointing to the right address when RTS is executed. Many programmers feel it is best not to use the stack during subroutines, except as appropriate in communications between main program and subroutine.

So, the stack can be used for an orderly implementation of subroutine linkage. What about the passing of *parameters?* Can the stack be used to simplify calling sequences? It turns out that is not a simple question. Before investigating the use of the stack, we need to look at the procedures presented in Chapter 12 in more depth. In Chapter 12 we saw two methods of passing parameters. One was to put them in registers, such as **AC** or **X**. This is the simplest method, but it is limited to a small number of parameters. The second method is to put the parameters into a *parameter area,* a reserved segment of memory, usually on page-0. We must make two important decisions in using parameter areas. The first is whether the parameters are data, or addresses. We saw examples of both in Chapter 12. One obvious factor in this choice is the amount of data to be passed. If there are only a few variables, it is probably simplest to pass their values. For many variables, passing the addresses is usually simpler. Other rather subtle factors are involved in the second decision, whether the parameter area is to be loaded by the main program or the subroutine.

To gain an understanding of some of these factors, let us consider some examples. First, consider the subroutine CMPR from Section 12.5. In this case the parameters are data, the numbers to be compared, and the main program loads the numbers to be compared into a parameter area at locations 0060 and 0061, and the subroutine always works with those locations. As an alternative approach, assume that at different times we may have two words in consecutive locations anywhere on page-0, and we would like to compare them without having to move them to a specific parameter area. Assume that at a particular time the two words are located in 0082 and 0083. The calling sequence would be

```
        .
        .
        .
LDX  #$82
JSR  CMPRA
        .
        .
```

The modified compare routine would be as shown here.

```
*   SUBROUTINE CMPRA
*    COMPARES TWO POSITIVE INTEGERS
*     UPON ENTRY:
*       TWO INTEGERS IN CONSECUTIVE LOCATIONS
*       ON PAGE-0, PAGE ADDRESS OF FIRST IN X
*     UPON EXIT:
*       LARGER INTEGER IN AC
*
CMPRA   LDA  0,X      /GET NUMBER
        CMP  1,X      /COMPARE SECOND NUMBER
        BCS  DONE     /BRANCH IF 1ST NUMBER
*                     /  LARGER
        LDA  1,X      /SECOND NUMBER LARGER
DONE    RTS
```

Here there is no parameter area in the usual sense. Instead, the main routine just loads the index register to enable the subroutine to access the data directly. Since the calling sequence sets **X** = 82, any reference to 0,X refers to 0082, and a reference to 1,X refers to 0083. Next, let us consider the example of the BCADD subroutine in Chapter 12. Here we definitely have a parameter area, at ltcations 00E0 to 00E6, and this area is loaded by the calling sequence of the main program. At different calls of the subroutine, the data strings may be in different places, but the parameters, the addresses of the arrays, are always passed through this same area. The heading comments for that subroutine are repeated here.

```
*    SUBROUTINE TO ADD TWO STRINGS OF BCD DIGITS
*    UPON ENTRY
*      00E0 = CNTR, SET TO NUMBER OF DIGITS
*      00E1,00E2 = APTR, POINTER TO FIRST STRING
*      00E3,00E4 = BPTR, POINTER TO SECOND STRING
*      00E5,00E6 = SUMPTR, POINTER TO SUM STRING
```

To use this routine, the calling sequence must load all these locations. A typical calling sequence in the main program, assuming the operand strings are at 0300 and 0360 and the result string at 03C0, is also repeated for convenience.

```
0307   LDA   CNT      /GET DIGIT COUNT
030A   STA   *E0      /LOAD DIGIT COUNT
030C   LDA   #$00     /LOAD PAGE ADDRESS OF
030E   STA   *E1      /FIRST OPERAND
0310   LDA   #$60     /LOAD PAGE ADDRESS OF
0312   STA   *E3      /SECOND OPERAND
0314   LDA   #$C0     /LOAD PAGE ADDRESS
0316   STA   *E5      /OF SUM
0318   LDA   #$03     /
031A   STA   *E2      /LOAD PAGE NUMBERS
031C   STA   *E4      /OF ALL POINTERS
031E   STA   *E6      /
0320   JSR   0200     /JUMP TO SUBROUTINE
0323   xxx   xxxx     /NEXT INSTRUCTION
```

This is conceptually simple enough, but it requires the insertion of 12 instructions into the main program every time the subroutine is to be called, and it requires the user of the subroutine to know all these addresses. Another approach is to let the subroutine do the work involved in getting the parameters to the parameter area. Assume that as a result of prior operations in the main program, such as input conversion from ASCII to BCD, we have the number of digits in 0092, the pointer to one input string in 0093 and 0094, the pointer to the other input string in 0095 and 0096, and the pointer to the sum string in 0097 and 0098. The calling sequence will consist of

```
LDX  #$92
JSR  BCDAD1
```

We then add steps at the beginning of the BCADD subroutine to move the parameters into the working area of the subroutine.

```
*    SUBROUTINE BCDAD1
*      ADDS TWO STRINGS OF BCD DIGITS
*      UPON ENTRY:
*        NO. OF DIGITS, POINTERS TO  OPERAND
*        STRINGS, AND POINTER TO RESULT STRING
*        IN 7 CONSECUTIVE LOCATIONS ON PAGE-0,
*        PAGE ADDRESS OF NO. OF DIGITS IN X.
*        CNTR, OP1PTR, OP2PTR, SUMPTR ARE
*        WORKING COUNTERS AND POINTERS
*        AT 00E0 TO 00E6.
*
BCDAD1       LDA   0,X          /GET DIGIT COUNT
             STA   *CNTR        /LOAD COUNTER
             LDA   1,X          /
             STA   OP1PTR       /
             LDA   2,X          /
             STA   OP1PTR+1     /
             LDA   3,X          /
             STA   OP2PTR       /COPY POINTERS
             LDA   4,X          /TO WORKING
             STA   OP2PTR+1     /LOCATIONS
             LDA   5,X          /
             STA   SUMPTR       /
             LDA   6,X          /
             STA   SUMPTR+1     /
             LDX   #0           /CLEAR X FOR
*                               /INDEXING ARRAYS
             CLC
             SED                /SET DECIMAL MODE
LOOP         as before
             .
             .
             .
```

This procedure certainly makes the calling of the subroutine much simpler. The subroutine is more complicated, but the subroutine only has to be written once, while the main program may call the subroutine many times. The list of parameters does have to be established in the main program, and there would clearly be no advantage if that had to be done every time the subroutine was called.

However, as we assumed here, a list of parameters may be established for other purposes in the main program, in which event this process of letting the subroutine transfer the parameters is clearly preferable. Another situation is one in which there are several sets of arrays, each with its own set of parameters on page 0, as shown in Fig. 15.6. If BCDAD1 was to be used for the first set of arrays, *X* would be set to $92 before JSR. If BCDAD1 was to be used for the second set of arrays, *X* would be set to $99, and so forth.

A variant on this method is the use of an *in-line parameter area,* in which the parameters are stored in the body of the main program, immediately following the JSR command. Consider again that the BCD add subroutine is to be called, with the operand string at 0300 and 0360, the result string at 03C0, 36 digits (24 hex) in the strings. The calling sequence in the main program would appear as follows.

|          |          | 0092 | 08 |
|          |          | 0093 | 00 |
| LDX | #$92 | 0094 | 0E |
| JSR | BCDAD1 | 0095 | 04 |
|          |          | 0096 | 0E |
|          |          | 0097 | 08 |
|          |          | 0098 | 0E |
|          |          | 0099 | 07 |
|          |          | 009A | 10 |
| LDX | #$99 | 009B | 1B |
| JSR | BCDAD1 | 009C | 14 |
|          |          | 009D | 1B |
|          |          | 009E | 18 |
|          |          | 009F | 1B |
|          |          | 00A0 | 1A |
|          |          | 00A1 | 10 |
| LDX | #$A0 | 00A2 | 1C |
| JSR | BCDAD1 | 00A3 | A2 |

*Figure 15.6   Page 0 parameter storage areas.*

The subroutine call comes at 0232. Immediately following this three-byte instruction, the parameters are loaded in-line with the program at 0235 to 0239, with the program proper continuing at 023A.

```
0232      JSR   BCDAD2      /JUMP TO BCD ADD SUBR.
0235            24          /NUMBER OF DIGITS
0236            00          /PAGE OFFSET, 1ST OPND
0237            60          /PAGE OFFSET, 2ND OPND
0238            C0          /PAGE OFFSET OF SUM
0239            03          /PAGE NMBR OF OPRNDS
023A      continue
```

The portion of the calling sequence in the subroutine is based on the fact that the JSR program pushes the address of the next location following that instruction onto the stack. Normally this address is the return address, but here it serves as a pointer to the parameter area. The portion of the subroutine for getting these parameters into the subroutine is shown in the following. We assume the subroutine proper will be unchanged from the version shown in Chapter 12. That is, it will work with locations 00E0–00E6.

The address pushed on the stack by JSR will be 0235, which is the address of the location containing the number of digits. The first four steps pull this address off the stack and load it into a pointer at 0050 and 0051. An indirect LDA through 50 then fetches the digit count, which is moved to 00E0, the working location. We then

```
*    SUBROUTINE BCDAD2
*      ADDS TWO STRINGS OF BCD DIGITS.
*      OPERAND STRINGS AND SUM STRINGS TO BE
*      ON SAME PAGE.   NO. OF DIGITS, PAGE
*      OFFSETS OF OPERAND STRINGS, PAGE OFFSET
*      OF RESULT STRING, AND PAGE NUMBER OF
*      STRINGS IN BODY OF MAIN PROGRAM IN
*      THAT ORDER, IMMEDIATELY FOLLOWING JSR.
*
BCDAD2   PLA                    /GET LOW BYTE OF
*                               /  RETURN ADDRESS
         STA  *50               /STORE AS POINTER
         PLA                    /GET HIGH BYTE OF
*                               /  RETURN ADDRESS
         STA  *51               /STORE AS POINTER
         LDA  (50)              /GET DIGIT COUNT
         STA  *E0               /STORE
         INC  *50               /INCREMENT POINTER
         LDA  (50)              /GET PAGE OFFSET OF
*                               /  1ST OPERAND
         STA  *E1               /STORE IN POINTER
         INC  *50
         LDA  (50)              /GET PAGE OFFSET OF
*                               /  2ND OPERAND
         STA  *E3               /STORE IN POINTER
         INC  *50
         LDA  (50)              /GET PG OFFSET OF SUM
         STA  *E5               /STORE IN POINTER
         INC  *50
         LDA  (50)              /GET PAGE NUMBER
         STA  *E2               /STORE
         STA  *E4               /  IN POINTER
         STA  *E6               /  HIGH BYTES
         INC  *50
         LDA  *51               /GET HIGH BYTE
*                               /  OF RETURN ADDRESS
         PHA                    /PUSH ON STACK
         LDA  *50               /GET LOW BYTE
*                               /  OF RETURN ADDRESS
         PHA                    /PUSH ON STACK
                   .
                   .
         body of subroutine
                   .
                   .
         RTS
```

increment 0050, to point of 0236, and we retrieve the page address of the first oper-
and and move it to 00E1, the first byte of the working pointer. Subsequent increments of 0050 and indirect loads fetch the remainder of the pointer bytes. The final
increment of 0050 brings it 40, the page address of the return point, and we then
push the updated return address onto the stack in preparation for the final RTS. The
remainder of the routine is unchanged from the first version in Chapter 12.

We saw that the stack was useful in organizing the linkage procedure, and we
used the stack to a minor degree for the calling sequence for the in-line parameter
area, for passing the address of the parameter area. Another approach is to use the
stack itself as the parameter area. Let us consider yet another variation of the BCD
addition routine.

```
*   SUBROUTINE BCDAD3
*   ADDS TWO STRINGS OF BCD DIGITS.
*    OPERAND STRINGS AND SUM STRINGS TO BE
*    ON SAME PAGE.  BEFORE CALLING PUSH NO.
*    OF DIGITS, PAGE OFFSETS OF OPERAND
*    STRINGS, PAGE OFFSET OF RESULT STRING,
*    AND PAGE NUMBER OF STRINGS ON STACK,
*    IN THAT ORDER.
```

So, all we are going to do is push the necessary information onto the stack. We need to know the order to put it on the stack, but we do not need to worry about any specific addresses. The calling sequence using the stack to pass the parameters is shown here.

```
0307    LDA    CNT      /GET DIGIT COUNT
030A    PHA             /PUSH ON STACK
030B    LDA    #$00     /PUSH PAGE ADDRESS OF
030D    PHA             /   FIRST OPERAND
030E    LDA    #$60     /PUSH PAGE ADDRESS OF
0310    PHA             /   SECOND OPERAND
0311    LDA    #$C0     /PUSH PAGE ADDRESS
0313    PHA             /   OF SUM
0314    LDA    #$03     /GET PAGE NUMBER
0316    PHA             /PUSH PAGE NUMBER
0317    JSR    BCDAD5   /JUMP TO SUBROUTINE
031A    xxx    xxxx     /NEXT INSTRUCTION
```

The situation on the stack after executing the JSR is shown in Fig. 15.7a, assuming the stack was initialized to 01FF. We can pull the parameters off the stack,

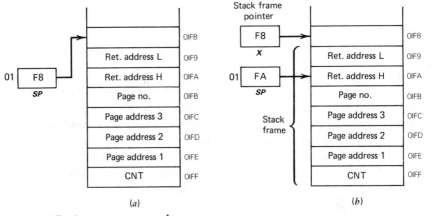

(a)                                                    (b)

*Figure 15.7   Parameters on stack.*

but we first have to do something about the return address, which is on top of the stack. One simple approach is to save the return address in a temporary location and then restore it before executing RTS. This approach is illustrated here.

```
*    SUBROUTINE BCDAD3
*    ADDS TWO STRINGS OF BCD DIGITS.
*     OPERAND STRINGS AND SUM STRINGS TO BE
*     ON SAME PAGE.  BEFORE CALLING PUSH NO.
*     OF DIGITS, PAGE OFFSETS OF OPERAND
*     STRINGS, PAGE OFFSET OF RESULT STRING,
*     AND PAGE NUMBER OF STRINGS ON STACK,
*     IN THAT ORDER.
*
BCDAD3    PLA                /SAVE LOW BYTE OF
          STA    RETL        / RETURN ADDRESS
          PLA                /SAVE HIGH BYTE OF
          STA    RETH        / RETURN ADDRESS
          PLA                /GET PAGE NO. OF
          STA    *E2         /  STRINGS, STORE
          STA    *E4         /   IN HIGH BYTES
          STA    *E6         /   OF POINTERS
          PLA                /PULL, LOAD PAGE
          STA    *E5         / OFFSET OF SUM
          PLA                /PULL, LOAD PAGE
          STA    *E3         / OFFSET, 1ST OPRND
          PLA                /PULL, LOAD PAGE
          STA    *E1         / OFFSET, 2ND OPRND
          PLA                /PULL, LOAD
          STA    *E0         / COUNT OF DIGITS
          .
          .
          .
     Body of subroutine
          .
          .
          .
          LDA    RETH        /PUSH HIGH BYTE
          PHA                / OF RET ADDRESS
          LDA    RETL        /PUSH LOW BYTE
          PHA                / OF RET ADDRESS
          RTS                /RETURN
```

This approach is certainly simple enough, but it does have the disadvantage of requiring that two locations in memory be set aside for saving the return address. We can avoid the use of extra memory locations by saving the setting of the stack at the beginning of the subroutine in the **X** register, which is used as a *stack frame pointer*. This procedure is shown in BCDAD4.

The situation immediately after execution of the subroutine jump is shown in Fig. 15.7a. The first step in BCDAD4 copies the stack pointer to **X**, which serves as the stack frame pointer. The indicated area in the stack, containing the return address and all parameters associated with the calling of the subroutine, is known as the *stack frame*. The function of the stack frame pointer is to preserve the address of this area while the stack pointer is used in the subroutine. The two "dummy" pulls

```
*    SUBROUTINE BCDAD4
*      ADDS TWO STRINGS OF BCD DIGITS.
*      OPERAND STRINGS AND SUM STRINGS TO BE
*      ON SAME PAGE.  BEFORE CALLING PUSH NO.
*      OF DIGITS, PAGE OFFSETS OF OPERAND
*      STRINGS, PAGE OFFSET OF RESULT STRING,
*      AND PAGE NUMBER OF STRINGS ON STACK,
*      IN THAT ORDER.
*
BCDAD4   TSX               /COPY SP TO X
         PLA               /DUMMY PULLS TO MOVE SP TO
         PLA               /    START OF PARAMETER AREA
         PLA               /GET PAGE NO. OF STRINGS
         STA   *E2         /    STORE IN
         STA   *E4         /    POINTERS TO OPERAND
         STA   *E6         /    AND RESULT STRINGS
         PLA               /PULL, LOAD PAGE
         STA   *E5         /    ADDRESS OF SUM STRING
         PLA               /PULL, LOAD PAGE
         STA   *E3         /    ADDRESS OF OPERAND STRING
         PLA               /PULL, LOAD PAGE
         STA   *E1         /    ADDRESS OF OTHER OPERAND
         PLA               /PULL, LOAD
         STA   *E0         /    COUNT OF DIGITS
         TXS               /RESTORE STACK POINTER TO INITIAL VALUE
          .
          .
          .
  Body of subroutine
          .

         RTS               /RETURN TO CALLING PROGRAM
```

move the stack pointer back two addresses, leaving the stack in the condition shown in Fig. 15.7*b*.

Note that, since memory reads are nondestructive, the return address is still intact at locations 01F9 and 01FA. The stack pointer is now in the right position to access the parameters, which is done in exactly the same manner as in BCDAD3. Following the copying of the parameters into the working area, the contents of the stack frame pointer are transferred back to the stack pointer, restoring the stack to the original condition shown in Fig. 15.7*a*.

We have now considered a variety of approaches to subroutine parameter passing and the hapless reader is no doubt hopelessly confused. Let us see if we can sort out the significant characteristics of these methods and draw some conclusions as to relative advantages and disadvantages. For convenience we shall number the methods, Method 1, Method 2, and so on, using letters A or D to indicate addresses or data are passed, for example, method *x*-A, method *x*-D.

**Method 1  Use of Registers**   Clearly this is the simplest method, but the number of parameters that can be handled is limited. There may or may not be a calling sequence, because the parameters may be in the registers for other purposes. Examples of Method 1-D are ABS in Section 12.5 and PRINT in Section 13.5.

**Method 2  Parameter Area Loaded by Main Program**   This is the simplest method next to Method 1. The subroutine uses a fixed parameter area, and the calling sequence in main program loads this area. The disadvantage is a complicated calling sequence in the main program every time the subroutine is called. CMPR in Section 12.5 is an example of Method 2-D. BCADD in Section 12.5 is an example of Method 2-A.

**Method 3  Parameter Area Loaded by Subroutine**   The main program passes only address of parameters, commonly in an index register. The advantage is a simpler calling sequence in the main program, but this advantage exists only if the parameter areas in main program can be used for several purposes, or if there are multiple parameter areas. If the main program has to load a parameter area every time the subroutine is called, this is essentially equivalent to Method 2. In this section, CMPRA is an example of Method 3-D, and BCAD1 is an example of Method 3-A.

**Method 4  In-Line Parameter Area**   The parameters are found in the body of the main program, immediately following the JSR instruction. The address of the parameter area is pushed on the stack by JSR. The parameters are usually fixed pointers, inserted when the program is written (or at compile time), so that no calling sequence in the main program is required. A complex calling sequence is needed in the subroutine to retrieve parameters. This method can be used only for fixed parameters, since the program is not subject to alteration during execution. An example is BCDAD2 in this section, which passes both data (digit count) and addresses.

**Method 5  Parameters on Stack**   The calling sequence in the main program pushes parameters on the stack prior to JSR. The calling sequence in the subroutine recovers parameters from the stack. The calling sequence in the main program is somewhat simpler than for Method 2, since no addresses are needed, but the procedure in the subroutine is complicated by the need to preserve the return address. Examples are BCDAD3 and BCDAD4 in this section.

It would be convenient if one of these methods were clearly superior to the others so we could concentrate on it and ignore the others, but that is not the case. Except in those cases where Method 1 can be used, most readers will find Method 2 conceptually the simplest. There are no tricky addressing methods needed, no problems in keeping track of the stack pointer. Where the same person is writing both the main and subs at the same time, and there are not too many subroutines, this is probably the best method. Where there are many subroutines, especially if written by more than one person, keeping track of a lot of parameter areas, particularly keeping them from interfering with each other, may be difficult. For such situations, one of the latter three methods will usually be preferable.

Methods 3, 4 and 5 may seem to have the disadvantage of complex sections of the calling sequence in the subroutine. However, it should be noted that the subroutine has to be written only once, whereas a calling sequence in the main program must be written every time the subroutine is used. The whole idea of a subroutine is that it will be used many times, so that it is preferable to make the calling process as seen from the main program as simple as possible, even at the expense of making the subroutine itself more complicated. From this point of view, Method 4 is attractive, since it is only necessary to insert the parameters into the program, but it is limited to cases where the parameters are fixed. It may seem that this is normally the case, since we have passed fixed parameters in most of the examples, but there is no reason we cannot pass variable parameters. For example, when we push the page address of the sum string on the stack for BCDAD3, instead of pushing C0 we could push SUMPTR, which might have different values on different executions of the main program. But with Method 4 this is not possible. For one thing, programs are frequently stored in ROM and cannot be altered. Even if the program is in RAM, making it theoretically possible to alter it, for example, to change the value of SUMPTR, it is a general principle of programming that a program should never alter itself.

So, we come to Method 5, for which the calling sequence in the main program is reasonable, but the calling sequence in the subroutine tends to be rather messy. This is probably the preferred method when writing "canned" routines for use by others. It is simple for the main program to use, requires a minimum of documentation, does not require reserving a lot of parameter areas for various subroutines, and is applicable to any computer that has a stack and that includes just about all of them. Stack processing of parameters is also advantageous when subroutines are nested. Without going into details, we can see that when several subroutines can call each other, each with its own parameter area, it will be very difficult to keep track of the different areas. If the parameters are all pushed on the stack, however, each subroutine can work off the top of the stack, without concern for how many sets of parameters are further down in the stack.

Stack handling of parameters is also virtually essential when subroutines must be *reentrant*. Reentrancy is an advanced concept that cannot be treated in depth in this book, but you should be aware of the problem. A subroutine is said to be *reentered* if is executing in response to one call, is interrupted before completion, and then is called by the interrupting routine. If the subroutine can continue correctly after servicing the interrupt and then returning to the original execution, it is said to be *reentrant*. If the subroutine works from parameters placed in fixed locations by the calling routine, the parameters passed by the original call will be wiped out by the subsequent call from the interrupt routine, making it impossible to resume the original execution. If the parameters are on the stack, however, the second call will just push them down in the stack, where they will still be available when the original execution is resumed. The handling of the stack for reentrant subroutines is so important that advanced microprocessors often include special stack commands to facilitate such procedures.

In Section 13.8 we considered a subroutine SCAN for controlling seven-segment display. We characterized this as a "background" routine, one that is called at regular intervals to update the display. It requires no parameters and should have no effect on the main program. The first thing we did in that program was to save *AC* and *X* in temporary locations so they could be returned to the main program unchanged. A simpler way to accomplish the same result is to push the registers to be saved onto the stack and then pull them off before returning to the main routine. The modified routine is shown here.

```
*   SUBROUTINE SCAN
*     UPON ENTRY:
*       DIGITS TO BE DISPLAYED STORED IN
*       SIX WORDS STARTING AT SCNBUF.
*       CURRENT DIGIT COUNT (IN RANGE 0-5)
*       IN DIGCNT. DIGCNT INITIALIZED TO
*       00 WHEN SCNBUF LOADED.
*
SCAN      PHA              /SAVE CURRENT
          TXA              / VALUES OF AC
          PHA              / AND X
           .
           .
  no change
           .
           .
SAVE      STX   DIGCNT     /SAVE DIGIT COUNT
          PLA              /RESTORE CONTENTS
          TAX              / OF X
          PLA              / AND AC
          RTS              /DONE
```

This may not seem to be any improvement. Because we cannot push or pull **X** directly, it requires more instructions than saving in temporary locations. However, it requires fewer bytes since all the instructions are one byte, and it does not require the programmer to keep track of any temporary locations. This last point is particularly important. If a subroutine requires any working locations, care must be taken to see that they do not overlap areas used by the main program. This can be done by reserving areas for the use of the subroutine, but, as discussed earlier, this can be a bother with many subroutines or canned subroutines. The big advantage of the stack is that this is a permanently reserved area available to be used as needed, in main or subs. The only precautions are that we must stay out of stack areas used by other programs and must restore the stack properly before executing RTS. The basic rule is to always do a push before a pull and do as many pushes as pulls. In this case we do two pushes onto the stack, and then two pulls from the stack, so the stack is in its original condition at RTS.

The problem of saving values for return to the main routine is not restricted to registers. Consider the use of BCDAD3, in which the parameters are passed on the stack. In the subroutine these parameters are moved to working locations 00E0 and 00EG and the return address is saved in memory. But this procedure requires that the main program reserve those locations for the use of the subroutine, which may be a nuisance. In BCDAD4, we eliminated the special areas for preserving the address through the use of the stack frame pointer. The stack frame pointer can also be useful in preserving the original values in the working locations on the stack, to be returned intact on return to the main program. Here again we can use the stack. Figure 15.8a shows the condition of the stack after executing JSR. Now we want to push the contents of the registers and 00E0 to 00E6 on the stack and then pull the parameters off the stack. We also assume that the contents of **AC** and **Y** are to be preserved. Obviously this is going to require some care to avoid losing parameters or the return address. The portion of the subroutine for handling this is shown here.

```
TSX                 /SET STACK FRAME PNTR
PHA                 /PUSH AC
TYA                 / AND Y
PHA                 / ON STACK
LDA    *E0          /
PHA                 /
LDA    *E1          /
PHA                 /
LDA    *E2          /   PUSH CONTENTS
PHA                 /
LDA    *E3          /   OF E0 TO E6
PHA                 /
LDA    *E4          /   ONTO STACK
PHA                 /
LDA    *E5          /
PHA                 /
LDA    *E6          /
PHA                 /
LDA    0103, X      /GET PAGE NUMBER
STA    *E2          /LOAD PAGE
STA    *E4          / NUMBERS OF
STA    *E6          / STRING POINTERS
LDA    0104, X      /GET PG OFFSET OF SUM
STA    *E5          /LOAD IN SUM POINTER
LDA    0105, X      /GET PG OFFSET, OPND2
STA    *E3          /LOAD IN POINTER
LDA    0106, X      /GET PG OFFSET, OPND1
STA    *E1          /LOAD IN POINTER
LDA    0107, X      /GET DIGIT COUNT
STA    *E0          /LOAD CNTR
       .
       .
no change
       .
       .
       .
PLA                 /RESTORE
STA    *E6          / E6
PLA                 /RESTORE
STA    *E5          / E5
       .
etc
       .
PLA
TAY                 /RESTORE Y
PLA                 /RESTORE AC
RTS
```

The first step establishes **X** as the stack frame pointer. The next three steps push **AC** and **Y** on the stack just as in the SCAN example just shown. The situation at this point is shown in Fig. 15.8b, with **X** pointing to the top of the stack frame. The next 14 commands simply push the contents of E0 and E6 onto the stack. The situation at the completion of this portion of the routine is shown in Fig. 15.8c. Note that **X** still preserves the location of the stack frame even though the stack pointer was incremented some arbitrary number of times. We now use indexing to retrieve the param-

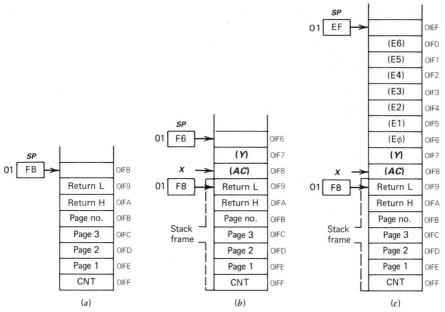

*Figure 15.8   Use of stack frame in subroutine.*

eters pushed on the stack by the calling sequence. This approach allows us to preserve the contents of **SP** for use in restoring the original values of E0 to E6, **AC**, and **Y** that have been pushed on the stack.

The next steps, starting with

**LDA 0103,X**

retrieve the parameters from the stack frame area, and move them to the working area, E0 and E6. Note that **Y**= F8. The parameter area starts three locations lower, at 01FB, which is the address generated by adding 0103 + F8. Similarly,

**LDA 0104,X**

will access 01FC, and so on.

When the subroutine is complete, the original values of E0 and E6 and **Y** and **AC** are restored, leaving **SP** in the proper position for the RTS. Study this procedure carefully. It is complex and possibly confusing, but it is very important. It is a standard method for working with the stack in subroutines, so every programmer should know it. Many newer processors include a separate stack frame pointer that can be used in push and pull operations, eliminating the need for indexing to access the stack frame and leaving the index register available for other purposes.

The final matter we want to mention in this section is the need for proper documentation of subroutines. We have already discussed this, but it cannot be emphasized too much. Clearly, we mst indicate what the subroutine does, where it expects to find the input parameters, and where it will return the output parameters. All the examples we have presented have included this information in the heading comments. If the subroutine may be used by others or with other main programs, the comments should also state how much memory the subroutine occupies and which registers and memory locations are used, but not preserved by the subroutine. Shown here is an appropriate set of comments for the original BCDADD routine, which works with a fixed parameter area.

```
*     SUBROUTINE BCDADD
*      ADDS TWO STRINGS OF BCD DIGITS.
*      LENGTH = 30 BYTES;
*      USES AC AND LOCATIONS 00E0 TO 00E6.
*      UPON ENTRY:
*        00E0 = CNTR, SET TO NUMBER OF DIGITS
*        00E1,00E2 = APTR, POINTER TO 1ST STRING
*        00E3,00E4 = BPTR, POINTER TO 2ND STRING
*        00E5,00E6 = SUMPTR, POINTER TO SUM
```

# 15.3  Interrupts

In our discussion of input/output in Chapter 13, we saw several examples of routines that called for periodically checking, or polling, some device, such as a keyboard, to determine if it is ready to send or receive data. We saw that polling of a slow device such as a keyboard could be done without interfering with other operations significantly, but that might not be the case with faster devices, or a large number of devices, requiring more frequent polling. The very nature of polling is such that, in order to avoid missing data, we usually have to poll much more frequently than we expect to find data. A typical polling operation might find data ready only 1 time out of 100, so that 99% of the polling time is wasted in some sense.

A better way to deal with situations where peripherals require service at irregular and unpredictable intervals is to provide a means whereby the peripheral can *interrupt* the processor to request service. That is, the processor can go along doing other things until a peripheral indicates that it has data, is ready to receive data, or otherwise needs to interact with the processor. At this point the processor "interrupts" what it is doing, deals with the peripheral, and then returns to its previous tasks.

The TB6502 processor will be provided with an active-low interrupt request line (*IRQ*). As long as this line is high, it will have no effect on the processor. Each time the TB6502 enters the fetch cycle it will check *IRQ*. If *IRQ* = 1, the TB6502 will proceed to fetch the next instruction; if *IRQ* = 0, the TB6502 will enter an interrupt cycle, provided that interrupts have not been disabled. It is clearly important that the programmer be able to control whether or not various devices will be allowed to interrupt. In Section 15.1 we mentioned the *I* bit, which is bit 5 of the processor status register. This is the interrupt disable flag. When this flag is set, interrupts are disabled and **IRQ** cannot affect the processor. When $I = 0$, the processor will be interrupted if **IRQ** goes low.

We can best explain what happens in an interrupt cycle in terms of the RTL sequence of the processor. First we insert a step 1A immediately ahead of the fetch cycle to check **IRQ** and the $I$ flag (**PS**[5]).

```
1A   → ((IRQ ∨ PS[2]), (IRQ ∧ PS[2]))/(1,100)
1    MA ← PC
```
continue with normal fetch cycle

If **IRQ** is high or interrupts are disabled, the sequence continues on to the normal fetch cycle. If **IRQ** is low and interrupts are enabled, the sequence branches to 100, which is the start of the interrupt cycle.

```
100   M⟨SP⟩ ← PC[15:8]; SP ← SP − 1
101   M⟨SP⟩ ← PC[7:0]; SP ← SP − 1
102   M⟨SP⟩ ← PS; SP ← SP − 1; PS[2] ← 1
103   PC[7:0] ← M(FFFE)
104   PC[15:8] ← M(FFFF)
      → (1)
```

The interrupt cycle is basically a jump to a subroutine. Steps 100 and 101 push the contents of **PC**––the address of the next instruction in the interrupted program, onto the stack; and 102 pushes **PS** onto the stack to preserve the status of the interrupted program. Step 102 also sets the interrupt disable flag to ensure that the processor will not be interrupted again until it has had a chance to deal with this interrupt. Steps 103 and 104 load the program counter with the contents of locations FFFE and FFFF and return to step 1. These locations must have previously been loaded with the address of a program or subroutine to deal with the interrupt. Thus, the return to step 1 with this address in **PC** will start execution of the interrupt program or subroutine. When the interrupt program is complete, there must be a way to get back to the interrupted program. This is provided by the RTI (return from interrupt) instruction, for which the steps are shown here.

```
110   SP ← SP + 1
111   PS ← M⟨SP⟩; SP ← SP + 1
112   PC[7:0] ← M⟨SP⟩; SP ← SP + 1
113   PC[15:8] ← M⟨SP⟩
      → (1)
```

Note that RTI is an assembly-language instruction. Thus execution arrives at RTL step 110 by virtue of the decoding of the opcode. This is in contrast to the jump to step 100, which is done in response to an interrupt. Step 111 pulls the reserved processor status of the stack and restores **PS**. Steps 112 and 113 pull the return address off the stack to **PC** and return to step 1 to resume execution of the reserved program. This is the same as RTS except that it also restores **PS**.

It is important to recognize that the interrupt cycle does not "do anything" about the interrupt except provide for a jump to a routine to deal with it. The programmer must supply that routine. The programmer must decide what is supposed to happen when an interrupt occurs, write a program to do that, store that program in memory, and enter the address of that program, low byte in FFFE, high byte in FFFF. This address is known as the *interrupt vector*.

Suppose that the only device in the system that can interrupt is a keyboard. Then the interrupt program could be a routine such as KYBD or KEYSCN in Sections 13.7 and 13.8, with suitable modifications, including changing RTS to RTI. The programmer might decide to locate that program in ROM at 0F40, in which case he or she would load 40 in FFFE and 0F in FFFF to form the interrupt vector. Note that FFFE and FFFF in the address space must be implemented. If you worked Problem 13.2 or 13.3, you may have wondered why they called for implementing the upper 4K or 2K in the address space. This is the reason. It is virtually standard to implement the upper locations in the address space in ROM to provide a place for the interrupt vector.

When we say that these addresses must be implemented, we mean that there must be unique locations that will respond when these addresses are issued. For example, consider the memory map of Fig. 13.2 in which the upper four address bits are not used. If addresses FFFE or FFFF are issued, address bits 4 and 5 will select ROM 2, and 0FFE or 0FFF will be accessed.

Since interrupts, by definition, occur at unpredictable points in the main program, it is especially important to see that the status of the program is preserved. The interrupt cycle saves **PS** automatically. In addition, **AC** should certainly be saved and **Y** or **X** must be saved if it is to be used in the interrupt program. The simplest way is to push them on the stack, just as we did in SCAN in the last section. Thus, the first instructions in a typical interrupt routine will be

```
PHA        /PUSH AC ON STACK
TYA        /GET Y
PHA        /PUSH ON STACK
```

And the last instructions will be

```
PLA        /PULL OLD Y
TAY        /RESTORE Y
PLA        /RESTORE AC
RTI
```

This switching from one program into another is sometimes referred to as *context switching*. The preservation of program status in context switching is so important that some processors push all the registers onto the stack automatically during the interrupt cycle and restore them all automatically when executing RTI. The choice between storing all the registers automatically or just the minimum possible is not an obvious one. If you are going to store all the registers anyway, it will certainly be faster to store them automatically than to execute separate instructions for that purpose. On the other hand, if the interrupt program does not need all the reg-

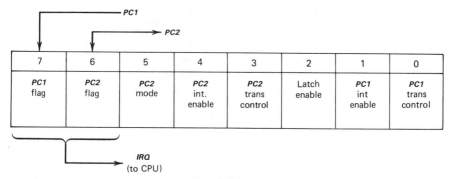

*Figure 15.9   Bit assignments in interface **PCR** register.*

isters, time spent storing unused registers is wasted. Some processors do it one way, some the other. At least one gives you a choice, with a normal interrupt (***IRQ***) input that stores all registers and a fast interrupt (***FIRQ***)that stores only ***PC*** and ***PS***.

Once the status of the interrupted program has been safely preserved, the next step is to identify the source of the interrupt. A situation such as suggested here, with only one source of interrupt, would be unusual. The normal situation is to have a number of devices that can interrupt to request service. If we assume that all devices connect to the prcessor through interface adapters such as discussed in the last chapter, the connection of interface lines is very simple. The bit assignments in the ***PCR*** register are shown in Fig. 15.9. As shown, there is an ***IRQ*** line connected to the ***PC1*** and ***PC2*** flags. This line will be normally high. If the ***PC1*** flag is enabled for interrupt (***PCR***[6] = 1), the ***IRQ*** line will go low when the ***PC1*** flag is set. If the ***PC2*** flag is enabled for interrupt (***PCR***[2] = 0 and ***PCR***[3] = 1), the ***IRQ*** line will go low when the ***PC2*** flag is set.

The electronics of the ***IRQ*** lines on the processor and interface chips is such that they can all be wired together in a wired-AND configuration, as shown in Fig. 15.10. If all the interface ***IRQ*** outputs are high, the common line will be high; if any of them go low, the common line will go low to signal an interrupt.

So, when the CPU finds ***IRQ*** low, it knows that at least one of the devices out there has interrupted. To respond, it obviously has to determine which one. The simplest method is software polling. Assume we have three interface units, as shown in Fig. 15.10, with the ***PCR*** registers denoted as ***PCR1, PCR2, PCR3***. We also assume that both interrupt flags are enabled in all three, so that there are six possible sources of interrupt. The routine shown here will check them one at a time by testing the ***PC1*** and ***PC2*** flags in each interface.

```
POLL      LDA   #$C0      /LOAD 11000000 IN AC
          AND   PCR1      /TEST FLAGS IN #1
          BEQ   CHK2      /IF ALL 0,
  *                       /NO INTERRUPT IN #1
          BMI   SUB1      /PC1 SET, BRANCH
          BPL   SUB2      /PC2 SET, BRANCH
```

```
CHK2      LDA   #$C0
          AND   PCR2        /TEST FLAGS IN #2
          BEQ   CHK3        /IF ALL 0,
*                          /NO INTERRUPT IN #2
          BMI   SUB3        /PC1 SET, BRANCH
          BPL   SUB4        /PC2 SET, BRANCH
CHK3      LDA   #$C0
          AND   PCR3        /TEST FLAGS IN #3
          BEQ   ERR         /ALL 0, ERRONEOUS IRQ
          BMI   SUB5        /PC1 SET, BRANCH
          BPL   SUB6        /PC2 SET, BRANCH
```

In each interface we want to check both flags, which are located in the first two bits of **PCR**. We start by loading 11000000 into **AC** in preparation for testing the first two bits in each **PCR**. If the result of the bit test of **PCR1** is 0, neither flag is set and we branch to check **PCR2**. If the result is nonzero, one or both are set. Recalling that the BIT command copies bit 0 of the tested location (the **PC1** flag in this case) into the **N** flag, the BMI command will take us to the appropriate subroutine if the **PC1** flag is set. If it is not set, the **PC2** flag must be, so the BPL takes us to the subroutine to deal with that interrupt. This process is then repeated for the other two interface units. If we find 0 as the result of the bit test of interface 3, something is wrong. None of the flags is set but **IRQ** was low. The error routine might notify the operator or might just go back to the main routine on the assumption that the error was transient.

This polling routine tests the interrupts in an order determined by the programmer, and this order sets the *priority* of the interrupts. Since the timing of interrupts is unpredictable, there is nothing to prevent several devices interrupting at about the same time. The polling order determines the order in which they will be served, independently of the order in which they occur. For example, suppose that the fifth device initiates the interrupt process by setting the **PC1** flag of interface 3. Then, during the interrupt cycle, but before the polling starts, the second device sets the **PC2** flag of interface 1. This will have no effect on the **IRQ** line since it is already low, but that device will be serviced first since its flag is tested first. That does not mean that the fifth device will be ignored. Sometime during the servicing of the second device, its flag will be reset. When control returns to the main program, **IRQ** will still be low due to the fifth device, so the interrupt process will begin again, this

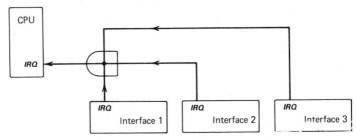

*Figure 15.10   Interconnection of **IRQ** lines.*

time servicing the fifth device, unless some other higher-priority device has raised its flag in the interim. A low-priority device can sometimes wait a long time.

The matter of priority is of great importance in designing a system. When there are a variety of devices that can interrupt, the programmer must decide which are most important, which have to be serviced most quickly to avoid some nonrecoverable error. For example, suppose we have an TBSIA receiving data at 9600 baud and a keyboard. Each time the TBSIA assembles a byte it loads it into the buffer register and starts assembling the next byte. If that first byte is not picked up in about 1.1 msec, it will be lost. When a key is pressed, the CPU can probably wait as long as 100 msec without losing the data.

Obviously, the TBSIA must have higher priority, but care must be taken to see that the keyboard is not blocked out entirely. One way to do that is to continue the scanning routine after servicing the first flag found. In this example, after servicing device 2, we could continue the scan from that point rather than returning to the main program. This can be accomplished by ending SUB2 with an RTS rather than RTI. The priority can also be controlled by use of the interrupt enable bits in the *PCR* registers. If you want to give certain devices special priority at certain times, you can disable the interrupts of other devices. The long and short of it is that you, as programmer, have absolute control of interrupt priority, but geat care must be exercised in analyzing the timing requirements of the various devices in the system if priority is to be properly assigned.

One problem with programmed polling is that it takes time. Where speed is of the essence in interrupt handling, *vectored interrupt systems* may be used. A vectored system is one in which there are separate interrupt lines for each source and the interrupt cycle directly generates an interrupt vector pointing to the service routine for the interrupting device. A system with only one interrupt line, as discussed here, is considered to be a *nonvectored* system; even though an interrupt vector is obtained (from FFFE and FFFF), it does not identify the source of the interrupt. Some processors have multiple *IRQ* inputs, with each causing a reference to a different location to find the appropriate interrupt vector. Single-*IRQ* processors such as the TB6502 can be converted to vectored interrupt with external hardware, and processor manufacturers often provide special chips to implement vectored interrupt systems.

The block diagram of a typical interrupt adapter chip accommodating up to eight interrupt sources is shown in Fig. 15.11. The separate *IRQ* lines are combined in an actual AND gate rather than wired-AND, so that their separate identities are retained. The *IRQ* lines also go to the vectoring logic, along with the address lines from the processor. The upper 11 bits of address go to *ADBUS* in the normal manner and also to vectoring logic along with the lower 5 bits of address. When any addresses other than FFFE or FFFF are issued, the lower 5 bits of address are passed along to *ADBUS* without alteration, so the interrupt adapter has no effect. If either FFFE or FFFF (the interrupt addresses) are issued, the vectoring logic modifies the lower 5 bits of address to point to a location associated with the source of the interrupt. This gives the CPU access to an interrupt vector area of 32 bytes, from FFE0 to FFFF. This area of memory is depicted in Fig. 15.12. Notice that one pair of bytes is set aside for RESET, an interrupt-like operation that will be discussed in the next section.

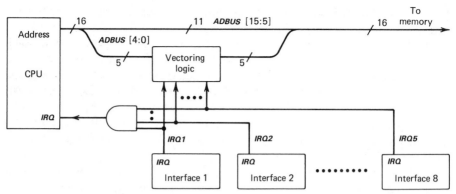

*Figure 15.11   Typical vectored interrupt adapter.*

Since two bytes are required for the address of each interrupt routine, the least significant address bit will not be changed by the network. It will be 0 for the low-order byte and 1 for the high-order byte just as in FFFE and FFFF. The following RTL statements describe a network like that in Fig. 15.11 that allow for only four interrupts with no provision for multiple interrupts.

**ADBUS**[15:5] = **ADDRESS**[15:5]
**ADBUS**[0] = **ADDRESS**[0]
**ADBUS**[4:1] = (**ADDRESS**[4:1]!0000!0001!  0010!0011)∗
    (∧/**ADDRESS**[15:1], (**IRQ1,IRQ2,IRQ3,IRQ4**) ∧
    (∧/**ADDRESS**[15:1]))

|  |  |  |
|---|---|---|
| FFE0 | L | Address of interrupt |
| 1 | H | routine for **IRQ1** |
| 2 | L | } For **IRQ2** |
| 3 | H | |
| 4 | L | } For **IRQ3** |
| 5 | H | |
| 6 | L | } For **IRQ4** |
| 7 | H | |
| . | . | |
| . | . | |
| FFFC | 00 | } For RESET |
| D | FF | |
| E | L | } For nonvectored |
| F | H | interrupt |

*Figure 15.12   Interrupt address area.*

Simple enough in principle, but the vectoring logic can become much more complex if it is required to establish the priorities in the event of multiple interrupts. The priorities may be fixed, that is, an interrupt on **IRQ1** will be serviced first, **IRQ2** next, and so on. In that event the only control the programmer has over priorities will be in the initial connections of devices to the interrupt adapter and in the enable control of the **PCR** flags. More complex interrupt adapters may include registers in which the processor can store control words to determine priority. We shall not concern ourselves further with vectored interrupt here. The main effect it has on the programmer is to eliminate the need for polling.

Now we know what device interrupted, so what do we do about it? There is no general answer to that question. It depends on what the device is, how it fits into the overall system, and the situation in the system at that time. One decision that has to be made is whether to allow interrupts of the interrupt. Step 102 in the interrupt cycle disables the interrupt line. This is essential. If the computer were interrupted again before it had a chance to preserve the interrupted program and identify the source of interrupt, chaos would result. However, that does not necessarily mean that no other interrupts can be handled until the current one has been resolved. Consider a process control system in which one interrupt is from a keyboard through which a supervisor can enter new instructions, and the other is from a pressure-monitoring device. An interrupt from this device indicates that a pressure vessel has reached its pressure limit and a relief valve must be opened. The keyboard routine might be one such as discussed in Section 13.7 that waits for key bounce to settle and for the operator to release the key, in which event it might tie up the processor for several seconds. In that case, we would probably want to enable the interrupt system during the keyboard routine, so that a pressure interrupt would be responded to immediately.

Fortunately, the use of the stack for preserving program status in interrupt routines makes it simple to nest interrupts. As we have noted, an interrupt cycle is just a special type of subroutine call, and we have already seen how the stack enables us to nest subroutines. As soon as we have preserved the status of the interrupted program, determined the identity of the interrupt source, and cleared the flag that caused the interrupt, we can enable the interrupt system with the CLI command, which clears the interrupt disable flag. If a second interrupt occurs before the first interrupt routine is complete, the status of that routine will be pushed onto the stack and the second routine will take over. When it is complete, RTI pulls the status of the first interrupt routine off the stack and it continues until another RTI restores the main program. Just as an interrupt while in the process of context switching to an interrupt routine cannot be allowed, interrupts cannot be allowed while switching back. If interrupt has been enabled during an interrupt routine, the closing steps of the subroutine should be of the form

```
RETRN    SEI        /SET INTERRUPT DISABLE
         PLA        /PULL OLD X
         TAX        /RESTORE X
         PLA        /PULL OLD AC
         RTI        /RETURN
```

Before starting any steps that restore the old program status, use SEI to set the interrupt disable flag. When RTI is executed, it will restore the old **PS**, which will enable interrupts (**I** = 0 in the old **PS** or there could have been no interrupt in the first place) so that the processor may be interrupted again immediately on returning to the main program. You may want to prevent interrupts for a while, to give the main program a chance to get on with things. This can be done by pulling the old **PS** off the stack with a PLA, setting the **I** flag, and then returning it to the stack before executing RTI.

Another step that will be common to any interrupt routine is clearing the flag that caused the interrupt in the first place. If you are using the interface adapters discussed in Chapter 13, this will require a read or write of the peripheral port, because that is the only way the **PC1** and **PC2** flags can be cleared. Even if the nature of the interrupt is such that no exchange of data is needed, you will need to do a read or write to clear the flags. Note that, while the setting of the flags is edge sensitive, that is, occurs only on transitions, the **IRQ** line is a level line, the interrupt will occur if **IRQ** = 0. This is why the disabling of the interrupt system during the context switch is essential. The **IRQ** line will remain low until the flag is cleared, so the interrupt system must not be reenabled until the flag has been cleared. If you wish to enable interrupts of the interrupt, be sure to clear the original flag first, or the CPU will respond again to the original input and you will be hopelessly lost.

As an example of the application of interrupts, let us consider again the process interaction discussed in connection with Fig. 13.26. Here we have a computer exercising on-line control of a process, and we have a keyboard by means of which an operator can enter new commands. In the approach taken in that example, the program checked the keyboard about 100 times a second to see if a new character had been typed. Even though the time involved is not large, this is very inefficient since new commands will be typed in only occasionally. We shall instead use an interrupt approach. Each time a key is pressed it will generate an interrupt. The interrupt routine will not only input characters but will also assemble the commands in a command buffer. When a carriage return is received, the routine will set a flag indicating

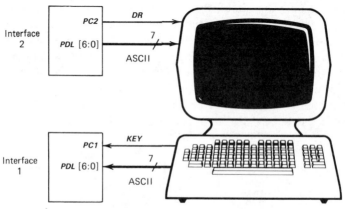

*Figure 15.13   Interface connections to CRT terminal.*

a complete command waiting for interpretation. The main program will check this flag at appropriate intervals, perhaps every 10 sec or so.

We shall assume communications through a complete CRT terminal as shown in Fig. 15.13. The keyboard will be the same one shown in Fig. 13.25, putting out a 7-bit ASCII code and line *KEY* that goes high when any key is pressed. It will communicate through interface 1, which will be set to generate an interrupt on a rising transition on *PC1*. The CRT display will be used to echo the typed characters. It will receive 7-bit ASCII characters via interface 2, set to generate a pulse dataready signal on *PC2*. We shall not use interactive signaling since the keyboard is so slow that there is no need to check for a datataken from the CRT display.

Communications between main and interrupt will be through a fixed parameter area on page 0. NEWCOM will be initially set to 0 by the main program and will be set to $80 (sign bit = 1) by the interrupt routine when a complete command is assembled. COMPTR will be the pointer to the command buffer area, initialized by the main program, used by the subroutine. Shown here are the initialization routines.

```
PINIT    LDA   #$FF
         STA   DDR2      /SET INTF. #2 FOR OUTPUT
         LDA   #0
         STA   DDR1      /SET INTF. #1 FOR INPUT
         LDA   #$28      /SET INTF. #2 FOR
         STA   PPR2      / PULSE HANDSHAKE
         LDA   #$03      /SET INTF. #1 FOR INTERRUPT
         STA   PPR1      / ON PC1, RISING EDGE
CINIT    LDA   #$40      /
         STA   $FFFE     /SET INTERRUPT
         LDA   #$06      /  VECTOR
         STA   $FFFF     /   TO 0640
         LDA   #0
         STA   NEWCOM    /CLEAR NEW COMMAND FLAG
         LDA   #$20
         STA   COMPTR    /SET LOWER BYTE OF POINTER
         LDA   #$04
         STA   COMPTR+1  /SET UPPER BYTE OF POINTER
```

The PINIT portion initializes the interfaces as described here. The CINIT portion initializes the parameters, assuming the command buffer is to start at 0420. The PINIT portion would normally be executed only when the system is started up. The CINIT portion would be executed whenever the system is ready to receive a new command.

We shall assume that the interupt routine is to be located at 0640, in which case FFFE = 40 and FFFF = 06. The following program starts by saving *AC* and *X* and then starts the poll. Since the keyboard is connected to *PC1*, it is the first device tested, so control will jump to SUB1. We have omitted the rest of the POLL routine, since it is irrelevant to this application.

```
*   INTERRUPT ROUTINE AT LOC. 0640
*
INTPRO    PHA                /SAVE
          TXA                /REGISTERS
          PHA                /
POLL      LDA   #$C0         /START
          AND   PCR1         / POLLING
          BEQ   CHK2         / POSSIBLE
          BMI   SUB1         / INTERRUPTS
            .
            .
            .
SUB1      BIT   NEWCOM       /COMMAND IN BUFFER?
          BNE   RETRN        /YES, IGNORE KEYBOARD
          JSR   DELAY        /NO, WAIT FOR KEY
                             /BOUNCE
          LDA   PP1          /READ CHARACTER
          AND   #$7F         /MASK OFF UPPER BIT
          STA   (COMPTR)     /STORE IN BUFFER
          STA   PP2          /ECHO TO CRT
          CMP   #$0D         /CARRIAGE RETURN?
          BNE   RETRN        /NO, RETURN
          LDA   #$80         /YES,
          STA   NEWCOM       /SET NEW COMMAND FLAG
          INC   COMPTR       /INCREMENT POINTER
RETRN     PLA                /RESTORE
          TAX                / REGISTERS
          PLA                /
          RTI                /RETURN TO MAIN
```

At SUB1 the program first checks to see if the buffer is already loaded with a command that has not yet been interpreted. If so, the new character from the keyboard will be ignored. For the operator, the fact that the key is not echoed on the display will be the signal that the computer is not yet ready for a new command. If the buffer is not full, the program jumps to DELAY to allow key bounce to settle. The character is then read in, stored, echoed, and then checked for CR. If CR, NEWCOM is set and control returns to main; if not CR, COMPTR is incremented in preparation for the next character before returning to main.

Since we are using interrupt to eliminate the need for regular checking of the keyboard, why not go one step further and interrupt when a complete command has been received, thus eliminating the need for regular checks of NEWCOM? Seems reasonable, but we have to consider carefully the problems of getting back to the main program. When we execute this program, control will return to main in a few milliseconds, depending on the debounce delay, and the main program can go right on from where it was. A return from a routine to interpret a new command might be more difficult. Suppose the new command called for some activity that is quite different from what was going on before? There is no information in the interpretation routine about what the main was doing before interrupt. Can we be assured of making a smooth transition from the old activity to the new activity? Halting right in the middle of doing one thing and switching to something different will not always work. It may be necessary to bring the original process to some stopping point before going on to something new. It is like shifting a car into reverse; you cannot do it just any old time.

You will find references in the literature to "interrupt-driven systems." The basic idea is that you have a computer monitoring a lot of different signals from the outside world, waiting to be interrupted and directed to some course of action in response to that interrupt. If the computer is really doing nothing but waiting between interrupt processes, this approach will work fine. When the interrupt process is over, the computer just drops back into the waiting loop. But if the computer is doing nothing but just waiting for interrupts, then it really is not being interrupted. It might just as well spend the "waiting time" polling the various inputs. It is truly an interrupt only if the computer is actively doing something and has to break away from that to do something else in response to the interrupt. But there are few meaningful activities that can be successfully terminated at any arbitrary time. Interrupt techniques are most likely to be useful where they are responding to temporary situations, following which the main program will continue as though nothing had happened. When changes in the course of program activity are to have a more lasting effect, such changes should generally come on a scheduled basis.

## 15.4  Start-Up and Reset

The problem of getting a computer started in the first place is an interesting and challenging problem, but one that few computer users have to worry about. Most of us come in contact with microprocessors as parts of complete computer systems that are configured for easy use. When we turn on the power, the computer comes up all ready to go, ready to accept input from a keyboard or terminal. If a program gets hung up in a loop or otherwise gets messed up, we can push the RESET button, and the computer comes back in a stable condition, all ready to try again. But how does this happen? Basically, a computer can do just one thing, read an instruction from memory and execute it. It has no built-in capability for doing such things as reading keyboards. We have written some programs for reading keyboards, but we need to get such programs into memory and start the computer executing them. How do we do that? Loading memory requires executing instructions, a JUMP to a certain location to execute is an instruction itself. We have a sort of chicken and egg problem here, how do we get things going in the first place?

In older computers, the process of getting a "blank" computer under way in the first place was known as "bootstrapping," conveying the idea of "pulling oneself up by one's own bootstraps." Typically, the computer would have a control panel with switches by means of which the operator could enter addresses and instruction words. When reset, the computer would be able to accept information through the switches. The operator would enter an address and an instruction (in binary, one switch per bit!) and then push the STORE button and the instruction would be stored in the specified location in memory. In this manner, the operator could (very laboriously) enter a start-up program in memory. This program would often be a program to enable the operator to communicate via a terminal, eliminating the use of binary switches. Once the program was entered, the operator would set the address switches to the address of the first instruction and then push the GO button, and the computer would start execution. If all this had to be done every time the computer was started, life would be very difficult. However, these computers usually had a special form of RAM, magnetic-core memory, that retains information even though the power is turned off. The next time the computer was turned on, the initialization program would still be in memory, so the operator had only to enter its address and press the GO button.

In microprocessor systems we have no permanent read/write memory, but we have something better, ROM. We store a start-up program in a ROM chip that becomes a permanent part of the system. Whenever the system is started or reset, the processor automatically refers to this ROM to get underway. For the TB6502, we have a **RESET** input. When this line is pulled low for at least two clock periods and then goes high, the processor enters the RESET cycle, which is basically a special interrupt procedure. In the RESET cycle the processor goes to locations FFFC and FFFD to obtain the address of the start-up routine. It then disables interrupts by setting the *I* flag and goes to the fetch cycle to fetch the first instruction of the start-up routine. For all this to work, the programmer must write a start-up program and put it in ROM and put the address of this program in FFFC (low byte) and FFFD (high byte).

As we said, most users do not have to worry about these problems because they work with systems in which all this has been taken care of. If you are designing a system from scratch, however, you cannot ignore the start-up problem. First, there are some electronic problems. Following initial application of power, you can require the operator to press a RESET button to drive the **RESET** line low. A better approach for most situations is to have some external electronics that will automatically hold the **RESET** line low for a short time after power is applied, and then set it high. The nature of the start-up program will, of course, depend on the function of the computer system, but there are a few things that almost always need to be done, principally initialization of registers. The RESET cycle will set the *I* flag and load *PC* but will leave all other registers in a random power-on condition. The usual first step in a start-up routine is to initialize the stack pointer at FF.

```
LDX   #$FF
TXS
```

The next step may be to initialize peripheral devices. The same line that drives the **RESET** input on the processor will also probably be connected to the **RESET** lines of various interface chips. In the case of the interface chips we have discussed, **RESET** will clear all registers. The start-up program will include appropriate PINIT segments, such as we have discussed earlier, to initialize the interfaces. The next step may be to execute CLD if binary operations are desired, or SED for decimal. If interrupts are to be enabled, CLI must be executed. At this point, control will usually jump to the regular operating program. If it is a monitor program, it will probably display some sign-on messages and then wait for instructions from the keyboard.

## 15.5   Tables and Other Data Structures

In our treatment of programming to this point we have focused primarily on the structures of the programs, the use of loops and subroutines, counters and pointers, and so on. This is important, of course, but it is not all of programming. Programs have to work with data, and the manner in which the data are organized is at least as important as the way the program is organized. The broad topic of data structures is properly the subject of entire books, and we can do little more than scratch the surface here. However, we feel it will be useful to present a few techniques for handling data that are likely to be of value in microprocessor applications.

Possibly the most common data structure in microprocessor programs is the *table,* and *table look-up* procedures occur frequently. We discussed table look-up in Chapter 13 in connection with the table of seven-segment codes, which is repeated here as Fig. 15.14.

For purposes of a more general discussion of tables, we need to establish some nomenclature. The input to a table look-up procedure is the *argument,* the output is the *result.* In the table look-up in subroutine DISP7, the BCD code for a digit is the argument, the seven-segment code is the result. The terms *argument* and *result* are completely general and are independent of the organization of the table or the look-up procedures. The table look-up problem is always the same—given the argument, find the result, by looking it up. The table itself consists of a set of *entries,* stored in a *range* of sequential locations in memory. The address of the first entry is the *base address,* the distance into the table to a particular entry is the *offset.* In the table of

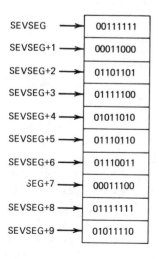

| | |
|---|---|
| SEVSEG | 00111111 |
| SEVSEG+1 | 00011000 |
| SEVSEG+2 | 01101101 |
| SEVSEG+3 | 01111100 |
| SEVSEG+4 | 01011010 |
| SEVSEG+5 | 01110110 |
| SEVSEG+6 | 01110011 |
| SEG+7 | 00011100 |
| SEVSEG+8 | 01111111 |
| SEVSEG+9 | 01011110 |

*Figure 15.14   Table for seven-segment code.*

Fig. 15.14, the entries are the seven-segment codes, the base address is SEVSEG, and the range is 10. The offsets range from 0 to 9. The entry with offset of 5, for example, is 01110110.

In the look-up procedure in DISP7, the argument, the BCD code, was used as the offset to access a particular entry, which was the result. Recall that this was done with index addressing. We loaded the BCD code into **X** and then executed

**LDA SEVSEG,X**

which accessed the table with the desired offset. This simple procedure works because the arguments are a set of sequential numbers. Suppose we want to go the other direction, given the seven-segment code as the argument, find the BCD code as the result. If we tried to use the seven-segment codes as addresses, the entries would be in 10 locations scattered randomly over a range of 128 locations, which would represent an intolerable waste of memory. We will store the table as shown in Fig. 15.14, and the look-up procedure will consist of comparing the argument with each of the entries until we find a match. The offset of the matching entry will be the result. A routine for this look-up process is shown here.

```
*   SUBROUTINE SEVBCD
*     CONVERTSFROM 7-SEGMENT CODE TO BCD
*       LENGTH = 12 BYTES
*       USES AC AND X
*       UPON ENTRY:
*         TABLE OF 7-SEGMENT CODES AT SEVSEG;
*         7-SEG CODE TO BE CONVERTED IN AC;
*       UPON EXIT:
*         BCD CODE IN AC
*
SEVBCD    LDX    #0           /SET X = 0
COMP      CMP    SEVSEG,X     /CMP ARG. TO ENTRY
          BEQ    MATCH        /BRANCH IF MATCH
          INX                 /INCREMENT X
          BNE    COMP         /COMPARE NEXT ENTRY
MATCH     TXA                 /MOVE OFFSET TO AC
          RTS                 /RETURN
```

Again we use indexing, this time stepping through the table until we find a match, at which point the offset in **X** is the desired BCD code.

In comparing these two procedures, note that we have reversed the relationship between arguments and results, and offsets and entries. It is important to understand this, that these terms are invariant, but that relationships between them depend on the procedures, and that there are many ways of using tables.

The procedures just used, involving a single table, work because either the arguments or results are a set of consecutive numbers. In the more general case, neither the arguments nor the results have any numeric significance, and two or more

| Mark | ASCII | EBCDIC |
|------|-------|--------|
| , | 10101100 | 01101011 |
| . | 10101110 | 01001011 |
| — | 10101101 | 01100000 |
| + | 10101001 | 01001011 |
| $ | 10100100 | 01011011 |
| # | 10100011 | 01111011 |

*Figure 15.15   ASCII and EDBCDIC codes for punctuation marks.*

parallel tables will be needed. As an example, consider the conversion between ASCII code and EBCDIC code. The ASCII code, as we have discussed, is the most widely used alphanumeric code. EBCDIC is an alphanumeric code used by IBM. The two codes are, except for parity, identical for alphabetic characters and numerals, but they are totally different for punctuation marks and special control characters. Figure 15.15 shows the ASCII and EBCDIC codes for some of the punctuation marks.

As you can see, the codes are quite different and have no numeric significance. Why IBM thought it had to be different from everybody else in the world on this matter is a good question, to which we have no answer. In any event, let us assume that we are receiving data from an IBM system in EBCDIC and have to convert it into ASCII for use in another (non-IBM) system. We set up two tables as shown in Fig. 15.16, one for ASCII, the other for EBCDIC, with corresponding codes having equal offset.

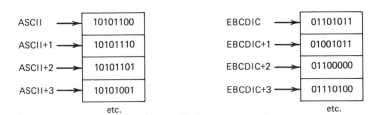

*Figure 15.16   Tables for ASCII and EBCDIC.*

The program that follows will find the ASCII code corresponding to a given EBCDIC code. You will note that this is virtually identical to the previous program. The only difference is that, when we find the match, we use the offset to go into the parallel table instead of using the offset directly as the result.

This general approach of using indexing to step through a table to find a match and then using the offset to enter parallel tables is very useful. Another application is in converting mnemonics to opcodes in assemblers. When you type in LDA, for example, the assembler searches a table of mnemonics for a match and then enters a parallel table to find the opcode. This approach can be used with any number of

```
*   PROGRAM EBCASC
*     CONVERTS EBCDIC CHARACTER TO ASCII
*       LENGTH = 14 BYTES
*       USES AC AND X
*       UPON ENTRY:
*         TABLE OF ASCII CODES AT ASCII
*         TABLE OF EBCDIC CODES AT EBCDIC;
*         EBCDIC CHAR. TO BE CONVERTED IN AC
*       UPON EXIT:
*         ASCII CHARACTER IN AC
*
EBCASC    LDX   #0          /CLEAR X
COMP      CMP   EBCDIC,X    /CMP ARG. TO ENTRY
          BEQ   MATCH       /BRANCH IF MATCH
          INX               /NO MATCH, INCR. X
          BNE   COMP        /CHECK NEXT ENTRY
MATCH     LDA   ASCII,X     /GET ASCII ENTRY
          RTS               /RETURN
```

parallel tables. You might have one table of names, a second table of corresponding addresses, a third table of telephone numbers, a fourth table of social security numbers, and so on. To find the address and phone number of a certain person, search the name table for a match, and then obtain the address and phone number from the parallel tables.

Note that the last process is exactly what you do when you look up an address or phone number in the telephone book. In the telephone book, names are arranged in alphabetical order to speed up the search. Can computer searches be similarly speeded up by arranging the entries in the table in certain ways? Very definitely. If the data are numeric or alphabetic, it probably will be useful to arrange the entries in numeric or alphabetic order. Actually, the two are the same in computers. If you will look at the hex equivalents for the letters of the alphabet in ASCII (Fig. 6.3), you will see that they are in numeric order. This is, of course, no accident, because it permits the same ordering procedures to be used for numeric or alphabetic data. If data are arranged in numeric order, there are search procedures that are much more efficient than a top to bottom linear search as used in these examples.

Another useful approach is to arrange entries in order of frequency of usage. For example, if we are converting codes for letters of the alphabet, we could arrange the codes in order of the frequency of occurrence of the letters in English text, e, a, i, . . . , at the beginning of the table, . . . , x, q, z, at the end of the table. Over a large number of letters, we would thus minimize the average number of comparisons to find a match. The most used letters would be found with only a few comparisons, while the seldom used letters would require the largest number of comparisons. We could follow this approach with the conversion of mnemonics to opcodes. Operations such as LDA, STA, and ADC occur far more often in typical programs than TSX, PLP, and SEI. The frequently used instructions would be at the beginning of the table, the seldom used ones at the end of the table.

When using a monitor, you will type in commands such as ERASE, PRINT, DUMP, which are executed immediately. These commands are not simply translated into an opcode, as in the case of an assembler, but must cause an immediate jump into a routine to carry out the specified action. Special tables called *jump tables* are

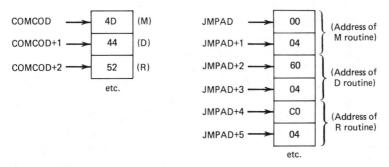

*Figure 15.17    Jump table.*

sometimes used for the command translation. Let us assume a very simple monitor that uses one-letter commands, such as M for memory display, D for deposit, R for register display. During the input routine, these code letters (in ASCII) will be loaded into location COMREG. We shall set up two tables, one containing the ASCII codes for the various commands, the other the addresses of the routines for executing the commands, as shown in Fig. 15.17. (The entries are shown in hex.) The table COMCOD contains the ASCII command codes. We assume that the routines for executing M, D, and R start at 0400, 0460, and 04C0, respectively. These addresses, which are the addresses to which the program must jump, are in table JMPAD. Table JMPAD has two entries for each entry in COMCOD, but we shall see that this is no problem. Shown here is a program segment for switching control to the appropriate command routine.

```
COMSW   LDA   COMREG     /GET COMMAND
        LDX   #$0        /CLEAR X
COMP    CMP   COMCOD,X   /COMPARE COMMAND TO COMMAND TABLE
        BEQ   MATCH      /BRANCH IF MATCH
        INX              /NO MATCH, INCREMENT X
        BNE   COMP       /GO BACK, CHECK NEXT COMMAND ENTRY
MATCH   TXA              /MOVE OFFSET TO AC
        ASL   A          /DOUBLE OFFSET
        TAX              /RETURN TO X
        LDA   JMPAD,X    /GET LOW BYTE OF JUMP ADDRESS
        STA   JMPTR      /PUT IN JUMP POINTER, LOW
        INX              /INCREMENT X
        LDA   JMPAD,X    /GET HIGH BYTE OF JUMP ADDRESS
        STA   JMPTR+1    /PUT IN JUMP POINTER, HIGH
        JMP   (JMPTR)    /JUMP TO ACTION ROUTINE
```

The first part of the program, down to MATCH, is the same as in the previous examples, searching COMCOD for a match. Because the jump table has two entries per entry in COMCOD, we must double the offset. This is done by moving the offset to *AC*, shifting it left, and returning it to *X*. We then access the low and high bytes of the jump address, move them into a jump pointer, and then make an indirect jump to the routine to execute the desired command. There are many ways of handling jump table routines: the preferable method is often determined by the available addressing modes for the jump commands. This routine is based on the assumption that the only addressing modes for JUMP are absolute and indirect.

This is not a particularly efficient approach for only a few commands. Another method is to test the command for all possibilities successively and branch accordingly. First, test for M; if it is M, branch to the M routine, otherwise test for D. If it is D, branch to the D routine, otherwise test for R, and so on. With this method, the number of tests, and therefore the length of the program, depends on the number of possible commands. On the other hand, the jump table program can handle any number of commands, although the length of the tables depends on the number of commands. Probably the biggest advantage of the jump table approach is that command routines can be added or deleted or moved around in memory to make room for modifications, simply by changing the tables. This is a general advantage of table-driven routines over procedural routines. It is usually easier to change tables than to change procedures.

Another very useful data structure is the *stack,* which we have already considered in terms of its mechanization with the hardware stack pointer (**SP**) and the PHA and PLA instructions. There are many situations in which it is useful to have more than one stack, and additional stacks can be implemented with software. Let

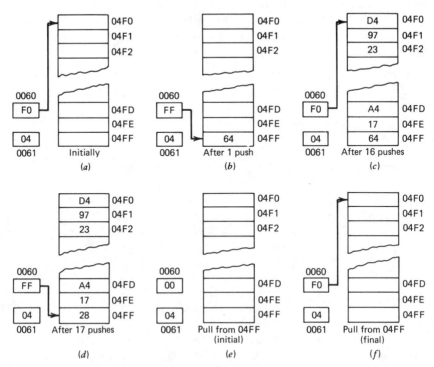

*Figure 15.18   Sixteen-word software stack.*

us assume we wish to implement a 16-word stack from 04FF to 04F0, as shown in Fig. 15.18. Locations 0060 and 0061 will be used as the pointer for this stack. We shall implement a decrementing stack in which the pointer points at the top of the stack, rather than the next empty location, as is the case with **SP**. This will facilitate a program to be discussed later. A PUSH will first decrement the stack pointer and then store; a PULL will read the top of the stack and then increment the stack pointer. It should be recognized that a limited stack is circular. In this case it will run from 04FF to 04F0, and then back to 04FF. This is nothing new, the hardware stack runs from 01FF to 0100 and then starts over at 01FF. Since this stack decrements before pushing, it should be initialized with the pointer at the location before the desired beginning of the stack. Thus, to start the stack at 01FF, we initialize the pointer to 01F0 (Fig. 15.18a). We shall check the pointer before decrementing on each PUSH. If 0060 contains F0, we shall reset it to 00 an then decrement it to bring it back to 04FF (Fig. 15.18b). After 16 pushes, the pointer will again point to 04F0 (Fig. 15.18c). On the seventeenth push, we reset 0060 to 00 and decrement it, thus overwriting 04FF (Fig. 15.18d). We reset it to F0, thus bringing the top of the stack back to 04F0 (Fig. 15.18f). PUSH and PULL subroutines for this stack are shown here.

```
*    SUBROUTINES PUSH, PULL
*     FOR HANDLING 16-WORD SOFTWARE STACK
*      LENGTH = 34 BYTES
*      USES 0060,0061 AS STACK POINTER,
*        INITIALIZED TO F0,04
*      USES 0062 AS TEMPORARY STORAGE
*      STACK RUNS FROM 04FF TO 04F0
*      STACK POINTER POINTS TO TOP OF STACK
*      UPON ENTRY FOR PUSH:
*        BYTE TO BE PUSHED IN AC
*      UPON EXIT AFTER PULL:
*        PULLED BYTE IN AC
*
PUSH       STA    *62        /SAVE ENTRY WHILE
*                            / CHECKING POINTER
           LDA    #$F0       /STACK POINTER
           CMP    *60        / AT 004F0?
           BNE    DECR       /NO, BRANCH TO DECR
           LDA    #$00       /YES, SET POINTER
           STA    *60        / TO 00
DECR       DEC    *60        /DECREMENT POINTER
           LDA    *62        /RETRIEVE ENTRY
           STA    (60)       /PUSH ON STACK
DONE1      RTS               /DONE WITH PUSH
PULL       LDA    (60)       /PULL ENTRY OFF STACK
           INC    *60        /INCREMENT POINTER
           BNE    DONE2      /IF NOT 00, DONE
           STA    *62        /SAVE ENTRY
           LDA    #$F0       /RESET POINTER
           STA    *60        / TO 04F0
           LDA    *62        /RETRIEVE ENTRY
DONE2      RTS               /DONE WITH PULL
```

There are actually two subroutines, PUSH and PULL, written together to save duplicating the header comments. PUSH first saves the entry while it checks the pointer. If the pointer is at 04F0, 0060 is reset to 00 before decrementing, so that the push will be to 04FF. The entry is then retrieved and pushed on the stack. PULL first gets the entry off the top of the stack and increments the pointer. If this increment does not bring 0060 to 00, the pull is done. If 0060 = 00, the entry is saved while 0060 is reset to F0 and then retrieved for return to the main routine. This stack routine is fairly simple, but it is inflexible in that the stack location and size are fixed. A better routine would permit the programmer to specify the stack size and location. We shall leave this as an exercise for you.

As we have seen, stacks are very useful in subroutine handling, for saving registers, passing parameters, and so forth, and such applications will be the major use of stacks for most programmers. However, stacks can be used in many ways, for many purposes. Indeed, there is a class of computers known as *stack machines,* because they do practically everything with stacks. Let us look at one rather ingenious use of stacks, for sorting a string of numbers into numerical order. Assume we have a string of decimal numbers in the range 1 to 99 to be sorted into order. Let

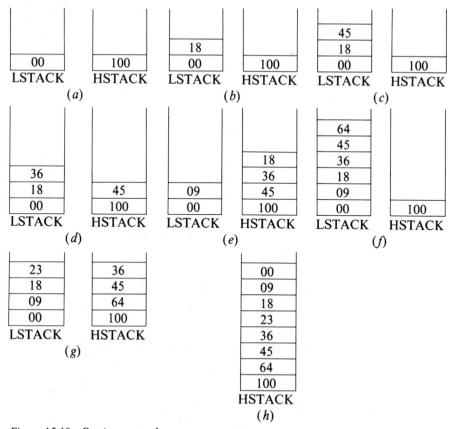

Figure 15.19   *Sorting on stacks.*

the entries in the string be

$$18, 45, 36, 09, 64, 23$$

We use two stacks, LSTACK and HSTACK, which are initialized to 0 and 100, as shown in Fig. 15.19a. Numbers in LSTACK will be arranged so that each entry is always larger than the previous entry, whereas HSTACK will be arranged with each entry smaller than the previous entry. When a new number, NUM, is to be entered, it will be compared to the top of each stack, which we will refer to as LTOP and HTOP. Numbers will be moved from one stack to the other until

$$LTOP < NUM < HTOP$$

and NUM is then pushed on LSTACK. Starting with our example string, 0 ⟨ 18 ⟨ 100, so we push 18 on LSTACK (Fig. 15.19b). Next, we have 18 ⟨ 45 ⟨ 100, so we push 45 on LSTACK (Fig. 15.19c). The next entry, 36, is not larger than 45, so we move 45 to HSTACK and then push 36 on LSTACK (Fig. 15.19d). In a similar manner, we move 36 and 18 to HSTACK before pushing 09 on LSTACK (Fig. 15.19e). The next entry, 64, is larger than 09, but not less than 18, 36, or 45, so we move these back to LSTACK before pushing 64 (Fig. 15.19f). The last number, 23 is not greater than 64, 45, or 36, so we move them back to HSTACK before pushing 23 on LSTACK (Fig. 15.19g). The final step is to move LSTACK to HSTACK (Fig. 15.19h). Ignoring 0 and 100, which are needed to initialize the process, we now see we have the string of numbers in order.

The program segment that follows assumes two stack routines similar to the preceding, with HPUSH and HPULL for handling HSTACK, and LPUSH and LPULL for LSTACK. The two stack pointers will be HPTR and LPTR.

```
*    PROGRAM DECSORT
*     SORTS STRING OF DECIMAL NUMBERS IN RANGE
*      01 TO 99 INTO NUMERICAL ORDER.
*     UPON ENTRY:
*      DPTR POINTS TO STRING OF NUMBERS
*      CNT CONTAINS COUNT OF NUMBERS
*     UPON EXIT:
*      NUMBERS IN NUMERICAL ORDER IN HSTACK
*
SINIT     LDA    #0          /INITIAL ENTRIES
          JSR    LPUSH       /
          LDA    #100        / TO STACKS
          JSR    HPUSH       /
GET       LDA    (DPTR)      /GET ENTRY, NUM
LCHK      CMP    (LPTR)      /CMP NUM TO LTOP
          BPL    HCHK        /NUM ) LTOP, CHK HTOP
          STA    TEMP        /SAVE NUM
          JSR    LPULL       /MOVE LTOP
          JSR    HPUSH       / TO HSTACK
          LDA    TEMP        /RETRIEVE NUM
          BNE    LCHK        /CHECK NEW LTOP
HCHK      CMP    (HPTR)      /NUM - HTOP
          BMI    PUSH        /NUM ( HTOP, BRANCH
          STA    TEMP        /SAVE NUM
          JSR    HPULL       /MOVE HTOP
```

```
          JSR  LPUSH      / TO LSTACK
          LDA  TEMP       /RETRIEVE NUM
          BNE  HCHK       /CHECK NEW HTOP
   PUSH   JSR  LPUSH      /PUSH NUM ON LSTACK
          INC  DPTR       /INCR. POINTER
          DEC  CNT        /DECR. COUNT
          BNE  GET        /IF NOT DONE, GO BACK
     *                    / FOR NEXT NUMBER
                 .
                 .
                 .
      finish by moving LSTACK to HSTACK
```

With this explanation, you should be able to follow the program without further comment. We shall leave the final steps, the consolidation of the two stacks into
one, as an exercise for you.

Another important data structure is the *queue,* also known as the first-in, first-
out (FIFO) stack. A queue is a one-dimensional list, in which elements are entered

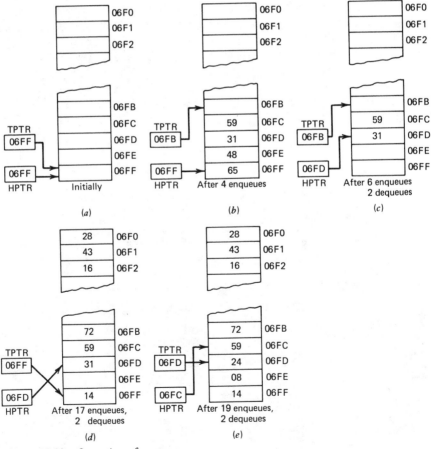

*Figure 15.20   Operation of a queue.*

at one end, the *tail,* and removed from the other end, the *head.* As described in Problem 15.17, queues can be realized in hardware using shift registers. As each word is entered at the tail, it is moved forward toward the head until it encounters the previous entry. When a word is removed, all remaining entries are shifted forward toward the head. This approach is not practical in programming, because of the time required to shift all the words every time a word is entered or removed. Instead, the queue is implemented in a manner similar to a stack, but using two pointers, one for the head (HPTR) and one for the tail (TPTR). Initially, both will point at the same location, as shown in Fig. 15.20a, where we assume a 16-word queue. As we enqueue (enter) a word, the tail pointer decrements, at all times pointing to the next unused space in the queue (Fig. 15.20b). When we dequeue (remove) a word, we likewise decrement the head pointer, so it always points to the head of the queue (Fig. 15.20c). If the head pointer catches up to the tail pointer, the queue is empty and a dequeue operation should return 0 or some sort of empty signal.

Queues are considered to be circular structures. Assuming the queue is initialized with the tail at 06FF, after 16 enqueues, loading the queue from 06FF to 06F0, the tail will return to 06FF (Fig. 15.20d). If the number of enqueues exceeds the number of dequeues by more than 16, the tail pointer will catch up with the head pointer and overwrite the head. In that case the head entry has been bumped out, and the head pointer must be pushed ahead one (Fig. 15.20e). It can be seen that the queue is a very dynamic data structure, with the two pointers moving in a circular path through an area of memory, defining the boundaries of a list.

```
*    SUBROUTINE FOR HANDLING 16-WORD
*      SOFTWARE QUEUE
*        LENGTH = 59 BYTES
*        USES 0070,0071 AS TAIL POINTER,
*          INITIALIZED TO FF,06
*        USES 0072,0073 AS HEAD POINTER,
*          INITIALIZED TO FF,06
*        USES 0074 AS TEMPORARY STORAGE
*      QUEUE RUNS FROM 06FF TO 06F0
*      UPON ENTRY FOR ENQUEUE:
*          BYTE TO BE ENQUEUED IN AC
*      UPON EXIT AFTER DEQUEUE:
*          DEQUEUED BYTE IN AC
*      DEQUEUE OF EMPTY QUEUE RETURNS 00
*
DEQU      LDA   *70        /CHECK FOR
          CMP   *72        /EMPTY QUEUE
          BNE   DEQ1
          LDA   #$00       /QUEUE EMPTY, RTRN 0
          RTS
DEQ1      LDA   (72)       /DEQUEUE HEAD ENTRY
          STA   *74        /SAVE
          LDA   #$F0       /CHECK FOR
          CMP   *72        /F0 IN HPTR
          BNE   DECH       /
          LDA   #$00       /RESET HPTR
          STA   *72        / TO 00
DECH      DEC   *72        /DECR. HEAD POINTER
          LDA   *74        /RETRIEVE ENTRY
          RTS              /RETURN FROM DEQUEUE
```

```
ENQU    STA   (TPTR)    /ENQUEUE ENTRY
        LDA   *72       /COMPARE POINTERS,
        CMP   *70       / IF NOT EQUAL,
        BNE   DECT      / DECR. TAIL POINTER
        CMP   #$F0      /HPTR AT F0?
        BNE   DECH1     /NO, DECREMENT HPTR
        LDA   #$00      /YES, RESET TO 00
        STA   *72
DECH1   DEC   *72
DECT    LDA   #$F0      /CHECK TPTR FOR F0
        CMP   *70
        BNE   DECT1     /NO, DECR. TPTR
        LDA   #$00
        STA   *70       /YES, SET TPTR TO 00
DECT1   DEC   *70
        RTS             /RETURN FROM ENQUEUE
```

The program should be self-explanatory given the previous discussion.

## Problems

**15.1**  Indicate the condition of the stack after executing the following sequence of instructions. Also indicate the contents of the memory locations 0050 and 0051. Assume **PS** = 08 initially.

```
0260   LDX   #$C8        0440   TSX
62     TXS                 41   INX
63     INX                 42   NOP
64     INX                 43   TXA
65     TXA                 44   PHA
66     PHA                 45   LDA   103,X
67     LDA   #0            48   STA   *50
69     PHA                 4A   LDA   104,X
6A     PHP                 4D   STA   *51
6B     JSR   0440
.      .     .
.      .     .
```

**15.2**  TABLE is an array of 20 positive integers. Write a subroutine to place a new entry in the table if the entry is larger than any element presently in the table. If an element is added, it will replace the smallest element in the table, and the smallest element will be returned to the main program. The main program will pass the address of TABLE on the stack and the new entry in **AC**. On return, **AC** will contain the new entry if it was not entered, or the deleted entry if it was entered. If the new entry was not entered, **C** = 0 on return; if it was entered, **C** = 1. Assume that **X** must be returned intact, but that 0050–0060 is available as a working area.

**15.3** A useful subroutine in programming is a block-move routine. Suppose you have written a program and find you need to insert some extra instructions in the middle of the program. To do this you need to move all the instructions from that point to the end down enough locations to make room for the new instructions. Assume that BEGIN and END point to the first and last locations in the block to be moved, and NUM is the number of locations down the block that is to be moved. These will be passed to the subroutine by Method 2 of Section 15.2.

**15.4** Repeat Problem 15.3, but use Method 3 to pass parameters.

**15.5** Repeat Problem 15.3, but use Method 5 to pass parameters, assuming all registers have to be preserved but that 0050–0060 are available as a working area.

**15.6** A subroutine is to be written that will select the largest of five data bytes. In this case, the parameters will be the five data bytes themselves and the result. Before calling the subroutine, the main routine will push the five data bytes on the stack. Following the return from the subroutine, the main routine should find the largest data byte on top of the stack. (The other bytes will have disappeared from the stack.) Write the subroutine just described, which will begin at symbolic address MAXBYT. Use symbolic addresses for all temporary storage used within the subroutine. The subroutine must restore the original contents of **AC** before returning, since this value may be needed by the main routine.

**15.7** You are to write a subroutine OUTCHR that will function as follows. The main routine will pass to it, in **AC**, either an ASCII code or a byte of two packed BCD digits. The **C** flag will be used to indicate which it is, **C** = 0 for an ASCII code, **C** = 1 for packed BCD digits. If the character is an ASCII code, it will be in the form of Fig. 6.3, that is, with the most significant bit equal to 0. The ASCII characters are to be sent, by means of subroutine PRINT, from Section 13.6 to a printer that requires ASCII characters with odd parity, that is, with the leading bit set to 1 or 0 so that the total number of 1's in the character will be odd. Subroutine OUTCHR should pass the ASCII character as received, in **AC**, to another subroutine, ODDPAR, which you are also to write, which will insert the correct parity bit and return the character, also in **AC**, after which it should be sent to PRINT. If the byte passed from the main routine is packed BCD, OUTCHR should convert it [upper byte (bits 0–3) first] to two ASCII characters with odd parity, to be sent to the printer. This subroutine may use **AC**, **X**, and locations 50 to 59 on page 0.

**15.8** Given an array $P$ of data elements, some of which may be duplicates. Write a subroutine to create a second array, $Q$, containing no duplicates, that is, only one copy of each distinct element in $P$. The main program should pass the starting addresses of $P$ and $Q$, and the number of elements in $P$. The subroutine should return the number of elements in $Q$. Start the subroutine at 0400. Show the calling sequence portion of the main program.

(a) Pass the parameters by Method 2 of Section 15.2. Put the address of $P$ at 0050,51, the address of $Q$ at 0052,53, the number of elements in $P$ at 0054, and the number of elements in $Q$ at 0055.

(b) Pass the parameters by Method 3 of Section 15.2.

(c) Pass the parameters by Method 5 of Section 15.2. Assume locations 0050–0055 available as a working area to be used by the subroutine.

(d) Pass the parameters by Method 5 of Section 15.2, but now use the stack-frame approach to preserve all registers and working areas.

**15.9**   The process interaction program of Section 15.3 is to be modified so that the CRT of Fig. 15.13 can be replaced by an RS-232 CRT as described in Problem 14.2. The computer will communicate with this CRT via a TBSIA, also as described in Problem 14.2. The main process should be interrupted either by *RDF* in the TBSIA being set, indicating that the first character of a new command has been received, or by the overrun flag being set, indicating that a new character has been sent before the last one was picked up. First, modify PINIT and CINIT as necessary. The new INTPRO will be modified to determine the source of the interrupt. If overrun, it should send the message WAIT to the CRT and then return to the main. If the interrupt is due to *RDF*, it should then proceed as in the original INTPRO, assembling a new command, with modifications as necessary for RS-232 signaling.

**15.10**  A particular TB6502 has only one device that can cause an interrupt, and the interrupt will be caused only by a 1 in *PCR*[7]. Suppose the main program is interrupted just before executing an instruction with its Opcode located at 0F0A. Suppose the stack pointer was initialized to 01FF, the stack was empty, *PS* = 00, *AC* = 11, and *Y* = 01 at the time of the interrupt. The first part of the interrupt routine is given as follows:

```
INTROUT     PHA
            TXA           / AC ← X
            PHA
            JSR     SUB
```

If INTROUT = FF00, indicate the complete contents of the stack following execution of these four instructions of the interrupt routine.

**15.11**  A part of any assembler is a routine to convert a mnemonic to a hex opcode. You are to write a routine to decode just a portion of the command set of the TB6502, consisting of the instructions AND, ASL, ADC, LDA, LDX, LSR, SBC, STA, STX. Assume that the three letters of the mnemonic are stored (in ASCII) in locations 50, 51, 52. The routine will end when the correct hex code, for absolute addressing, is in the accumulator. Use a table approach as suggested in Fig. 15.17, modifying as necessary to reflect the fact that three-letter codes have to be checked. Show the contents of all tables, and write the routine.

**15.12** Although table-driven routines are very useful, they are not always the best approach. Sometimes a combination of table-driven conversion and procedural conversion will be effective. As an example, let us continue with the problem of converting mnemonics to opcodes, to include provision for the addressing modes. The first step will be to add a character to each mnemonic to indicate the addressing mode—space for absolute, # for immediate, * for page 0 ( for indirect, X for indexed, and A for accumulator, which applies only to shift and rotate instructions. For example, if the instruction is

    **ADC    BAKER,X**

the assembler will place

    **ADCX**

in locations 50–53 for conversion to an opcode.

    We could handle this by another level of table look-up, but the table will be large, because there will be an entry for every addressing mode for every instruction. Instead, we note that there are fixed relationships among the opcodes for various addressing modes. For example, if we consider just page 0 and absolute addressing, the opcodes for AND are 25 and 2D, for ASL they are 06 and 0E, for ADC they are 65 and 6D, and so on. Thus, the opcode for page 0 addressing can be obtained from the opcode for absolute addressing simply by adding $-8$. The codes for the other addressing modes are similarly related.

    Extend the program of Problem 15.11 to accommodate the addressing modes available for the same set of instructions. Use table-driven routines to determine the opcode for absolute addressing and then apply the appropriate correction to generate the opcode if the addressing mode is not absolute.

**15.13** Modify the software stack routine of Section 15.5 as necessary to include an initialization section that will permit the user to specify the location and size of the stack. The beginning of the stack should be passed in a pointer at locations 60, 61, and the number of words in location 63.

**15.14** Modify the software stack routine of Section 15.5 to implement a postincrement stack running from 02F0 to 031F. In a postincrement stack the stack pointer is incremented on PUSH after storing the word.

**15.15** Complete the stack sort routine of Section 15.5.

**15.16** Modify the queue routine of Section 15.5 to implement a 32-word incrementing queue that runs from 0500 to 051F.

**15.17** Microprocessors are sometimes used in applications where they must communicate with other devices that send or receive in short bursts of data too fast for the computer to keep up. Assuming there is enough time between these bursts for the computer to handle the data, a possible solution is the use of a buffer memory, a high-speed memory that can accept data from a peripheral at a high rate and then pass it on to the computer at a slower rate, or vice versa. One such device is the FIFO (first-in, first-out) memory, shown in Fig. P15.17a in an 8 × 32 configuration. This is a hardware realization of the queue discussed in Section 15.5. It consists of eight 32-bit shift registers to store 32 8-bit words. Words are read in at the "bottom" of the memory and read out from the "top" of the memory. Data written in at the bottom automatically ripple through until they reach the output or another data word. When data are read out from the top, all other data words automatically shift upward one location toward the output. Thus, however many words are in the memory, they are always at the top, that is, the output end.

There are four control lines on this device. When the bottom location is available to accept a word, line **IR** will be low. When line **LOAD** goes high, the data word will be read into bottom location and **IR** will go high. When

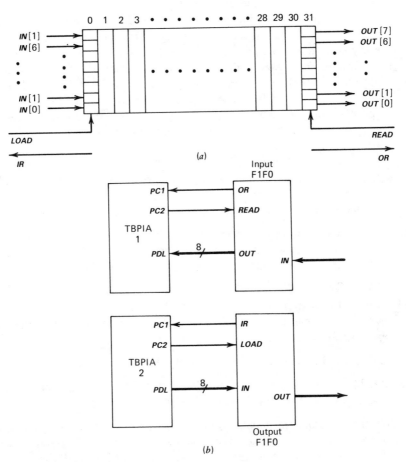

*Figure P15.17*

*LOAD* returns low, the word will start to ripple toward the top and *IR* will go low again, indicating that another word can be accepted. If the word loaded fills the memory, *IR* will stay high and further *LOAD* pulses will be ignored until words are read out, making room for more to be entered. When there is a word at the top available to be read, line *OR* will be high. When the line *READ* goes high, line *OR* will go low, indicating that the top location has been cleared and no data are available. When *READ* returns low, all words will be shifted up one location and *OR* will again go high to indicate that data are available. If the READ pulse empties the memory, *OR* will remain low and no data will be available until more are loaded.

We wish to use two FIFOs, one for input via interface 1 (Fig. 13.9) and one for output via interface 2, as shown in Fig. P15.17*b*. First, write device drivers for these two units. The input driver will simply read a word if available and return it through *AC*. The output driver will write a word received through *AC* if there is room. Both drivers will simply wait if data or space are not available. Also include initialization routines for both interfaces. Then write an input routine that will empty the input FIFO into 32 locations on age 0 starting at INBUF. Finally, write an output routine that will load 32 words into the output FIFO from 32 locations on page 0 starting at OUTBUF.

# References

1. Coats, R. B., *Software Engineering for Small Computers,* Reston Publishing, Reston, Va., 1982.
2. Khambata, A. J., *Microprocessors/Microcomputers,* Wiley, New York, 1982.
3. Leventhal, L. A., *Introduction to Microprocessors: Software, Hardware, Programming,* Prentice-Hall, Englewood Cliffs, N.J., 1978.
4. Leventhal, L. A., *6502 Assembly Language Programming,* Osborne/McGraw-Hill, Berkeley, Calif., 1979.
5. Peatman, J. B., *Microcomputer-Based Design,* McGraw-Hill, New York, 1977.
6. Wakerly, J. F., *Microcomputer Architecture and Programming,* Wiley, New York, 1981.

# Microprocessor-Based Systems Design

## 16.1 Introduction

Up to this point in our consideration of microprocessors, we have been primarily concerned with programming. This is appropriate, as we obviously have to be able to program a computer before we can apply it to any useful tasks. However, we should note that the primary applications of microprocessors are not in "conventional" computing, that is, using computers to solve mathematical problems or for bookkeeping and other types of business data processing. Rather, microprocessors are used as parts of larger systems, often for instrumentation and control. In such situations, the job of the engineer goes far beyond programming. The engineer must determine what parts of the system operations can be handled by a computer, select a microprocessor appropriate for the task, determine how much RAM and ROM will be needed, determine what imputs and outputs are needed, select peripheral devices, and so forth. Only when these things have been done will programming come into the picture. The engineer must design a complete system, of which the programs form only a part. In this chapter we wish to consider an example of such a complete system design, in the use of a microprocessor to control a vending machine.

## 16.2 Vending Machine Operation

You are certainly familiar with the simple form of vending machines used for soft drinks or candy, in which there are a small number of items available, all at the same

**457**

price, and the customer simply deposits the required amount and pulls a knob under the desired item. In the simplest versions of these machines, the operation is wholly mechanical, with the falling coin simply triggering a release that permits one of the knobs to be pulled out, dispensing the desired item. However, many vending machines are now much more sophisticated, permitting sale of a number of different items with different prices, keeping track of the amount deposited, making change, and so forth. Some even "talk" to the customer, giving verbal instructions on how to operate the machine.

We shall consider a machine that which can handle up to 100 different items. These will be displayed in a case, each with a corresponding number, from 00 to 99, and a price, from $0.25 to $9.95. To purchase an item, the customer will enter the number of the desired item on a keyboard and then deposit the necessary amount of money, following which the item will be dispensed and change made if any is required. We wish to design a unit that will receive the instructions and money from the customer and control the vending machine. We shall not be concerned with the mechanical design of the vending machine, since this will depend on the nature of the items to be sold. Our objective is to design a control unit that can be used with any vending machine meeting the general specifications given here, that is, up to 100 items, with prices in the range of $0.25 to $9.95.

The control panel of the unit will appear to the customer as shown in Fig. 16.1 There will be a 12-key keyboard consisting of 10 decimal keys and two keys marked "SELECT" and "CANCEL." There will be a two-digit display marked "ITEM

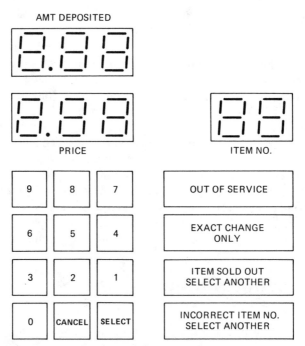

*Figure 16.1   Vending machine control panel.*

NO.," a three-digit display marked "PRICE," and a three-digit display marked "AMT DEPOSITED." There will also be four panels that will display messages when lit up. The messages will be "OUT OF SERVICE," "EXACT CHANGE ONLY," "ITEM SOLD OUT, SELECT ANOTHER," and "INCORRECT ITEM NO., SELECT ANOTHER." There will also be a set of instructions for use, that will read as follows.

---

## Instructions for Use

Items for sale are displayed in the case, together with the price and a two-digit item number.

To purchase an item proceed as follows:

1. Push the button marked "SELECT."
2. Enter the two-digit number of the desired item on the decimal keyboard. The number of the selected item will be displayed.

If the item you have selected is sold out, a message stating this will be displayed, and you may push the "SELECT" button and select another. If the item is available, the price will be displayed.

3. Deposit coins. You may deposit nickels, dimes, or quarters in any combination equaling or exceeding the purchase price. As you deposit coins, the amount deposited will be displayed.

When the correct amount has been deposited, the selected item will be dispensed. If the amount deposited exceeds the price, you will receive change, unless the "EXACT CHANGE ONLY" message is lit.

If you make a mistake at any point, press the "CANCEL" button. Any coins deposited will be returned, and you may start over.

---

# 16.3   System Specifications

The description of the last section specifies the operation of the system as seen by the customer. The next step is to develop technical specifications of the system, with emphasis on the basic system structure. It is convenient to divide the machine into four sections: (1) the vending machine itself; (2) the coin mechanism that receives the coins and issues change; (3) the control panel, consisting of the keyboard and displays; (4) the control unit, consisting of the microprocessor, memory, and associated logic. These sections, together with the signal lines among them, are shown in Fig. 16.2.

All signal lines between sections of the system will use active-low signaling. This is usually advantageous in communications between computers and various peripheral devices. For example, there will be outputs from the computer to the vend-

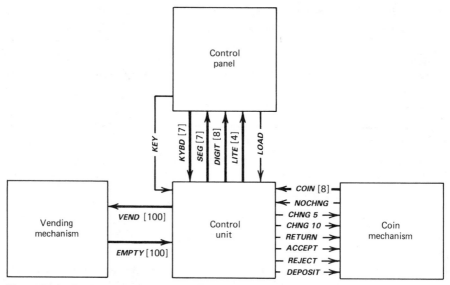

*Figure 16.2   Basic system organization.*

ing machine to activate the mechanism to dispense an item. This might require energizing a solenoid (electromagnet) to activate the mechanism. One terminal of the solenoid coil would be connected to the appropriate power source and the other terminal would be connected to a line to the computer. The computer can then activate the device simply by grounding the line through a driver of adequate current capacity, that is, the computer does not have to provide the power to activate the solenoid. In a similar manner, there will be inputs to the computer to indicate when various items are sold out. With active-low signaling, these lines need only be connected to switches that will ground the lines when items are sold out, that is, the computer need not provide any signaling voltages.

The signal lines will be as follows;

(A)   Between control unit and vending mechanism
      **VEND**[100]    100 lines to dispense items.
      **EMPTY**[100]    100 lines indicating that items are sold out.
(B)   Between control unit and control panel
      **KYBD**[7]    Seven lines for the 3 × 4 keyboard array.
      **KEY**   Line indicating that a key has been pressed.
      **SEG**[7]    Seven lines for the segments of the price and item number displays.
      **DIGIT**[8]    Eight lines for the eight digits in the price and item number displays.
      **LITE**[4]    Four lines for the four lighted message displays.
      **LOAD**   Input from switch to allow operator of vending machine to load new prices.
(C)   Between control unit and coin mechanism
      **COIN**[8]    Eight lines indicating type of coin deposited.
      **CHNG5**   Line to dispense nickel change.

*CHNG10*   Line to dispense dime change.

*NOCHNG*   Line to indicate that change dispenser is empty.

*ACCEPT*   Line to move coin to holding box.

*REJECT*   Line to return incorrect or counterfeit coin.

*DEPOSIT*   Line to deposit coins in coin box.

*RETURN*   Line to return coins.

The meaning of the *VEND* and *EMPTY* lines is fairly obvious. The 100 *VEND*-lines to the vending mechanism each control the dispensing of a single item. Each of the 100 *EMPTY* lines indicates that a specific item is sold out. Since cost is clearly important and speed is no problem, it is appropriate to have the microprocessor handle as many tasks as possible, so we use an unencoded keyboard with the 12 switches arranged in a 4 × 3 array, requiring seven bits of input over lines *KYBD*[7]. The ITEM NO., PRICE, and AMT DEPOSITED displays will require eight seven-segment indicators, which will be multiplexed in the same manner as discussed in connection with Fig. 13.24. For this purpose we have lines *SEG*[7] and *DIGIT*[8[] to control the segments and digits. The *LITE* lines will turn on the lighted messages, OUT OF SERVICE, EXACT CHANGE ONLY, ITEM SOLD OUT, SELECT ANOTHER, and INCORRECT ITEM NO., SELECT ANOTHER. The last message is provided so that this controller can be used with vending mechanisms with less than 100 items. It will be displayed if the customer selects a nonexistent item.

The *LOAD* input relates to a feature of the vending machine not yet discussed. It is obviously necessary that the operator of the vending machine be able to set the prices of the various items. The main operating programs of this system will be in ROM, but it would be very inconvenient for the operator to have to replace a ROM anytime it was necessary to change prices. We will provide RAM to hold prices, of the type used in "constant memory" pocket calculators. This is basically an electronic RAM with a very low current requirement. It will normally be powered by the regular power supply of the control unit, but if power should be lost, it will switch over to a standby battery that can retain the contents for several days. The machine will thus be operative even if it should inadvertently lose power for short periods of time.

The *LOAD* switch will be on the back of the control panel. When the operator throws the switch, prices of items can be changed by using the keyboard to enter the items number and then a new price. If the machine has less than 100 items, the operator will enter 0.00 as the price corresponding to any unused item numbers. To provide protection against the machine being without power for so long that the contents of the RAM are lost, we will set aside two test locations in memory. Every time the LOAD routine is entered, these two locations will automatically be loaded, one with all 1's, the other with all 0's. Then, whenever a customer selects an item, the computer will first check the two test locations. If they do not contain all 1's and all 0's, this will be an indication that the RAM contents are no longer valid. The computer will turn on the OUT OF SERVICE message, and the machine will be inoperative until the price RAM is reloaded.

The nature of the signals to the coin mechanism will, of course, depend on the design of that mechanism. We assume that the coins first move down a slot to a position where the type of coin—nickel, dime, quarter, or other (counterfeit)—is

| Code | | Meaning |
|------|---|---------|
| | 0000 | No coin |
| | 0001 | Coin diameter of dime |
| **COIN**[7:4] | 0011 | Coin diameter of nickel |
| | 0111 | Coin diameter of quarter |
| | 1111 | Coin larger than quarter |
| | 0000 | No coin |
| | 0001 | Coin weight of dime |
| **COIN**[3:0] | 0011 | Coin weight of nickel |
| | 0111 | Coin weight of quarter |
| | 1111 | Coin heavier than quarter |

*Figure 16.3   Coding of coin diameter and weight.*

determined. If the coin is counterfeit or otherwise unacceptable (e.g., a penny) it will be returned. If it is acceptable, it will be moved to a temporary holding box. The holding box will collect the coins until the purchase price has been deposited, at which time the coins will drop into the coin box. Anytime the customer pushes the CANCEL button, any coins in the holding box will be returned. The coin changer will have two dispensing cylinders, one for nickels and one for dimes. A pulse on **CHNG5** will dispense one nickel, and a pulse on **CHNG10** will dispense one dime. One or both cylinders empty will cause the line **NOCHNG** to go active.

We shall assume a coin detector system using eight photocells to measure the diameter and weight of the coin. The four photocells indicating diameter will drive lines **COIN**[7:4]; the four photocells indicating weight will drive lines **COIN**[3:0]. The coding of these lines is tabulated in Fig. 16.3.

This is a very unsophisticated coin detector, checking only the weight and diameter of the coin. If they agree, the coin is genuine; if not, it is counterfeit.*

## 16.4   Control Unit Design

The first step in the design of the control unit would normally be to choose a microprocessor. In this case, since the TB6502 is the only one we are familiar with, we shall use it. In an actual industrial design situation, it would be desirable to consider other processors, although familiarity with a particular unit is often sufficient reason for using it. Learning about a new unit takes time and therefore costs money, so it may be cheaper in the long run to use the old unit, even though the new one might be superior in some respects. In this case, the control tasks to be performed are

---

*A vending machine of this complexity would likely use a more sophisticated scheme for detecting counterfeit coins than presented here. As might be expected, the methods actually used are not widely publicized. The simple two-parameter system described here will serve to illustrate typical design approaches.

relatvely simple, so that the TB6502 should have ample capacity, and there should be no need to go to a more powerful processor.

Since this is a fairly simple system, a fairly small memory may be sufficient. It might thus be possible to use a single-chip system, namely, a system with processor and a small amount of memory on a single chip. Unfortunately, determining the amount of memory is not something that can be done until some work has been done on the programming. It may not be necessary to do the detailed coding, but some estimate of the size of the programs must be made. But the size of the programs is itself affected by the choice of processor, so we have a sort of "chicken and egg" problem here. This is not unusual, design is usually a circular process. You make some tentative choices of equipment, then proceed with the design. As you proceed you may find that some of your initial choices were bad, so you make some changes and refine the later steps accordingly, which may in turn indicate the need for still more changes, and so on.

Having settled on the TB6502, we next consider the interface between that processor and the various peripherals. Since the item numbers are two-digit decimal numbers, it seems logical to handle them in BCD form, with one byte of two packed BCD digits being used to specify the item number. The vending mechanism has two sets of 100 lines. When an item is selected by a customer, the computer must first check the corresponding **EMPTY** line to see if that item is available, and then select the corresponding **VEND** line to dispense the item. Obviously, we cannot have 100 inputs and 100 outputs coming directly into the computer. The selection of 1 line out of 100 on input requires multiplexing; the selection of 1 line out of 100 on output requires demultiplexing. A possible hardware arrangement to provide the required multiplexing is shown in Fig. 16.4

The item number in BCD is loaded into the **POR** register of an interface unit. The 100 **EMPTY** lines are applied to ten 10-input multiplexers, the outputs of which drive another 10-input multiplexer. The control inputs of the first 10 multiplexers are driven by the lower digit of the item number, the output multiplexer is driven by the upper digit. Assume the item number is 16. The lower digit (5), applied to the 10 first-level multiplexers, will route the values on **EMPTY**[05], **EMPTY**[15], **EMPTY**[25], and so on through their respective multiplexers. The upper digit (1) applied to the output multiplexer will select the output of MUX1, in this case **EMPTY**[15]. This output will be applied to **PC1** for testing. The decoding/demultiplexing of an item number into a signal on one of 100 **VEND** lines can be accomplished in an analogous manner using eleven 4-line-to-10-line decoders. We shall leave this as an exercise for you. The **PC2** line will be used to enable the **VEND** demultiplexer.

Three interface units will be required to interface to the control panel. One will be used to connect to the keyboard in exactly the same manner as shown in Fig. 13.20, except that line **PDL**[7] will not be used for the keyboard, since there are only three rows. We shall also assume that this keyboard is equipped with an additional line KEY that goes low any time any key is pressed. This will be connected to the PC1 input of the interface. The input from the **LOAD** switch will be connected to **PC2**. The other two interface units will be connected to drive the eight seven-segment indicators in exactly the manner shown in Fig. 13.24, except that all eight outputs

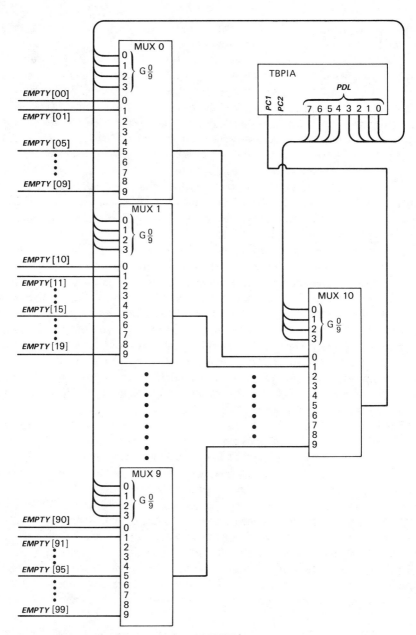

*Figure 16.4   Multiplexing to select EMPTY line.*

of the three- to eight-line decoder will be used to control the eight digits. The four lines **PDL**[7:4] from the interface controlling the digit lines will be used to control the four message lights.

Two interface units will be required to interface to the coin mechanism. One will be connected to the eight **COIN** inputs and the other will be used for the other

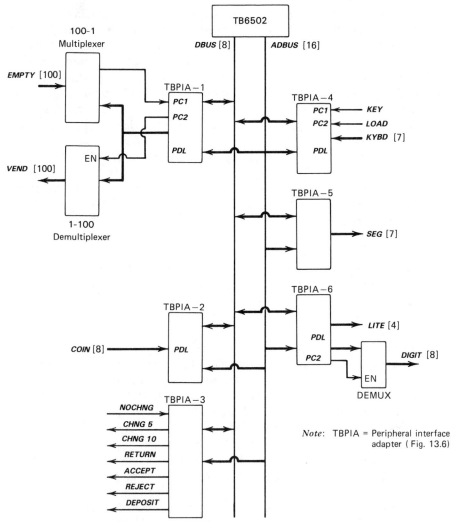

*Figure 16.5   Control system block diagram.*

seven lines to the coin unit. The connections described here are summarized in the block diagram of Fig. 16.5.

# 16.5   Planning the Program

At this point we are ready to plan the programs for the controller. In Section 11.5 we discussed the programming process. The first step is to define the problem. The preceeding sections have accomplished that step. We now know what it is we are going to control and what it is supposed to do. The second step is to plan the program.

Note that this does not involve coding. Instead, we develop the structure of the program, breaking the problem up into parts, specifying what is to be accomplished by each part of the program, and determining the relationships among the various parts. The importance of this step cannot be overemphasized. Novice programmers (and experienced ones, too) often skip this step and start right in coding. Unless the program is trivial, such a practice will inevitably result in poor programs. There are many techniques and styles for planning programs. The authors favor the use of flowcharts in this part of the programming process, and flowcharts will be used extensively in the following.

We first note that the controller will have two distinctly different modes of operation. In the NORMAL mode, the controller is interacting with customers, accepting item numbers, accepting cash, and dispensing items. In the LOAD mode, the operator will enter or change prices. It is apparent that the LOAD mode should totally override the NORMAL mode. Therefore, we will use the main program to implement the NORMAL mode, and the LOAD mode will be handled by an interrupt. When the operator presses the LOAD switch, the interrupt line will be pulled active and the computer will jump to an interrupt routine to handle the loading of new prices. The return from the interrupt program will go back to the main program at that point where it is waiting for a customer to start a transaction. We shall concentrate on the main program, leaving the interrupt program for the LOAD mode as an exercise for you.

In planning the main program, we shall break it up into sections. First, let us consider the power-on section, the flowchart of which is shown in Fig. 16.6. You will recall from Section 15.4 that, when power is first applied to the TB6502, it automatically obtains a reset vector that points to an initialization routine. This routine will initialize the various registers and the interface adapters. Since there are six PIAs in this system, there is quite a bit of initializing to be done.

There are two basically different power-on situations to be considered. In one case, the unit has been without power for a considerable time and is being started "from scratch." In the other case, power has been inadvertently lost for a relatively short time, and the contents of the RAM are still valid. Following initialization, the program checks the contents of the two test locations in RAM. If they are correct, control branches to Ⓒ , awaiting a customer transaction. If the test locations do not check out, the program turns on the "OUT OF SERVICE" message and goes into a loop that can be broken only the the operator entering the LOAD mode.

It might seem, at first glance, that the power-on section could simply dead-end after turning on the "OUT OF SERVICE" message. But we must consider the possibility of a customer ignoring the instructions, ignoring the "OUT OF SERVICE" message, and depositing coins. When nothing happens, he probably reads the instructions and pushes the CANCEL button to get his money back. It will be a first principle in the design of this program that there must be no circumstance in which the CANCEL signal can be ignored. A customer presses that button when he wants his money back. When people cannot get their money back from a vending machine, they get angry and sometimes vent their anger on the defenseless machine. We do not want that to happen, so we put a loop in the power-on section that constantly monitors the keyboard for a CANCEL signal. If one appears, the unit refunds any

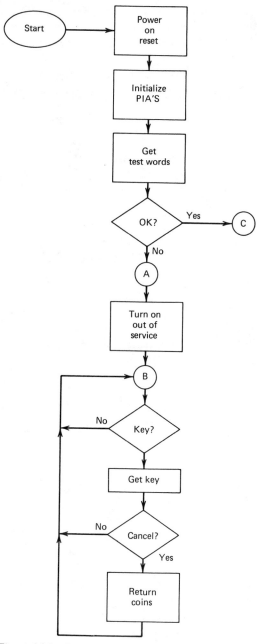

*Figure 16.6    Power-on section of main program.*

coins that may have been deposited. This is an example of a general principle. In designing a program that will interact with humans, you must try to anticipate and provide for all the dumb things they will do. In the vernacular, try to make the program "idiot-proof."

The next part of the program, the transaction initialization phase, starts at **○** (Fig. 16.7). This point is reached from the power-on phase if the RAM con-

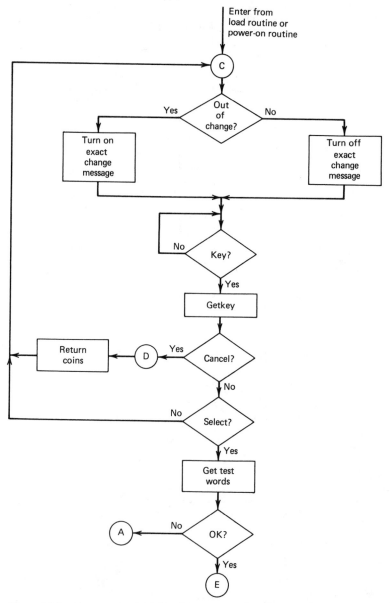

*Figure 16.7   Transaction initialization section of main program.*

tents are valid, on return from the LOAD routine, or after completion of a transaction. The first step is to check to see if there are coins available for making change and to turn the EXACT CHANGE ONLY message on or off accordingly. The program next waits for a key to be pressed and then checks for CANCEL or SELECT. If it is neither, the program ignores it and waits for another closure. If the key pressed is CANCEL, the program returns any coins deposited and returns to wait for another closure. Note again that we check for CANCEL even though there is no reason for it to be pressed at this point if the customer is following instructions. If SELECT was pressed, indicating that a customer is starting a transaction, the program checks the two test words in memory. If they are not correct, control returns to point Ⓐ to turn on the OUT OF SERVICE message and waits for the LOAD routine to be executed. This check is redundant, since we presumably cannot reach point Ⓒ unless the RAM is operating correctly. However, it provides an extra degree of reliability at a very small cost in additional program steps. If SELECT was pressed, control proceeds to the next phase to accept the item number.

The item selection phase starts at point Ⓔ , waiting for the item number to be entered (Fig. 16.8). If either CANCEL or SELECT is pressed at this point, it will be assumed that customer wants to start over, so control returns to point Ⓓ to return coins and wait for a new transaction to start. If a digit is entered, it is saved. Item numbers will be handled in BCD, so the digit will be shifted to the upper half of the location if it is the first digit, and the second digit will be loaded into the lower half. When the complete item number is entered, the program will look up the price. If the price is zero, this will indicate that an item number has been entered that is not in use, as in the case where the actual vending mechanism is designed for fewer than 100 items. In this case the INCORRECT ITEM message will be turned on and control will return to Ⓒ to await the start of new transaction.If the price is nonzero, the program will then check the appropriate EMPTY line, turning on the SOLD OUT message if the EMPTY line for that item is active. Assuming everything is all right, the program will display the price and set the amount deposited to 0.00 and go to the next phase to wait for coins.

The coin deposit and item vend phase of the program is shown in Fig. 16.9. The program alternately checks for a key closure or the deposit of a coin. The checking for key closure again is needed to permit the customer to cancel the transaction at any time.

When a coin is received, it is first checked to see if it is a good coin. If not, the coin is rejected and the program waits for another coin. If the coin is good, it is accepted, that is, moved to the holding box, and the amount deposited is updated. The amount deposited is then checked against the price, with the program returning to await additional coins if the required amount has not yet been received. If the amount is correct, the item is dispensed. If too much has been deposited, change is issued and the item dispensed. When the item has been dispensed, control then returns to point Ⓒ to await a new transaction.

This completes the planning of the program, except for the LOAD routine. The program as a whole is complex, but planning is not difficult when the problem is broken into reasonably simple segments. This approach to program planning is very important. Never treat a complex programming task as one big chunk. Analyze the problem in terms of isolating the component tasks. Plan the sections to accomplish

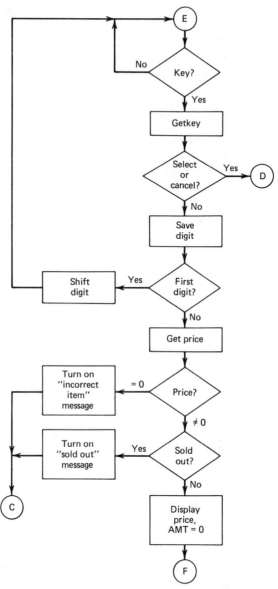

*Figure 16.8   Item selection section of main program.*

each task, keeping these sections simple enough that you can be reasonably sure that you have taken all pertinent factors into account. Where appropriate, treat the sections as subroutines, so that the same code can be used at various points in the overall program. Not only does this approach make the initial planning and programming simpler, it also simplifies debugging. When something does not work as expected, you can usually isolate it to one particular section and identify the source of the problem quickly. Flowcharts are particularly helpful in determining the relationships

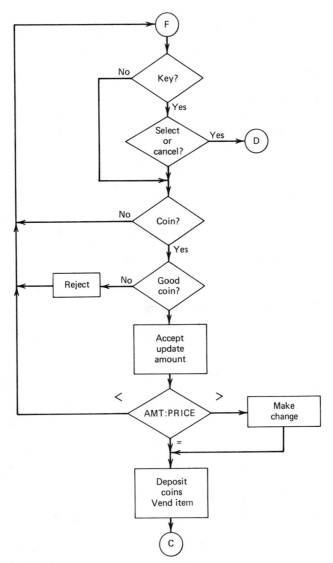

*Figure 16.9   Coin deposit and item vend section of main program.*

between various sections, as you can see right on the chart where you need to come into a section to accomplish a desired function.

## 16.6   Memory Map Layout

At this point we are in a position to make some estimates of the amount of memory required for various sections of the program and to set up an appropriate memory map. First, we note that the price table must be stored in RAM. Since the prices range from $0.25 to $9.95, we shall need three BCD digits for each price. The prices can thus be packed in two bytes each, with four bits left over in the upper byte. We could pack the prices with two prices per three bytes, but this would complicate the table look-up unduly. Thus, we need 200 bytes for the price table. We also need RAM to implement the stack, which is always on page 1 for the TB6502. We shall use 256 bytes of RAM to implement page 1, the first 202 bytes for the price table and the two test words, the remaining bytes for the stack. Page-0 must be implemented if we are to do any indirect addressing, and page-0 adressing is useful in reducing program size. We shall use ROM for the first 128 bytes of page-0, for permanent pointers for indirect addressing, and for constants, and RAM for the upper 128 bytes, which can be used for temporary storage and for variable pointers.

Two subroutines will be used frequently: one to scan the keyboard, the other to scan the seven-segment display. Subroutines for this purpose were written in Chapter 14, with the key-scan routine requiring 36 instructions, the seven-segment routine, 18 instructions. Since instructions in a typical program are mostly two-byte and three-byte instructions, with a few one-byte instructions, the average number of bytes per instruction is probably about 2.4. However, when estimating memory requirements for programs not yet written, it is desirable to allocate more than probably needed to allow a margin for error. For that reason, we shall use three bytes per instruction in our estimates, so that we shall allocate 108 bytes for the key-scan routine, 54 bytes for the seven-segment routine.

In the power-on section, there are 6 PIAs with two registers each (**DDR** and **CCR**) to be initialized. It will require two instructions for each, a load immediate and a store, or 24 instructions for all six PIAs. We shall also initialize the stack pointer and, perhaps, a few counters or pointers. Let us allow for 10 altogether, another 20 instructions. We then check the two test words, about six instructions. The rest of this section will probably require about 10 more instructions. The total is 60 instructions, or 180 bytes. This does not include the key-scan subroutine, which has already been accounted for. The transaction initialization section is quite simple. We allow six instructions to check for change, eight to check the keyboard for CANCEL or SELECT, and six to check the test locations in RAM. This gives a total of 20 instructions, or 60 bytes.

In the item selection section, we again allow eight instructions to check for CANCEL or SELECT, and six instructions to save and shift the digits. Getting the price will involve a table look-up routine, one that will be somewhat more complicated than those studied earlier since the item number is in BCD. This number will probably first have to be converted to binary to provide a convenient offset into the price table. We shall allow 30 instructions for this function. The checks for zero price and item sold out will require another 10 instructions. This gives a total of 54 instructions, or 162 bytes.

In the deposit and vend section, we again allow eight instructions to check for CANCEL or SELECT, and eight instructions to check for a good coin. We shall allow six instructions to update the amount deposited, eight instructions to compare the amount to the price, and eight instructions to compute the change and dispense the item. This gives a total of 36 instructions, or 108 bytes. Summarizing, we come up with a total of 672 bytes. Since we have not yet planned the LOAD section, we cannot make any detailed estimate. However, it seems likely that it will be much less complex than the main program, so we shall arbitarily allow 300 bytes for that routine. This gives a grand total of 972 bytes, so it appears that a 1K ROM will be adequate. This will also allow for the six bytes necessary for the interrupt and reset vectors.

The final step in this phase of planning is to assign locations in the address space to the various memory elements. We have 128 bytes of ROM to go in the lower half of page 0, 128 bytes of RAM in the upper half of page 0, and 256 bytes of RAM to implement page 1. The 1K of ROM for the program and the 24 locations for the six PIAs should be located in the address space in such a manner as to minimize the amount of external address decoding required.

There are many possible arrangements. We shall start the layout by noting that the the division of page 0 into two 128-byte segments suggests that page 1 be similarly divided. We shall, therefore, implement these two pages with one 128-byte ROM chip and three 128-byte RAM chips.

The main ROM will be most economically realized in the form of a single 1K chip. On this basis we will use the lower 2K of the address space, with ADBUS[10] dividing this space into two 1K segments, as shown in Fig. 16.10. The 1K ROM is in the upper 1K segment, with the lower segment accommodating the four 128-byte ROM and RAMchips and the six PIAs.

We then use a three- to eight-line decoder to divide the lower 1K into eight 128-byte segments, with ROM 1 and the RAMs in lower four of these. The six PIAs are allocated to the remaining four segments on the basis of their function. PIA 1, interfacing with the vending mechanism, is in segment 4. PIA 2 and PIA 3, interfacing with the coin mechanism, are in segment 5. PIA 4, interfacing with the keyboard, is in segment 6. PIA 5 and PIA 6, interfacing with the display, are in segment 7. Note that an additional bit is needed to select PIA 2 or PIA 3 in segment 5, and PIA 5 or PIA 6 in segment 7.

The complete map, wih decoding logic, is shown in Fig. 16.10. The resultant address ranges for each device in the address space are listed in Fig. 16.11. Note that the ROM occupies the highest addresses in the implemented address range. This is very important for implementing the interrupt and address vectors. Since the upper five bits of the addresses are not used at all, addresses in the range FFFA–FFFF, the addresses of the interrupt and reset vectors, will access locations 07FA–07FF in ROM 2.

You may find it surprising that such a small amount of memory is needed for a fairly complex system. This is characteristic of control problems. Such problems normally involve mostly programs and relatively little data. And programs do not require as much memory as data in most cases. As seen here, 1000 bytes can accommodate quite a lot of program, but 1000 bytes of data are not much data. This low memory requirement suggests two possibilities. One is to use a single-chip microcomputer, one with a microprocessor and a small amount of RAM and ROM on a single chip, eliminating the need for separate wiring to memory chips. A number of such units are available on the market, and they have appeared in response to a need for economical realization of control systems such as this vending machine controller.

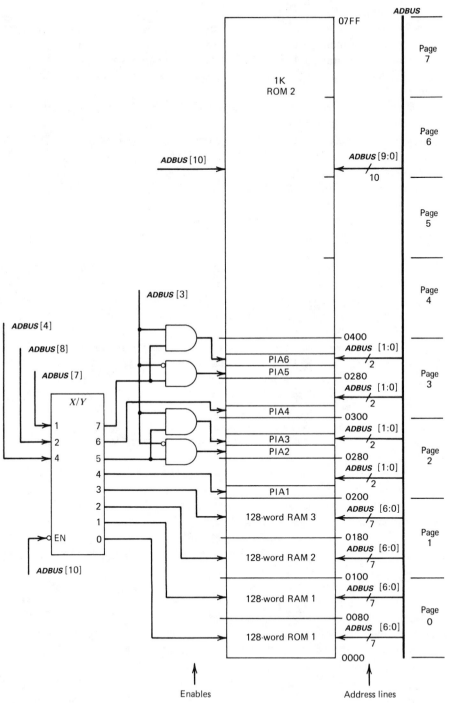

*Figure 16.10    Memory map for vending machine controller.*

| Device | Address range |
|--------|---------------|
| ROM 1 | 0000–007F |
| RAM 1 | 0080–00FF |
| RAM 2 | 0100–017F |
| RAM 3 | 0180–017F |
| PIA 1 | 0200–0202 |
| PIA 2 | 0280–0282 |
| PIA 3 | 0284–0286 |
| PIA 4 | 0300–0302 |
| PIA 5 | 0380–0382 |
| PIA 6 | 0384–0386 |
| ROM 2 | 0400–07FF |

Figure 16.11  Address ranges for vending machine controller.

A second possibility is to take advantage of this unused memory space to expand the capabilities of the system. For example, the controller could keep track of the inventory and have access to a telephone line or other communication link. The central office of the vending machine operator could then remotely interrogate the controller to determine if the machine needed restocking or other service. Another possibility is to have the controller compile data with regard to purchasing patterns, which could be very useful to the operator. For example, if it turned out that customers purchased item B only when item A was sold out, this would probably indicate that item B was a second choice, and more business would be obtained by replacing item B with a larger stock of item A. This sort of thing has happened in many cases when microprocessors have been used to carry out functions that were previously done in other ways. The designers have found that the microprocessor could not only implement the original functions economically but could also add additional capabilities at little or no additional cost.

## 16.7  Coding the Program

At this point we are ready to start on the actual coding of the program. Note again how much effort has to be expended on problem definition and planning before we are ready to start with coding. After all this work, the coding is almost anticlimactic. And that is the way it should be. The coding should be mostly a routine task. The "real thinking" should be done before coding starts.

This first step in the coding is to set up a tentative symbol table. This is very important because we are treating the program as a set of smaller sections and subroutines, all of which will be referring to a shared set of variables and branch points. We need to assign names, or symbols, for these variables and branch locations so we can refer to them in a uniform manner as we code the various sections of the problem. We could assign the name as we run across them, but it is more efficent to make as many assignments as possible all at one time. It is unlikely that we shall anticipate all the variables and branch points, and we shall add to the symbol table as we code, but it is still advantageous to set up as much of the symbol table in advance as possible. It is not necessary to assign addresses at this time, but we shall make some tentative address assignments in order to get an initial picture of the manner in which the program will be arranged in memory.

The tentative symbol table is shown in Fig. 16.12. Comments are not a part of the usual symbol table, but we include them here because this symbol table is to be used in the coding process. We need the comments to remind us of the meanings of the various symbols and to ensure that we shall use them properly and consistently

| Symbol | Address | Comments |
|---|---|---|
| PRICE | 0100–01C7 | Price table in RAM; two bytes for each price; two low-order digits in even-numbered byte, high-order digit in odd-numbered byte; price for item 00 in 0100,0101, running to price for item 99 in 01C6,01C7 |
| TEST1 | 01C8 | Test word = 0000 |
| TEST2 | 01C9 | Test word = FFFF |
| KEYSCAN | 0750 | Subroutine to scan keyboard |
| DISPLAY | 07C0 | Subroutine to scan seven-segment displays |
| START | 0400 | Start of main program |
| POINTA |  | Start of OUT OF SERVICE routine in power-on section |
| POINTB |  | Start of key-check loop in power-on section |
| POINTC |  | Start of transaction initialization section |
| POINTD |  | Coin return loop for CANCEL |
| POINTE |  | Start of item selection section |
| POINTF |  | Start of deposit/vend section |
| LOAD |  | Start of interrupt routine to load new prices |
| PRICEH | 00FF | Temp for high and low bytes of price of item |
| PRICEL | 00FE | being vended |
| AMTH | 00FD | Temp for high and low bytes of amount |
| AMTL | 00FC | deposited |
| ITEM | 00FB | Temp for item number |
| CHNG | 00FA | Temp for change due |
| SCNBUF | 00F0–00F7 | Array for seven-segment codes for current display |
| DIGCNT | 00F8 | Digit count for seven-segment display |
| N | 00EF | Temp for variable in KEYSCAN |
| SWNO | 00EE | Temp for variable in KEYSCAN |
| SEVSEG | 0000–0009 | Table of seven-segment codes in ROM |

*Figure 16.12  Tentative symbol table for vending machine program.*

as we code the various sections of the program. We have put the two subroutines at the high end of ROM, just below the reserved locations for interrupt and reset vectors. We can locate these as we are using routines already written and can thus be fairly sure of the space needed. For the main routine, however, since we have only rough estimates of length, we specify only the starting location, at the start of the main ROM.

Note that we have named the various points in the program in accordance with the flowchart, POINTA, POINTB, and so forth. This will greatly simplify coding, making it easy to see exactly where we are in reference to the flowcharts. The LOAD routine will go between the main program and the subroutines. We have specified the locations of the various constants, tables, and data, as there is n reason not to do so at this time. The names and addresses of the locations in the PIAs will also have to be included in the final symbol table, but we shall omit them for now, because they can be easily determined from Fig. 16.11.

Since we shall be testing and setting values on various peripheral lines, a table showing which peripheral lines are connected to which output lines of the interface adapters will be needed. Figure 16.5 shows which lines are connected to which PIAs, but it does not show the exact line numbers in every case. Figure 16.13 provides this information.

With all this information tabulated, and the program planned in the form of flowcharts, the actual coding, as noted earlier, should be routine. We shall code one

| PIA 1 | HIGH DIGIT OF ITEM NO. = *PDL*[7:4] |
| | LOW DIGIT ID ITEM NO. = *PDL*[3:0] |
| | *EMPTY* INPUT FROM MULTIPLEXER = *PC1* |
| | ENABLE FOR *VEND* DEMULTIPLEXER = *PC2* |
| PIA 2 | *COIN*[7:0] = *PDL*[7:0] |
| PIA 3 | *NOCHNG* = *PDL*[7] |
| | *CHNG5* = *PDL*[5] |
| | *CHNG10* = *PDL*[4] |
| | *DEPOSIT* = *PDL*[3] |
| | *ACCEPT* = *PDL*[2] |
| | *REJECT* = *PDL*[1] |
| | *RETURN* = *PDL*[0] |
| PIA 4 | KYBD[6:0] = *PDL*[6:0] |
| | *KEY* = *PC1* |
| | *LOAD* = *PC2* |
| PIA 5 | *SEG*[6:0] = *PDL*[6:0] |
| PIA 6 | DIGIT NUMBER FOR 7-SEG DISPLAY = *PDL*[2:0] |
| | INCORRECT ITEM MESSAGE = *PDL*[4] |
| | SOLD OUT MESSAGE = *PDL*[5] |
| | EXACT CHANGE MESSAGE = *PDL*[6] |
| | OUT OF SERVICE MESSAGE = *PDL*[7] |
| | ENABLE FOR DIGIT DRIVER = *PC2* |

*Figure 16.13   Terminal numbers of peripheral lines.*

section of the program, leaving the rest as exercises for you. The section we shall code is the transaction initialization section, for which the flowchart is shown in Fig. 16.7.

The program starts at POINTC. The first step is to fetch the peripheral data from PI-3, and test bit 7, which corresponds to NOCHNG. If this is 1, we turn on the EXACT CHANGE message; if it is 0, we turn off the message. At KEYC we read in the PCR of PIA-4 and check PC1, which is controlled by KEY. We hold in a loop at KEYC until KEYgoes high. We then jump to a 2-msec delay to allow key bounce to settle and then jump to KEYSCAN to determine which key was pressed.

```
*TRANSACTION INITIATION SECTION OF VENDING
* MACHINE MAIN PROGRAM
*
POINTC    LDA    PP3         /GET NOCHNG
          BPL    CHANGE      /IF 0, GO TO CHANGE
          LDA    #$02        /TURN ON
          ORA    PP6         /   "EXACT CHANGE"
          STA    PP6         /   MESSAGE
          BNE    KEYC        /GO TO CHECK KEY
CHANGE    LDA    #$FD        /TURN OFF
          AND    PP6         /   "EXACT CHANGE"
          STA    PP6         /   MESSAGE
KEYC      LDA    PCR4        /READ KEY LINE
          BPL    KEYC        /NOT SET, READ AGAIN
          JSR    DELAY       /WAIT FOR KEY BOUNCE
          JSR    KEYSCAN     /GET KEY NUMBER
          CMP    #$01        /CANCEL? IF NO,
          BNE    SEL1        / GO CHCK FOR SELECT
          LDA    #$03        /TURN ON
          ORA    PP3         /   REJECT AND RETURN
          STA    PP3         /   SIGNALS
          LDA    #$FC        /TURN OFF
          AND    PP3         /   REJECT AND RETURN
          STA    PP3         /   SIGNALS
          JMP    POINTC      /RETURN TO POINT C
SEL1      BNE    POINTC      /NOT SELECT, POINT C
          LDA    TEST1       /GET ALL-0 TEST WORD
          BNE    POINTA      /NOT ALL-0, POINT A
          LDA    TEST2       /GET ALL-1 TEST WORD
          EOR    #$FF        /
          BNE    POINTA      /NOT ALL-1, POINT A
POINTE    .      .           /   .    .
```

We assume that the key numbers range from 0 to 11, with KEY 0 = SELECT, KEY 1 = CANCEL, KEY 2 = 0, KEY 3 = 1, and so on. If the key is CANCEL, we issue both the REJECT and RETURN signals since there might be coins in either the coin slot or the holding box. If the key is neither CANCEL nor SELECT, it is ignored, since SELECT must be pressed to initiate a transaction. If SELECT is pressed, the two RAM test words are checked, and control returns to POINTA if either is wrong. If the test words check out, control advances to POINTE, the start of the item selection section of the program.

Some general comments about this program are appropriate. First, we find that it consists of 28 instructions, requiring 73 bytes, compared with our estimate of 20 instructions, 60 bytes. So our estimate was low, and if all our estimates are similarly low, we may need more memory. Next, we note that the delay routine for handling key bounce was not included in the original flowchart. This was simply overlooked in the initial planning, and this will require some memory space that we did not take into account in the estimates used to set up the memory map. Next, note that the RETURN and REJECT signals are turned on and then off. Obviously, we do not want to leave them on, or all coins will be rejected. However, this short segment will turn them on for only a few microseconds, which may not be long enough to trigger the mechanisms. If we were actually building this unit, at this point we would check the specifications for the coin mechanism, to determine the required duration of the various activation signals. It may be necessary to insert a delay here to hold the signals on for the required time.

It may be discouraging to find these things overlooked in spite of all the planning that was done before coding started. But this is perfectly normal. Design is a circular, iterative process. No designer, no matter how skillful or experienced, will ever anticipate all the aspects of a complex design right from the start. You do what we have done here. You start in with the design process, include those features you recognize at the beginning, and move ahead. When you reach a point where you realize that you have overlooked something, go back and add what is needed, make the adjustments required by these additions, and then continue. In this case we probably would not write the delay routines or modify the memory map at this point, since we do not know yet whether additional ROM will be needed. We would make a note of the fact that a delay routine will be needed and continue with the coding. When all the coding other than the delay routine was completed, we would see if we had enough room left for the delay routine. If not, we would adjust the memory map accordingly and make such other changes as this might require. It is perfectly normal to have to make several passes through the design process before converging on a working design.

As noted earlier, we will leave the remaining coding as an exercise for you. When that coding is done and the memory map adjusted if necessary, the next step would be a verification of the design. This would probably be done first by emulation of the system on a microprocessor development system. Next, the system might be built in the lab and tested. When it has been verified that the control system works properly, it will then be time to start on the actual production design of the vending machine. This is a process far beyond the scope of this book. In this chapter we have endeavored to illustrate the design of a microprocessor control system in terms of the logic design, system organization, and programming, with specifications expressed in terms of required interactions with a given set of input and output lines.

## Problems

**16.1** Draw a block diagram similar to Fig. 16.4 for a 1–100 demultiplexer to drive the **VEND** lines.

**16.2**  Modify the KEYSCN routine of Chapter 13 as required to accomodate the 12-key keyboard used with the vending machine. Assume the keys are numbered from 00 to 0B, with key 00 = *SELECT*, key 01 = *CANCEL*, and so on up to key-0B = 9. Modify Fig. 13.27 as necessary to reflect this arrangement, and rewrite the program. Note that there is no need to use *C* to indicate if a key has been pressed, since there is a separate KEY line for that purpose. Also note that the debounce delay is no longer needed, since that will be included when the main program checks for closure on *KEY*.

**16.3**  Modify subroutine SCAN of Chapter 13 to accommodate the eight-digit display of the vending machine.

**16.4**  Code the power-on section of Fig. 16.6.

**16.5**  Code the item selection section of Fig. 16.8.

**16.6**  Code the deposit/vend section of Fig. 16.9.

**16.7**  Write the LOAD routine for the vending machine. This routine will be an interrupt routine, accessed when the operator presses the *LOAD* button. You need not concern yourself with this part of the process. Just write the routine and assume the interrupt system is properly set up. After pressing the *LOAD* button, the operator should then enter the number of the first item for which a new price is to be entered. The system should then display the item number and the price currently stored in RAM. The operator should then enter the three-digit price and press *SELECT* again. The system should enter this price in memory and display it so the operator can verify that it has been entered properly. The operator may then repeat this process, that is, enter an item number, press *SELECT*, enter the price, press *SELECT*, and so on. If, after entering a price and pressing *SELECT*, the operaor presses *SELECT* again without entering an item number, the system should advance to the next sequential item number. This will permit the operator to enter a whole series of prices without having to enter every item number. The operation will be terminated by pressing the *CANCEL* button, which will cause a jump to the main program, at the start of the transaction initialization section.

**16.8**  Assume that the coding has been completed and it has been determined that another 256 bytes of ROM are needed. Modify the memory map of Fig. 16.10 to move all PIAs onto page 2 and to add a 256-byte ROM chip to implement page 3. Also correct Fig. 16.11 to show the address ranges with this modification.

# The Clock-Mode Assumption Reexamined

## 17.1 Attention to Output Waveforms

In Chapter 9, when we designed the waveform generator of Section 9.4, we said little regarding the ultimate application of the circuit. Consider the following dialogue that might well have taken place prior to the design of this circuit.

Application engineer:  Can you design a circuit for me that can be controlled to generate any of the four output waveforms given in Figure 9.11?

Design engineer:  No problem! I assume that a clock-mode circuit will be satisfactory.

Application engineer:  I don't care what is inside your circuit package, but the output must conform faithfully to Figure 9.11 with no extraneous or transient output changes (glitches). In addition, the frequency (when $X_1 = X_2 = 1$) must be a very stable 5 MHz $\pm$ .0001%.

Design engineer:  Hmmm.

The design engineer doesn't let on that he is unsure whether a similar design that he had already completed for a less exacting application will satisfy the requirement of no transient output transitions. Given the extreme requirement for frequency stability, he is certain that a clock-mode design with a crystal-controlled clock input is the only reasonable option. He is confident that he will find some way to eliminate extraneous output changes, if indeed any appear.

**481**

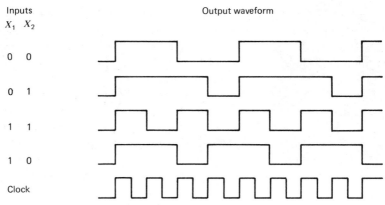

Inputs
$X_1$  $X_2$

Output waveform

0  0

0  1

1  1

1  0

Clock

*Figure 17.1   Waveform generator output and clock.*

To see if the design engineer has any reason for concern, let us look again at the desired output waveforms as shown in Fig. 17.1, together with the driving clock that itself must have a frequency of 10 MHz. The circuit will employ two rising edge-triggered $D$ flip-flops. The two flip-flops must be capable of changing state at any of the transitions (rising or falling) of the output waveform. There must be a rising clock edge at each one of these transitions, hence the clock frequency of 10 MHz.

Next, let us consider the output logic network for the waveform generator as shown in Fig. 17.2. This network is a realization of the output expression given by Eq. 9.2. Any unwanted output changes must, of course, be changes in the logical value of the OR gate. In general, the existence of unanticipated output changes will be a function of the relative gate delays of the various gates in the network. To show that such a transition can occur in the network of Fig. 17.2, let us consider a particular combination of gate delays and a particular point in the output waveform.

Although the gates from the same logic family used to construct a circuit will usually have approximately the same gate delay, variations in the delay in individual

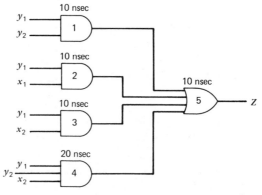

*Figure 17.2   Waveform generator output network.*

gates can occur. In the situation that we are about to consider, a problem can occur if the delay in gate 4 is greater than the delay in the other AND gates. Consequently, let us postulate a delay of 20 nsec for gate 4 and delays of 10 nsec for the other gates. These delays are indicated in Fig. 17.2.

Let us assume that the desired output waveform is the one corresponding to $X_1 = 0$ and $X_2 = 1$. The transition table for the waveform generator is repeated in Fig. 17.3$a$. In the column corresponding to $X_1 X_2 = 01$, the states $y_1 y_2$ change in the order $00 \rightarrow 01 \rightarrow 11 \rightarrow 10 \rightarrow 00$. That is, only one of the two flip-flops changes state each clock period. Thus the circuit behaves as a counter, counting in the Gray code. Eventually we shall see that under certain circumstances it may not be possible to avoid transient output changes in the presence of multiple changes in inputs or state vari-

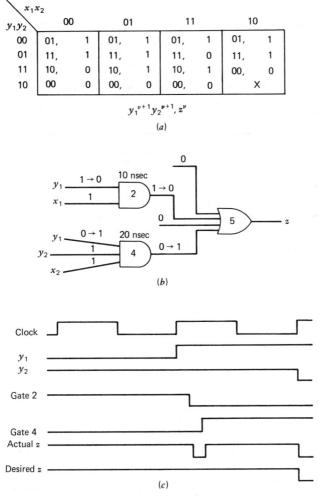

Figure 17.3   Malfunction in the waveform generator.

ables. For now let us confine our attention to the case where $y_2$ remains constant at 1 and $y_1$ changes from 0 to 1. Under these circumstances gates 1 and 3 have a constant output of 0 and, therefore, do not affect the circuit output, $z$. This is emphasized in Fig. 17.3b where the assumed input and state variable values are indicated.

In Fig. 17.3c we depict the actual behavior of the waveform generator, as $y_1$ changes from 0 to 1 under the input and delay conditions stated previously. Notice that the output of gate 2 goes to 0 10 nsec before gate 4 goes to 1. Thus there is a 10 nsec period during which the outputs of all four AND gates are 0. The output, $z$, will correspondingly go to 0 for a short period following the delay in the OR gate. (The actual output pulse will be less than 10 nsec wide due to rise-time effects, but this is of no help.) The actual behavior on line $z$ is depicted just above the desired output waveform in Fig. 17.3c.

The phenomenon that caused the malfunction just described above is called a *hazard*. A hazard occurs in a two-level AND–OR network wherever the output is to remain 1 before and after a single input change but no single AND gate has output 1 both before and after the change. Similar hazards occur in two-level OR–AND, NAND–NAND, NOR–NOR, and multilevel networks as well. It is possible to avoid hazards in the design of two-level networks subject to single input changes, as will be discussed in the next section.

Recalling our opening conversation between the design engineer and the applications engineer, we must inquire into the relation between hazards and the clock-mode assumption. Is the waveform generator a clock-mode circuit? The answer is yes. If a circuit operates strictly in the clock mode, hazards are of no concern. Suppose some such glitch as shown on the output line in Fig. 17.3c appears at the output of the network driving the $D$ input to one of the flip-flops. As in this case, all such transient circuit values occur in response to a rising clock transition. As in this case, the transients disappear and all gate outputs settle to their proper values before the next clock transitions. Thus the hazards have no effect within the clock-mode circuit.

Do we conclude then that the applications engineer was merely an alarmist in demanding that no transient output transtions be allowed? No! His circuit, which is to be driven by the waveform generator, may not have the same clock or may not be clock mode! There are innumerable special circuits that must be connected to digital systems and that will respond to every input transition anticipated or not. The waveform generator acceptable to the application engineer must, then, be only internally clock mode with a hazard-free output network. We shall complete this design in the next section.

# 17.2   Elimination of Hazards

Fortunately, the existence or nonexistence of hazards is a property of the combinational logic realization of a Boolean function. As we shall see, it is possible to eliminate these hazards by altering the realization. Figure 17.4a is a reproduction of the Karnaugh map of the output function of the waveform generator originally given in

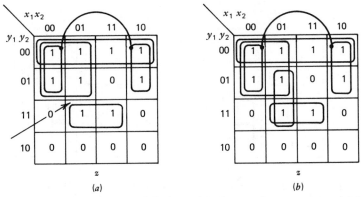

*Figure 17.4* *Elimination of the hazard in the waveform generator:* (a) Z, (b) Z with hazard eliminated.

Fig. 9.14. The realization of this map that led to the network of Fig. 17.2 is given as Eq. 17.1.

$$z = \bar{y}_1\bar{y}_2 \lor \bar{y}_1\bar{x}_1 \lor \bar{y}_1 x_2 \lor y_1 y_2 x_2 \tag{17.1}$$

This expression was first given as Eq. 9.2.

The arrow in Fig. 17.4 points to the source of the problem discussed in the previous section. This point is the transition of $y_1$ from 0 to 1 in the $X_1 X_2 = 01$ column while $y_2 = 1$. Notice that not all the 1's in this column are contained in a single product grouping. The state of the circuit moves from the $\bar{y}_1 \bar{x}_1$ product to the minterm $y_1 y_2 x_2$ when $y_1$ goes from 0 to 1. No product is 1 both before and after the transition.

In Fig. 17.4*b* we have added a redundant product to those selected in Fig. 17.4*a*. Note that this product will be 1 both before and after the transition of $y_1$ from 0 to 1 in the $x_1 x_2 = 01$ column. This leads to a new expression for the output of the waveform generator given by Eq. 17.2.

$$z = \bar{y}_1 x_2 \lor \bar{y}_1 \bar{x}_1 \lor \bar{y}_2 x_2 \lor y_1 y_2 x_2 \lor \bar{x}_1 x_2 y_2 \tag{17.2}$$

When the network of Fig. 17.2 is modified to incorporate the new product term and agree with Eq. 17.2, the extraneous output pulse of Fig. 17.3*c* will disappear.

Will the output of the waveform generator satisfy the specification provided by the application engineer once this modification is included? In arguing that the answer is yes we note that changes in the circuit inputs are infrequent. In this case only transient outputs arising from state changes are of concern. Therefore, the Karnaugh map of Fig. 17.4*b* need only be examined on a column-by-column basis to determine if any additional hazards exist. Clearly the first and last columns present no problems, since in each case all the 1's are included in single-product terms. The third column is not a problem, since $y_1 y_2$ will never change directly from 00 to 11 or vice versa in a single clock period.

Our analysis of the hazard in the waveform generator leads us to the following theorem, which we shall state without further proof.

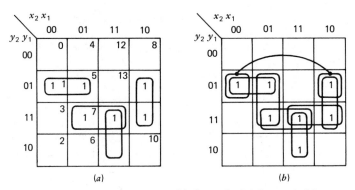

*Figure 17.5    Input and state variable hazards: (a) Original, (b) hazard free.*

**Theorem:**   *Transient output transitions in sum of product networks caused by hazards may be eliminated by overlapping the hazard transition with redundant product terms provided that no two outputs or state variables are allowed to change simultaneously.*

## Example 17.1

Obtain a hazard-free realization of the output network given by the Karnaugh map of Fig. 17.5a subject to any change in a single input or a single state variable.

Solution:

We note that a single input change can cause a transition between minterms $m_1$ and $m_9$ as well as between $m_{11}$ and $m_{15}$. A single change in a state variable can cause a transition between $m_5$ and $m_7$. None of these pairs of minterms are included in the same cubes in the Karnaugh map of Fig. 17.5a. This situation is remedied by adding three redundant products as shown in the map of Fig. 17.5b.

The sum of products network realization of the final Karnaugh map of Fig. 17.5b is given as Eq. 17.3.

$$z = \bar{y}_2 y_1 \bar{x}_2 \vee y_2 y_1 x_1 \vee y_2 y_1 x_2 \vee \bar{y}_2 y_1 \bar{x}_1 \vee y_1 \bar{x}_2 x_1 \vee y_1 x_2 \bar{x}_1 \vee y_2 x_1 x_2 \qquad (17.3)$$

■

# 17.3  Synchronizing Inputs to Clock-Mode Circuits

Design in the clock mode (subject to the clock-mode assumption) is by far the easiest and most convenient approach to the design of sequential circuits. The large majority of existing digital systems have been designed this way. In Section 17.1 we saw that occasionally it is necessary to impose additional conditions on the outputs of systems designed in the clock mode. The purpose of this section is to consider possible problems associated with inputs to clock-mode systems.

The clock-mode assumption may be restated from Chapter 8 as follows.

In a circuit operating in the clock mode all input transitions and all state-variable transitions must be synchronized by a clock pulse and all the variables must be stable before the next clock pulse. Usually these transitions are in synchronism with either a rising or a falling edge of the clock.

Designers are able to ensure that transitions on state variables are clock synchronized. They often have no such control of inputs that are generated in other systems. Suppose that system D to be designed is to receive inputs from system A. There are many reasons why an input from system A might not be synchronized with the internal clock of system D: (1) system A might be driven by an internal clock not synchronized with the clock of system D; (2) system A might not be a clock-mode system; (3) there might be a significant transmission line delay between systems A and D, so that the inputs would be delayed and, therefore, unsynchronized even if it were possible to drive the two systems with the same clock.

Even with these reasons in mind, the designer of system D should probably not give up without a fight. If steps can be taken to adjust specifications or physical location so that the two systems can be driven by the same clock, the designer of system D is well advised to make the attempt. The engineering time saved and the greater confidence in the reliability of the final design are well worth some time spent on interpersonal relations.

Try as he might, the system designer may finally have no choice but to deal with some unsynchronized inputs. In Fig. 17.6 we see system D with examples of two possible kinds of inputs. As discussed in Chapter 10, we assume that system D is partitioned into a control section and a data section. The vector **DATA**[4] is an input to the data section or a data input while $x$ is a control input. If these signals arrive at system D unsynchronized with the clock, then some steps must be taken to synchronize them so that system D can operate in the clock mode. As is typically the case, the function of the control input, $x$, is to indicate when the data input lines, **DATA**, are stable and can be accepted by system D. If system D monitors line $x$ to determine when **DATA** is stable before clocking it into the register, **CR**, there is no need to provide for synchronizing the lines, **DATA**, within system D. A similar argu-

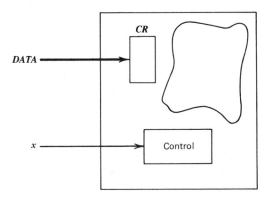

*Figure 17.6   Clock-mode system D with unsynchronized inputs.*

ment with respect to other input data vectors reduces the problem of input synchronization to synchronizing the control inputs.

To make our example more specific, let us assume that the only appearance of **DATA** within the RTL description is in step 10 as follows.

> 10.   **CR ∗ x ← DATA**;
>       → (*x̄,x*) / (10,11).

Since the transfer of **DATA** into **CR** is conditional on *x*, we need not worry about synchronizing the data lines; but let us examine the problems that result if we fail to synchronize the control input, *x*. Figure 17.7 is a realization of step 10 using falling edge *D* flip-flops. Suppose that *x* is an unsynchronized control input that goes to 1 immediately ahead of the trailing edge of a clock pulse. A 1 signal appears at point *a* one gate delay after the transition on line *x*. The level at point *a* is used to gate a clock pulse that will cause the transfer, **CR ←DATA** to take place. The resultant control pulse may be a narrow spike as shown, which may or may not be sufficient to effect the desired transfer. Even in a case in which the step 10 transfer did not occur, the signal at point *a* may have appeared at the input of control flip-flop 11 in time for the trailing edge of the clock to set this flip-flop to 1. This situation is depicted in Fig. 17.7 by the shaded area on the step 11 waveform, which indicates that the value is uncertain. It is also possible that gate delays will be such that point *b* will still be 1 at the time of the trailing edge of clock pulse one even though *a* reached the value 1 prior to this edge. This could result in 1's appearing simultaneously in control flip-flops 10 and 11 These two 1's could then propagate independently within the control unit causing a sequence of erroneous transfers. On the other hand, a different combination of gate delays could cause 0's to appear at the input of both flip-flops 10 and 12 coincident with the trailing edge of the clock pulse. In this case the control level would be lost. Again, these various possibilities are represented by the shaded areas for steps 10 and 11.

Now that we have observed the potential danger associated with an unsynchronized control input, let us consider what might be done to *synchronize* that input. First, let us transfer *x* into a *"synchronizing"* flip-flop, *r*, in a statement following ENDSEQUENCE and modify step 10 as follows.

> 10.    **CR ∗ r ←** DATA;
>        → (*r̄,r*) / (10,11).
>        .
>        .
>        .
>
> ENDSEQUENCE
>    *r ← x*
> END

Now the branch and transfer in step 10 are controlled by *r* rather than *x*. This means that *r* must be set to 1 by *x* one clock period before the transfer in step 10 is accomplished and control can continue to step 11. The argument that this measure will solve the synchronization problem is the following.

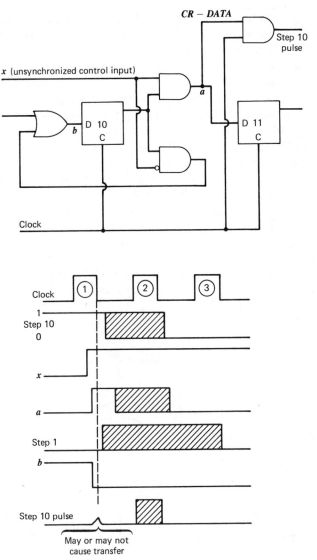

*Figure 17.7 Effect of unsynchronized control input.*

**Argument for a Synchronizing Flip-Flop** It is assumed that whenever $x$ goes to 1, it will remain there for several clock periods. If a transition of $x$ from 0 to 1 occurs very near a clockig transition, this transition may or may not set $r$ to 1. No matter! If $r$ is set, the transfer in step 10 and the branch to step 11 will be accomplished during the next clock period. If not, $r$ will be set one period later; and step 10 will be executed two periods later. Either behavior is acceptable.

If you are tempted to breathe more easily at this point, don't! Unfortunately, the flip-flop, $r$, may not behave precisely as assumed in the preceding argument. A number of researchers have verified empirically and argued theoretically that simultaneous transitions on the clock and data inputs of a flip-flop can leave the output gates of the device in the linear region. See, for example, Ref. [6]. The result is that the output may oscillate for an entire clock period without settling to either the old or a new logical value. For an illustration of this behavior, consider Fig. 17.8, which shows the pair of cross-coupled gates that connect to the flip-flop output. Every type of bipolar flip-flop will have this pair of gates at the output. The rest of the flip-flop, which will vary with the type, is represented by the box. A similar argument can be made for the MOS flip-flops within LSI parts. Suppose the flip-flop output is initially 0. Point $a$ in the circuit will be normally 1 but must be driven to 0 for a period of time to cause the output, $z$, to got to 1. The third waveform of Fig. 17.8 shows without verification what can happen when the $D$ input goes to 1 too close to the triggering clock transition. This tiny zero-going pulse may not have sufficient energy to drive $z$ to $+5$ V or even past the threshold of about 2.5 V for logical 1. In this case the output of both gates in the cross-coupled pair will be between 1 and 2.5 V when the pulse on $a$ has disappeared. The result is the oscillation of $z$ shown in the last waveform that can persist until the next triggering clock edge.

We have not shown that synchronization is hopeless. What has been demonstrated by the research in the area is that there will always be finite probabilty of

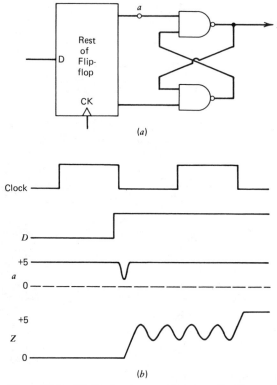

*Figure 17.8   Oscillation in a synchronizer flip-flop.*

failure in the synchronization process. It is up to the designer to reduce this proba-biltity of failure to a point that it is no more significant than the probability of other failures such as shorts and opens within the system. This is done by reducing the minimal interval between a data transition and the clock transition within which trouble will result. Even without synchronization, the interval might, for example, be only 1 nsec out of a 500-nsec clock period for a failure probability of 0.2%.

Computation of this probability is beyond the scope of this book (see Ref. [9]), but it is improved by merely adding the fallible synchronizing circuit just described. The probability of oscillation is less than the probability of one of the forms of erro-neous behavior of Fig. 17.7 given an unsynchronized $x$. Also, the oscillation may die out in one clock period depending on the magnitude of the pulse at point $a$.

Other steps that might be taken to further reduce the risk of synchronization failure are

1. Modify the circuit design of the synchronizing flip-flop, usually by increasing gain, so that oscillations will die out more quickly.
2. Connect the input of a second synchronizing flip-flop, $s$, between the input of the unsynchronized input, $x$, and the D input of the flip-flop, $r$.
3. Extend the length of time allowed for the oscilltions to die out before the syn-chronized input is used.

We can apply points 2 and 3 to reduce the risk of synchronization problems in the system of Fig. 17.6 by modifying the statements after ENDSEQUENCE as follows.

```
ENDSEQUENCE
r * (∧/CNT) ← s: s * (∧/CNT) ← x;
    CNT ← INC(CNT).
END.
```

Now $r$ and $s$ are updated less frequently. If the counter, **CNT**, is modulo-8, for exam-ple, then $s$ and $r$ are triggered only once every eight clock periods. Whenever $s$ goes into oscillation, these oscillations have eight clock periods to die out before $r$ is affected. The disadvantage is, of course, a possible eight-clock-period delay in the acceptance of **DATA** after this vector is available. If this response time is too slow for a particular application, the only choice is circuit enhancement or toleration of a larger probability of failure.

If you are uncomfortable with the complexity of synchronization, heed our ear-lier advice and make every effort to arrange to drive all systems with the same clock.

# 17.4  Clock Skew on Edge-Triggered Flip-Flops

There is yet another problem that can stand in the way of satisfaction of the clock-mode assumption. This problem concerns the clock itself. If all flip-flops are to change state synchronized with the edge of a clock source, this edge must actually reach the clock input of all flip-flops at approximately the same time. Sometimes delays on the clock lines can cause this not to happen. We shall illustrate this problem in the following paragraphs using rising edge-triggered $D$ flip-flops.

Sometimes a particular realization of a system will introduce delay in the clock line and cause this clock-mode assumption to be violated. Consider the clock-mode sequential circuit of Fig. 17.9 Each flip-flop shown will be assumed to be rising edge-triggered, so that any state change will take place immediately as the rising edge of a clock pulse reaches its $ck$ input terminal. In effect, the circuit is a simple 2-bit shift register. Assume that, because of some peculiarity in wiring the clock distribution lines, a delay has appeared between the clock line entering flip-flop 1 and the clock line entering flip-flop 2. This may have been caused by a long wire or by a pair of inverters inserted to help handle heavy loading on the clock line. In high-speed circuits, wire length can become quite critical.

The initial values of the flip-flops are shown in Fig. 17.9$b$ to be $y_1 = y_2 = 1$, and the input is 0. If the circuit were working properly, the next clock pulse would place a 0 in flip-flop 1, and the second pulse would propagate the 0 on to $y_2$. In Fig. 17.9$b$, we observe the problem caused by the delay in the clock line. Note that $y_1$ quite properly goes to 0 on the rising edge of clock pulse 1. This value is propagated directly to the input of flip-flop 2 before the clock pulse, $CK2$, reaches this device. Thus the positive transition on $CK2$ occurs when $D_2 = 0$, so that $y_2$ goes to 0. Thus, a state change has taken place at clock pulse 1 that should not have occurred until pulse 2.

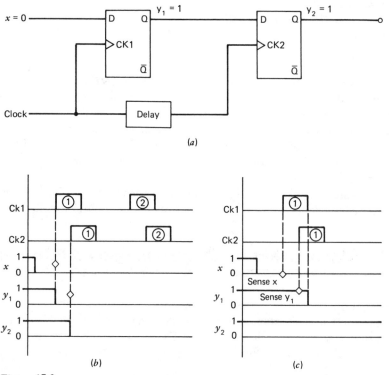

(a)

(b)                                      (c)

*Figure 17.9*

Care should be taken to assure that delays from the clock source to the clock inputs of all flip-flops are approximately the same. In slower systems this can be accomplished by merely balancing the use of any inverters required in the clock lines to increase fanout and keeping wires as short as possible. For faster systms the transmission line delay on wires, and therefore the physical layout, will become increasingly important.

A more complicated flip-flop can be of help when it is impossible to balance the delays on clock lines. Suppose that the data input was sensed at the leading edge of the clock pulse but the outputs did not change until the trailing edge, as shown in Fig. 17.9c. In that case the original value of $y_1$ would be sensed by the input of flip-flop 2 by pulse 1 and the circuit would perform properly provided the clock pulse width exceeded the *clock skew*. The design of this more complicated leading-edge-sensing trailing-edge-executing flip-flop is discussed in Ref. [4].

# 17.5   Is That Clock Really Necessary?

In the past four sections we probed some of the difficulties associated with clock-mode design. Perhaps we have lead you to wonder if it is possible to design sequential circuits without the benefit (or perhaps you are thinking detriment) of the clock. Indeed it is possible to design without a clock. Sequential circuits without an explicit clock will be called *level-mode sequential circuits* (another commonly used but probably misleading term is asynchronous). In the remaining sections of this chapter we shall analyze some circuits designed in the level mode. In the process we shall encounter a number of new problems and restrictions associated with this approach to design. In short, if you thought that clock-mode design was plagued with problems, wait until you have seen level mode!

As a beginning, let us see if we can remove a clock from a clock-mode sequential circuit and expect it to continue to function. The circuit shown in Fig. 17.10 was deliberately designed so that this would be the case, so let us start there. Clearly the flip-flops in Fig. 17.10 must not be edge-triggered. If the clock source were disconnected from this type of flip-flop, it would be impossible for the device to change state. We therefore assume that our storage elements are clocked latches as shown in Fig. 17.11a. Removing the clock from this circuit is the same as setting it to a permanent logical 1, so that the outputs $Y_1$ and $Y_2$ of the logic network will always propagate through the clock control NAND gates directly to the inputs to the cross-coupled pair of gates. Thus, removing the clock is the same as replacing the NAND gates with inverters, resulting in the circuit shown in Fig. 17.11b.

We have previously argued that multiple-state changes in a single clock period can be the result in the circuit of Fig. 17.11a, if the clock for this type of latch is anything but a sequence of very narrow pulses. Setting the clock to 1 can be interpreted as making the clock pulses infinitely wide. The circuit of Fig. 17.11b will exhibit the same behavior as the clock-mode sequence checker only if it is somehow "immune" to multiple state changes.

It is easy to see that the combinational logic networks in Fig. 17.11b implement the following Boolean functions.

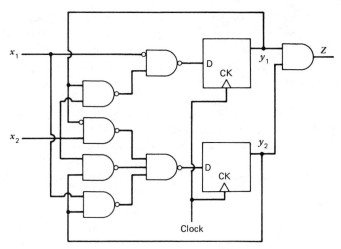

*Figure 17.10   Clock-mode sequence checker.*

$$Y_1 = x_1 \lor y_1 x_2 \tag{17.4}$$
$$Y_2 = y_1 x_2 \lor y_2 x_2 \lor y_2 x_1 \tag{17.5}$$
$$z = y_1 y_2 \tag{17.6}$$

Substituting all possible combinations of the four variables in these three expressions yields the transition table of Fig. 17.12. $Y_1$ and $Y_2$ are the next state values for present states defined by the values of $y_1$ and $y_2$. For level-mode circuits the $y_1$'s are called *secondaries*, whereas the $Y_1$'s are called *excitations*.

In the transition table of Fig. 17.12 there are several entries for which the two excitations (next state values) are the same as the two secondaries. All such entries

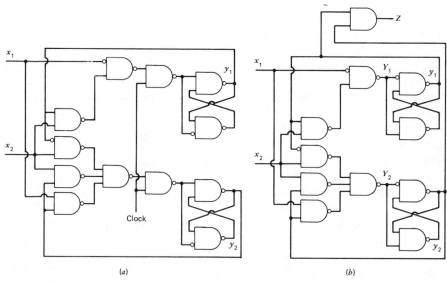

(a)                                                          (b)

*Figure 17.11   Removing the clock.*

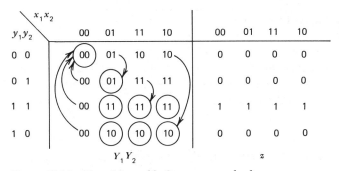

Figure 17.12    Transition table for sequence checker.

are circled in the figure. Suppose now that something has initiated a change of state in the circuit. We can see from Fig. 17.11$b$ that anytime that $y_1$ differs from $Y_1$ it will change value until it agrees with $Y_1$. This will not happen instantaneously due to the two gate delays in the cross-coupled pair of NAND gates. Similarly $y_2$ will change whenever it differs from $Y_2$. If the values of $Y_1$ and $Y_2$ do not agree with the values specified by Eqs. 17.4 and 17.5 for the present values of $y_1$ and $y_2$, the input logic network will cause $Y_1$ and $Y_2$ to change. As we argued when we made the clock pulses infinitely wide, the values of $y_1$, $y_2$, $Y_1$ and $Y_2$ could go on changing (oscillating) indefinitely. In fact, this oscillating will not terminate until $y_1$ agrees with $Y_1$ and $y_2$ agrees with $Y_2$. As mentioned, this agreement will exist for each of the circled pairs of values for $Y_1$ and $Y_2$ shown in Fig. 17.12. Since the circuit will stabilize at these pairs of values, they are termed *stable states*. Values of $Y_1Y_2$ that do not agree with the values of $y_1y_2$ for that row are called *unstable states*.

Notice that for the circuit of Fig. 17.11$b$ there is a stable state on each row of Fig. 17.12. Thus each state of the clock-mode network of Fig. 17.10 corresponds to a stable state in the network of Fig. 17.11$b$. If this were not the case for some row, the level-mode circuit could not exist at those values of $y_1$ and $y_2$ in the steady state. In addition, each next state or unstable state entry is the same as some stable state in the corresponding column. For example, the unstable state 01 is entered in the first row of the column $x_1x_2 = 01$. In the same column on the next row is the stable state 01. Suppose that the circuit is in stable state 00 (the first row) when the inputs $x_1x_2$ change from 00 to 01. First $Y_1Y_2$ will change to 01 and then $y_1y_2$ will change to 01 where the circuit will stabilize. This activity is indicated by the arrow in the second column of Fig. 17.12. For each unstable state in the transition table there is an arrow to the target stable state suggesting similar activity.

We may generalize this discussion with the following statement.

If a sequential circuit is to be oprated without a clock (in the level mode), its transition table must satisfy the following two conditions.

1. Each row in the transition table corresponding to state variable values at which the circuit must be able to rest for a finite period of time must include a stable state.
2. Each unstable state entry in the table must cause the circuit to assume a stable state in the same column.

A consequence of these two conditions is that a level-mode circuit can change state only if one of the inputs changes value.

Despite the fact that we chose as an introductory example a clock-mode circuit that would function equally well in the level mode, this is not usually the case. A typical clock-mode circuit will not satisfy these two conditions, if the clock is removed. The following is an example.

### Example 17.2

Construct a level-mode equivalent of the waveform generator considered in Section 17.1.

### Solution

Before we examine the circuit itself, let us consider the transition table that is given once more as Fig. 17.13.

The first step is to circle the stable states, that is those pairs of excitation values that agree with the pair of secondary values for the same row. We have already done that in Fig. 17.13. Notice that there are no circles. Therefore, there would be no stable states, so conditions 1 and 2 cannot be satisfied. Any attempted level-mode implementation of this table would oscillate indefinitely at a frequency determined by circuit parameters. The state variables might oscillate between 0 and $+5$ V or drop into the linear region resulting in a small magnitude sinusoidal output.

Since the behavior of any level-mode realization of the waveform generator would not be a predictable function of its inputs, it would be of no interest as a digital circuit.                                                                          ■

It should not be surprising that the waveform generator cannot function without a clock. Each column functions independently as a Gray code counter incrementing with each clock pulse. No stable state at all is an extreme example, but most clock-mode transition tables will include some row that does not satisfy condition 1 or some combination of next state values that fail to satisfy condition 2.

Just as with clock-mode circuits, we may assign symbols to each stable state or combination of state variable values. We may then replace each stable state in the transition table with the corresponding symbol enclosed by a circle. Next, we may replace each combination of values for each unstable state with the corresponding symbol without a circle. The result is a *flow table*. It is often less confusing to follow circuit activity through a flow table than through a transition table.

| $y_1 y_2$ \ $x_1 x_2$ | 0 0 | 0 1 | 1 1 | 1 0 |
|---|---|---|---|---|
| 00 | 01, 1 | 01, 1 | 01, 1 | 01, 1 |
| 01 | 11, 1 | 11, 1 | 11, 0 | 11, 1 |
| 11 | 10, 0 | 10, 1 | 10, 1 | 00, 0 |
| 10 | 00  0 | 00, 0 | 00, 0 | X |

$$Y_1 Y_2, z$$

*Figure 17.13   Transition table for waveform generator.*

| $y_1y_2$ \ $x_1x_2$ | 00 | 01 | 11 | 10 | 00 | 01 | 11 | 10 |
|---|---|---|---|---|---|---|---|---|
| 0 0 | ①  | 2  | 4  | 4  | 0 | 0 | 0 | 0 |
| 0 1 | 1  | ②  | 3  | 3  | 0 | 0 | 0 | 0 |
| 1 1 | 1  | ③  | ③  | ③  | 1 | 1 | 1 | 1 |
| 1 0 | 1  | ④  | ④  | ④  | 0 | 0 | 0 | 0 |

Figure 17.14   *Flow table for sequence checker.*

## Example 17.3

Obtain a flow table for the sequence checker from the transition table of Fig. 17.12. Using this flow table, obtain a timing diagram that relates output behavior to input behavior for this circuit.

### Solution

First we assign state ① to state variable values $y_1y_2 = 01$, state ② to $y_1y_2 = 01$, state ③ to $y_1y_2 = 11$, and state ④ to $y_1y_2 = 10$. Substituting these states for the corresponding stable and unstable state variable combinations in Fig. 17.12 yields the flow table of Fig. 17.14.

As its name indicates, this circuit was designed to go to ① if and only if a particular sequence of input combinations occurs. This sequence is 00 followed by 01 followed by 11. The circuit will return to zero when both inputs are 0. In the timing diagram of Fig. 17.15 we assume the circuit to be initially in state 1 with output 0. Observe from the flow table that the circuit will go to stable state ④, if $x_1$ goes to 1 before $x_2$. Notice that stable state ④ occupies three columns in the last row of the flow table. This holds the circuit at state ④ until both inputs return to 0 preventing any output of 1 between points $t_1$ and $t_2$ in Fig. 17.15. Between points $t_2$ and $t_3$ we show the inputs going first to 01, taking the circuit to state ②, and then to 11, taking the circuit to state ③ and the output finally to 1. The last part of the timing

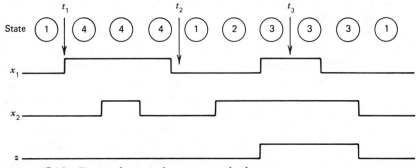

Figure 17.15   *Timing diagram for sequence checker.*

diagram shows the circuit holding at state ③ with the output 1 until both inputs return to 0. ∎

## 17.6 Synthesis of a Level-mode Sequential Circuit from a Flow Table

In the analysis examples in the last section we followed the sequence of steps listed here in obtaining a flow table of a level-mode sequential given a logic network diagram.

*Analysis*

1. Obtain Boolean expressions for the network state variables and outputs from the diagram.
2. Obtain a transition table by substituting all combinations of values in the expressions found in step 1.
3. Assign a state symbol to each combination of variable values.
4. Substitute state symbols for the combinations of values in the transition table to form the the flow table.

Our primary interest throughout the book has been design or synthesis of new systems rather than the analysis of existing systems. Typically, synthesis is the reverse process of analysis. In the case of level-mode design, this relationship is especially precise. With one addition, the step-by-step process for the synthesis of sequential circuits is essentially the four steps just listed in reverse order.

*Synthesis*

1. Derive a flow table representation from an initial statement of the problem (usually a natural language statement).
2. Assign a combination of state variable values to each stable state.
3. Substitute these state variable combinations for both the stable and unstable states in the flow table to form a transition table.
4. Use Karnaugh maps to obtain Boolean expressions for each excitation and output.
5. Obtain a network realization of these Boolean expressions.

The flow table derivation in the first step, which is most critical, was not considered in the analysis process. In clock-mode design the first unambiguous formulation of a system was an ASM chart or RTL description. For level mode this first unambiguous description is the flow table. A variety of reasoning processes can lead to correct flow tables. This correctness of the result is the ultimate criteria. Reference [4] presents a standard process for flow table derivation useful over a broad range of applications. In this section we shall obtain flow tables by "hook or crook" and concentrate on the last four steps of the synthesis process.

The second step of assigning stable states to state variable combinations has, of course, a counterpart in clock-mode design. In our clock-mode discussion we noted briefly that the particular state assignment could affect the number of gates required in the memory element input logic, but we concluded that the most cost effective approach was probably arbitrary state assignment. State assignment is more critical for level-mode design. We shall see in the next section that for some flow tables the circuit resulting from one state assignment will work while that from another will not.

### Example 17.4

Obtain a level-mode realization of a modulo-2 counter. This circuit has a single input, $T$. When $T$ is 0, the circuit output is to stay constant at 1 or 0. Each time $T$ changes from 0 to 1 the output is to change values from 0 to 1 or 1 to 0. This simple circuit will function nicely in the clock mode. A clock-mode state table is given in Fig. 17.16.

Solution:

In Fig. 17.17a we see a direct translation of the clock-mode state table to a level-mode flow table by replacing each present state in the state table by a stable state in the transition table. This flow table cannot be used to construct a functioning level-mode circuit. The first of the two necessary conditions set forth in the previous section is satisfied, but the second is not. There is a stable state in each row, but there is none in the second column. Therefore, the unstable states in this column call for continuous transitions back and forth between the two rows as dictated by circuit parameters.

To avoid this problem we add two rows, each of which has a stable state in the $T = 1$ column to form the flow table of Fig. 17.17b. From state ① the circuit now goes to state ④, which has output 1 when $T$ goes to 1. Similarly, there is a transition from stable state ② with output 1 to stable state ③ with output 0, when $T$ goes to 1. There remain two blank entries in the first column of Fig. 17.17b. The initial problem statement indicated that the output of the modulo-2 counter must not change while $T$ is 0. We now take this to mean that the output must not change when $T$ changes from 1 to 0. Since stable states ① and ③ both have output 0, the circuit must go from ③ to ① when $T$ goes to 0. Similarly the circuit must go from ④ to ② when $T$ goes to 0, so that the output remains 1. This is accomplished by adding the unstable 1 in the first row of the first column and unstable 2 in the third row, as shown in Fig. 17.17c. ∎

Let us pause briefly to examine what has really been accomplished with the development of the flow chart of Fig. 17.17c before we proceed with its implementation. We have described a device that will change state on the leading edge of every

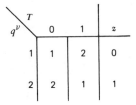

*Figure 17.16   State table for modulo-2 counter.*

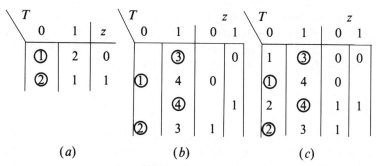

Figure 17.17   Evolution of a flow table for T flip-flop.

pulse on line $T$ and retain its value until the advent of the next leading edge. This activity closely resembles that of the $J$-$K$ flip-flop with the $J$ and $K$ inputs tied together, as used in the design of counters in Chapter 8. Figure 17.18$a$ shows a leading edge-triggered $J$-$K$ flip-flop connected in this manner. The box in Fig. 17.18$b$ with the single input labeled $T$ is intended to represent an implementation of the flow table of Fig. 17.17$c$. To this input we have connected the output of an AND gate with inputs at a level labeled $TL$ and the clock. If we restrict changes on $TL$ to occurring while the clock is 0 and a sufficient period of time before a rising transition on the clock, we claim that the function of the circuits in Figs. 17.18$a$ and $b$ are exactly the same.

We have just succeeded in designing a *clocked T flip-flop* using level-mode techniques. We accomplished this by adding states to the flow table until conditions 1 and 2 of the previous section were satisfied. By adding stable states as necessary, it is possible to design the other types of edge-triggered or master–slave flip-flops. A level-mode development of $D$ and $J$-$K$ flip-flops may be found in Ref. [4]. More complex clock-mode sequential circuits containing two or more flip-flops could also be designed in the level mode by adding sufficient stable states to ensure satisfaction of conditions 1 and 2. This would be considerably more work than regarding the flip-flops as primitive elements and working in the clock mode as we have done in previous chapters. Perhaps the most important application of level-mode techniques is the development of primitive elements for use in clock-mode design.

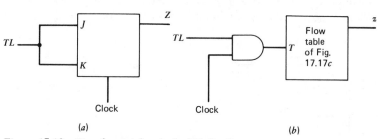

Figure 17.18   Two forms of a clocked T flip-flop.

**Example 17.4 (continued)**

We are now ready to complete the final four steps of the synthesis process for the $T$ flip-flop flow table of Fig. 17.17c.

Solution

One state assignment that will work for this circuit follows the row order in the figure. Thus we assign $y_1 y_2 = 00$ to ③, 01 to ①, 11 to ④, and 10 to ②. Step 3 of the synthesis process results in the transition table of Fig. 17.19a. This is translated directly to the Karnaugh maps for the individual state variables given in Fig. 17.19b.

From the Karnaugh maps of Fig. 17.19b we obtain the following expressions for the state variables.

$$Y_1 = \overline{T} y_1 \vee T y_2 \tag{17.7}$$

$$Y_2 = \overline{T} \overline{y_1} \vee T y_2 \tag{17.8}$$

∎

So far nothing has been said about the output in the realization of the $T$ flip-flop proceeding as Example 17.4. Let us look again at the flow table of Fig. 17.17. Notice that when unstable state 1 was entered in the first row of the first column that a 0 was entered in the corresponding square of the output portion of the table. Suppose for a moment that a 1 had been entered in this output square, as shown in the partial flow table of Fig. 17.20a.

Now assume the circuit to be initially in state ③ when the input $T$ changes from 1 to 0. The circuit will go eventually to stable state 12 where the output is 0. Briefly, however, the circuit will be in unstable state 1 ($y_1 y_2 = 00$ with $T = 0$). During this short period of time the output will be 1, as specified by the flow table. The original problem specification called for the output to remain 0 during 1 to 0 transitions on $T$. As was made clear in the discussion of hazards in Section 17.1, unexpected transient output pulses such as the one in Fig. 17.20b can cause unpleasant consquences. We, therefore, conclude that the flow table of Fig. 17.17c, which

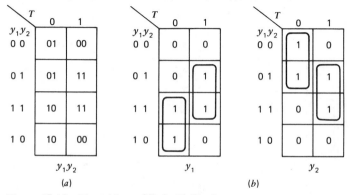

*Figure 17.19 Transition table for T flip-flops.*

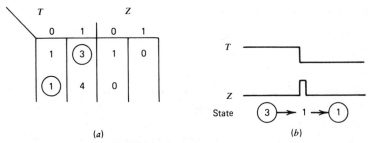

(a)    (b)

*Figure 17.20   Spurious output for unstable state.*

also avoids a transient output during the transition from state 4 to state 2, is the proper one. There are two don't-care output entries in this table, they correspond to unstable states in transitions between two stable states for which the output is not the same. When the output is to change from 0 to 1, for example, it rarely matters whether it changes to 1 slightly earlier when the circuit is in the unstable state or slightly later after the circuit reaches the new stable state. Thus these output entries may be left unspecified to facilitate optimization of the combinational logic.

We may summarize this discussion with the following rule for the specification of outputs corresponding to unstable states.

Rule   Output entries corresponding to unstable states must be specified as necessary to prevent extraneous output transitions.

Hazards can also cause unwanted output transients in level-mode circuits. Still worse, they can cause improper state transitions. Notice that the Karnaugh maps of Fig. 17.19b for the $T$ flip-flop reflect no provision for eliminating hazards. In fact, a realization of Eqs. 17.7 and 17.8 would build hazards into both state variables. Suppose that in such a realization $y_1 y_2$ are initially 01 when $T$ changes from 0 to 1. This would cause one of the AND gates in the realization of $Y_2$ to go to 0 and the output of the other to go to 1 as depicted in Fig. 17.21. This could cause $Y_2$ and then $y_2$ to go to 0 and perhaps stick there, depending on relative delays of gates in the realization. This could in turn cause $y_1$ to return to 0 or remain there, since $Y_1 Y_2 = 00$

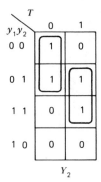

$Y_2$    *Figure 17.21   State variable hazard.*

is specified in both squares of the transition table for which $y_2 = 0$. In either event the circuit would stabilize at ⓪⓪ rather than at ①① as intended.

Hazards must be eliminated in realizations of excitations for level-mode circuits by adding redundant AND gates as discussed in Section 17.2. It is important to note that **hazards in state variable realizations do not cause problems in clock-mode circuits.** In the clock mode there is always time for the excitations to settle down at the values specified in the transition table before the next clock pulse arrives.

### Example 17.4 (continued)

Complete the realization of the edge-triggered $T$ flip-flop so that it is free from hazards and output transients.

Solution
Karnaugh maps for the $z$, $Y_1$, and $Y_2$ are given in Fig. 17.22.

Boolean expressions taken from these Karnaugh maps are give by Eqs. 17.9, 17.10, and 17.11

$$z = y_1 \tag{17.9}$$

$$Y_1 = \overline{T}y_1 \vee Ty_2 \vee y_1y_2 \tag{17.10}$$

$$Y_2 = \overline{T}\overline{y_1} \vee Ty_2 \vee \overline{y_1}y_2 \tag{17.11}$$

The expressions for the hazards differ from Eqs. 17.7 and 17.8 in that in each case a redundant product has been added to eliminate a hazard. A logic network diagram realizing these expressions is given in Fig. 17.23.    ∎

You will note that the cross-coupled pairs of NAND gates that have up until now separated the secondaries from the excitation are not included in Fig. 17.23. The only effect of these pairs is to add additional delay in the loop for each state variable. The delay in the combinational logic network generating the excitations is sufficient to permit the storage of information, so the function of the network is logically unaffected by removing these cross-coupled pairs. In considering this form of

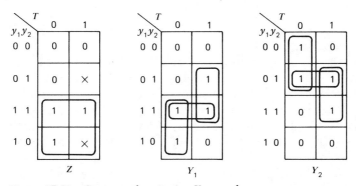

*Figure 17.22   Output and excitation Karnaugh maps.*

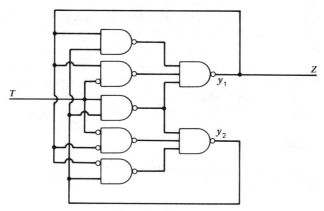

*Figure 17.23    Network realization of edge-triggered* T *flip-flop.*

network it may be helpful to think of the excitations as specified by the map as next values for the secondaries that are not represented physically in the network.

# 17.7    Races in State Variable Transitions

In this section we shall investigate still another problem that appears in level-mode but not clock-mode circuits. We first obtain a straightforward realization of the flow table of Fig. 17.24$a$ by assigning $y_2y_1 = 00$ to ⓐ, 01 to ⓑ, and 11 to ⓒ. The resulting transition table is given in Fig. 17.24$b$. Translation of the transition table to two separate Karnaugh maps is left to you.

The Boolean expressions given by Eqs. 17.12 and 17.13 would be taken from these maps. The output expression is evident from the flow table.

$$Y_2 = \bar{x}Cy_1 \tag{17.12}$$

$$Y_1 = x \lor \bar{y}_2\bar{y}_1 \lor Cy_1 \tag{17.13}$$

$$z = y_1 \tag{17.14}$$

A realization of these expressiosn is given in the logic network diagram of Fig. 17.25, but it may not work.

Let us assume that the circuit is in the stable state $xCy_2y_1 = 0111$, corresponding to the stable state ⓒ in the second column, third row of the state table. The corresponding values at various points in the circuit are shown in Fig. 17.25. Note that the 1 values for $Y_2$ and $Y_1$ are produced by gates A1 and A3, respectively. Now assume that $C$ goes from 1 to 0, as shown. This should take the circuit through the unstable a entry in the third row to the stable ⓐ in the first column, first row, with both state variables changing to 0.

Recalling that a NAND gate will have a 1 out if any input is 0, we see that the $C = 0$ signal will drive the outputs of A1 and A3 to 1. Providing A2 does not

| xC | 00 | 01 | 11 | 10 | 00 | 01 | 11 | 10 |
|---|---|---|---|---|---|---|---|---|
|  | (a) | (a) | b | – | 0 | 0 | 0 | – |
|  | (b) | c | (b) | (b) | 0 | – | 0 | 0 |
|  | a | (c) | – | – | – | 1 | – | – |

$z$

(a)

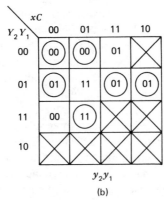

(b)

Figure 17.24   Flow table and transition table.

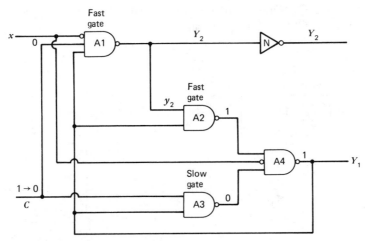

Figure 17.25   Realization of the table in Fig. 17.24b.

change too fast, the 1 at the output of A3 will then drive $Y_1$ to 0 at the output of A4. This value of $Y_1$, fed back to the input of A2, will lock its output at 1, in turn stabilizing $Y_1$ at 0, as required.

But now assume that the gates have unequal delays, as shown in Fig. 17.25. Then the change in $C$ will first drive the output of A1 to 1, which will in turn drive A2 to 0 *before* the output of A3 goes to 1. As a result, $Y_1$ will not change and the circuit will make an erroneous transition to $Y_2Y_1 = 01$, the stable b state in the first column.

The difficulty just described occurred because both secondaries were required to change simultaneously following a single input change but did not change at the same time due to unequal circuit delays. Such a situation is termed a *race* because the nature of the transition may depend on which variable changes fastest. Sometimes races may be "fixed" by introducing extra delays, but this approach adds to the complexity of the circuit and doesn't always work. Let us consider some possible solutions to the problem by manipulating the flow table itself.

It is possible for a circuit to assume more than one unstable state in the process of moving to a new stable state. If for a given initial state and input transition such a sequence of unstable states is the same each time the transition is executed, then it is termed a *cycle*. For example, in the flow table and corresponding transition table of Fig. 17.26 the circuit will cycle through three unstable states during a transition from ① to ④. Note in the transition table that the next state for $y_2y_1x_2x_1 = 0001$ is not $y_2y_1 = 10$. Instead, the machine proceeds from 00 to 01 to 11 to 10. Each of the transitions involves a change in only one secondary. Thus, the circuit proceeds reliably through each of the unstable states to the final stable state. The cycle may be entered from stable states ② and ③ as well.

Where the simultaneous change of more than one secondary is specified by the transition table, the condition is termed a race. The race illustrated in Fig. 17.27 is a noncritical race. Suppose the inputs change to 01 while the machine is in state $y_2y_1 = 00$. If $y_2$ changes first, the machine goes to unstable state 10. If $y_1$ changes first, the machine goes to 01. For either of these unstable states, the $Y$ map shows the next state to be stable ②. Thus, no matter what the outcome of the race, the stable state ②, as designated by the $Y$ map, is reached.

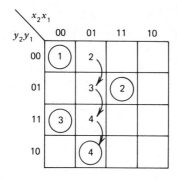

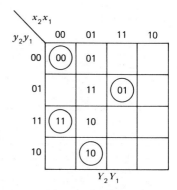

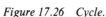

*Figure 17.26    Cycle.*

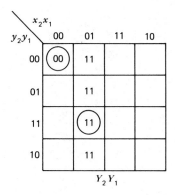

Figure 17.27   Noncritical race.

The situation encountered in the control delay, as discussed in the last section, is an example of a *critical* race, a situation in which there are two or more stable states in a given column, and the final state reached by the circuit depends on the order in which the the variables change. In other words, the desired result might occur or might not occur depending on the actual circuit delays.

Another critical race is illustrated in Fig. 17.28. There the circuit is supposed to move from stable ② ($y_2y_1 = 01$) in the $x = 1$ column to stable ③ ($y_2y_1 = 10$) in the $x = 0$ column. If $Y_1$ changes first, the excitations will be $Y_2Y_1 = 00$, leading to an erroneous transition to stable ①. If $Y_2$ changes first, the excitations will be $Y_2Y_1 = 11$, leading to unstable 1 in the third row and thence to stable ①. Thus, no matter what the delays, the wrong transition will occur.

It might seem that the first type of critical race would be less troublesome since it can, in theory, be eliminated by proper adjustment of delays. In practice, however, constraining delays is difficult and expensive, so both types of critical races are to be avoided. In some problems, this may be accomplished by a judicious assignment of state variables. In others, it becomes necessary to use more than a minimal number of states.

### Example 17.5

Modify the sequential circuit of Fig. 17.25 to eliminate the critical race.

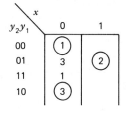
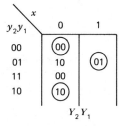

Figure 17.28   Critical race.

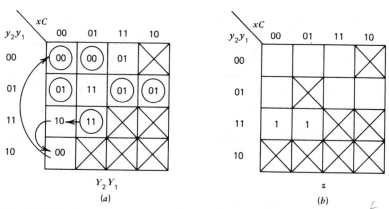

**Figure 17.29**   *Revised transition table for Fig. 17.24.*

## Solution

It is only necessary to eliminate the critical race in column 00 of Fig. 17.24 and replace it with a cycle through the spare state. A new transition table is given in Fig. 17.29. The only changes are in the lower two squares of column $xC = 00$. Thus, from stable state $\textcircled{11}$, the circuit will cycle as indicated by the arrows when $C$ goes to 0. In terms of the excitation maps, the only change is an additional 1 in the third row. The new equation for $Y_2$ is

$$Y_2 = y_2 y_1 + \bar{x} C y_1 \qquad (17.15)$$

There is no change in the equation for $Y_1$. To avoid a possible output pulse during the cycle, an additional 1 is added to the output map as shown in Fig. 17.29*b*. This does not change the equation for $z$. The transition table of Fig. 17.29 leads to the reversed realization in Fig. 17.30    ∎

In Section 17.1 we stated two restrictive conditions on transition tables that must be satisfied if the tables were to be realizable as reliable level-mode circuits. To these we add the following third condition.

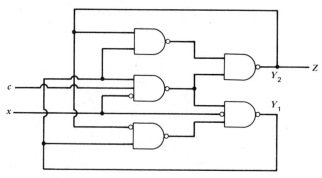

**Figure 17.30**   *Realization of flow table of Fig. 17.25.*

Third Condition for Level Mode    The transition table must be free from critical races.

As with the first two conditions, this one need not be satisfied by transition tables for clock-mode realization. Any number of flip-flops may be required to change state simultaneously in a clock-mode circuit. At the beginning of a clock period the excitations will begin to change in response to the previous clock edge. These new excitation valves will have the entire clock period to stabilize at the flip-flop inputs before the next clock edge allows the flip-flop outputs (secondaries) to change again.

To eliminate critical races, we must assign the state variables in such a manner that transitions between states require the change of only one variable at a time. To this end, it is desirable that stable states between which transitions occur be given adjacent assignments, that is, assignments differing in only one variable. To find such assignments, it is convenient to make a *transition diagram* on the Boolean hypercube. Consider the state table shown in Fig. 17.31$a$ and the transition diagram in Fig. 17.31$b$. To set up the transition diagram we start by assigning ⓐ to 00. Noting that ⓐ has transitions to ⓒ, we give ⓒ an adjacent assignment, at 01, and indicate the transition by an arrow from ⓐ to ⓒ. Next, the transition from ⓒ back to ⓐ is indicated by an arrow from ⓒ to ⓐ. Since there is a transition from ⓒ to ⓑ, ⓑ is given an adjacent assignment at 11, leaving 10 for ⓓ. The diagram is then completed by filling in arrows for all the other transitions. In this case, all transitions are between adjacent states, so this assignment is free of critical races.

A slightly different situation is illustrated in Fig. 17.32. Now ⓓ has a transition to ⓒ, resulting in a diagonal transition, that is, a change of two variables. This diagonal transition cannot be eliminated since ⓓ has transitions to three other states and cannot be adjacent to all of these with only two state variables. However, this diagonal transition is seen to represent a noncritical race and will cause no problems. Whenever there is only one stable state in a column, transitions into that column need not be considered in choosing a state assignment since critical races cannot

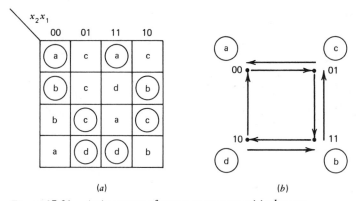

*Figure 17.31    Assignments of states to prevent critical races.*

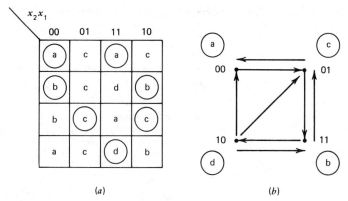

(a)                                                          (b)

Figure 17.32    Assignment with noncritical race.

occur unless there are at least two stable entries in a column. Thus, the same assignment will be valid for this state table.

In Fig. 17.33, we see a more difficult situation. Here we have several critical races, as indicated by the diagonal transitions, and it is clear that no permutation of the assignment can eliminate all diagonal transitions. For example, the first row requires both Ⓑ and Ⓓ to be adjacent to Ⓐ, while the second row requires Ⓑ and Ⓓ to be adjacent. Clearly, not all these requirements can be satisfied.

In situations such as that in Fig. 17.32, the critical races can be eliminated only by the use of spare states. If the number of states in the original state table is not a power of 2, spare states will automatically be available. This was the case in the control delay, where there were only three states and the spare state was used for a cycle to avoid the critical race. If the original number of states is a power of 2 or if a satisfactory assignment is not found using available spare states, spare states must be created by adding extra state variables.

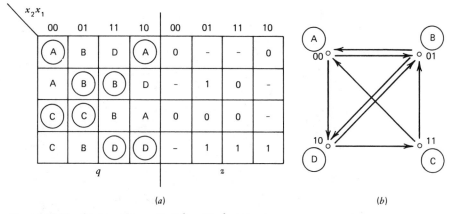

(a)                                                          (b)

Figure 17.33    State assignment with critical races.

# 17.8   Critical Race-Free State Assignments

Where more than two state variables are required, a variety of techniques have been devised for determining critical race-free state assignments. For more complete discussions see Refs. [5,11]. The most straightforward approach and the one that usually leads to a minimal gate realization is the shared row assignment method. A few four stable state flow tables such as the one given in Fig. 17.29 require three state variables. Flow tables with five and six stable states can usually also be realized using three state variables. With five state variables there are three spare rows that can be share by cycles in one or more of the columns as necessary. A cycle in one column will not conflict with the activity in another column even though it may share a common row. With six state variables there will be two spare rows for cycles.

A useful concept in finding such assignments is the *destination set*. In any column of a state table, a destination set consists of any stable state and any rows that make transitions into that state. For example, in the state table of Fig. 17.34, in the first column, the stable states are ①, ③, ⑤; and row 4 leads to ①, row 2 to ⑤. The destination sets for this column are thus (1,4)(2,5)(3). Similarly, the destination states for the second, third, and fourth columns are (1,3)(2,4)(5), (1,4)(2,3)(5), and (1,3)(2)(4,5), respectively. The destination sets are important because, to avoid critical races, the members of each set must either be adjacent or so located in relation to the spare states that cyclic transitions for all sets in a given column may be made without interference.

A useful assignment for five-state tables is shown in Fig. 17.35a. It will work for almost every table but will fail for a few rare cases, as we shall see. For this case we arbitrarily assign the five states as shown in Fig. 17.35b on the K-map, and, Fig. 17.35c on the Boolean hypercube. In the first column of the state table, the destination sets are (1,4)(2,5)(3), requiring transitions from 4 to 1 and from 2 to 5. We see in Fig. 17.35c that this is impossible with this assignment because the two cycles must use the same spare rows enroute to differnt destinations. Rows may be *shared*

| $x_2 x_1$ | 00 | 01 | 11 | 10 | 00 | 01 | 11 | 10 |
|---|---|---|---|---|---|---|---|---|
| 1 | ① | 3 | 4 | ① | 0 | – | 0 | 0 |
| 2 | 5 | 4 | ② | ② | – | 1 | 1 | 0 |
| 3 | ③ | ③ | 2 | 1 | 1 | 1 | 1 | – |
| 4 | 1 | ④ | ④ | ④ | – | 1 | 0 | 1 |
| 5 | ⑤ | ⑤ | ⑤ | 4 | 0 | 1 | 1 | 1 |

*Figure 17.34   Example five-state table.*

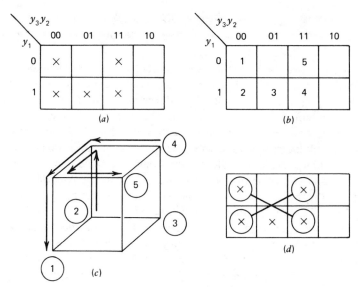

Figure 17.35   *Five stable state shared row assignments.*

only in the sense that they may be used for different cycles in different columns. In any single column, a transitional row may be part of a cycle to only one destination.

This problem occurs only in columns having (2, 2, 1) destination sets, that is, columns having 2, 2, and 1 members in three destination sets. It can be eliminated by permuting the assignment so that the two two-member sets do not occupy diagonally opposite vertices of the hypercube. In terms of the assignment map, the two-member sets must not occupy the connected rows as shown in Fig. 17.35*d*. There are 15 possible (2, 2, 1) column configurations and, as long as all 15 do not occur in the same table, it is possible to avoid assigning any (2, 2, 1) sets in the "forbidden" pattern. This assignment will thus fail only in the unlikely event of a state table with 15 or more columns, including all 15 possible (2, 2, 1) columns [10].

The easiest way to use this assignment is to construct any aribtrary (2, 2, 1) group of destination sets that does *not* occur in the destination sets of the state table and assign them to the "forbidden" pattern. This guarantees that none of the (2, 2, 1) columns that are in the state table can be assigned to that pattern.

For the table of Fig. 17.34, we see that the (2, 2, 1) pattern (1, 2),(3, 5)(4), for example, does not appear among the destination sets, so we assign (1, 2) and (3, 5) to the diagonal locations, as shown in Fig. 17.36*a*. For ease of reference, the spare rows are labeled, $\alpha$, $\beta$, $\gamma$. To set up the augmented state table (Fig. 17.36*b*), we start by filling in the stable states and the corresponding outputs, just as for a multiple-row assignment.

The transitional entries are filled in with reference to the map of the assignment (Fig. 17.36*a*) to determine if the transition is direct or whether a cycle through spare states must be provided. State ① makes transitions to ③ in column 01 and to 4 in column 11. The transition to ③ is direct, but the transition to ④ uses the cycle ① → $\alpha$ → 4 → ④, as seen in the third column of Fig. 17.36*b*. Transitions

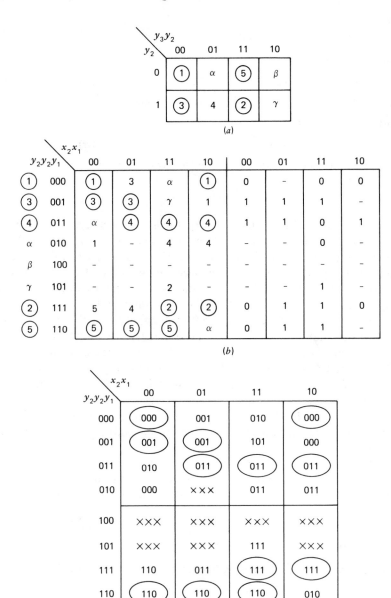

Figure 17.36   Valid five-state assignment for Fig. 17.34.

② → ④ and ② → ⑤ are both direct. Transition ③ → ① is direct, but ③ → γ → 2 → ② cycles through γ. The other entries are filled in similarly. We note here an advantage of the shared-row method. It is rare that all the spare row entries are needed, resulting in many don't-cares. By contrast, the multiple-row method does not provide any don't-cares other than those in the original state table.

So far we have not addressed the problem of assigning outputs to unstable

| | | | | | | | |
|---|---|---|---|---|---|---|---|
| States: | ④ | → | α | → | ① | → | 1 |
| Output: | 1 | | – | | – | | 0 |

(*a*)

| | | | | | | | |
|---|---|---|---|---|---|---|---|
| States: | ④ | → | α | → | 1 | → | ① |
| Output: | 1 | | 0 | | 1 | | 0 |

(*b*)

| | | | | | | | |
|---|---|---|---|---|---|---|---|
| States: | ④ | → | α | → | ① | → | 1 |
| Output: | 1 | | 1 | | – | | 0 |

(*c*)

| | | | | | | | |
|---|---|---|---|---|---|---|---|
| States: | ④ | → | α | → | 1 | → | ① |
| Output: | 1 | | – | | 0 | | 0 |

(*d*)

*Figure 17.37 Transitional output assignment in cyclic transition.*

states found in cycles. Again, spurious output pulses must be avoided. If the output does not change in between the initial and final stable states of a cycle, this same output value must be assigned to all intermediate unstable states in the cycle. If the output is to change, it must change only once during a cycle. Therefore, it can be a don't-care for only one transitional state during the cycle. Consider the transition from ④ in the second column to ① in the first column, in Fig. 17.36*b*, a transition that cycles through $\alpha$. If we let both transitional outputs be don't-cares, as shown in Fig. 17.37*a*, the result, depending on groupings of don't-cares, might be a momentary 1 pulse, as shown in Fig. 17.37*b*. To prevent this, either the first transitional state must be specified as 1, or the second as 0, as shown in Figs. 17.37*c* and *d*, respectively. Then choice between the latter two should be made on the basis of which seems likely to result in the simplest output equations.

The transition table follows directly from the augmented state table by replacing each state by the corresponding state variable combination, as shown in Fig. 17.36*c*.

There are no generally applicable three-variable assignments for six and seven state tables, but suitable assignments can often be found through careful analysis of the destination sets. This may be illustrated by the following example.

## Example 17.6

Obtain a critical race-free state assignment for the six stable state flow table of Fig. 17.38.

| $x_1x_2$: 00 | 01 | 11 | 10 |
|---|---|---|---|
| 3 | 6 | ① | ① |
| ② | 3 | 1 | ② |
| ③ | ③ | ③ | 5 |
| 3 | ④ | ④ | 1 |
| 2 | ⑤ | 6 | 5 |
| ⑥ | ⑥ | ⑥ | 2 |

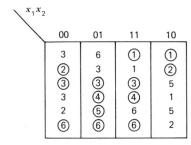

*Figure 17.38 Example six-state table.*

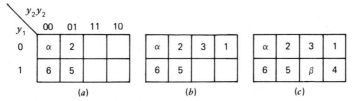

Figure 17.39 Development of race-free assignment for Fig. 17.38.

## Solution

The destination sets are (1, 3, 4)(2, 5)(6), (1, 6)(2, 3)(4)(5), (1, 2)(3)(4)(5, 6), and (1, 4)(2, 6)(3, 5). First we note a destination set with more than two members, (1, 3, 4), in the first row. In such cases, it is often possible to cycle within the destination set, for example, ① → 4 → ③ or ④ → 1 → ③. In view of the flexibility

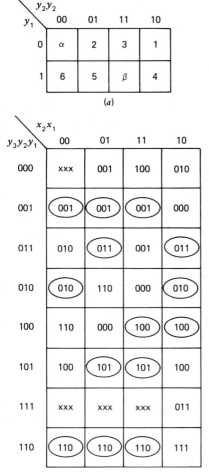

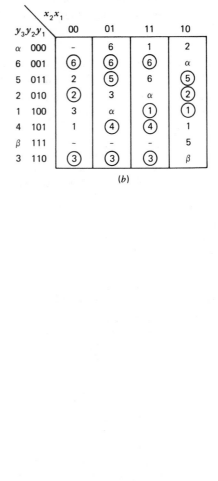

Figure 17.40 Six stable state shared row tables.

offered by such cycles, it is usually best to consider the two-member destination sets first. In this case, we note that state 2 is a member of four such sets (2, 5), (2, 3), (1, 2), and (2, 6). In a three-variable assignment, a single state can be adjacent to only three other states, so these four destination states require that ② be adjacent to one of the spare states, which we shall designate $\alpha$. Of the four states paired with ② in destination sets, ⑤ and ⑥ are not part of the (1, 3, 4) set, so we place them adjacent to the (2, $\alpha$) pair, giving the intiial assignment shown in Fig. 17.39a. The remaining state paired with ②, ①, and ③ must be adjacent to ② or $\alpha$, so we locate them as shown in Fig. 17.39b. Finally, noting that state ④ can cycle to ③ through state ①, we place 1 and the remaining spare state, $\beta$, as shown in Fig. 17.39c. We then check the destination sets for all columns to see that all transitions can be made without critical races with the assignment of Fig. 17.39c. The development of the augmented state table and transition table is shown in Fig. 17.40. ■

## 17.9  More Constraints on Level-Mode Realizations

In all the level-mode examples considered so far, there have been sufficient constraints in the original problem statement to assure that no two input lines ever changed state at the same time. As we shall soon see, this greatly simplified their realization. A fourth condition is almost always imposed on level-mode realizations.

> **Fourth Level-Mode Condition**   Only one circuit input is allowed to change value at a time. All state variables that change in response to a change on one input line must have stabilized at their final values before another change on an input line can occur.

Reference [8] calls circuits that satisfy this condition *fundamental mode* and considers some approaches to the design of level-mode circuits that are not fundamental mode. Few design problems are most effectively pursued in the level mode allowing multiple input changes, so our discussion of this topic will be limited to the few observations that follow.

The partial flow table of Fig. 17.41 illustrates what may happen when specific behavior is specified for a simultaneous change on more than one input that differs from the behavior expected when just one of those inputs changes. Suppose $x_1x_2$ are initially 00. Note that the circuit is expected to remain in state ⓐ , if the two inputs

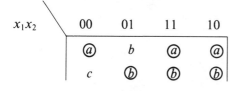

| $x_1x_2$ | 00 | 01 | 11 | 10 |
|---|---|---|---|---|
| | ⓐ | b | ⓐ | ⓐ |
| | c | ⓑ | ⓑ | ⓑ |

*Figure 17.41   Problem multiple input change.*

change simultaneously to 11. In practice, the circuit cannot consistently distinguish between this multiple input change and a change in first one input and then the other a very short time later. If, for example, the circuit of Fig. 17.41 thought $x_2$ changed first, it would go to state ⓑ. Otherwise it would remain in state ⓐ. Random gate delays might cause one copy of a circuit to behave one way and another copy differently. Sequential circuit designs with behavior dependent on gate delays are never acceptable.

Occasionally it does not matter whether a set of inputs change at exactly the same time or inputs within the set change at slightly different times. If the resultant behavior is always the same, entries corresponding to multiple changes can appear in the flow table. An example is given in the partial flow table of Fig. 17.42. Here the circuit will go to state ⓑ whenever the inputs change from 00.

Sometimes a flow table will include entries that seem to correspond to multiple input changes even though such changes cannot actually occur. The flow table of Fig. 17.41 would actually satisfy condition 4, if multiple input changes between 00 and 11 or from 10 to 01 were not allowed by the specification. That these entries are not "don't cares" follows from the fact that there is more than one stable state on the first row.

Some of the more tempting level-mode design problems that do not satisfy condition 4 effectively include a need for some form of synchronization. That is, certain output changes are required to be synchronized with changes on some specific input, even though these output changes are caused by changes on some other input. It is possible to realize a flow table for such a circuit, but its realization will always include some cross-coupled pair susceptible to being driven by a partial pulse with resultant oscillation as depicted in Fig. 17.8. That is, level-mode techniques by themselves will not solve the synchronization problem.

Before leaving this chapter, let us point out one additional problem that could affect the operation of some properly designed level-mode circuits. The transition table from which the edge-triggered $T$ flip-flop was implemented is repeated as Fig. 17.43. The realization eliminated all hazards in the combinational logic. Suppose the circuit is initially in stable state 01 when $T$ changes to 1. Imagine that there is very

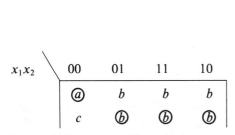

Figure 17.42   *Manageable multiple input change.*

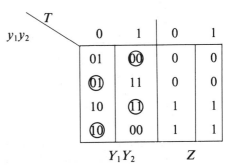

Figure 17.43   *Essential hazard.*

little delay in the loop realizing $Y_2$, so that this secondary immediately changes to 1 before the effect of the new value of $T$ has propagated into the network realizing $Y_1$. If the circuit thinks that both $y_1$ and $y_2$ are 1 while $T$ is 0, $y_1$ would proceed to 0 as suggested by the arrow in Fig. 17.43. Once the circuit fully realized that $T = 1$, it would continue to state 00, the wrong stable state.

The phenomenon just discussed has been called on *essential hazard*. The term essential is probably used because the problem appears in the flow tables of a number of circuits of fundamental importance and is a characteristic of the flow table itself and not its particular realization.

> Definition   An essential hazard exists in a state table if there is a stable state from which three consecutive changes in a single input variable will take the circuit to a different stable state than the first change alone.

Essential hazards are easily identified by an examination of the flow table. The obvious remedy is the inclusion of sufficient delay in all feedback loops. If the delay of individual gates is fairly uniform, extra gates in each loop for the sole purpose of increasing delay will not be required. In most technologies a direct implementation of the edge-triggered $T$ flip-flop circuit of Fig. 17.23 would work.

## Problems

**17.1**   Construct a logic block diagram of the logic network described by the Karnaugh map realization in Fig. 17.5$a$. Assume that each gate in your realization has a delay of 5 nsec, except for the gate covering minterms 1 and 5 in the map, which has a delay of 20 nsec. Let the initial input values be $y_2 = 0, y_1 = 1, x_2 = 1$, and $x_1 = 0$. Plot a timing diagram showing the output value of each gate in the network for 50 nsec after input, $x_2$, changes from 1 to 0.

**17.2**   Construct the logic block diagram of a hazard free two level NAND–NAND realization of the Boolean function defined by the Karnaugh map of Fig. P17.2

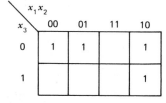

*Figure P17.2*

**17.3**   Consider a synchronizing network consisting of two falling edge-triggered flip-flops, *r* and *s*, and a three-bit counter, CNT, as described at the end of Section 17.3. Let the initial value of input, *x*, and both flip-flops be 0. Assume a $0 \rightarrow 1$ transition on *x* occurs at a time when **CNT** = 1, 1, 1 and close enough to the falling edge of the clock to cause oscillations that last for 6 clock periods. Sketch a timing diagram showing $x$, $r$, and $s$ for 10 clock periods following the transition on $x$ just described and consistent with the preceeding discussion.

**17.4** Consider again the synchronizing network described in Problem 17.3. Suppose it is driven by a clock of frequency 10 MHz. Assume that a transition on $x$ will cause oscillations on the output of flip-flop $s$, which last for more than 800 nsec if and only if that transition occurs at least 1 nsec before a falling clock transition (when $CNT = 1, 1, 1$) but no more than 2 nsec before that clock transition. Assume that transitions on $x$ (always widely separated in time) occur at random and will occur with equal likelihood at any point in the clock period. What is the probability that a particular transition on $x$ will cause oscillations that are still active at the time the new value after the input transition is to be transferred into flip-flop $r$.

**17.5** Consider the level-mode sequential circut given in Fig. P17.5.
   (a) Obtain a transition table for the circuit.
   (b) Translate the transition table into a flow table assuming states $a$, $b$, $c$, and $d$ represent state variable combinations $Y_1 Y_2 = 00, 01, 11,$ and $10$, respectively. Circle all stable states in the flow table.
   (c) Indicate all critical races that exist within the circuit.

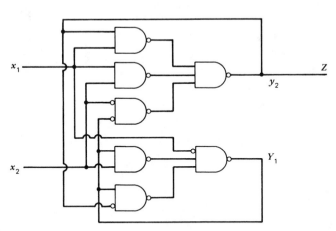

*Figure P17.5*

**17.6** Assign secondaries to the minimal flow table of Fig. P17.6. Construct a transition table that avoids critical races. Obtain Boolean expressions for each excitation variable.

**17.7** Make a secondary assignment for the flow table of Fig. P17.7. Construct a transition table without critical races.

$q$

| | $x_2x_1$ | | | | | $z_2z_1$ | | |
| :---: | :---: | :---: | :---: | :---: | :---: | :---: | :---: | :---: |
| | | | | | | $x_2x_1$ | | |
| 00 | 01 | 11 | 10 | 00 | 01 | 11 | 10 |
| ① | 2 | ① | 3 | 01 | — | 01 | — |
| 1 | ② | ② | ② | — | 00 | 00 | 00 |
| 1 | ③ | 2 | ③ | — | 10 | — | 10 |

*Figure P17.6*

$q$

| | $x_2x_1$ | | | | | $z$ | | |
| :---: | :---: | :---: | :---: | :---: | :---: | :---: | :---: | :---: |
| | | | | | | $x_2x_1$ | | |
| 00 | 01 | 11 | 10 | 00 | 01 | 11 | 10 |
| ⓐ | ⓐ | d | b | 0 | 1 | | |
| a | ⓑ | c | ⓑ | 0 | | | 0 |
| ⓒ | a | ⓒ | d | 1 | | 1 | |
| c | b | ⓓ | ⓓ | | | 1 | 1 |

*Figure P17.7*

**17.8**   For the circuit and secondary assignment given in Fig. P17.8 find expressions for $Y_2$, $Y_1$, and $z$ that are free from all static hazards.

$q$

| $Y_2Y_1$ | $x_2x_1$ | | | | | $z$ | | |
| :---: | :---: | :---: | :---: | :---: | :---: | :---: | :---: | :---: |
| | | | | | | $x_2x_1$ | | |
| | 00 | 01 | 11 | 10 | 00 | 01 | 11 | 10 |
| 00 | a | a | a | b | 1 | 1 | 0 | 0 |
| 01 | a | b | c | b | 1 | 1 | 1 | 0 |
| 11 | c | c | c | d | 0 | 0 | 1 | 1 |
| 10 | c | – | a | d | 0 | – | 0 | 1 |

*Figure P17.8*

**17.9**   Determine the flow table of a level-mode sequential circuit with two inputs, $x_2$ and $x_1$. The single output, $z$, is to be 1 only when $x_2x_1 = 10$ provided that this is the fourth of a sequence of input combinations 00 01 11 10.

Otherwise the output is to be 0. Include a tabulation of outputs assuring that no spurious 1 outputs will occur during transitions between two states with 0 outputs. Both inputs will not change simultaneously.

**17.10** For the flow table of Fig. P17.10, determine state variable assignments free of critical races and determine excitation equations free of all combinational logic hazards.

$x_1 x_2$

| 00 | 01 | 11 | 10 |
|------|------|------|------|
| ①, 0 | 2,− | ①,1 | 5,− |
| 1, − | ②,0 | ②,0 | 4,− |
| ③, 0 | 5 ,− | 4,− | ③,1 |
| 3, − | 5,− | ④,0 | ④,0 |
| ⑤, 0 | ⑤,0 | 1,− | ⑤,1 |

$q, z$

*Figure P17.10*

**17.11** A circuit has a single-input line on which pulses of various widths will occur at random. Construct the primitive flow table of this circuit with a single output, $z$, on which a pulse will occur coinciding with every fourth input pulse.

**17.12** A troublesome problem associated with a switch is contact bounce. There are generally many switches and relays in the console and input–output sections of computers. It is very difficult and expensive to construct switches and relays so that the contacts do not bounce. This problem is illustrated in Fig. P17.12. At $t_1$, the switch moves from $A$ to $B$. It makes initial contact at $B$ and then bounces several times, producing a series of several short pulses before settling permanently at $B$. We may assume it does not bounce back far enough to contact $A$. At $t_2$, the switch is moved back to $A$ with similar results.

Design a "bounce eliminator" to produce a $z$ output as shown following $B$ but eliminating the multiple transitions.

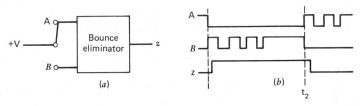

*Figure P17.12*

**17.13**  A fundamental mode circuit has two inputs and two outputs. The two outputs should indicate which of the inputs has changed most recently and whether the number of times it has changed is odd or even.
(a)  How many secondaries will be required?
(b)  Obtain a minimal flow table of such a circuit.
(c)  Obtain a race-free transition table.
(d)  Design a circuit to realize the transition table of part c that is free of static and dynamic hazards.
(e)  If any essential hazards exist, indicate them on the flow table.

**17.14**  A fundamental mode circuit is to be designed that will identify a particular sequence of inputs and provide an output to trigger a combination lock. The inputs to the circuit are three switches labeled, $X_3$, $X_2$, and $X_1$. The output of the circuit is to be 0 unless the input switches are in the position $X_3X_2X_1$ = 010 where this position occurs at the conclusion of a sequence of input positions $X_3X_2X_1$ = 101 111 011 010. Note that two switches are not required to be switched simultaneously.

Using only NAND gates, design a fundamental mode circuit that will have an output of 1 only at the conclusion of this sequence of switch positions. A new correct sequence may begin every time the switches are set to $X_3X_2X_1$ = 101. Be sure that the circuit is free from hazards and critical races.

# References

1. Chaney, T., and C. E. Molnar, "Anomalous Behavior of Synchronizer and Arbiter Circuits," *IEEE Trans. Computers, C-22:* April 1973, pp. 421–422.
2. Couranz, G. R., and D. F. Wann, "Theoretical and Experimental Behavior of Synchronizers Operating in the Metastable Region," *IEEE Trans. Computers, C-24;* June 1975, pp. 604–616.
3. Evans, D. J., "Accurate Simulation of Flip-flop Timing," *Proc. of Design Automation Conference,* No. 15, June 1978, p. 393.
4. Hill, F. J., and G. R. Peterson, *Introduction to Switching Theory and Logical Design,* 3rd ed., Wiley, New York, 1981.
5. Huffman, D. A., "The Design and Use of Hazard-Free Switching Networks," *JACM, 4,* 1957, p. 47.
6. Liu, B., and N. C. Gallagher, "On the 'Metastable Region' of Flip-Flop Circuits," *Proc. of the IEEE,* April 1977, pp. 581–583.
7. Marino, L. R., "General Theory of Metastable Operation," *IEEE Trans. Computers, C-30,* No. 2; February 1981, pp. 107–115.
8. McCluskey, E. J., "Fundamental and Pulse Mode Sequential Circuits," *Proc. IFIP Congress, 1962,* North-Holland, Amsterdam, 1963.
9. McCluskey, E. J., *Introduction to the Theory of Switching Circuits,* McGraw-Hill, New York, 1965.
10. Saucier, G. A., "Encoding of Asynchronous Sequential Networks," *IEEE Trans. Computers, EC 16,* 1967, p. 3.
11. Unger, S. H., "Hazards and Delays in Asynchronous Sequential Switching Circuits," *IRE Trans. Circuit Theory, CT-6,* 1959, p. 12.
12. Unger, S. H., *Asynchronous Sequential Switching Circuits,* Wiley, New York, 1969.

# Powers of Two

| | | | | |
|---|---|---|---|---|
| $2^{0}$ | 1 | | $2^{-1}$ | 0.5 |
| $2^{1}$ | 2 | | $2^{-2}$ | 0.25 |
| $2^{2}$ | 4 | | $2^{-3}$ | 0.125 |
| $2^{3}$ | 8 | | $2^{-4}$ | 0.0625 |
| $2^{4}$ | 16 | | $2^{-5}$ | 0.03125 |
| $2^{5}$ | 32 | | $2^{-6}$ | 0.015625 |
| $2^{6}$ | 64 | | $2^{-7}$ | 0.0078125 |
| $2^{7}$ | 128 | | $2^{-8}$ | 0.00390625 |
| $2^{8}$ | 256 | | $2^{-9}$ | 0.001953125 |
| $2^{9}$ | 512 | | $2^{-10}$ | 0.009765625 |
| $2^{10}$ | 1,024 | | | |
| $2^{11}$ | 2,048 | | | |
| $2^{12}$ | 4,096 | | | |
| $2^{13}$ | 8,192 | | | |
| $2^{14}$ | 16,384 | | | |
| $2^{15}$ | 32,768 | | | |
| $2^{16}$ | 65,536 | | | |
| $2^{17}$ | 131,072 | | | |
| $2^{18}$ | 262,144 | | | |
| $2^{19}$ | 524,288 | | | |
| $2^{20}$ | 1,048,576 | | | |

# APPENDIX B

# Summary of TB6502 Instruction Set

| Command | Mnemonic | Operation[a] | Flags[b] N Z C V |
|---|---|---|---|
| Or | ORA | $AC \leftarrow AC \vee M\langle MA \rangle$ | x x – – |
| And | AND | $AC \leftarrow AC \wedge M\langle MA \rangle$ | x x – – |
| Ex-OR | EOR | $AC \leftarrow AC \oplus M\langle MA \rangle$ | x x – – |
| Add with carry | ADC | $AC \leftarrow AC + M\langle MA \rangle + C$ | x x x x |
| Subtract with carry | SBC | $AC \leftarrow AC - M\langle MA \rangle - C$ | x x x x |
| Store accumulator | STA | $M\langle MA \rangle \leftarrow AC$ | – – – – |
| Load accumulator | LDA | $AC \leftarrow M\langle MA \rangle$ | x x – – |
| Jump | JMP | $PC \leftarrow M\langle MA \rangle$ | – – – – |
| Subroutine jump[c] | JSR | $M\langle SP \rangle \leftarrow PC;$ $PC \leftarrow M\langle MA \rangle$ | – – – – |
| Decrement | DEC | $M\langle MA \rangle \leftarrow M\langle MA \rangle - 1$ | x x – – |
| Increment | INC | $M\langle MA \rangle \leftarrow M\langle MA \rangle + 1$ | x x – – |
| Compare | CMP | Set flags on $AC - M\langle MA \rangle$ | x x x – |
| Bit test AC with memory | BIT | Set Z flag on $AC \wedge M\langle MA \rangle$ | – x – – |
| Load Y | LDY | $Y \leftarrow M\langle MA \rangle$ | x x – – |
| Store Y | STY | $M\langle M \rangle \leftarrow Y$ | – – – – |
| Compare Y | CPY | Set flags on $Y - M\langle MA \rangle$ | x x x – |
| Load X | LDX | $X \leftarrow M\langle MA \rangle$ | x x – – |
| Store X | STX | $M\langle MA \rangle \leftarrow X$ | – – – – |
| Compare X | CPX | Set flags on $X - M\langle MA \rangle$ | x x x – |

| Command | Mnemonic | Operation[a] | Flags[b] | | | |
|---|---|---|---|---|---|---|
| | | | N | Z | C | V |
| Increment *Y* | INY | $Y \leftarrow Y + 1$ | x | x | – | – |
| Decrement *Y* | DEY | $Y \leftarrow Y - 1$ | x | x | – | – |
| Transfer *AC* to *Y* | TAY | $Y \leftarrow AC$ | x | x | – | – |
| Transfer *Y* to *AC* | TYA | $AC \leftarrow Y$ | x | x | – | – |
| Increment *X* | INX | $X \leftarrow X + 1$ | x | x | – | – |
| Decrement *X* | DEX | $X \leftarrow X - 1$ | x | x | – | – |
| Transfer *AC* to *X* | TAX | $X \leftarrow AC$ | x | x | – | – |
| Transfer *X* to *AC* | TXA | $AC \leftarrow X$ | x | x | – | – |
| Shift left | ASL | $\boxed{C} \leftarrow \boxed{M\langle MA\rangle} \leftarrow 0$ | x | x | x | – |
| Rotate left | ROL | $\boxed{C} \leftarrow \boxed{M\langle MA\rangle} \leftarrow$ (loop) | x | x | x | – |
| Shift right | LSR | $0 \rightarrow \boxed{M\langle MA\rangle} \rightarrow \boxed{C}$ | x | x | x | – |
| Rotate right | ROR | (loop) $\boxed{M\langle MA\rangle} \rightarrow \boxed{C}$ | x | x | x |  |
| Shift *AC* left | ASL A | $\boxed{C} \leftarrow \boxed{AC} \leftarrow 0$ | x | x | x | – |
| Rotate *AC* left | ROL A | $\boxed{C} \leftarrow \boxed{AC} \leftarrow$ (loop) | x | x | x | – |
| Shift *AC* right | LSR A | $0 \rightarrow \boxed{AC} \rightarrow \boxed{C}$ | x | x | x | – |
| Rotate *AC* right | ROR A | *AC*       *C* | x | x | x | – |
| Clear *C* | CLC | $C \leftarrow 0$ | – | – | 0 | – |
| Set *C* | SEC | $C \leftarrow 1$ | – | – | 1 | – |
| Set decimal | SED | $D \leftarrow 1$ | – | – | – | – |
| Clear decimal | CLD | $D \leftarrow 0$ | – | – | – | – |
| Set interrupt | SEI | $I \leftarrow 1$ | – | – | – | – |
| Clear interrupt | CLI | $I \leftarrow 0$ | – | – | – | – |
| Branch on plus | BPL | Branch if $N = 0$ | – | – | – | – |
| Branch on minus | BMI | Branch if $N = 1$ | – | – | – | – |
| Branch if not equal to 0 | BNE | Branch if $Z = 0$ | – | – | – | – |
| Branch if equal to 0 | BEQ | Branch if $Z = 1$ | – | – | – | – |
| Branch if carry clear | BCC | Branch if $C = 0$ | – | – | – | – |
| Branch if carry set | BCS | Branch if $C = 1$ | – | – | – | – |
| Branch if no overflow | BVC | Branch if $V = 0$ | – | – | – | – |
| Branch if overflow | BVS | Branch if $V = 1$ | – | – | – | – |
| Push *AC* on stack | PHA | $M\langle SP\rangle \leftarrow AC$; $SP \leftarrow SP - 1$ | – | – | – | – |
| Push status on stack | PHP | $M\langle SP\rangle \leftarrow PS$; $SP \leftarrow SP - 1$ | – | – | – | – |
| Pull *AC* from stack | PLA | $SP \leftarrow SP + 1$; $AC \leftarrow M\langle SP\rangle$ | – | – | – | – |
| Pull status from stack | PLP | $SP \leftarrow SP + 1$; $PS \leftarrow M\langle SP\rangle$ | – | – | – | – |
| Transfer *SP* to *X* | TSX | $X \leftarrow SP$ | x | x | – | – |

| Command | Mnemonic | Operation[a] | Flags[b] N Z C V |
|---------|----------|--------------|------------------|
| Transfer $X$ to $SP$ | TXS | $SP \leftarrow X$ | - - - - |
| Subroutine return[c] | RTS | $PC \leftarrow M\langle SP\rangle$ | - - - - |
| Return from interrupt[c] | RTI | $PC \leftarrow M\langle SP\rangle$ | - - - - |
| No operation | NOP | | - - - - |

[a]For addressed instructions, $M\langle MA\rangle$ indicates the addressed location in memory. For immediate addressing, the second operand is not $M\langle MA\rangle$; it is the second byte of the instruction. For instructions involving the stack, $M\langle SP\rangle$ indicates the location pointed at by the stack pointer.

[b]In the Flags column, an x indicates that the flag is affected by the operation, a dash (-) that it is not affected.

[c]For JSR, RTS, and RTI, we have indicated only that the return address is pusehd onto or pulled from the stack. The changes of the stack pointer are not indicated. See Section 15.2 for details on stack operatiosn in JSR and RTS, Section 15.3 for details on RTA.

# C

# TB6502 Instruction Set: Opcodes, Bytes, Cycles

## Nonaddressed Instructions   (All One Byte)

| Command | Mnemonic | OP | N |
|---|---|---|---|
| Increment $Y$ | INY | C8 | 2 |
| Decrement $Y$ | DEY | 88 | 2 |
| Transfer $AC$ to $Y$ | TAY | A8 | 2 |
| Transfer $Y$ to $AC$ | TYA | 98 | 2 |
| Increment $X$ | INX | E8 | 2 |
| Decrement $X$ | DEX | CA | 2 |
| Transfer $AC$ to $X$ | TAX | AA | 2 |
| Transfer $X$ to $AC$ | TXA | 8A | 2 |
| Shift $AC$ left | ASL A | 0A | 2 |
| Rotate $AC$ left | ROL A | 2A | 2 |
| Shift $AC$ right | LSR A | 4A | 2 |
| Rotate $AC$ right | ROR A | 6A | 2 |
| Clear $C$ | CLC | 18 | 2 |
| Set $C$ | SEC | 38 | 2 |
| Set decimal | SED | F8 | 2 |
| Clear decimal | CLD | D8 | 2 |
| Set interrupt disable | SEI | 78 | 2 |
| Clear interrupt disable | CLI | 58 | 2 |
| No operation | NOP | EA | 2 |
| Subroutine return | RTS | 60 | 6 |
| Return from interrupt | RTI | 40 | 6 |
| Push accumulator | PHA | 48 | 3 |
| Pull accumulator | PLA | 68 | 4 |
| Push processor status | PHP | 08 | 3 |
| Pull processor status | PLP | 28 | 4 |
| Transfer stack pointer to $X$ | TSX | BA | 2 |
| Transfer $X$ to stack pointer | TXS | 9A | 2 |

**Addressed Instructions[a]**

| Command | Mnemonic | Immed 2 bytes | | Page 0 2 bytes | | Abs 3 bytes | | Indir.[b] 2 bytes | | Index[c] X[c] 3 bytes | |
|---|---|---|---|---|---|---|---|---|---|---|---|
| | | OP | N | OP | N | OP | N | OP | N | OP | N |
| Or | ORA | 09 | 2 | 05 | 3 | 0D | 4 | 11 | 5 | 1D | 4 |
| And | AND | 29 | 2 | 25 | 3 | 2D | 4 | 31 | 5 | 3D | 4 |
| Ex-OR | EOR | 49 | 2 | 45 | 3 | 4D | 4 | 51 | 5 | 5D | 4 |
| Add w/carry | ADC | 69 | 2 | 65 | 3 | 6D | 4 | 71 | 5 | 7D | 4 |
| Subtract w/$C$ | SBC | E9 | 2 | E5 | 3 | ED | 4 | F1 | 5 | FD | 4 |
| Store $AC$ | STA | | | 85 | 3 | 8D | 4 | 91 | 6 | 9D | 5 |
| Load $AC$ | LDA | A9 | 2 | A5 | 3 | AD | 4 | B1 | 5 | BD | 4 |
| Jump | JMP | | | | | 4C | 3 | 6C | 5 | | |
| Sbrtne Jmp | JSR | | | | | 20 | 6 | | | | |
| Decrement | DEC | | | C6 | 5 | CE | 6 | | | DE | 7 |
| Increment | INC | | | E6 | 5 | EE | 6 | | | FE | 7 |
| Compare | CMP | C9 | 2 | C5 | 3 | CD | 4 | D1 | 5 | DD | 4 |
| Bit test | BIT | | | 24 | 3 | 2C | 4 | | | | |
| Load $Y$ | LDY | A0 | 2 | A4 | 3 | AC | 4 | | | BC | 4 |
| Store $Y$ | STY | | | 84 | 3 | CE | 4 | | | | |
| Compare $Y$ | CPY | C0 | 2 | C4 | 3 | CC | 4 | | | | |
| Load $X$ | LDX | A2 | 2 | A6 | 3 | AE | 4 | | | | |
| Store $X$ | STX | | | 86 | 3 | 8E | 4 | | | | |
| Compare $X$ | CPX | E0 | 2 | E4 | 3 | EC | 4 | | | | |
| Shft left | ASL | | | 06 | 5 | 0E | 6 | | | 1E | 7 |
| Rot left | ROL | | | 26 | 5 | 2E | 6 | | | 3E | 7 |
| Shft right | LSR | | | 46 | 5 | 4E | 6 | | | 5E | 7 |
| Rot right | ROR | | | 66 | 5 | 6E | 6 | | | 7E | 7 |

[a]Op, hex opcode; N, number of clock cycles.

[b]See Appendix D for a discussion of indirect addressing.

[c]Add 1 to $N$ if indexing crosses page boundary. The entries for the numbers of clock cycles are those published for the 6502. Where the number of clock cycles is important in programming, entries from the table should be used. The TB6502, as first described in Chapter 11, is not cycle for cycle equivalent to the 6502.

## Branch Instructions    (All Two Bytes)[a]

| Command | Mnemonic | OP | N |
|---|---|---|---|
| Branch on plus | BPL | 10 | 2 |
| Branch on minus | BMI | 30 | 2 |
| Branch if not equal to 0 | BNE | D0 | 2 |
| Branch if equal to 0 | BEQ | F0 | 2 |
| Branch if carry clear | BCC | 90 | 2 |
| Branch if carry set | BCS | B0 | 2 |
| Branch if no overflow | BVC | 50 | 2 |
| Branch if overflow | BVS | 70 | 2 |

[a]For all branches, add 1 to $N$ if branch is taken, add 2 to $N$ if branch crosses page boundary.

**APPENDIX**

# D

# Mapping the TB6502 into the 6502

Ideally, courses using this book will include an assembly-languge programming laboratory. If the microcomputer used in the laboratory is 6502 based, students will have no difficulty in applying the material in this text to their laboratory exercises. The TB6502 instructions used throughout the book and summarized in this appendix are a subset of the complete 6502 instruction set. With one important exception, all programs given in the book will run directly on the 6502. In order that the 6502 laboratory experience can follow the book as closely as possible, the handling of this exception is explained in the following.

Indirect addressing as described in Chapter 12 is the "classic" form of indirect addressing, in which the address in the instruction is not the address of the data, but the address of the address of the data (see Figure 12.4.) In Chapter 12 we also introduced a different form of indirect addressing, indirect indexed addressing, which is shown in Figure 12.9. The 6502 does not have the regular indirect addressing, but it does have indirect indexed addressing. However, by comparing Fig. 12.4 and 12.9, you will see that indirect indexed addressing will produce the same results as regular indirect addressing if **Y** contains 00.

This is the key to using indirect addressing as described in the text on a 6502. The opcodes given in this appendix for indirect addressing are, in fact, the opcodes for indirect indexed addressing, which, if **Y** = 00, is equivalent to indirect addressing. When running any program in this text using indirect addressing on the 6502, insert the command

   **LDY #00**

at the beginning of the program. Every program in the text will then run exactly as described if entered in hex. When using an assembler, an additional correction must be made. Because simple indirect addressing does not exist on the 6502, assemblers for the 6502 generally will not recognize the form

   LDA (50)

Instead, use the form for indirect indexed addressing,

   LDA (50), Y

This will assemble to the opcodes given in this appendix for indirect addressing and will produce the same results if *Y* is set to 00.

   The 6502 also has another form of indirect addressing, indexed indirect. Indexed indirect uses the *X* register, and the contents of *X* are added to the page 0 address contained in the second byte of the instruction. For example, in

   LDA (50, X)

the contents of *X* will be added to 50, to refer to a different pair of locations on page 0 to obtain the pointer, as illustrated in Fig. A.1.

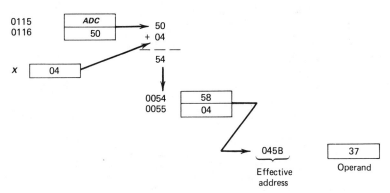

*Figure A.1    Indexed indirect addressing in the 6502.*

We have omitted indexed indirect addressing from the TB6502 in the interests of simplicity. The 6502 also permits regular indexing with the *Y* register and indexing on page 0 with *X* and, for a few instructions, with *Y*. These modes have not been used in any programs in the text. They add little to the power of the machine and would complicate the presentation needlessly for an introductory text. Regular indexing with *X* is the only index addressing mode included in Appendix C.

# Index

**535**